Urban-Rural-Partnerships: Sustainable and Resilient

Urban-Rural-Partnerships: Sustainable and Resilient

Editors

Stephan Bartke
Sigrun Kabisch

MDPI • Basel • Beijing • Wuhan • Barcelona • Belgrade • Manchester • Tokyo • Cluj • Tianjin

Editors

Stephan Bartke
German Environment Agency, President's Office PB1—Planning and Process Management, Research Coordination
Germany

Sigrun Kabisch
Helmholtz Centre for Environmental Research—UFZ, Department of Urban and Environmental Sociology
Germany

Editorial Office
MDPI
St. Alban-Anlage 66
4052 Basel, Switzerland

This is a reprint of articles from the Special Issue published online in the open access journal *Land* (ISSN 2073-445X) (available at: https://www.mdpi.com/journal/land/special_issues/UPR2020).

For citation purposes, cite each article independently as indicated on the article page online and as indicated below:

LastName, A.A.; LastName, B.B.; LastName, C.C. Article Title. *Journal Name* **Year**, *Volume Number*, Page Range.

ISBN 978-3-0365-5141-8 (Hbk)
ISBN 978-3-0365-5142-5 (PDF)

Cover image courtesy of Stadt-Land-Plus

Contents

About the Editors

Stephan Bartke

Stephan Bartke (Dr.) is an economist by training. Since August 2021, Stephan has been working at the German Environment Agency (UBA) in the President's Office, PB1, on planning, process management and research coordination, with a focus on third-party-funded projects and European research networking. His own inter- and transdisciplinary projects, including the synthesis project Stadt-Land-Plus funded by the German Federal Ministry of Education and Research (BMBF), have aimed at facilitating more sustainable land use and governance in urban and rural environments.

Sigrun Kabisch

Sigrun Kabisch (Prof. Dr. habil.) is the head of the Department of Urban and Environmental Sociology at the Helmholtz-Center for Environmental Research, UFZ, in Leipzig. Sigrun has been teaching sociological urban geography at Leipzig University. In her interdisciplinary and international research projects, she deals with urban transformation processes in an environmental context. Social concerns and questions of environmental justice are central elements of her complex, practice-oriented projects on sustainable and resilient urban development.

Preface to "Urban-Rural-Partnerships: Sustainable and Resilient"

How do we want to live in the future? How will we want to reside, move and conduct business? Answers to these questions have to be collaboratively found on the ground. With over 450 participants from close to 50 different countries at the URP2020 conference in November 2020, concepts for urban–rural partnerships were debated upon and scientific results were exchanged with on-the-ground practical experience and knowledge.

The "Sustainable & Resilient Urban-Rural Partnerships – URP2020" conference was organized as part of Germany's presidency of the Council of the European Union in 2020. By focusing on sustainable and resilient urban–rural partnerships, the international URP2020 conference aimed at discussing novel urban–rural imaginaries, integrating strategies and projects which explore present and future potential in terms of sustainability and resilience. It contributed to the implementation of the UN Sustainable Development Goals, the UN Habitat Urban–Rural Linkages Guiding Principles, the New Leipzig Charter and the Territorial Agenda 2030.

The conference took place under the auspices of the German Federal Minister for Education and Research, Anja Karliczek. It was sponsored by the BMBF-funded project "Stadt-Land-Plus" (FKZ 033L200), and took place at the Helmholtz-Centre for Environmental Research, UFZ, in Leipzig, Germany. Full documentation is available online at www.URP2020.eu.

At the conference, it was generally recognized that strong regions with equal living conditions for all, simultaneously providing desirable environmental qualities, rely, at least in part, upon sustainable and resilient urban–rural partnerships. Impulses for this can already be found in the New Leipzig Charter and the Territorial Agenda of the EU 2030 documents. Urban–rural partnerships are of critical importance for implementing the UN 2030 Agenda, including the Sustainable Development Goals, the HABITAT III objectives and the European Green Deal.

This book is a reprint of the Special Issue "Urban-Rural-Partnerships: Sustainable and Resilient" in Land, which, based on the URP2020 objectives, invited original contributions dealing with interactions in regional systems, particularly between urban and rural actors, institutions and projects to tackle great societal challenges. The 16 contributions published included conceptual and methodological papers, as well as case studies dedicated to striking examples and providing transferable knowledge and solutions.

First, Simon offers an insightful general introduction by focusing on "Co-Productive Tools for Transcending the Divide: Building Urban–Rural Partnerships in the Spirit of the New Leipzig Charter". Next, Knickel et al. investigate the implications for rural–urban relations by transitioning towards a sustainable wellbeing economy. The contributions by Martens et al. on short food-supply chains, Kosow et al. on policy mixes addressing goal conflicts of commercial land use management, Sponagel et al. introducing expert-based maps as a regional planning tool; and Greiving, Becker and Gall Rochrig et al. examining structural change in the Rhenish lignite mining region all present research developed with the BMBF funding measure Stadt-Land-Plus. The contributions by Toc et al. and Borseková also shed light on regional change focusing on Central Eastern European areas. The following papers, namely, by Banzhaff et al., widen the focus to China and India, with thematic investigations of nature-based solutions for resilience, as well as resilience and migration (Liu et al. and Link et al.), e-commerce and regional inequality (Wang et al.), the urban–rural income gap (Zhang et al.) and experimenting with urban–rural partnerships for sustainable sanitation in India by Dasgupta and colleagues.

The Guest Editors hope that the contributions stimulate learning processes on various levels, i.e., cross- and transdisciplinary, as well as from the local level to entire regions and to the broader European and international levels, in order to foster an understanding of integrated regional and urban–rural development.

Acknowledgements:
We are grateful for the support of the reviewers contributing to the quality of the published research in this special issue. Partly, the guest editorial work has been funded by the German Federal Ministry of Education and Research (BMBF) through the project "Stadt-Land-Plus – Wissenschaftliches Querschnittsvorhaben" (FKZ033L200) and by the hosting institutions of the editors.

Stephan Bartke and Sigrun Kabisch
Editors

land

Article

Co-Productive Tools for Transcending the Divide: Building Urban–Rural Partnerships in the Spirit of the New Leipzig Charter

David Simon

Department of Geography, Royal Holloway, University of London, EGHAM, Surrey TW20 0EX, UK; d.simon@rhul.ac.uk

Abstract: The outdated and discredited notion of a binary urban–rural divide remains stubbornly widely used. However, it both sets up and reflects oppositional politics and processes between the two supposedly mutually exclusive categories of space and place, which hamper urban–rural partnerships. Empirical reality on the ground is far more complex. Just as more appropriate conceptualisations and approaches have evolved, so new research methods and tools have been developed to overcome the different institutional barriers and stakeholder priorities in the face of contemporary real-world complexities and the urgency of tackling the 'wicked' challenges of sustainability, which also underpin the New Leipzig Charter. The focus here is on co-production and related methods, which can be considered as representing the top-most rungs of Arnstein's (1969) Ladder of Participation. The relevance and application of these methods are exemplified from the work of Mistra Urban Futures in relation to transcending conventional European urban–rural divisions and forming partnerships, with due attention to problems and limitations. Such methods have considerable potential, including for addressing unequal power relations, but are time-consuming and require careful adaptation to each situation.

Keywords: urban–rural continuum; transdisciplinarity; deep participation; co-production; co-creation; co-design; sustainability; unequal power relations; New Leipzig Charter; URP2020 conference

Citation: Simon, D. Co-Productive Tools for Transcending the Divide: Building Urban–Rural Partnerships in the Spirit of the New Leipzig Charter. *Land* **2021**, *10*, 894. https://doi.org/10.3390/land10090894

Academic Editor: Stephan Bartke

Received: 29 July 2021
Accepted: 23 August 2021
Published: 25 August 2021

1. Introduction

Urban–rural partnerships of diverse kinds are increasingly important and urgent to help address today's profound global challenges, many of which are transboundary in nature. This requires an appropriate vision as well as tools capable of overcoming the obstacles to effective governance and management posed by the often outdated and inadequate conceptualisations, approaches, policies and methods on which most practitioners and academics still rely. This paper analyses these issues in turn, drawing on global literature and experience to explain their relevance to implementation of the New Leipzig Charter, adopted in late 2020, in the European context.

Since the 1990s, more appropriate conceptualisations of urban–rural relationships than the simplistic and discredited traditional urban–rural dichotomy have centred on a spectrum or continuum of conditions across space. This may vary in gradient rather than being isotropic—i.e., having an even gradient—between places that everyone would unequivocally characterize as urban at one extreme and as rural at the other. The notion of a peri-urban transition or interface zone within that continuum is now widely accepted and is being applied in many parts of the world [1–10].

In contexts of rapid urbanisation and urban growth, peri-urbanisation usually means a process of transition whereby a previously rural area beyond a city progressively acquires a mix of land uses and functions conventionally associated with both rural and urban areas. The process is dynamic, with the proportion of urban activities increasing, and often ends in such a zone becoming physically and visually urban, even if its legal and institutional status remains unchanged. Meanwhile, the urban fringe or edge moves further out into previously rural land. Depending on wider politico-economic conditions, however, the process may stall, so that the mixture of land uses becomes a more enduring feature. Even in the relatively stable urban areas of Europe, North America and Japan, abrupt urban edges or fringes are relatively rare nowadays, and longstanding peri-urban transition zones more accurately describe the mix of land uses now found there [5,6,10–12].

The precise legal/administrative situations vary by region and country, but large cities almost invariably comprise multiple local governments (or authorities), such as municipalities or districts. Those established in or for urban areas are defined as urban, while those established in previously rural or peri-urban areas but now absorbed into a city have often not evolved to match their current situations. Hence the constituent local government units in a city—and even more in a larger functional city region—commonly possess substantially different sets of powers, responsibilities, capabilities and revenue sources. This underlines the importance of having a strategic metropolitan or city regional governance structure to co-ordinate and ensure a coherent and effective operation in relation to transboundary infrastructure, services and policies among the constituent local authorities (e.g., [13]). Similar issues arise regarding the division of powers, responsibilities and resources between national, regional and local authorities to ensure effective multi-level (or multi-/cross-scale) governance.

Official governance structures and processes are generally rigid and unable to keep pace with changing situations and needs. This is particularly true nowadays, with a growing number of important and often intractable or 'wicked' transboundary challenges because they are so difficult to tackle effectively. Key examples include sustainability (including biodiversity) and climate/environmental change; the unprecedented mobility of people, goods, services and finance; and global health epi- and pandemics—such as avian influenza, Ebola, dengue fever and currently COVID-19. When these become linked, they are especially challenging. One particularly relevant challenge arising from the foregoing relates to the establishment and development of urban–rural partnerships to address these transboundary issues in the context of unequal powers, resources and capacity among the many organisations, stakeholder groups and different categories of local government.

The European Union context: The URP2020 conference in Leipzig in November 2020 highlighted the importance of urban–rural partnerships to meet current sustainability and other 'wicked' problems within the European Union (EU). The New Leipzig Charter (EU 2020), which was launched at the same time, seeks to provide a way forward to meet such urban sustainability challenges within a changing Europe characterised by diverse and dynamic urban–rural conditions, within which some areas are urbanising rapidly, while others are stable, 'mature' or even declining. The recent experience of the city of Leipzig itself exemplifies what effective and coherent action can achieve in terms of regeneration in the wake of industrial obsolescence and political change by bringing diverse stakeholders together to discover that, despite antagonisms and even conflicting priorities, there is much to be gained by co-operating and co-creating [14,15].

The tools required to achieve such progress and to embrace different stakeholders, regardless of what kind of local government unit they live within, are very different from the conventional planning tools associated with technocratic, modernist and generally top–down planning and management. Despite local variations, existing formats of public consultations about proposed (re-)development schemes illustrate the point well. These are widely felt to be inadequate, often merely serving to confirm or validate limited choices among a small set of alternatives pre-determined by professional planning officials and/or elected representatives. Residents and users—and, importantly, different sub-groups and minorities among them defined in different ways—are rarely asked about their aspirations and priorities for the future of their neighbourhoods or other parts of the city. Hence public participation rates are generally low, and the exercises are perceived as having little relevance or even as being patronising.

To transcend this kind of impasse and give meaning to the transformative and participatory spirit, as well as letter, of the New Leipzig Charter [16], it is essential to refresh and update existing partnerships and to forge new single- and multi-purpose alliances, including those addressing urban–rural relations. This necessity is emphasized in all elements of the 2015–2016 global sustainability agenda, including the Sendai Framework for Disaster Risk Reduction; 2030 Agenda for Sustainable Development (of which the Sustainable Development Goals (SDGs) are the key monitoring and evaluation toolkit); COP 21 Paris Agreement of the United Nations Framework Convention on Climate Change; and the New Urban Agenda (NUA). Underpinning all these is the principle of inclusivity and of 'leaving no one behind'. The SDGs and NUA will be returned to below.

The *New Leipzig Charter; The transformative power of cities for the common good* [16] has been updated from the original 2007 version [17]—which focused on measures to promote 'the sustainable European city'—to ensure consistency with these recent global agendas and newer perspectives and approaches, including urban transformation. To this end, it explicitly recognizes the importance of responsive and interdependent multi-level governance that integrates individual neighbourhoods with local authorities and functional city regions, which embody urban–rural dynamics and relations:

In parts this covers a metropolitan area or a combination of other territorial entities. In order to adapt urban policies to people's daily lives,

> … *towns and cities need to cooperate and coordinate their policies and instruments with their surrounding suburban and rural areas on policies for housing, commercial areas, mobility, services, green and blue infrastructure, material flows, local and regional food systems and energy supply, among others* [16] (p. 3; my emphasis added).

Here, then, is explicit recognition of both vertical and horizontal functional and governance interdependencies, and the complexity of collaboration and co-ordination in practice across very different kinds of politico-administrative jurisdictions. The Charter goes on to explain the vision of urban transformation as comprising the social, ecological and economic dimensions of sustainable development—also termed the just, green and productive dimensions [16] (pp. 3–5).

The Charter's explicitly normative vision of future European cities is wholly consistent with contemporary perspectives on integrated and holistic sustainable urban transformations, which embrace social justice and equity/redistribution issues alongside environmental and economic productive sustainability (e.g., [18,19]). All these concepts and requirements are very challenging, even once they have crossed the thresholds of political acceptability and legitimacy by being incorporated into such high-level strategic documents and commitments.

One additional challenge is that SDG11, on sustainable cities and communities, is naturally geared to urban conditions. While, like Agenda 2030 as a whole, it does recognise the integral nature of functional relations and interdependencies beyond a city's limits, how the Goal's targets and indicators will be measured in peri-urban and rural localities within a functional city region that fall within non-urban local government jurisdictions, still requires research and experimentation. Europe is ideally placed to undertake such experimentation, which will also help to progress implementation of the Leipzig Charter. The NUA also explicitly recognises that cities form integral components of sustainable regions and countries [20]. Indeed, its three transformative commitments—namely, social inclusion and ending poverty; inclusive urban prosperity and opportunities for all; and environmentally sustainable and resilient urban development—apply equally well at this scale and to the urban–rural partnerships that are in focus here. Moreover, because the NUA lacks specific means of implementation, monitoring and evaluation of its own—a role to be fulfilled by the SDGs—no additional challenges in the context of urban–rural partnerships arise from the NUA.

The rest of this paper sets out and assesses the very different toolkit required to tackle these challenges. The next section explains the limitations of conventional methods in greater detail, while the third section delves into co-creation, co-design and co-production methods, providing relevant examples from the experience of Mistra Urban Futures (2010–2019), the international centre on urban sustainability headquartered in Gothenburg, Sweden [21]. Section 4 assesses the potential of such methods, including various challenges and limitations, while the concluding section pulls together the various threads and underlines the importance of using sustainable methods to achieve sustainability.

2. Rising to the Challenge: The Need for an Appropriate Conceptual Approach

If conventional urban planning and management tools are already inadequate and lacking in credibility, they are certainly not up to the job of implementing such bold visions and bringing about substantially different, more sustainable, integrated and equitable urban, peri-urban and rural places. Diverse groups of inhabitants and economic actors need to be involved through substantively increased active participation at all stages from conceptualization to implementation. To this end, 'deep' (as opposed to superficially) participatory methods and especially co-productive or co-creative tools are particularly appropriate. These correspond to the top-most rungs (Nos. 6 to 8) of a slightly modified version of Sherry Arnstein's Ladder of Citizen Participation [22] (Figure 1), which identifies the full spectrum from 'manipulation' at the bottom to 'citizen control' at the top. These are categorised in different ways on the basis of direct research experience, as indicated by the labels to the left and right of the ladder.

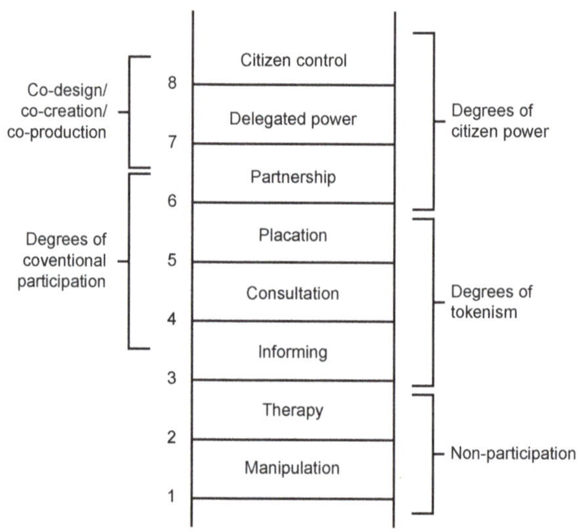

Figure 1. Author's modified version of Arnstein's (1969) Ladder of Participation (Adapted from [22]).

The challenge should not be underestimated because, as Tony Bebbington [23] (p. 281) put it,

> ... theorising participatory development [and hence, by extension, even more so co-production] *necessarily* requires an engagement with practices that pose awkward questions about attitudes and behaviours ... unexpected outcomes, and normative commitments. Meanwhile, practising participation *necessarily* requires engagement with theories that pose difficult questions and challenges, that force the practitioner never to lose sight of the wider picture ... .

Acknowledging the shortcomings of the conventional tools and methods is important as the rationale for developing a fundamentally different toolkit. Existing processes and methods have an overwhelming short- or medium-term timeline of generally not more than five years. This reflects the institutionalisation of linked electoral, planning and budget cycles in most societies, which often triggers rapid changes when political control shifts at elections. These planning and budget cycles are also usually fundamentally sectoral in nature, being aggregated to produce an overall plan or budget, rather than being undertaken on an integrated basis. However, such approaches, and the associated gaps, intersectoral and policy disjunctures and volatility, are particularly unsuited to the far longer integrated strategic planning time horizons of perhaps 20–30 years required to tackle the wicked societal challenges such as economic restructuring, climate change and urban or territorial sustainability.

Such institutional timelines and cycles will be extremely hard to alter, if only because of the inertia and interests vested in long-established political and bureaucratic processes and procedures. A different way of working across sectoral and political party lines is therefore essential if the challenges are to be met, so that the foundations laid during one cycle and are not then ripped out or ignored in subsequent cycles but instead are built upon. This requires important elements of consensus or alliance building that are often unfamiliar in relation to how most current systems operate. A different, co-creative approach has the potential to overcome this immediate problem, as I argue in the next section.

The value of participatory research and service design/delivery has long been recognised and numerous approaches and methods have been developed since at least the 1980s, perhaps most prominently in relation to rural but also some urban development interventions in poor countries. They have also been adapted to diverse rural and urban contexts worldwide, although explicitly urban–rural participatory engagement has been rare. These approaches differ in the depth and extent of citizen or beneficiary participation, with the most sustainable interventions usually being those where participation has been sufficient to generate a sense of buy-in or 'ownership'. This was/is not a specific school of thought but a diverse movement, inspired by figures including Robert Chambers [24–26] and the late Manfred Max-Neef, the Chilean economist, who advocated human-scale development [27,28]. Such approaches therefore became *de rigeur* and were required by many funding and donor agencies. However, 'quick and dirty' versions of slow and deeply participatory methods were often deployed in the interests of rapid delivery to meet donors' budget or planning cycle deadlines. This devalued and discredited such approaches, leading to accusations that a 'tyranny of participation' had arisen and spawning efforts to chart new paths to more substantive transformatory approaches [29,30] based on co-creation/-production, to which the paper now turns.

3. Methods and Materials: Transdisciplinary Co-Creation, Co-Design and Co-Production

Alongside such participatory initiatives arose what are now termed methods of co-creation, co-design and co-production, which are fundamentally different approaches from conventional methods. They seek to establish a level playing field for all participants, regardless of which stakeholder group they belong to and what level of professional or lay expertise they possess. As such, these approaches share a normative philosophical position with poststructuralist theories and methods, namely, that all forms of knowledge and experience have value and can contribute to solving the problem(s) at hand. Hence these approaches aim to decentre (usually western) expert knowledge in favour of multiple, plural (including indigenous and lay) and particularly hybrid forms of knowledge that combine elements of different knowledges into new forms.

Working with such diverse forms of knowledge held by different stakeholder groups makes such approaches, by definition, also transdisciplinary. This distinguishes them from inter- or multi-disciplinary engagements, which bring together different academic or professional disciplines. Transdisciplinarity also has diverse roots, ranging from 'alternative' and heterodox economics and development, e.g., [31], to joint service planning and delivery with local authorities, and is now an increasingly important movement in relation to research and policy [32–41]. As with other participatory methods, they have to date usually been applied in either urban or rural settings, but conceptually they are well suited to spanning urban–rural relations, as I will explore further below.

Linked to their profoundly different epistemologies of knowledge—i.e., assumptions and starting points—compared to conventional methods, transdisciplinary co-creation, co-design and co-production methods (henceforth collectively referred to here as co-production for simplicity) do not produce blueprints or templates for replication. Instead, each situation and research team is regarded as distinctive and requires a bespoke process to be undertaken by the participants to work out their common ground, priorities and the methods appropriate to the problem at hand.

There is a wide range of methods, developed and adapted to diverse situations but all sharing these key features [33,35,38,42–45]. This provides the substantive active engagement required for buy-in and shared ownership of the process and outcomes, which increases substantially the prospects for successful implementation and sustainability of the results. Conversely, this certainly adds considerably to the time requirements and uncertainties of the process. It may also create difficulties in maintaining active engagement and participation by all team members, especially for those for whom participation raises direct opportunity costs and trade-offs. The tools to help maintain involvement and momentum are discussed below.

Accordingly, such methods are held to be appropriate and to provide new opportunities for tackling complex and intractable 'wicked' societal challenges. Their transformatory potential in this regard appears promising on the basis of experiences to date in diverse contexts worldwide, straddling both the global North and South [33,35,39,41–44,46–50]. However, building up a broader evidence base—including on urban–rural partnerships—is urgent. Although there is no specific limit to the number of participants and stakeholders in a project team, the complexity of the process does increase with team size. Another key advantage of such approaches is that they can serve as effective tools for addressing transboundary service delivery and research issues within a city region or across the urban–rural spectrum because they are appropriate to building teams that straddle administrative, political or other jurisdictional boundaries. The essential prerequisite is that the respective communities, local authorities and other organisations, institutions and key individuals are willing to commit to such a process; but, careful consideration by the initiators of whom to include in the initial discussions on the basis of relevance and inclusivity is important [35,43].

For the chances of success in co-productive research projects to be maximised, it is therefore important that no single or subset of participating organisations 'owns' or is perceived to own the process and its outcomes. Experience shows that if that happens, other parties may lose momentum or interest, jeopardising the rationale and potential outcomes. This can be avoided by hammering out ground rules at the outset, including agreement regarding shared ownership of the intellectual property produced by all participating bodies, regardless of size, effective power or resources contributed to the process. That said, local authorities have sometimes led successful co-production exercises to improve the relevance and appropriateness of their service delivery [42,51], while there are examples of top–down co-creation experiments led by local authorities [52].

It is also important to acknowledge that both subtle and overt power relations do exist and must be mediated to ensure that all voices are heard and are not silenced or drowned out by vocal participants or representatives of more powerful bodies. Furthermore, most, if not all, participants are likely to be unfamiliar with the co-production approach and having to negotiate and collaborate with people from diverse organisations, including some that are likely to be in structural opposition to their own. Hence recognising that most participants will be outside their normal comfort zones and making efforts to put them at ease through the initial stages is important.

For these reasons, it is often advantageous to utilise an external—ideally professional—facilitator. This is one of the building blocks or forms of 'supporting infrastructure' found valuable in safeguarding and supporting transdisciplinary co-production processes [38,43,53,54]. Devoting adequate time and effort in the inception stages of transdisciplinary co-production projects to establish the ground rules with which everyone is comfortable is essential to build confidence and mutual trust as the basis for developing later momentum. Key issues in this regard include agreeing to share ownership of the intellectual property to be created, and identifying and understanding the often-conflicting rationalities of different professions and stakeholders [43,55,56].

An example of conflicting rationalities would be the diverse ways in which different professions and other urban and non-urban actors perceive the same problem of river flooding in urban areas, as has become an increasingly dramatic climate change challenge in many European (and other) countries:

- An urban planner might see the heart of the problem as being inappropriate construction within a floodplain without remedial and diversionary measures.
- By contrast, a hydraulic engineer would focus on the inadequacy of embankments, stormwater drainage capacity and the presence of old bridges or other structures that impede river flow or trap logs and other material carried downstream by the flood.
- A biogeographer or urban ecologist would focus on the inadequacy of blue and green infrastructure and permeable surfaces in the city to increase rainfall penetration into the soil and reduce run-off.
- A disaster risk reduction or emergency response professional would focus on the adequacy of early warning systems, rescue and evacuation protocols and stocks of emergency food and relief supplies.
- On the other hand, neighbourhood residents' associations would be concerned with some or all of these issues as they affect their particular community rather than having a strategic focus on the city as a whole.
- Farmers and forestry officials up- and downstream of the urban area would focus on flood mitigation and remediation affecting their lands, livelihoods and resources.
- Hotel and leisure industry operators up- and downstream of the urban area would be concerned primarily with flood mitigation and the security of road/rail infrastructure to provide access to their facilities, and hence their investments and livelihoods.
- A regional official will be concerned with the upstream causes and downstream consequences of both the existing flooding and remedial action taken to address it.

Another and more durable form of facilitation or 'active intermediation' can be provided by a 'boundary' or boundary-crossing' organisation that is separate from the respective stakeholders, and the offices of which provide a more neutral or 'safe' space in which team members can engage. Experience shows that being in their normal work environments, literally embedded within the day-to-day workplace pressures and power relations, can be inhibiting to thinking and engaging 'outside the box'. Hence, having such a resource organisation, which can also provide direct facilitation, served Mistra Urban Futures and various other transdisciplinary co-production teams well [33], [39] (pp. 6–10), [43] (pp. 31–47) and [54].

Diverse Examples of Appropriate Co-Production Methods

As explained above, it is important to develop or adapt methods to suit the particular situation. To assist readers in understanding the range of possibilities, Table 1 contains a selection of different types of methods utilised by at least one Europe-based team within the Mistra Urban Futures international research centre on urban sustainability. These are a combination of single-city and comparative cross-city methods, as explained briefly in the following text. All share the objective of bringing together and building a shared understanding of the common goal and ethos among a diverse transdisciplinary group in order to produce integrated knowledge [43] (Chapter 4). A priori, especially given the need for adaptation to particular local circumstances in any event, there is no reason to expect that they would not be appropriate to urban–rural or regional contexts. Accordingly, their respective key features are explained as encouragement for readers to experiment and build up relevant experience in practice.

Table 1. Selection of co-production methods utilised by Mistra Urban Futures.

• Symmetrical leadership and participation for cross-learning
• Joint problem formulation and solution—iterative design thinking
• Study circles, co-writing of profound 'changes in outlook'
• Exchange visits—reflective translocal learning
• An inverted citizens' jury
• Comparative transdisciplinary co-production around the SDGs and NUA

Symmetrical leadership and participation worked well in relation to some projects among a set of local and regional government bodies, national research agencies and universities in Gothenburg, Sweden, where the principal differences were between academic and various professional knowledges, in both cases spanning different disciplines and sectors. Without minimising the differences that existed, it was felt that the group was relatively symmetrical in terms of these knowledges and their epistemologies. Hence the joint leadership by a senior university researcher and senior city practitioner from the municipality worked effectively, starting with intensive ground-breaking and familiarization activities and continuing throughout the project. Rotating chairmanship, regular meetings and ongoing feedback activities served the project well in building mutual respect and enabling production of appropriate deliverables [43] (pp. 101–103). This method could therefore be utilised effectively in the context of urban–rural partnerships where professionals of comparable expertise levels and experience are involved from across diverse organisations and institutions, provided that the respective institutional contexts are adequately explored during the early stages.

Iterative design thinking was used successfully by the group of stakeholder institutions working to develop a joint formulation of the problem and then a solution in the forms of a funding application and later a work plan to establish a node of Mistra Urban Futures in Stockholm, Sweden. This challenge-led approach originates from the field of design and has been developed to address practical problems through a 5-stage process, starting with the specific, then generalising, before reverting again to the specific in a sequence resembling a 'double diamond' [43] (pp. 103–105). Like the previous method, this works well in a situation of relatively symmetrical power relations and some shared understandings. Perhaps, therefore, the opportunities to deploy it in urban–rural partnership contexts may again be greatest among professionals from different stakeholder groups or organisations, where comparable levels of expertise and experience might exist. Beyond that, it probably has more limited relevance.

Study circles and co-writing is another method used successfully in Gothenburg that brought together academics, the local authority and other public sector officials and was also jointly chaired by an academic and civil servant. The objective was to address the social dimensions of sustainability in the city, a particularly urgent challenge in the context of the increasing segregation, inequalities of income and health and falling political participation. Three substantial workshops with the interested participants helped to frame the project around understanding the underlying drivers and conflicting goals of the problems identified. Theoretical and practical experience were contributed by the respective participants en route to defining key questions and then three sub-projects to address them. Working groups comprising both academic researchers and civil servants iterated ideas and co-wrote eight mental shifts or changes in outlook required to address the challenges. Gaining an understanding of the working environment and conditions of other stakeholders was also an important ingredient [43] (pp. 106–111). This last point highlights that the method should be able to work effectively in different contexts, including urban–rural partnerships, though some of the elements might require more time in view of the wider diversity of contexts and working environments.

As indicated above, and discussed further in the next section, many co-production methods are very time consuming and hence difficult to implement in contexts where time and staff capacity are often key constraints. Mistra Urban Futures' Sheffield–Manchester team has experimented with methods to address this through academic researchers working with local government officials and some civil society members. One pertinent example utilised three carefully organised and structured complementary international study visits within a year to the international Mistra Urban Futures conference in Cape Town, to an international participatory democracy conference in Barcelona, and to their municipal counterparts in Gothenburg. Active co-productive learning was pursued during and after each visit through organised meetings and reflection in groups, individually and by means of interviews, to use the insights gained from other contexts to improve their understanding of the situations, constraints and other possibilities in their own city. The main foci were citizen involvement in decision-making and deep reflection on policy and practice in their own context. These led in turn to discussions about what the prerequisites would be for implementing alternatives that were deemed desirable and possible. This was dubbed 'learning from the outside in' to co-produce practicable conclusions and tools. It built on longstanding co-productive relations and the associated trust between the academic researchers and local authority gatekeepers but, even so, the former needed to be active intermediaries in promoting shared learning within the local authority organization [43] (pp. 129–134) [56]. This method could easily be adapted to help establish or deepen urban–rural partnerships by including participants from different urban, peri-urban and rural jurisdictions/local authorities, or working in different localities for a utility firm or other agency, just as this team comprised participants from different boroughs within Greater Manchester.

In another experiment by the Sheffield–Manchester team, the well-known model of a citizens' jury, in which a random, diverse group of local residents hears presentations by a set of experts and deliberate on recommendations, which are then disseminated to the wider public and key stakeholders, designed to shift official practice, was turned inside out for the purpose of improving the care at home of elderly relatives. This was designed to coincide with an official review of such care and intended to mobilise the experience of diverse healthcare professionals with the existing model as a way of co-producing alternative visions. Hence, healthcare professionals from different fields and levels of seniority were recruited to deliberate as a jury through six sessions on testimony from home care givers and service users in a context removed from their normal work environments and with the assistance of two facilitators. It proved successful, demonstrating the value of such co-productive processes within professional organisations and bodies. The process and its recommendations were both taken up by the senior officials responsible for healthcare [43] (pp. 150–154). Such an inverted jury process could readily be established, comprising professionals located within different jurisdictions within an urban region or other context to combine urban–rural relations and facilitate partnership formation.

My final example is a comparative project across seven cities on four continents—including three in Europe (Gothenburg and Malmö in Sweden, and Sheffield in the UK)—in which Mistra Urban Futures worked. It was designed to understand and facilitate engagement with and uptake of the targets and indicators of Sustainable Development Goal (SDG) 11 and the New Urban Agenda (NUA). Given the global nature of this agenda, the project was designed by the Centre's Secretariat but with the essential flexibility to enable local co-productive implementation according to the circumstances and level of buy-in and interest within each municipality to work with an academic researcher. In Cape Town, the researcher was embedded within the municipality as part of a longstanding collaborative arrangement, while the researcher in Malmö held joint appointments in the university and municipality. There was considerable variation across the cities in terms of the level of municipal engagement, relating to both local factors (such as use of existing indicator sets felt to be comparably appropriate or appreciation of the value of the SDGs in governance for sustainability as well as just in monitoring) or having a well-placed 'champion', and national considerations, such as awaiting national guidance on local authority reporting. This underlined the importance of a multi-level governance perspective [57], which is highly pertinent to the issue of urban–rural partnerships (see below). The researchers were able to add research capacity, to provide in-depth assessments of how various municipal policies could be mapped onto the SDGs, and to provide comparative insights from the other cities. Establishment of a peer-to-peer learning network among the city teams, which met face to face annually and online, also proved particularly valuable for information sharing and enhancing their understanding about the different approaches to the same problems being adopted in the various cities [43] (pp. 122–128), [58–61] (See Supplementary). This is an excellent example of an approach eminently suitable for use with both individual and groups of rural as well as urban local/regional governments and other organisations. In the regional or even national sectoral contexts, these would span the urban, peri-urban and rural categories of government and community institutions, as required in terms of meaningful urban–rural partnerships.

4. Discussion and Limitations

In the context of this Special Issue from the URP2020 conference [62] and implementation of the New Leipzig Charter, one notable benefit of transdisciplinary co-production methods is their flexibility to different contexts and situations, provided that appropriate adaptations are made. Hence, although the examples above were developed for urban applications, such methods are equally amenable to use across jurisdictions of different types and with any number of stakeholder groups within urban, peri-urban and rural contexts, or combinations of these. They could be used to bring together stakeholders in one sector/industry or from several to promote integrated thought, research and action.

The brief summaries in the previous section are intended merely to illustrate some of the relevant methods of transdisciplinary co-production that have been utilised in different contexts to address specific sustainability challenges that are not amenable to conventional methods and solutions. In each case, a process of discussion and reflection led to the identification of several potential methods that might appropriate and hence enable progress in that context by breaking down barriers, forging shared understandings and providing opportunities to learn and formulate suitable ways forward jointly. Final decisions followed detailed discussion of the pros and cons of each in relation to the particular situation and what level of adaptation would be required to make them useful and acceptable. Sometimes short trials were also used in making the final choice. These are important steps to follow since "[c]ontext is everything and it is therefore essential to start by reading and reflecting on the contextual information and guidelines provided regarding what the authors see as key factors or attributes that make the methods successful where they were developed" [43] (p. 167).

Provision of training at the outset is also crucial so that participants understand and can gain sufficient appreciation of the tools not to become frustrated and distracted from the purpose. This is most effectively done by a facilitator or boundary-crossing/active intermediary organization. Applying such a method in the context of research is itself experimental and a learning process—which adds novelty and originality as well as enhancing a sense of bridge building among participants from different organisations and stakeholder groups and developing that crucial sense of joint ownership of both the process and outcomes. Including regular time for reflection and modification if necessary is an important part of any implementation process.

Inevitably, however, such methods will not suit all situations or participants. This underlines the importance of adequate initial thought and discussion. It is also important to acknowledge several challenges and limitations in relation to these methods, which may be encountered individually or jointly.

First, as already explained, the essence of transdisciplinary co-production approaches is inclusivity and appropriateness. This militates against the use of blueprints or templates to formalise, regularise or speed up projects but requires that each initiative or project should be individualised to reflect the particular context and circumstances. This puts a premium on effective leadership and facilitation (see the sixth limitation below).

Second, and following directly from this, the start-up and ongoing transaction costs are higher than with conventional projects and methods where greater replicability is possible. Some of the inception challenges regarding whom to include, the need to negotiate and understand conflicting rationalities, the priorities and relative power of the respective stakeholders, hammering out a joint methodology, and of maintaining active participation and engagement by all participants have already been explained.

Third, it follows logically that such processes make project timelines and the eventual outcomes more uncertain and difficult to predict accurately. This applies even when due care and attention are applied during the start-up phase, and means that some stakeholders will not wish to embark on such a process. Adjustments to expectations and budget, project and reporting cycle deadlines will need to be negotiated—which will not be equally easy or even feasible for all stakeholders.

One way that Mistra Urban Futures addressed this problem was to operate by means of formal multi-stakeholder agreements (memoranda of understanding or agreement, contracts and the like, depending on the situation), which provided assurance of institutional moral, political and resource support to the individual participants. These took time to negotiate and update or renew for successive programme funding phases but did then facilitate and speed up project-specific formalities and the design of co-production processes under their aegis [39], ([43] Chapter 4). Having brought numerous projects to remarkably successful conclusion in recent years, both within Mistra Urban Futures and other programmes [33,35], illustrates how far we have come in the few years since Voorberg et al. [44] highlighted the lack of attention in the co-production literature to outcomes.

Fourth, a key challenge underpinning the previous limitations is the imperative of overcoming short-termism and self-interest by participating individuals and their respective organisations. Bureaucratic practices and procedures, such as the duration of budget and planning cycles, are often long established and are widely embedded in regulations and legislation, including those that govern the respective powers, responsibilities and financial resourcing of the respective central, regional and local government authorities. They are therefore commensurately difficult to revise. The same applies to current electoral cycles at all levels of government. If commitments to effecting substantive and transformative change to achieve sustainability are therefore to have real meaning, elected representatives will need pressure from their electorates to forge cross-party agreements on certain fundamental commitments in this regard that can be embedded as being 'above party politics' in order to serve as building blocks for sustainability across successive electoral cycles. Achieving this may well be easier in systems with proportional representation and/or traditions of coalition governance than in adversarial, first-past-the-post systems.

Fifth, the collaborative relations required among participating stakeholders and measures already mentioned to ensure and maintain active participation can lead to frustration among key professional officials in local authorities or other public bodies and agencies who feel that their training and expertise are undervalued relative to all the other voices and perspectives having to be accommodated. They sometimes also feel that the duration of co-production is an unnecessary luxury and/or that the outcomes and outputs may be sub-optimal in terms of their professional training and judgements. Some initial training, as for all participants, in relation to key parameters of the approach and how to manage expectations is often very helpful, while an element of choice as to which professional officers join a project also provides flexibility for those well disposed towards the approach to volunteer or agree to serve.

Sixth, as already identified above, the duration and challenges of co-production processes put a premium on facilitation and leadership to retain interest and participation by all team members. Identifying, understanding and accommodating the various voices, disciplines and competing rationalities during the inception phase as the basis for working forwards is crucial, along with ongoing facilitation and active intermediation [43] (pp. 31–47), [54–56]—what one might call the 'zen of relationship maintenance'.

Finally, a corollary of the distinctiveness of each situation and project, along with the importance of building and maintaining interpersonal trust to underpin effective working relations, is that co-production approaches are unlikely to be scalable. They work best in the context of relatively small teams, where face-to-face and trustful relationships can develop and then be built upon. Experience shows that even when a member leaves the team due to being assigned a new role or leaving the organisation, the replacement may not fit in immediately and will have to work hard to 'catch up', forge effective interpersonal working relationships and build trust [39,43].

5. Conclusions

Pulling together the threads of this paper requires a few concluding observations. First, urban–rural partnerships, as elaborated in the URP2020 conference [62] and the New Leipzig Charter [16], exemplify the context of current sustainability challenges in that they involve many and diverse, unequal stakeholder groups and organisations and also straddle numerous politico-administrative boundaries. As such there are often no, or only inadequate, mechanisms and processes for seeking to address such transboundary issues, the number, scale and scope of which have been increasing lately and which constitute classic 'wicked societal problems'.

Second, another common reason for the intractability of urban regional sustainability problems is that many existing political institutions and processes operate along traditional lines and have lost credibility among at least some residents because they are perceived as being too bureaucratic, top-down and unrepresentative, often defending vested interests rather than being forward-looking. Most urban planning processes fit this picture because they are expert-led and do not engage residents or the wider public adequately. Even when public consultations are held, these tend to have limited scope, merely offering residents a choice among a few alternative schemes proposed by the planners as meeting mainly technical criteria. Alternative suggestions are not invited and there is no possibility to reject them all as not meeting local needs or priorities. Low voter turnouts in many local and regional elections reflect their perceptions of reduced relevance and powerlessness.

Third, such situations perpetuate the familiar 'we versus them' situations of frustration and low perceived relevance. Yet, the importance of finding a way forward in relation to the wicked problems implies the need for very different approaches that can break through these barriers and be truly inclusive of diverse forms of knowledge, 'ways of knowing', perceptions and priorities [33,35,39,41,43,44,47].

The diverse body of transdisciplinary co-production approaches—and the positive outcomes—explored in this paper provide a potentially valuable way through this impasse precisely because they are designed to bring together representatives of different community groups and other stakeholders and organisations to explore ways to solve mutual problems by building shared understandings and mutual respect on the basis that all forms and sources of knowledge have potential value. Planners and other technical specialists are one such group, not the only source of expertise. As such, these 'deep participatory' approaches constitute a very different epistemology and methodology from conventional procedures and toolkits. The mental shifts required by all parties to accept this and learn to work together to co-design better public services, more appropriate built environments or to find novel solutions to intractable societal problems are considerable. Moreover, and this is my fourth observation, transdisciplinary co-production methods are no magic or silver bullet, capable of solving all the problems straightforwardly. The mere fact of getting—and hopefully keeping—the various interested parties in a room to discuss, negotiate and work together does not mean that implicit or overt interpersonal or institutional power relations simply evaporate. They need to be managed collectively, perhaps steered by an external facilitator and/or boundary-crossing organisation. Even if everyone accepts intellectually that this is necessary and appropriate, ensuring that vocal or forceful individuals learn not to dominate and to ensure that all voices are heard and taken seriously does take time and effort. This is just one of the prerequisites for achieving effective active engagement and joint ownership of both the process and outcomes.

Fifth and a related point is that these methods work on the basis of negotiated and mutually acceptable processes and outcomes. This usually implies seeking consensus, which also takes time but can still prove elusive. It is therefore important that the initial team-building and procedure-defining phase includes discussion and agreement on a mechanism for breaking a deadlock if full consensus is unachievable. This could be a certain proportion of majority vote or a sequential process to agree successive stages of activity or recommendations until a stumbling block is reached, as a way of emphasising agreement over difference.

Finally, the very advantageous character of transdisciplinary co-production projects that requires tailoring to individual circumstances rather than having an off-the-shelf toolkit, also represents their Achilles Heel because of the time and effort required to achieve this and then to run the process to completion. This potential deterrent makes these approaches unsuited to all situations and sets of stakeholders. However, they do provide a fundamentally different way to approach intractable problems by creating a more level playing field, particularly if well facilitated by a skilled professional or a boundary-crossing organisation to create a more neutral space for thinking outside of the box and researching and negotiating away from the respective stakeholders' normal work environments. How far they do provide game-changing transformative potential to break through the constraints of existing institutional practices and procedures, thus enabling them to tackle long-term societal challenges, requires further research in diverse settings. This recognises the experimental nature of transdisciplinary co-production, and the challenges involved. In the words of Fokdal et al. [33] (pp. 18–19):

> [o]ften processes are messy, complex, difficult and time consuming, and there is a large risk of failure. The kind of knowledge produced through these processes, however, is what we need in order to be able to localize the Sustainable Development Goals and the New Urban Agenda (NUA) . . . .

Supplementary Materials: This paper has no direct Supplementary Materials beyond the various publications cited in the text. Those produced by Mistra Urban Futures are published open access at the URLs indicated; these and unpublished reports are available at www.mistraurbanfutures.org.

Funding: The writing of this paper received no funding. The research by Mistra Urban Futures on which the paper draws was funded by several grants from Mistra, the Swedish Foundation for Strategic Environmental Research; Sida, the Swedish International Development Cooperation Agency; and the Gothenburg Consortium of organisations (no grant numbers).

Institutional Review Board Statement: Not applicable.

Informed Consent Statement: Informed consent was obtained from all participants in the studies reported here.

Data Availability Statement: This paper does not draw on any new primary data; see Supplementary Materials for information on the underlying source reports and publications.

Acknowledgments: My thanks to the organizers of the URP2020 conference, Sigrun Kabisch and Stephan Bartke, who are also the guest co-editors of this Special Issue, for the kind invitation to deliver the closing plenary lecture. They also made the eventual online event accessible and enjoyable. Co-productive research by former colleagues within Mistra Urban Futures and the generosity of the core funders mentioned above is gratefully acknowledged.

Conflicts of Interest: The author declares no conflict of interest.

References

1. Bartels, L.E.; Bruns, A.; Simon, D. Towards Situated Analyses of Uneven Peri Urbanization: An (Urban) Political Ecology Perspective. *Antipode* **2020**, *52*, 1237–1258. [CrossRef]
2. Li, C.; Wang, M.; Song, Y. Vulnerability and livelihood restoration of landless households after land acquisition: Evidence from peri-urban China. *Habitat Int.* **2018**, *79*, 109–115. [CrossRef]
3. Lichter, D.T.; Ziliak, J.P. The rural–urban interface: New patterns of spatial interdependence and inequality in America. *Ann. Am. Acad. Political Soc. Sci.* **2017**, *672*, 6–25. [CrossRef]
4. Lin, G.C.S. Peri-urbanism in globalizing China: A study of new urbanism in Dongguan. *Eurasian Geogr. Econ.* **2006**, *47*, 28–53. [CrossRef]
5. Simon, D. Urban environments: Issues on the peri-urban fringe. *Annu. Rev. Environ. Resour.* **2008**, *33*, 167–185. [CrossRef]
6. Simon, D. Peri-urbanization. In *Palgrave Encyclopedia of Urban and Regional Futures*; Brears, R., Ed.; SpringerNature: Cham, Switzerland, 2021; chap. 192–1; p. 5. [CrossRef]
7. Simon, D.; Adam-Bradford, A. Hybrid planning in the peri-urban interface: Applying archaeology and contemporary dynamics for more sustainable, resilient cities. In *Balanced Urban Development: Options and Strategies for Liveable Cities*; Maheshwari, B., Singh, V.P., Thoradeniya, B., Eds.; Springer: Dordrecht, The Netherlands; Frankfurt, Germany; London, UK, 2016; pp. 57–83. [CrossRef]
8. Simon, D.; McGregor, D.; Nsiah-Gyabaah, K. The changing urban–rural interface of African cities: Definitional issues and an application to Kumasi, Ghana. *Environ. Urban.* **2004**, *16*, 235–247. [CrossRef]
9. Simon, D.; McGregor, D.; Thompson, D. Contemporary perspectives on the peri-urban zones of cities in developing countries. In *The Peri-Urban Interface: Approaches to Sustainable Natural and Human Resource Use*; McGregor, D., Simon, D., Thompson, D., Eds.; Earthscan: London, UK, 2006; pp. 3–17.
10. Sorensen, A. Periurbanization as the institutionalization of place: The case of Japan. *Cities* **2016**, *53*, 134–140. [CrossRef]
11. Garner, B. "Perfectly positioned": The blurring of urban, suburban, and rural boundaries in a southern community. *Ann. Am. Acad. Political Soc. Sci.* **2017**, *672*, 46–63. [CrossRef]
12. Qviström, M. Searching for an open future: Planning history as a means of peri-urban landscape analysis. *J. Environ. Plan. Manag.* **2013**, *56*, 1549–1569. [CrossRef]
13. FAO, RUAF Foundation. *A Vision for City Region. Food Systems; Building Sustainable and Resilient City Regions*; Food and Agriculture Organization of the United Nations: Rome, Italy, 2015. Available online:: http://www.fao.org/3/a-i4789e.pdf (accessed on 24 November 2020).
14. Kabisch, N. Transformation of urban brownfields through co-creation: The multi-functional Lene-Voigt Park in Leipzig as a case in point. *Urban Transform.* **2019**, *1*, 1–12. [CrossRef]
15. Stadt Leipzig 'Leipzig and the Leipzig Charter—On the Way to a Sustainable City'. 2020. Available online: https://www.youtube.com/watch?v=SbifTeLR4jc&feature=emb_logo (accessed on 23 November 2020).
16. European Union. *The New Leipzig Charter. The Transformative Power of Cities for the Common Good*; EU2020.de. European Commission: Brussels, Belgium, 2020. Available online:: https://ec.europa.eu/regional_policy/en/newsroom/news/2020/12/12-08-2020-new-leipzig-charter-the-transformative-power-of-cities-for-the-common-good (accessed on 19 July 2021).
17. European Union. *Leipzig Charta Zur Nachhaltigen Europäischen Stadt*; Bundesministerium für Umwelt, Naturschutz, Bau und Reaktorsicherheit (BMUB): Berlin, Germany, 2007. Available online:: https://www.bmu.de/fileadmin/Daten_BMU/Download_PDF/Nationale_Stadtentwicklung/leipzig_charta_de_bf.pdf (accessed on 19 July 2021).
18. Elmqvist, T.; Bai, X.; Frantzeskaki, N.; Griffith, C.; Maddox, D.; McPhearson, T.; Parnell, S.; Romero-Lankao, P.; Simon, D.; Watkins, M. (Eds.) *Urban Planet*; Cambridge University Press: Cambridge, UK, 2018. Available online:: http://www.cambridge.org/9781107196933 (accessed on 18 November 2018).
19. Simon, D. (Ed.) *Rethinking Sustainable Cities*; Policy Press: Bristol, UK, 2016.
20. United Nations. *New Urban Agenda*; Habitat III. UN: New York, NY, USA, 2017. Available online:: https://unhabitat.org/sites/default/files/2019/05/nua-english.pdf (accessed on 18 August 2021).
21. Mistra Urban Futures. Available online: www.mistraturbanfutures.org (accessed on 20 July 2021).
22. Arnstein, S.R. A ladder of citizen participation. *J. Am. Plan. Assoc.* **1969**, *35*, 216–224. [CrossRef]
23. Bebbington, A. Theorizing participation and institutional change: Ethnography and political economy. In *Participation: From Tyranny to Transformation? Exploring New Approaches to Participation in Development*; Hickey, S., Mohan, G., Eds.; Zed Books: London, UK, 2004; pp. 278–283.
24. Chambers, R. *Rural Development: Putting the Last First*; Pearson Education: Harlow, UK, 1983.
25. Chambers, R. *Whose Reality Counts? Putting the First Last*; Practical Action Publishing: Rugby, UK, 1997.
26. Chambers, R. *Can. We Know Better? Reflections for Development*; Practical Action Publishing: Rugby, UK, 2017.
27. Max-Neef, M.A.; Elizalde Hevia, A.; Hopenhayn, M. Human scale development: An option for the future. *Dev. Dialogue* **1989**, *1*, 5–80.
28. Max-Neef, M.A.; Elizalde, A.; Hopenhayn, M. *Human Scale Development: Conception, Application and Further Reflections*; Apex Books: New York, NY, USA, 1991.
29. Cooke, B.; Kothari, U. (Eds.) *Participation: The New Tyranny*; Zed Books: London, UK, 2001.

30. Hickey, S.; Mohan, G. (Eds.) *Participation: From Tyranny to Transformation? Exploring New Approaches to Participation in Development*; Zed Books: London, UK, 2004.
31. Max-Neef, M.A. Foundations of transdisciplinarity. *Ecol. Econ.* **2005**, *53*, 5–16. [CrossRef]
32. Bransden, T.; Honingh, M. Distinguishing different types of co-production: A conceptual analysis based on the classical definitions. *Public Adm. Rev.* **2016**, *76*, 427–435. [CrossRef]
33. Fokdal, J.; Bina, O.; Chiles, P.; Ojamäe, L.; Paadam, K. *Enabling City: Inter—And Transdisciplinary Encounters*; Routledge: Abingdon, UK, 2021.
34. Hoffmann, S.; Pohl, C.; Hering, J.G. Methods and procedures of transdisciplinary knowledge integration: Empirical insights from four thematic synthesis procedures. *Ecol. Soc.* **2017**, *22*, 27. [CrossRef]
35. Horvath, C.; Carpenter, J. (Eds.) *Co-Creation in Theory and Practice; Exploring Creativity in the Global North and South*; Policy Press: Bristol, UK, 2020.
36. Pohl, C. What is progress in transdisciplinary research? *Futures* **2011**, *43*, 618–626. [CrossRef]
37. Polk, M. Transdisciplinary co-production: Designing and testing a transdisciplinary research framework for societal problem solving. *Futures* **2015**, *65*, 110–122. [CrossRef]
38. Simon, D.; Palmer, H.; Smit, W.; Riise, J.; Valencia, S. The challenges of transdisciplinary co-production: From unilocal to comparative research. *Environ. Urban.* **2018**, *30*, 481–500. [CrossRef]
39. Simon, D.; Palmer, H.; Riise, J. (Eds.) *Comparative Urban. Research from Theory to Practice: Co-Production for Sustainability*; Policy Press: Bristol, UK, 2020. Available online:: www.oapen.org/search?identifier=1007874 (accessed on 23 July 2020).
40. Simon, D.; Palmer, H.; Riise, J. (Eds.) Assessment: Learning between theory and practice. In *Comparative Urban Research from Theory to Practice: Co-Production for Sustainability*; Policy Press: Bristol, UK, 2020; pp. 155–172.
41. Thompson Klein, J.; Grossenbacher-Mansuy, W.; Häberli, R.; Bill, A.; Scholz, R.W.; Welti, M. (Eds.) *Transdisciplinarity: Joint Problem Solving among Science, Technology, and Society—An Effective Way for Managing Complexity*; Birkhäuser Verlag: Basel, Switzerland, 2001; Springer: Basel, Switzerland, 2013.
42. Durose, C.; Richardson, L. *Designing Public Policy for Co-Production: Theory, Practice and Change*; Policy Press: Bristol, UK, 2016.
43. Hemström, K.; Simon, D.; Palmer, H.; Perry, B.; Polk, M. *Transdisciplinary Knowledge Co-Production. A Guide for Sustainable Cities*; Practical Action Publishing: Rugby, UK, 2021. Available online:: practicalactionpublishing.com/book/2544/transdisciplinary-knowledge-co-production-for-sustainable-cities (accessed on 24 May 2021).
44. Voorberg, W.H.; Bekkers, V.J.J.M.; Tummers, L.G. A systematic review of co-creation and co-production: Embarking on the social innovation journey. *Public Manag. Rev.* **2015**, *17*, 1333–1357. [CrossRef]
45. Wolf, G.; Mahaffey, N. Designing difference: Co-production of spaces of potentiality. *Urban Plan.* **2016**, *1*, 59–67. [CrossRef]
46. Klenk, N.; Meehan, K. Climate change and transdisciplinary science: Problematizing the integration imperative. *Environ. Sci. Policy* **2015**, *54*, 160–167. [CrossRef]
47. Mauser, W.; Klepper, G.; Rice, M.; Schmalzbauer, B.S.; Hackmann, H.; Leemans, R.; Moore, H. Transdisciplinary global change research: The co-creation of knowledge for sustainability. *Curr. Opin. Environ. Sustain.* **2013**, *5*, 420–431. [CrossRef]
48. Pohl, C.; Thompson Klein, J.; Hoffman, S.; Mitchell, C.; Fam, D. Conceptualising transdisciplinary integration as a multidimensional process. *Environ. Sci. Policy* **2021**, *118*, 20–26. [CrossRef]
49. Pohl, C.; Truffer, B.; Hirsch Hadorn, G. Addressing wicked problems through transdisciplinary research. In *The Oxford Handbook of Interdisciplinarity*, 2nd ed.; Frodeman, R., Klein, J.T., Pacheco, R.C.S., Eds.; Oxford University Press: Oxford, UK, 2017; pp. 319–331.
50. Tengo, M.; Hill, R.; Malmer, P.; Raymond, C.M.; Spierenburg, M.; Danielsen, F.; Elmqvist, T.; Folke, C. Weaving knowledge systems in IPBES, CBD and beyond-lessons learned for sustainability. *Curr. Opin. Environ. Sustain.* **2017**, *26*, 17–25. [CrossRef]
51. Joshi, A.; Moore, M. Institutionalised co-production: Unorthodox public service delivery in challenging environments. *J. Dev. Stud.* **2004**, *40*, 31–49. [CrossRef]
52. Pruvot, S. A top-down experiment in co-creation in Greater Paris. In *Co-Creation in Theory and Practice; Exploring Creativity in the Global North and South*; Horvath, C., Carpenter, J., Eds.; Policy Press: Bristol, UK, 2020; pp. 137–153.
53. Lux, A.; Schäfer, M.; Bergmann, M.; Jahn, T.; Marg, O.; Nagy, E.; Ransiek, A.-C.; Theiler, L. Societal effects of transdisciplinary sustainability research—How can they be strengthened during the research process? *Environ. Sci. Policy* **2019**, *101*, 183–191. [CrossRef]
54. Palmer, H.; Polk, M.; Simon, D.; Hansson, S. Evaluative and enabling infrastructures: Supporting the ability of urban co-production processes to contribute to societal change. *Urban Transform.* **2020**, *2*, 1–4. [CrossRef]
55. Smit, W.; Simon, D.; Durakovic, E.; Dymitrow, M.; Haysom, G.; Hemström, K.; Riise, J. The challenge of conflicting rationalities about urban development: Experiences from Mistra Urban Futures' transdisciplinary urban research. *Trialog* **2021**, *137*, 31–37.
56. Watson, V. Conflicting rationalities: Implications for planning theory and ethics. *Plan. Theory Pract.* **2003**, *4*, 395–407. [CrossRef]
57. Perry, B.; Russell, B. Participatory cities from the 'outside in': The value of comparative learning. In *Comparative Urban Research from Theory to Practice: Co-Production for Sustainability*; Simon, D., Palmer, H., Riise, J., Eds.; Policy Press: Bristol, UK, 2020; pp. 133–154.
58. Croese, S.; Oloko, M.; Simon, D.; Valencia, S.C. Bringing the global to the local: The challenges of multi-level governance for global policy implementation in Africa. *Int. J. Urban Sustain. Dev.* **2021**, 1–13. [CrossRef]
59. Hansson, S.; Simon, D.; Arfvidsson, H. The double function of SDG indicators: Governance for sustainable urban development. *Area Dev. Policy* **2019**, *4*, 217–235. [CrossRef]

60. Valencia, S.; Simon, D.; Croese, S.; Nordqvist, J.; Oloko, M.; Sharma, T.; Taylor Buck, N.; Versace, I. Adapting the Sustainable Development Goals and the New Urban Agenda to the City Level: Initial reflections from a comparative research project. *Int. J. Urban Sustain. Dev.* **2019**, *11*, 4–23. [CrossRef]
61. Valencia, S.C.; Simon, D.; Croese, S.; Diprose, K.; Nordqvist, J.; Oloko, M.; Sharma, T.; Versace, I. Internationally initiated projects with local co-production: Urban Sustainable Development Goal project. In *Comparative Urban Research from Theory to Practice: Co-Production for Sustainability*; Simon, D., Palmer, H., Riise, J., Eds.; Policy Press: Bristol, UK, 2020; p. 113.
62. URP2020 Conference. Available online: www.urp2020.eu (accessed on 18 August 2021).

land

MDPI

Article

Transitioning towards a Sustainable Wellbeing Economy—Implications for Rural–Urban Relations

Karlheinz Knickel [1], Alexandra Almeida [2], Francesca Galli [3], Kerstin Hausegger-Nestelberger [4], Bryonny Goodwin-Hawkins [5,†], Mojca Hrabar [6], Daniel Keech [5], Marina Knickel [3], Olli Lehtonen [7], Damian Maye [5], Irune Ruiz-Martinez [8], Sandra Šūmane [9], Hans Vulto [10] and Johannes S. C. Wiskerke [11,*]

1 HELSUS—Helsinki Institute of Sustainability Science, Department of Economics and Management, Faculty of Agriculture and Forestry, University of Helsinki, FI-00014 Helsinki, Finland; karlheinz.knickel@helsinki.fi
2 Comissão de Coordenação e Desenvolvimento Regional de Lisboa e Vale do Tejo (CCDR-LVT), 1269-053 Lisbon, Portugal; alexandra.almeida@ccdr-lvt.pt
3 Department of Agriculture, University of Pisa, Food and Environment (DAFE), I-56124 Pisa, Italy; francesca.galli@unipi.it (F.G.); marina.kobzeva@phd.unipi.it (M.K.)
4 Regionalmanagement Steirischer Zentralraum, A-8010 Graz, Austria; hausegger-nestelberger@zentralraum-stmk.at
5 Countryside and Community Research Institute, University of Gloucestershire, Cheltenham GL50 2RH, UK; bgoodwinhawkins@glos.ac.uk (B.G.-H.); dkeech@glos.ac.uk (D.K.); dmaye@glos.ac.uk (D.M.)
6 Oikos, Sustainable Development, Inc., SI-1241 Ljubljana-Kamnik, Slovenia; mojca.hrabar@oikos.si
7 Natural Resources Institute Finland (LUKE), FI-00790 Helsinki, Finland; olli.lehtonen@luke.fi
8 Institute of Local Development, University of València, ES-46010 València, Spain; irune.ruiz@uv.es
9 Baltic Studies Centre, LV-1014 Rīga, Latvia; sandra.sumane@gmail.com
10 Municipality of Ede, NL-6877 Ede, The Netherlands; hans.vulto@ede.nl
11 Rural Sociology Group (RSO), Section Space, Place and Society (SPS), Wageningen University, NL-6700 HB Wageningen, The Netherlands
* Correspondence: han.wiskerke@wur.nl
† Before: Department of Geography and Earth Sciences, Aberystwyth University, Aberystwyth SY23 3FL, UK.

Citation: Knickel, K.; Almeida, A.; Galli, F.; Hausegger-Nestelberger, K.; Goodwin-Hawkins, B.; Hrabar, M.; Keech, D.; Knickel, M.; Lehtonen, O.; Maye, D.; et al. Transitioning towards a Sustainable Wellbeing Economy—Implications for Rural–Urban Relations. *Land* **2021**, *10*, 512. https://doi.org/10.3390/land10050512

Academic Editors: Stephan Bartke and Sigrun Kabisch

Received: 28 March 2021
Accepted: 6 May 2021
Published: 11 May 2021

Abstract: This article focuses on the question of how a shift from a narrow economic perspective to a wider sustainable wellbeing focus in regional development strategies and actions might change rural–urban relations. A brief review of relevant research and discourses about economic development models provides the foundation for the analysis. The review leads to the development of an analytical framework that puts the notion of sustainable wellbeing at its center. The criteria included in the analytical framework are then used to assess the current situation, challenges and perceived ways forward based on data and analyses from 11 European regions. The focus of the analysis is on different expressions of a sustainable wellbeing economy, and aspects of territorial development that are consistent with the basic features of a wellbeing economy are identified. Development dynamics and tensions between different development goals and resource uses, strategies and actions that are in favor of sustainable wellbeing goals, and conditions for more mutually beneficial rural–urban relationships are discussed. The article concludes with the implications for local government, and governance and policy frameworks. Reference is made to current high-level strategic policy frameworks and the European Green Deal.

Keywords: sustainable wellbeing; rural–urban relations; land management; local government; governance; policy; case studies; Europe

1. Introduction

In 2012, the OECD contended in its Environmental Outlook to 2050 that business-as-usual is not an option anymore: "*Humanity has witnessed unprecedented growth and prosperity in the past decades, with the size of the world economy more than tripling [. . .]. This growth, however, has been accompanied by environmental pollution and natural resource depletion. The current growth model and the mismanagement of natural assets could ultimately undermine human development"* [1].

In 2015, the 193 countries of the UN General Assembly adopted the 2030 Agenda for Sustainable Development. The 17 Sustainable Development Goals (SDG) that concretize the agenda aim to build an *"inclusive, sustainable and resilient future for people and planet"* [2].

These high-level goals align, in terms of programmatic aims, with the European Commission's Europe 2020 strategy. The strategy explicitly demands an economy based on knowledge and innovation that is resource-efficient and greener, and that provides high employment and social cohesion [3]. The European Commission has recently introduced the European Green Deal. Pitched as *"a new EU growth strategy"*, the Green Deal promises to *"enable European citizens and businesses to benefit from a sustainable green transition ... by [supporting] investments in green technologies, sustainable solutions and new businesses"* [4].

1.1. Evolution of (Economic) Development Models

Together with the increasing recognition of the global challenges, societies are confronted with, there has been an evolution in thinking about development. In the following, five broad types of (economic) development will be briefly reviewed in order to derive an analytical framework that will be used later in the empirical analysis. The examined models start with conventional and end with, in many respects, more encompassing and advanced models:

- Classical development models focused on economic growth;
- Green growth, smart growth, and circular economy models;
- Collaborative (or sharing) economy and distributed economy models;
- Eco-economy and regenerative economy models;
- Foundational economy and sustainable wellbeing models.

Up to now, classical development models have used economic growth as the primary marker of progress and prosperity [5]. Strong economic growth tends to be connected with expansion, not only in markets but also in space, and it can literally be seen how cities and economic powerhouses expand, putting pressure on the surrounding peri-urban areas, as well as natural resources and buffer capacities in natural systems. Fast growing regions also leave less space for those regions that still need to grow their economies because of insufficient job and income opportunities, or low living standards [6,7]. This aspect is crucial as across Europe very divergent trends in economic development can be seen. In some stagnating (or shrinking) regions current and prospective growth is much lower than elsewhere [6], and these differences are even more pronounced at global level [2,8].

Due to the more and more evident downsides of classical economic models governments are increasingly seeking economic growth that is smart (innovation-led), inclusive, and sustainable [9]. Models like green growth and smart growth and, more recently, circular economy [3,10] attempt to mitigate negative effects without putting into question the basic mechanisms underlying growth-centered policies. The European Commission supports related investments in resource use efficiency and through its circular economy action plan adopted in 2016 [11]. The Ellen MacArthur Foundation (2021) argues that increases in resource use efficiency are not sufficient, emphasizing that circular models build economic, natural, and social capital, thereby encouraging a rather different approach to thinking about growth [12]. Isserman et al. (2009) and Van Leeuwen (2015) argue that even if growth as such occurs, it might not trickle down to the local economy and that its benefits do not necessarily accrue to the actors themselves [7,13]. An example of this could be a new logistics center that requires a lot of land but is largely automated with only a few remaining staff members often recruited from outside of the region. These few examples indicate that the different interpretations of a circular economy and smart growth (or smart specialization) might have implications for rural–urban relations.

Jackson (2009, 2016), Raworth (2017), the OECD (2012) and others go a step further arguing that, in the resource-intensive lifestyles world, and in aggregate (or overall), further economic growth is not possible [1,8,14,15]. Raworth (2017) adds that economic growth tends to be progressively less connected with job creation and securing income and wellbeing for a broader population [15]. The resulting question is what the key features of alternative development models and the related progress measures are.

The collaborative economy (sometimes also referred to as a sharing economy) represents a more profound departure from classical growth models. It focuses more on the shared creation, production, distribution, trade, and consumption of goods and services. Working collaboratively is seen as transformative for both the communities where exchanges are happening and for the individuals involved [16]. Collaborative economy services are rapidly emerging across Europe, and range from sharing houses and car journeys to domestic services. They can potentially provide opportunities for citizens and innovative entrepreneurs and offer greater choice to consumers. So far, however, they have also created significant tensions, for example between new service providers and existing market operators [17]. Mason (2015) warns that business models in the collaborative or sharing economy are often based on the privatization of socially produced information, and that these new approaches might reproduce the old order if guided by old principles [18].

Another model which is meant as an alternative to large-scale, centralized production units that tend to be associated with socio-economically unsustainable dynamics is the distributed economy [19]. In a distributed economy a significant share of production activities is organized in the form of smaller units that are synergistically connected with each other and that prioritize quality in their production. Information technology is facilitating these changes. Locally controlled bioenergy and smart distributed energy systems are other expressions of a distributed economy. Essentially, structuring an economy as a distributed network can more equitably spread income and wealth amongst all those who help to generate it.

The eco-economy and regenerative economy models are in many ways complementary to the distributed economy model. Both emphasize environmental sustainability and a regenerative use of natural resources [20,21]. An eco-economy requires the "*principles of ecology* [to] *establish the framework for the formulation of economic policy and economists and ecologists* [to] *work together to fashion the new economy*" [20], p.4. Marsden and Farioli (2015) argue that the framing of the eco-economy goes far beyond the framing of the bioeconomy, especially in its implications for environmental, social, *and* spatial development; while the bioeconomy focuses on renewables and resource efficiency, the eco-economy also comprises more place-based systems and a wider distribution of value added [21]. A regenerative economy puts particular emphasis on the balance between efficiency and resilience, collaboration and competition, diversity and coherence, and small, medium, and large organizations with their different needs [22]. Both, the eco-economy and the regenerative economy model emphasize multifunctional resource uses, including of land.

Finally, there is a set of development models which correspond with Trebeck's (2020) much more encompassing notion of a more humane and sustainable economy [23]. The related indicators include, for example, the EU's sustainable development indicators, OECD's better life indicators, the genuine progress indicator (GPI) and the human development index (HDI). The related quality of life goals are expressed in everyone's need for health, a reasonable living standard, close relationships, activities considered meaningful by people, and self-realization. Closely connected are demands for an equitable distribution of wellbeing at present, as well as securing the wellbeing of future generations [24]. The foundational economy concept which was originally introduced in the 'manifesto for the foundational economy' [25,26] brings together these different aspects. Like the model of 'doughnut economics' proposed by Kate Raworth (2017, 2019) [15,27] it highlights the importance of addressing environmental sustainability (with ecological ceilings that life depends on, etc.; or the 'planetary boundaries' as coined by Johan Rockström and colleagues [28]), and social justice (i.e., access to healthcare, education, etc.) goals together. In Kate Raworth's doughnut model an economy is considered prosperous when 12 key social foundations are met without overshooting any of the nine key ecological ceilings [15,27].

In this article, these different notions are subsumed under the overarching concept of a sustainable wellbeing economy (the development model and term introduced by Katherine Trebeck [23]). Table A1 provides a summary overview on the framing of each model and the related shifts in focus. Central in the analysis presented in this article is the encompassing concept of a sustainable wellbeing economy.

1.2. Research Questions

The evolution of development models briefly sketched out above, leads to the two research questions that this article is about:

1. To what extent and in what ways is the shift from a narrow economic perspective to wider sustainable wellbeing, explicitly or implicitly, expressed in regional development strategies and actions?
2. How could the shift towards a wider sustainable wellbeing perspective change rural–urban relations, and under what conditions can it lead to more mutually beneficial relations?

2. Methodology and Empirical Basis

The approach applied in this article is based on three steps:

- Elaboration of an analytical framework with key criteria for exploring the significance of sustainable wellbeing goals in regional level strategies and actions;
- Application of the analytical framework in 11 European regions to determine in how far a shift towards sustainable wellbeing can be recognized;
- Exploring how this shift could change rural–urban relations and under what conditions this could lead to more mutually beneficial relations.

The first step, elaboration of an analytical framework, builds on the review of models of (economic, sustainable) development presented in the previous section. Based on the review and related studies, policy and strategy papers, key criteria for exploring the significance of sustainable wellbeing goals are derived.

The second step involves the use of the framework and criteria for analyzing the current situation, trends, challenges, perceived ways forward, strategies, and actions in 11 European regions. The third step comprises a synthesis of observations and an analysis and discussion focused on and structured by the two research questions.

In the second and third step a multi-method approach to data collection and analysis is used, given that different kinds of data are available in the different case study regions (see section 'Empirical basis'). Multiple sources of evidence were used to ensure the internal validity of all analyses (e.g., pooling existing literature and data and including inputs from all relevant stakeholders). Generally, authors drew on primary and secondary data analyses and expert discussions.

The cooperation between practitioners and specialists from these different countries in this analysis and the elaboration of the article, involves integrating a range of disciplinary approaches (interdisciplinarity) and drawing on the experiential knowledge of practitioners (transdisciplinarity). This, in turn, provides deep insights into how regional development challenges and rural–urban relations are addressed in different national contexts. The two research questions and the criteria in the analytical framework were used to jointly select illustrative examples for this article.

2.1. Analytical Framework of Sustainable Wellbeing

The review presented in the introduction has highlighted a range of models, most of which represent a significant departure from classical economic development notions. The three main concerns reflected in the more recent models are:

- The limitedness of natural resources and of the buffer capacity of natural systems which are effectively expressed in the planetary boundaries work; related to this, demands for a more environmentally sustainable development;

- The limited recognition of socio-cultural and quality of life goals in orthodox development models. Quality of life is becoming more important especially for the younger generation, but also more broadly;
- The uneven distribution of income and access to goods and services, and the increasing discomfort regarding the wellbeing of future generations; related to this, demands for a more, equitable and inclusive economic development.

The analytical framework presented in Table 1 draws out 17 criteria that—based on the reviewed models and literature—appear most effective in describing these three concerns. The criteria are grouped into the three broad kinds of concerns (or dimensions): environmentally sustainable, socio-cultural and quality of life, and equitable and inclusive economic development. It is recognized that some aspects, like 'equitable and inclusive' or 'resilient' could also be understood as transversal and therefore equally relevant to any of the three dimensions.

Table 1. Key dimensions and criteria of sustainable wellbeing.

Environmentally Sustainable	Socio-Cultural and Quality of Life	Equitable and Inclusive Economic Development
(1) Climate-friendly production systems and lifestyles (2) Natural capital, natural resources integrity and resilience (3) Sustainable management of land, maintenance of high nature value areas and ecosystem services provision (4) More efficient use of finite resources (decoupling) (5) Transition to renewable energy (6) Sustainable mobility	(7) Social capital, diversity and resilience (8) Social justice and good living conditions for all (9) Activities considered meaningful by people, social recognition and security (10) Collaboration and coherence (11) Healthy food (12) Education and healthcare	(13) Decent, satisfying jobs and enough household income for all (14) Fair income distribution (15) Equitable access to resources and inclusive development (16) Strengthening of local economic relations, diversity, synergies and resilience (17) Maintaining the given resource base for future generations

Source: Authors' compilation based on the results of the review of models of (economic, sustainable) development and related studies, policy and strategy papers.

Linking back the analytical framework with the five broad types of development model introduced earlier, it is apparent that classical economic growth models only maintain some limited meaning in regions where there is a tangible lack of job and income opportunities (Criterion 13 in Table 1). Green growth, smart growth, and circular economy models are reflected above all in the framework's environmentally sustainable dimension (Criteria 1–6). Rebound effects (or take-back effects) often mean a reduction in expected gains from new technologies aimed at increasing the efficiency of resource use because of behavioral or other systemic responses. Criterion 2 'Natural capital, natural resources integrity and resilience' is, therefore, only partially met. Eco-economy and regenerative economy models go further in particular in meeting Criterion 2. Collaborative (or sharing) economy and distributed economy models are reflected well in Criteria 8–10 and 13–16. All 17 criteria in the analytical framework are covered well with foundational economy and sustainable wellbeing models. As the latter play an increasingly important role in urban and regional development, they are chosen as reference system for this article.

The basic assumption underlying the empirical analysis presented in this article is that a shift towards a sustainable wellbeing economy can be recognized if perceived ways forward and regional development strategies and actions are in correspondence with these 17 criteria.

2.2. Empirical Basis

The empirical basis for this article are 11 regions that were selected for the EU-funded Horizon 2020 ROBUST project on enhancing rural–urban relations. The 11 regions cover a broad range of scales, historical, institutional, socio-economic and cultural settings, regional development trends and dynamics, and geographical locations.

The use of a case study approach ensures that the trends, strategies and actions identified in each of the 11 regions are analyzed with close attention paid to the unique mix of resources, structures, activities, formal and informal institutions, aspirations, and development trajectories that each region features. In line with that it is the aim to discover the diversity of possible pathways and actions leading to sustainable wellbeing economies, and *not* to compare the regions.

As the set of 11 case study regions covers a broad range of settings, trends, and dynamics, it will be possible to illustrate potential synergies, as well as tensions in much needed transitions in very different circumstances.

A map of the 11 study regions is presented in Figure 1.

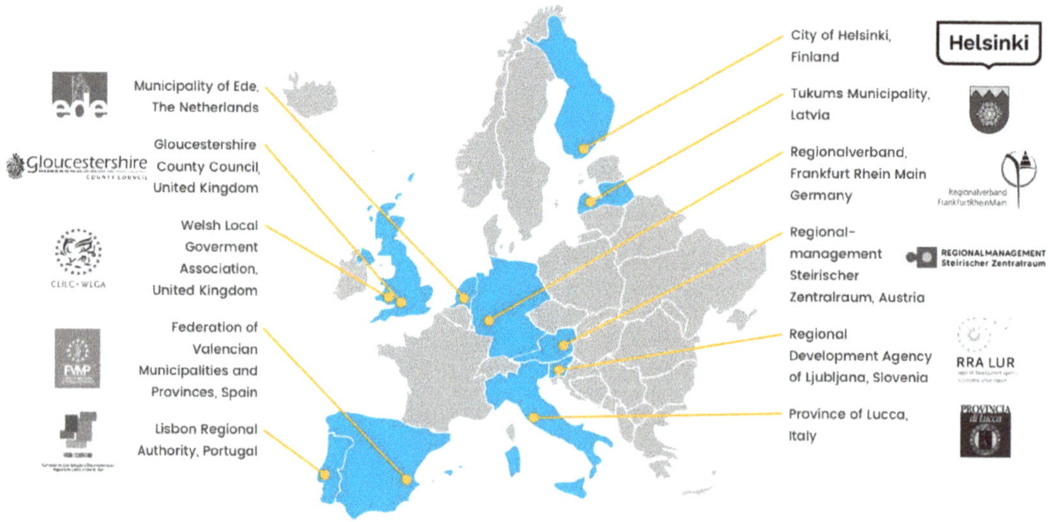

Figure 1. ROBUST consortium partners leading the project's 11 Living Labs.

3. Results

In this section, the analytical framework is used to examine the situation and trends in the 11 study regions, which are briefly introduced first.

3.1. Characterization of the Development in the 11 EU Case Study Regions

Ede is a municipality, the east of which abuts Netherlands' largest nature reserve, The Veluwe, while its west has an agricultural character and is part of the central Dutch plains. The key concerns of municipal decision-makers are promoting rural business models that build on farming diversification, boosting the provisioning of ecosystem services and increasing environmental quality. Consequently, regional development efforts are focused on integrating food, environmental, and planning policies, introducing novel forms of territorial cooperation and new types of offset mechanisms between public and private goods.

The Frankfurt/Rhein-Main region in Germany is known for its international airport, finance sector, and high-tech industry. Because the city of Frankfurt (Main) is economically successful with favorable employment opportunities, continuing population growth is foreseen. Municipal decision-makers and planners recognize the importance of quality of life and good living conditions but face the challenge of accommodating a rising demand for affordable housing while preserving remaining green spaces.

Gloucestershire, in the UK, also anticipates significant levels of growth. Key challenges in the future include the need to balance growth alongside ensuring the long-term protection of its environmental assets. Not all growth can be accommodated within the existing urban areas. Rural and peri-urban land will need to be sacrificed for new housing, supporting infrastructure and commercial enterprises. To maximize the benefits of growth and minimize its potential impacts, local decision-makers are drawn to championing the ideals of clean, green growth as an alternative to conventional practice. Support for the transition to a more circular economy is seen as key.

The population of Helsinki-Uusimaa region in Finland is expected to grow rapidly in the next 20 years. Development will differ within the region because around half of the rural and peri-urban municipalities in the region are expecting a drop in population numbers. In these shrinking municipalities, an ageing population will significantly affect cultural and economic dynamics, including basic services provisioning. The region's priority is to promote smart growth and adaptation by enabling knowledge networks and multi-locality for sustainable life, work, and entrepreneurship. Balancing the social impacts of these structural changes while enhancing the competitiveness of the region and not exceeding resource limits is a central issue for decision-makers.

Similarly, the Lisbon Metropolitan Area, Portugal, faces enormous peri-urban pressure, urban migration, and depopulation in rural areas. Expanding urban territories with high population density exist next to the places with prevailing rural lifestyles and primary sector activities, which face lower income opportunities and depopulation. Decision-makers recognize the need to strengthen mutually beneficial relationships while building on local assets, bridging metropolitan communities and economies for a harmonized and integrated territorial development.

The economic boom experienced in the 1990s–2000s in Slovenia's Ljubljana Urban Region, resulted in urban sprawl and suburbanization. As an employment center, Ljubljana attracts in-migration, which along with a steep increase in tourism results in high property prices and costs of living, leading in turn to the suburbanization of surrounding rural areas and pressure on agriculture and the natural environment. A municipal priority intended to foster more beneficial relations between the rapidly growing city and surrounding rural areas is to promote short food supply chains and increase local provisioning in public institutions.

Local and high-quality food products and wine play also a significant role for the local food system, the ecosystems and the cultural traditions of the Province of Lucca, Italy, characterized by dispersed settlements and peri-urban agriculture. Land fragmentation and abandonment, and the resulting deterioration of peri-urban agricultural areas are widespread. The province's goal is to foster a more sustainable spatial development through policies that mitigate urban sprawl, encourage multifunctional agriculture, and preserve environmental and cultural values.

In mid-Wales, UK, remoteness, poverty and depopulation are longstanding problems compounded by long distances from urban centers of economic development and market failures in service provision. As a predominantly rural region, mid-Wales has been structurally overlooked by national policies that focus on investment in city-regions. Local government priorities hence focus on strategies for fostering rural growth, while maintaining agricultural landscapes, natural resources, and the distinctive Welsh culture and language.

Tukums Municipality in Latvia is also predominantly rural with agriculture and food central to regional development. Tukums is facing depopulation and related social and economic challenges, including maintaining service provision, infrastructure and quality of life. The municipality expects that ensuring sustainable living and working conditions of high quality will retain young people, attract new residents and visitors, and encourage business development. Vibrant cultural life is seen as one key ingredient of quality of life and sustainable living conditions in the region. For that purpose, a joint cultural strategy for the municipality is developed to preserve and add value to the rich cultural and historical heritage of the region that can also boost economic and social activities.

The Metropolitan Area of Styria in Austria includes the vibrant city of Graz and its surroundings which benefit from in-migration, as well as small and remote rural municipalities to the west. Development models that valorize the region's heterogeneous conditions and benefit from sub-regional potentials and aspirations aim to counteract the widening gap between rural areas and urban Graz, and to contribute to a higher quality of life overall. Decision-makers are pooling existing resources in the different sub-regions, fostering interregional cooperation in public infrastructure, social services, and cultural activities, and creating synergies that can benefit the whole region.

The Province of Valencia, Spain, is illustrative of many Mediterranean regions. Most of the population is concentrated in the metropolitan area. Over time, unbalanced population growth and development has resulted in complex territorial, social and economic tensions. A key question for decision-makers is whether shifting from sector-based (mainly tourism), short-term growth to a territory-based, comprehensive long-term view could help the region better manage challenges in the future. Focus areas include fostering smart growth to improve rural–urban relations and overcoming the negative impacts of low-cost tourism.

Table 2 provides a summary overview of the 11 study regions, as well as a simple characterization based on area size (sqkm), population density (inh./sqkm) and population change (% p.a.).

Table 2. Brief characterization of the 11 regions examined in this paper.

Region	Area	Population	
		Density	Change *
	sqkm	inh./sqkm	%
Ede Municipality, Netherlands	318	364	+0.9%
Frankfurt/Rhein-Main, Germany	2458	960	+1.2%
Gloucestershire, United Kingdom	3150	239	+0.9%
Helsinki-Uusimaa Region, Finland	9568	176	+1.0%
Lisbon Metropolitan Area, Portugal	3015	944	+1.3%
Ljubljana Region, Slovenia	2334	237	+0.8%
Lucca Province, Italy	1773	220	−0.1%
Mid-Wales, United Kingdom	17,034	60	−0.2%
Metropolitan Area Styria, Austria	1890	261	+1.1%
Tukums Municipality, Latvia	1195	23	−1.2%
Province of Valencia, Spain	10,812	228	+1.0%

Source: Authors compilation based on information provided on the project website: http://rural-urban.eu/ (accessed: 26 April 2021). * last 5 years in % p.a. (i.e., approximately 2015–2020).

The distribution of regions by population density and population change is shown in Figure A1. The data indicate that there are two regions with a low population density and significant depopulation (Tukums, mid-Wales). Lucca stands out with an average population density and a slightly decreasing population. Two regions have a very high population density and significant increases in population (Frankfurt/Rhein-Main, Lisbon). All other regions have a population density of around 200–400 inhabitants/sqkm and population changes of around +1% p.a.

3.2. Do Sustainable Wellbeing Goals Play a Role in Regional Level Strategies?

Following the introduction of the 11 regions, the analytical framework is used to answer the two main research questions. The three related dimensions in the analytical framework structure the discussion:

- Environmentally sustainable development;
- Socio-cultural aspects and quality of life;
- Equitable and inclusive economic development.

Wherever relevant, for each dimension attention is paid to the related differences between urban, peri-urban, and rural areas, and the connections across dimensions and boundaries.

3.2.1. Environmentally Sustainable Development

The key issues examined in this sub-section are the maintenance of natural resources and ecosystem integrity, the sustainable management of land, nature conservation, and high nature value areas, the role of a more efficient use of finite resources (decoupling), and the importance of climate friendly production systems and lifestyles, the latter also in respect of sustainable mobility.

To contain the expansion of settlement areas and the spreading of cities and suburbs over more and more agricultural land and nature areas is a key issue in all regions except Tukums and mid-Wales. The challenge is to restrict the overuse and fragmentation of space due to demand for infrastructural, business, and housing developments, and to preserve natural resources, while at the same time accommodating growth in population numbers and jobs, and the associated housing and infrastructural development.

A central concern of regional planners in Frankfurt/Rhein-Main is the intense demand for settlement and business development areas and, related to this, the question how environmental quality can be maintained as the city grows, both spatially and economically. Policymakers are asking how much growth is (still) possible, and how the conflicting goals between the further expansion of economic activity and the associated demands for open space can be balanced. Open space is maintained in the Rhein-Main Regional Park as an asset for people to enjoy attractive landscapes on their doorstep. The main challenge is to balance competing land uses identified for green open areas [29].

The situation is similar in Ljubljana and Helsinki-Uusimaa. In both regions accelerated suburbanization contributes to the loss of open space and increasing environmental problems. At the same time, the urban area benefits from the ecosystem services provided by surrounding areas. The harmonization of economic growth and environmental requirements is increasingly seen as one of the key challenges to be addressed. Spatial planning is becoming more effective in incorporating environmental goals, partly due to extensive Natura 2000 network, improved protection of water resources and better flood prevention and mitigation. In Helsinki-Uusimaa, expanding built-up areas and increasing the use of natural resources for building, livelihoods, energy consumption, and recreation create pressures on ecosystems. To reduce urban sprawl and mitigate climate change, regional authorities are encouraging the eco-efficiency of cities and increased density of urban areas. The ongoing mapping and valuation of ecosystem services and green infrastructure is to support planning. The region's land use and business planners see biodiversity and a sustainable use of natural resources as prerequisites for wellbeing. This in turn generates pressure to develop regulatory frameworks, for instance in land use and building acts, that integrate ecosystem services into decision-making [30,31].

A closely related question in Ljubljana and Helsinki-Uusimaa is how overall resilience can be increased when connecting the dispersed rural settlements with the capital city. Decision-makers are exploring ways to enhance mutually beneficial relationships and foster sustainable economic development. Similarly, in Gloucestershire, greater attention is being given to the imposition of ever-progressive measures to protect and ensure the long-term resilience of the county's designated environmental assets and opportunities to secure beneficial enhancements in biodiversity and broader resource quality. Decision-makers argue that, ideally, this must occur in an integrated and coordinated fashion, in tandem with the ambition to deliver housing, employment and infrastructure growth and avoid unduly pressuring and degrading existing, resident local communities. Simultaneously, the adoption of circular business models throughout the local economy could help keep increasingly scarce and energy-intensive raw materials at maximum use for as long as possible and, thus, reduce their demand and the burden upon source locations, which are often rural areas.

In Ede, circular farming finds its expression in more traditional land-based farming practices, as well as in more high-tech inspired practices of urban rest-flow valorization. Circular systems are integrating renewable energy production and urban rest flow valorization. While they are clearly beneficial overall, locally they can be a source of conflicting interests. Windmills and solar fields, for example, are generally regarded as spoiling the landscape.

Decision-makers in the Province of Lucca aim to reduce land use conflicts resulting from urbanization and foster more sustainable spatial development. Building activities that lead to a loss of agricultural land are increasingly challenged as a waste of natural resources. A sustainable vision for the region has been agreed and policies put in place that mitigate urban sprawl, encourage multifunctional agriculture, and that preserve environmental and cultural values [32]. Closely related, and like Ede, policy also aims to encourage the strengthening of sustainable forms of agriculture, to the benefit of the environment, the landscape, and the provision of ecosystem services.

The apparently relentless growth in traffic volumes is a major problem in several regions (Frankfurt/Rhein-Main, Graz/Styria, Lisbon, and Ljubljana). Especially the exponential growth in transport and logistics related to online trade—often with only negligible local employment effects—is aggravating problems. Regional administrations are confronted with the question how further growth in mobility, transport and logistics can be sustainable. In many regions, less centralized, adaptive solutions are being developed which will also impact the relations between urban, peri-urban, and rural areas. In Valencia, the rapid growth of low-cost air travel is imposing growing pressures on environmental resources along the coast. Low-cost air travel and tourism have limited positive effects on local economies, while consuming local resources. The predominance of low-cost tourism is therefore increasingly perceived as detrimental to the realization of the region's potentials.

These regional examples indicate that the environmental dimension is increasingly considered as intrinsic to sustainable regional development. At the same time traditional approaches to natural resources management and nature conservation continue to play a significant role. Restructuring of (economic) activities to achieve a more efficient use of finite resources (decoupling) plays a much lesser role. Illustrations of an economic restructuring are Valencia where the costs and benefits of low-cost tourism are questioned, and Gloucestershire which is aspiring to a circular economy. The same applies to climate friendly production systems and lifestyles where it can be assumed that awareness has grown although few tangible efforts are visible. The only two regions not struggling with rapid expansion of economic activities and population growth are mid-Wales and Tukums. Both regions are confronted with a lack of job and income opportunities, as well as a decreasing and aging population, which is discussed subsequently.

3.2.2. Socio-Cultural Aspects and Quality of Life

In this sub-section good living conditions are the focus of analysis alongside activities considered meaningful by people, a socially more balanced development, access to healthy food, education, and healthcare, diversity and resilience, social recognition and security, and collaboration and coherence.

Socio-cultural potentials tend to play a limited role for decision-makers in the study regions in respect of both the development of regions and quality of life goals. Exceptions to this are Lucca, Graz/Styria, Tukums, and mid-Wales, where socio-cultural aspects are seen as providing significant opportunities. Ongoing discussions on how to develop sustainable cultural tourism in ways that do not exploit and undermine, but preserve and let local culture flourish, and that are beneficial both for visitors and locals, are an example. Tourism can boost local economies, social and cultural life, incite preservation and maintenance of environmental resources, provide recognition of local culture (identities, traditions, language, food, etc.), but it can also exploit and destroy local resources [33].

In the Province of Lucca, opportunities related to quality food products and culture play a central role. The intermunicipal food policy for the Lucca plain includes encouraging access to locally produced food in open air markets, shops, restaurants, school and company canteens, and the establishment of shared vegetable gardens, on public or private land. The Community for Food and Agro-biodiversity in Garfagnana combines the use of natural and cultural resources. This community-based initiative represents a multi-actor, cross-sectoral network that is driven by private sector actors and civil society organizations (CSO) who are mobilizing support from public actors, thereby also acquiring research and development funding. The initiative builds on the interests of tourists in consuming and buying local products, as well as in getting a sense of the place. The high and increasing reputation of Garfagnana is based on authenticity connected with rural and nature tourism. Local initiatives promote the conservation and valorization of local agro-biodiversity through the production, processing and marketing of quality food products and through tourism development, leading to new start-up firms [32].

An illustration of a new cooperative business model that is to improve access to resources and foster a more inclusive development is implemented in the more rural western part of the Metropolitan Area of Styria. The initiative aggregates products and services offered by female entrepreneurs. Particular attention is paid to women on their way to self-employment and to providing cooperation opportunities with female farmers. To secure livelihoods based on self-employment is seen as creating new regional jobs and an important measure for making the more rural parts of the region a more attractive place to live and to reduce out-migration [34].

Tukums' sustainable development strategy has four interrelated goals aimed at reviving the region's prosperity: educated and socially active and responsible residents; diversified smart businesses; attractive living space, including environmental conditions; and collaborative governance. Tourism development includes initiatives aimed at fostering the valorization of local natural and cultural resources. Examples include introducing visitors to local farming traditions, gastronomic tourism routes that offer the possibility of discovering local or traditional food and recipes, routes integrating local architectural sites like renovated castles and manor houses, and a range of creative workshops showcasing typical traditional handicrafts.

Over the past two decades, the focus in the development of Lisbon Metropolitan Area has evolved from infrastructure needs and the preservation of natural systems, to highlighting social cohesion, communities, citizenship and diversity, thereby projecting a metropolis based on solidarity, cosmopolitanism and interculturality.

The examples from the 11 study areas illustrate how socio-cultural and environmental can be effectively combined in place-specific strategies. Businesses and community groups using commercial activities to deliver social, cultural, and environmental benefits play a central role in this integration.

3.2.3. Equitable and Inclusive Economic Development

In the following, traits of a more equitable and inclusive economic development in the 11 regions are identified. The focus is on the provision of a sufficient income for all and in all parts of a region, the goal to achieve a more equitable income distribution, the strengthening of local economic relations, and the maintenance of the given resource base for future generations.

The new edition of the regional land use plan for Frankfurt/Rhein-Main aims at reducing the massive increase in settlement areas, thereby enhancing the quality of life in the wider region [29]. A rather different but potentially very effective strategy for achieving a more even spatial development could be the location of innovation clusters outside of the Frankfurt/Main conurbation. The resulting development would be more dispersed, bringing opportunities to reduce commuting and traffic into the city and contributing to a more even utilization of infrastructure [35].

Other examples show that not all regions are in the position of abandoning growth strategies. In regions like Mid Wales and Tukums, more growth in economic activities is seen as essential in providing jobs, income opportunities, social recognition, and security, especially for the younger generation. In these regions, economic growth is directly coupled with social needs. In mid-Wales, the *Growing mid-Wales* partnership arises from a policy context in which polycentric development through 'city deals' has emerged to ostensibly temper London's economic dominance. However, city deals still follow a model rooted in classical economics wherein the benefits of growth in an urban center are assumed to trickle down to a wider region. *Growing mid-Wales* seeks to wrest a city deal without a city, thereby challenging the dominant narrative of growth.

Championing good practice in the evolution of circular business models is a cornerstone measure in the climate change strategy for Gloucestershire. Circular economy principles are seen as presenting tangible opportunities to ensure that future growth will occur in an equitable way, particularly with regards to the scale and concentration of resource use and the realization of enduring benefits. Circular practice is to also help drive the resilience of existing local communities in the county, many of whom will be challenged by external forces, such as climate change and global economic competitiveness. The County Council is seen as having an important role to play as a local resource regulator (through planning for minerals, waste management, and public infrastructure), and simultaneously as an economic actor via its own procurement of goods and services. Related initiatives include the adoption of waste minimization strategies and the expansion and reinforcement of green infrastructure. Decision-makers expect that, in totality, these approaches will reduce some existing inequities for example by tackling the flow of wastes from urban (producing) to rural (disposal) locations.

The cooperative for female entrepreneurs in Styria shows that decent and satisfying jobs, healthy food, education, healthcare, and mobility can emerge from the strengthening of local economies. In Tukums, a regional food strategy was developed to reconnect local producers and consumers and promotes sustainable food consumption, thereby supporting the local economy, and contributing to regional environmental quality and health. The regional analysis shows that improved market access can increase farmers' professional satisfaction and living conditions, and that both is beneficial for rural communities and economies [36]. Related support measures include public food procurement, branding of local food, and food education of consumers and food professionals. Similar initiatives are pursued in the inter-municipal food policy in Lucca Province.

Across the 11 regions, the two main economic development drivers are the need to provide jobs and enough household income for citizens, and, loosely related, the strengthening of local economic relations, diversity and synergies. Other aspects like a fair income distribution, equitable access to resources and inclusive development, and the maintenance of the given resource base for future generations appear still rather distant from current strategies and actions.

3.3. A Synopsis of Key Findings

Table 3 provides a synopsis of how important specific sustainable wellbeing aspects are for each region. Criterion marked 'h' (high) indicate regions where corresponding strategies and actions are more pronounced, and 'm' (moderate) denotes regions where a criterion is somewhat important. The scoring is based on the information and examples provided in the previous section and the best professional judgement of the author(s) from the respective region.

Across the 11 case study regions, three aspects are hardly recognizable in development strategies and actions: Criteria 9 'Activities considered meaningful by people, social recognition and security' (only in Tukums), Criteria 14 'Fair income distribution' and Criteria 17 'Maintaining the given resource base for future generations'. One other aspect seems only important in Helsinki-Uusima and Lisbon Metropolitan Area: Criteria 8 'Social justice and good living conditions for all'. It is telling that these four aspects can be seen as more profound expressions of a more humane and sustainable wellbeing economy. This, in turn, shows how far the distance is between conventional growth models that tend to still dominate policymaking in many regions and the notions developed by Kate Raworth, Katherine Trebeck, and the Foundational Economy Collective [15,23,25–27].

Table 3. Summary overview on how important specific sustainable wellbeing aspects are for each region. ('**h**' = high; 'm' = moderate; empty cells denote that a criterion does not stand out).

Region	Environmentally Sustainable						Socio-Cultural and Quality of Life						Equitable and Inclusive Economic Development				
	1	2	3	4	5	6	7	8	9	10	11	12	13	14	15	16	17
Ede, NL	m	m	**h**	m	m	m	m			m	m				m	m	
Frankfurt/Rhein-Main, DE	m		**h**	m	m	m	m				m	m				m	
Gloucestershire, UK		**h**	m	**h**													
Helsinki, FI	m		m	m			m	m				m	**h**	m	**h**	**h**	
Lisbon Metropolitan Area, PT			**h**		m	m	m	m		m		**h**	m		m	**h**	
Ljubljana, SI			m			m	m			m	m		m		m	m	
Lucca Province, IT		m	m				m			m	**h**					**h**	
Mid-Wales, UK			m				m			m			m				
Metropolitan Area Styria, AT			m			**h**	m			m		m	**h**			**h**	
Tukums, LV	m		m				**h**		m	m	m		**h**				
Valencia, ES										m			m			**h**	

(1) Climate-friendly production systems and lifestyles; (2) Natural capital, natural resources integrity and resilience; (3) Sustainable management of land, maintenance of high nature value areas and ESS provision; (4) More efficient use of finite resources and decoupling; (5) Transition to renewable energy; (6) Sustainable mobility; (7) Social capital, diversity and resilience; (8) Social justice and good living conditions for all; (9) Activities considered meaningful by people, social recognition and security; (10) Collaboration and coherence; (11) Healthy food; (12) Education and healthcare; (13) Decent, satisfying jobs and enough household income for all; (14) Fair income distribution; (15) Equitable access to resources and inclusive development; (16) Strengthening of local economic relations, diversity, synergies and resilience; (17) Maintaining the given resource base for future generations.

4. Discussion

4.1. Can a Shift towards a Sustainable Wellbeing Economy Be Recognised?

Based on the evidence presented in the previous section, the extent and ways that sustainable wellbeing goals play a role in regional development strategies and actions can be assessed. Overall, many examples of thinking and acting could be identified that suggest that development trajectories are re-orienting towards strategies that aim at reconciling environmental, social, and economic goals.

However, we also found several examples that are still primarily about managing conflicting interests, and more specifically about dealing with land use conflicts at the urban fringe and the difficulty to maintain green spaces in urban and peri-urban areas. Frankfurt/Rhein-Main provides an illustration of a region with an enormous competition for land, and numerous pressures resulting from a very dynamic economic development. Ede, Gloucestershire, Helsinki, and Ljubljana are other regions also confronted with the enormous challenge of reconciling economic development and environmental goals.

Several other examples go beyond the management of conflicting interests, indicating that the related thinking and strategies are becoming more encompassing, cutting across economic, environmental, and socio-cultural goals. They show that integration goes beyond mitigating conflicts and can lead to development paths that are more mutually beneficial and sustainable. The development strategy for the Helsinki-Uusimaa region provides a notable illustration. The related strategic priorities include human wellbeing and competence, successful and responsible business, and being climate-aware and a diverse region [31]. The city of Helsinki uses some of the UN sustainable development goals as a frame for shifting to wider sustainable wellbeing (e.g., SDG4, SDG8, SDG10, SDG13). Implementation of the SDGs will take place locally through municipal cooperation and investments in education and learning opportunities, promotion of sustainable economic growth, significant investments in the reduction in inequality and becoming climate responsible [30].

That a more integrated approach commands the need for new approaches in urban planning and regional governance is shown by Lisbon regional authority (CCDR-LVT) and the Lisbon Metropolitan Area as a mandatory public association of 18 municipalities. Both are working together for the 2030 regional strategy to assure integration and better outcomes. Ljubljana Urban Region is another example where a considerable shift in development strategies can be seen. While the first regional development plan in 2002 prepared for EU accession focused on economic and infrastructural development, by 2014 regional development planning had shifted considerably. Economic development is still at the forefront, but more attention is now paid to sustainable mobility and inter-modality to cope with increased commuting and to reducing transport-related pollution, as well as to conserving and enhancing environmental services and green infrastructure.

Even more ambitious is the Welsh Government's *Wellbeing of Future Generations Act*, which came into force in 2015. It offers an explicit model of sustainable wellbeing in contradistinction to classical economic thinking. Rather than solely quantitative economic indicators, the act foregrounds seven wellbeing goals, including 'A healthier Wales' and 'A Wales of vibrant culture and Welsh language', which collectively envision 'the Wales we want' beyond a solely economic sphere. Instead of determining that so-called 'lagging' regions must 'catch-up' according to narrowly quantified criteria [37], the seven wellbeing goals offer "the art of the possible" (https://futuregenerations.wales/the-art-of-the-possible/ (accessed on 26 April 2021)). Notably, the act has provoked the publication by Oxfam Cymru, an NGO, of "*The Welsh Doughnut—A framework for environmental sustainability and social justice*" [38], which is an explicit attempt to link the new national policy context to the emerging theorization of doughnut economics.

The few examples provide vivid illustrations of what a shift from a narrow economic perspective to wider sustainable wellbeing can look like. The related strategies do not yet include the full spectrum of what a sustainable wellbeing economy comprises, but they still represent considerable progress.

4.2. *How Could the Shift towards a Wider Sustainable Wellbeing Perspective Change Rural–Urban Relations and under What Conditions Can It Lead to More Mutually Beneficial Relations?*

The examples from the 11 regions illustrate manifold interactions and functional relations between urban, peri-urban, and rural areas, including their dynamics. The interactions (and dependencies) that were identified relate to the distribution and use of local assets, such as social and natural capital, and the way new strategic orientations can contribute to delivering more sustainable, integrated, and inclusive forms of development.

Generally, relations can be seen as positive when they are mutually supportive, i.e., when all parties benefit. Especially the examples of Ede, Lucca, Ljubljana, and Styria feature traits of more beneficial relations between urban and rural areas. Typical features include mutually supportive relations between different sectors, activities and actors, and synergies between ecological, economic, and social motivations.

A key concept used in Ede (and the Netherlands as a whole) is *spatial quality*. The new Environmental and Planning Act (EPA) demands that municipalities develop a single 'physical vision document' for the whole area of the municipality, including both its rural and urban parts. Elaborating a single vision, forces planners to consider how certain spatial functions and policies affect both urban and rural areas. In Lucca province, attention is paid to strengthening public and private sensitivity for the value of available open space and rural land, and of its ability to provide ecosystem services. The valorization of more sustainable resource uses is seen as crucial [39]. In the Ljubljana Urban Area, the increasing importance of sustainable development goals has reinforced the role of peri-urban and rural areas in preserving a high quality of life and providing local high-quality food. Short food supply chains are being established, sometimes experimenting with new ICT-based business models and delivery systems. Public procurement plays an increasingly important role in stimulating change. New forms of public transport, inter-modality and carpooling are being tested in Styria to improve accessibility.

The examples from Ede, Lucca, Ljubljana and Tukums show that food can act as a catalyst in connecting conservation and valorization, and in integrating rural, peri-urban, and urban spaces and functions. The related strategies facilitate access to local food, foster knowledge and education on local food, its production and consumption, and encourage sustainable agriculture. In Lucca, the concept of 'smart rural' links the capacity of organic farming and local and high-quality products with sustainable consumption. In Ede, an active community is working on sustainable food, agriculture, and food waste topics involving Slow Food, local schools and families. A challenge for policymakers in Ede is to integrate the world market oriented intensive livestock farms in the region as the smaller scale diversified farming initiatives seem more motivated to build connections with local urban communities.

The process of rebalancing economic development and quality of life goals in Frankfurt-Rhein/Main represents an ambitious spatial objective. Policymakers and planners are rediscovering the polycentric nature of the region characterized by an intricate pattern of peri-urban centers and high-quality open space. The regional spatial plan for the Territory of Lisbon Metropolitan Area aims at connecting four broader priorities: environmental sustainability (including revitalizing the rural environment); measures to contain urban expansion; socio-territorial cohesion with a focus on eradicating precarious residential situations and guaranteeing equal opportunities and access in housing; and improving the metropolitan transport system [40]. All four priorities have the potential to foster a sustainable wellbeing economy while at the same time improving territorial relations.

In the Valencia region, the latest tourism strategy aims at a more balanced distribution of incomes in rural and urban areas. Technological advances like Wi-Fi networks, mobile technology, cloud computing, etc., are fostering rural and urban synergies and improving living conditions and sustainability at the same time. Valencia, for the same reason, has shifted towards pursuing a territorial longer-term perspective that connects the domains of business development, labor markets, public infrastructure, and sustainable food systems, and that connects activities across the urban, peri-urban, and rural parts of the region.

In the Helsinki-Uusimaa region, managing urban growth dominates regional development and planning. At the same time, counter-urbanization occurs in the region, manifested in seasonal mobility patterns and multi-locality. At the same time, multi-locality and rural–urban connections are hardly recognized by official development strategies and policies and transport systems remain strongly linked to the needs of permanent residents. Driven by growth corridor thinking, most investments in transport infrastructure are directed to strengthening the connections between the largest urban centers. Overall, Helsinki-Uusimaa is the only example where regional cohesion seems to play a lesser role despite differentiated development within the region.

Most of our examples illustrate the potential of cross-sectoral coordination and cooperation for more balanced rural–urban relations. That both are not easily achieved is a finding which is supported by Meijers and Van der Wouw (2019) who refer to hierarchical relationships in which urban areas tend to outcompete rural ones [41]. The resulting question is under what conditions more mutually beneficial rural–urban relations can be expected.

When discussing under what conditions rural areas might benefit from the shift towards a wellbeing economy, particular attention needs to be paid to the importance of networks and cooperation, and the new possibilities that ICT, digital infrastructures, and the circular economy offer. Powering local economies requires activating external networks for knowledge exchange, and new kinds of supply chains and markets. Online platforms that connect food producers with processors and consumers, combine technological with social and organizational innovation.

In Graz/Styria, Helsinki-Uusimaa, Ljubljana, and other regions, the sharing economy, intercommunal cooperation and the balancing of sub-regional developments are becoming increasingly important. Well-working governance arrangements are being elaborated that are to organize the sharing of functions and space in ways that benefit urban, peri-urban, and rural areas. The possibilities that digital infrastructures offer for overcoming the distance between rural places and urban markets, and for generating new employment opportunities, play an increasingly important role. In Helsinki-Uusimaa investments in high-speed internet connections are to enable place-independent living and working, and lead to a more dispersed spatial distribution of economic activity. Mid-Wales is exploring 'Gigabit Hubs' to improve digital infrastructure, offering high-speed internet connections and co-working spaces. Improving digital infrastructures and development of public e-services in health, culture, and other domains is also on the agenda in Tukums. The expectation is that this will improve the quality of life, especially for rural residents in more remote areas. In the Valencia region, it is expected that investments in sustainable tourism and digital infrastructure could improve the connections between environmental, cultural, and food sectors, as well as boosting regional economic benefits.

In the Metropolitan Area of Styria, shared economy approaches and intercommunal cooperation play a central role. The expectation is that providing high-quality digital access and smart mobility will improve the quality of life of residents, as well as business opportunities. A more recent car-sharing initiative in the region follows a multimodal approach, connecting the City of Graz to 10 suburban and rural municipalities in the metropolitan region. In this way, synergies between the city and the surrounding rural areas are fostered that bring many financial and structural benefits. The Municipal Master Plan of Sintra emphasizes rural tourism and urban logistics for local products as key investment areas [42]. Decision-makers expect more mutually beneficial rural-urban relations, a dynamic, innovative and competitive regional economy, and increased wellbeing overall.

The examples illustrate how decision-makers try to deliberately connect, and even blend, goals that formerly were thought to conflict with each other, and how this can lead to more beneficial territorial relations. That the changes towards a more sustainable economy are not always without conflict and beneficial for rural areas is illustrated by examples from Ede and Frankfurt/Rhein-Main.

5. Conclusions

> *"The economy is a means to an end, and should be helping us to live good lives, which means we need to redesign the economy. . . . A wellbeing economy is one that will deliver human and ecological wellbeing."* (Trebeck, 2020)

In consideration of the evidence gathered from the 11 regions, is there evidence of a paradigm shift in economic development? Clearly there are manifold indications of a reorientation in regional level strategies and action that represent a shift from a narrow economic perspective to wider sustainable wellbeing. The regional examples include features of circular economy models (Ede, Gloucestershire), distributed economy models (Ljubljana, Metropolitan Area Styria), eco-economy and regenerative economy models (Lucca Province), as well as sustainable wellbeing models and doughnut economics (Lisbon Metropolitan Area, mid-Wales). Especially the examples of mid-Wales, Tukums, Ljubljana, Valencia, Ede and Frankfurt/Rhein-Main show how these new strategies are blended with traits of classical growth models.

The analytical framework and aspects presented in Table 1 helped to go beyond these broad types and describe these shifts in more concrete terms. Many territorial initiatives can be described as innovative, but they tend to be local and often still in the initial stages. However, several impressive examples of initiatives were found that focus on making lifestyles and economic systems more environmentally sustainable (e.g., maintaining the natural resource base and ecosystem integrity, nature conservation and preservation of high nature value areas, or promotion of climate friendly production systems and lifestyles). Some examples illustrate how socio-cultural and quality of life goals can reinforce each other. Generally, changes are most often driven by civil society and or the private sector.

When scrutinizing development dynamics and strategies in the 11 regions, it becomes clear how much situations differ. Regional disparities in growth dynamics, employment and living conditions are significant. Some regions struggle with depopulation and a lack of jobs, while others are trying to cope with high growth rates and seemingly limitless expansion. It is therefore important to pay attention to contextualizing, and to being explicit about regional needs and aspirations. Our 11 examples also illustrate that the cross-sectoral activities that could enhance rural, peri-urban, and urban linkages are highly complex and differentiated.

Related to regional development strategies and actions, and their connection with rural–urban relations, three broad clusters of regions can be identified:

- In four regions (Ede, NL; Frankfurt/Rhein-Main, DE; Gloucestershire, UK; mid-Wales, UK) balancing urban growth and economic goals with environmental goals, an increased ecosystem services provision, environmental protection, and sustainable modes of mobility are dominant strategic goals. Mid-Wales pays particular attention to encouraging smart growth, while maintaining the distinctive Welsh culture and language;
- Four other regions emphasize strategies and actions that are to strengthen local economic relations, a more balanced (harmonized, integrated territorial) development and social and territorial cohesion (Helsinki, FI; Lisbon Metropolitan Area, PT; Valencia, ES; Ljubljana, SI). A particular aspect in Valencia is the shift from a sector-based, short-term view to a territory-based, long-term view. In Ljubljana, the related aim is to counteract suburbanization and the reduced availability of public services in rural areas, thereby fostering inclusion;
- Three regions emphasize in their strategies the valorization of social, environmental (landscape), and cultural values assets and cultural heritage (i.e., less related to balancing or rural–urban relations) (Lucca Province, IT; Tukums, LV; Metropolitan Area Styria, AT). Tukums connects sustainable living and working conditions of high quality, valorizing regional cultural capital and improving accessibility in its strategies. In Styria, particular attention is paid to fostering cooperation in public infrastructure, social services and cultural activities, thereby enhancing quality of life. Especially the regions in this last cluster point to a connection of more synergistic territorial relations with sustainable wellbeing.

More generally, it seems important to understand under what conditions synergies can be generated that result in a more balanced development. Related to new strategic policy frameworks like the European Green Deal, it became clear that many regional development strategies and plans are not playing out their potentials yet. One reason is that the realization of different objectives is not always perceived in their connectedness. Currently there are only few indications of a deliberate application of a more holistic, sustainable wellbeing perspective, and of strategies that address the environmental, social, cultural, and economic potentials of territories in a more integrated fashion. To link conventional development goals with the provision of social and environmental services and longer-term societal goals remains a major challenge. Overall, a major shift is still needed from sectoral, short-term goals in regional plans to strategies that are long-term, territorial, and comprehensive. The problem is that current policy frameworks do not yet incentivize the necessary shifts towards quality of life and sustainable wellbeing. One of the few instances, where mainstream sectoral policies, and perspectives, are the main driver, is the transition towards sustainable energy. Carbon emergency and zero carbon targets are likely to lead to a wider shift of the metrics for how regional economic development is planned.

More research is needed on how to encourage the implementation of new research-based concepts and models, such as the application of ecosystem services or valorization approaches, more systematically. At the same time, it is important to remain critical and reflexive regarding the way terms, such as the bioeconomy, the circular economy, or smart growth, are used. Great care needs to be taken not to confuse *means* and *goals*. The exploratory analyses presented in this article point to a need to fundamentally shift mindsets and perspectives, especially in policy development. The initiatives that were identified are important in this respect because they show why alternatives are desirable, and that they are feasible. They, in the words of Trebeck (2020, p.1), *"open up people's imagination that a different economy is possible"* [23].

Author Contributions: Conceptualization, K.K.; methodology, K.K.; analysis, All authors; investigation, All authors; writing—original draft preparation, K.K.; writing—review and editing, All authors. All authors have read and agreed to the published version of the manuscript.

Funding: This research has been carried out in the ROBUST project funded under the Horizon 2020 Framework Programme of the European Union under Grant Agreement no. 727988. The information and views set out in this article are those of the authors, and do not necessarily reflect the official opinion of the European Union.

Institutional Review Board Statement: Not applicable.

Informed Consent Statement: Not applicable.

Data Availability Statement: Data sharing not applicable.

Acknowledgments: The authors like to thank the European Union for the funding of this research and Alexia Rouby of DG AGRI for excellent support during project implementation. We also like to thank all colleagues in the project for the great cooperation.

Conflicts of Interest: The authors declare no conflict of interest.

Appendix A

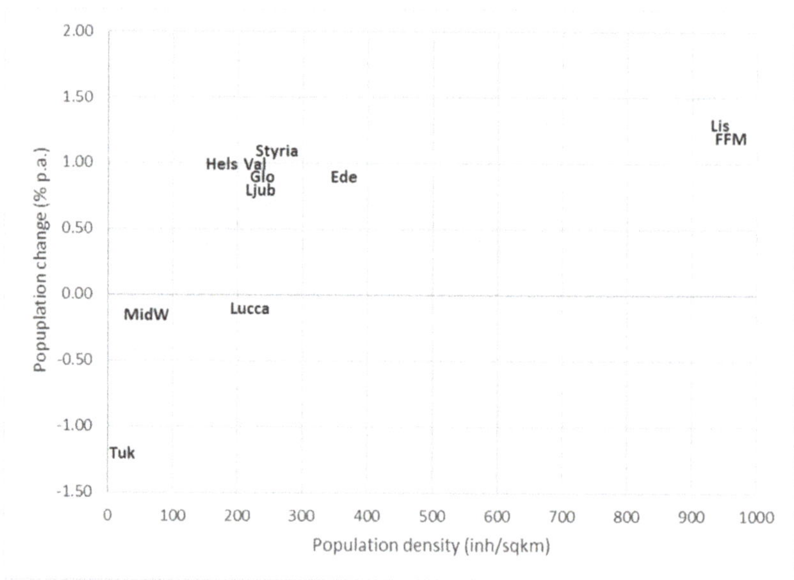

Figure A1. The 11 regions by population density and change.

Table A1. Framing of each model, key focus areas and key references.

Model	Framing	Key Focus and Mechanism	References
Classical economic growth	• Growth is central for obtaining wealth • Markets balance supply and demand	• Policies that promote economic growth • Curtailing regulation	Smith (1776) [43]
Smart growth, green growth and circular economy	• Decoupling • Recycling • Resource use efficiency	• Reducing use of limited resources • Reducing waste	COM (2010, 2017), Ellen MacArthur Foundation (2020)
Collaborative or sharing economy	• Direct interactions • Use of information and communications technology	• Wider access to goods, services • Wider distribution of resources • Socially produced information	Botsman and Rogers (2010) [44]
Distributed economy	• Economic structures and scale • Use of information and communications technology, smart coordination • Sharing	• Wider distribution of benefits and wellbeing among present generation	Johansson et al. (2005)
Eco-economy, regenerative economy	• Use of given resources • Resilience • Systemic health	• Renewal in natural, social and economic systems • Environmental sustainability and regenerative use of natural resources • Securing the wellbeing for future generations	Brown (2001), Marsden and Farioli (2015)
Foundational economy, sustainable wellbeing economy	• Wellbeing • Planetary boundaries • Social justice (standards, floors)	• Fostering a just and sustainable (safe) development • Sustainable wellbeing	Bentham et al. (2013) Raworth (2017, 2019)

Source: Authors' compilation based on the results of the review of studies, analyses, policy and strategy papers on alternative models of (economic, sustainable) development.

References

1. OECD. *OECD Environmental Outlook to 2050: The Consequences of Inaction*; OECD: Paris, France, 2012.
2. UN. Transforming Our World: The 2030 Agenda for Sustainable Development. UN Resolution A/RES/70/1 of 25 September 2015. Available online: https://www.un.org/ga/search/view_doc.asp?symbol=A/RES/70/1&Lang=E (accessed on 26 April 2021).
3. European Commission. Europe 2020: A Strategy for Smart, Sustainable and Inclusive Growth. 2010. Available online: http://eur-lex.europa.eu/LexUriServ/LexUriServ.do?uri=COM:2010:2020:FIN:EN:PDF (accessed on 26 April 2021).
4. European Commission. A European Green Deal: Striving to be the First Climate-Neutral Continent. 2020. Available online: https://ec.europa.eu/info/strategy/priorities-2019-2024/european-green-deal_en (accessed on 26 April 2021).
5. Sabir, I. Indigenous culture and the western concept of development. *Fountain* **2002**, *40*, 1–4.
6. Rauhut, D.; Humer, A. EU Cohesion Policy and spatial economic growth: Trajectories in economic thought. *Eur. Plan. Stud.* **2020**, *28*, 2116–2133. [CrossRef] [PubMed]
7. Isserman, A.M.; Feser, E.; Warren, D.E. Why Some Rural Places Prosper and Others Do Not. *Int. Reg. Sci. Rev.* **2009**, *32*, 300–342. [CrossRef]
8. Jackson, T. *Prosperity without Growth: Foundations for the Economy of Tomorrow*, 2nd ed.; Routledge: London, UK, 2016.
9. Mazzucato, M. Mission-oriented innovation policies: Challenges and opportunities. *Ind. Corp. Chang.* **2018**, *27*, 803–815. Available online: https://academic.oup.com/icc/article/27/5/803/5127692 (accessed on 26 April 2021). [CrossRef]

10. European Commission. *Strengthening Innovation in Europe's Regions: Strategies for Resilient, Inclusive and Sustainable Growth. Communication from the Commission to the European Parliament, the Council, the European Economic and Social Committee and the Committee of the Regions*; European Commission: Brussels, Belgium, 2017.
11. European Commission. *Circular Economy Action Plan, COM/2020/98 Final*; European Commission: Brussels, Belgium, 2016.
12. Ellen MacArthur Foundation. 2021. Available online: https://www.ellenmacarthurfoundation.org/circular-economy/concept (accessed on 14 April 2021).
13. Van Leeuwen, E. Rural-urban Synergies: An Explorative Study at the NUTS3 Level. *Appl. Spat. Anal.* **2015**, *8*, 273–289. [CrossRef]
14. Jackson, T. *Prosperity without Growth: Economics for a Finite Planet*; Sustainable Development Commission: London, UK, 2019; ISBN 9781844078943.
15. Raworth, K. *Doughnut Economics: Seven Ways to Think Like a 21st Century Economist*; Random House/Penguin: New York, NY, USA, 2017.
16. European Commission. Collaborative Economy. 2019. Available online: https://ec.europa.eu/growth/single-market/services/collaborative-economy_en (accessed on 16 September 2019).
17. Capital Institute. Eight Principles of a Regenerative Economy. 2019. Available online: https://capitalinstitute.org/8-principles-regenerative-economy/ (accessed on 16 September 2019).
18. Mason, P. *PostCapitalism: A Guide to Our Future*; Penguin Books: London, UK, 2015.
19. Johansson, A.; Kisch, P.; Mirata, M. Distributed economies—A new engine for innovation Archived September 28, 2007, at the Wayback Machine. *J. Clean. Prod.* **2005**, *13*, 971–979. [CrossRef]
20. Brown, L.R. *Eco-Economy: Building an Economy for the Earth*; Earthscan/Routledge: London, UK, 2001.
21. Marsden, T.; Farioli, F. Natural powers: From the bioeconomy to the eco-economy and sustainable place-making. *Sustain. Sci.* **2015**, *10*, 331–344. [CrossRef]
22. Andreucci, M.B.; Marvuglia, A.; Baltov, M.; Hansen, P. (Eds.) *Rethinking Sustainability towards a Regenerative Economy*; Springer: Berlin/Heidelberg, Germany, 2021; ISBN 978-3-030-71819-0.
23. Trebeck, K. Welcome to the Wellbeing Economy. Circular Conversations—An Interview with Katherine Trebeck. 2020. Available online: https://www.circularconversations.com/blog/wellbeingeconomy (accessed on 23 March 2020).
24. Finnish Environment Institute (Syke). Sustainable Wellbeing Remains a Distant Goal. 2021. Available online: https://www.syke.fi/en-US/Finland_and_sustainable_wellbeing/Sustainable_wellbeing_remains_a_distant_goal (accessed on 24 April 2021).
25. Bentham, J.; Bowman, A.; de la Cuesta, M.; Engelen, E.; Ertürk, I.; Folkman, P.; Froud, J.; Johal, S.; Law, J.; Leaver, A.; et al. Manifesto for the foundational economy. CRESC Working Paper 131. 2013. Available online: https://foundationaleconomycom.files.wordpress.com/2017/01/wp131.pdf (accessed on 26 April 2021).
26. The Foundational Economy Collective. *Foundational Economy: The Infrastructure of Everyday Life*; Manchester University Press: Manchester, UK, 2018.
27. Raworth, K. What on Earth is the Doughnut? 2019. Available online: https://www.kateraworth.com/doughnut/ (accessed on 22 July 2019).
28. Rockström, J.; Steffen, W.; Noone, K.; Persson, Å.; Chapin, F.S.; Lambin, E.F.; Lenton, T.M.; Scheffer, M.; Folke, C.; Schellnhuber, H.J.; et al. A safe operating space for humanity. *Nature* **2009**, *461*, 472–475. [CrossRef] [PubMed]
29. Henke, R. *Regional Land Use Plan*; Rapid Appraisal Report; Rural-Urban Governance Ar-Rangements and Planning Instruments; Rhein-Main: Frankfurt, Germany, 2018.
30. City of Helsinki. From Agenda to Action: The Implementation of the UN Sustainable Development Goals in Helsinki. 2019. Available online: https://www.hel.fi/static/helsinki/julkaisut/SDG-VLR-Helsinki-2019-en.pdf (accessed on 5 June 2020).
31. Helsinki-Uusimaa Regional Council. The Helsinki-Uusimaa Regional Programme 2.0. 2018. Available online: https://www.interregeurope.eu/fileadmin/user_upload/tx_tevprojects/library/file_1518441359.pdf (accessed on 5 June 2020).
32. Arcuri, S.; Galli, F.; Rovai, M. *Community for Food and Agro-biodiversity, Lucca, Italy. ROBUST, Rural-urban Governance Arrangements and Planning Instruments*; University of Pisa: Pisa, Italy, 2018.
33. Šūmane, S. Expressions of Urban—Peri-Urban—Rural Relationships: (New) Tourism in Tukums. 2018. Available online: https://rural-urban.eu/sites/default/files/S-TUK4%20New%20Tourism%20in%20Tukums.pdf (accessed on 26 April 2021).
34. Bauchinger, L. *Law on Planning and Development of the Province of Styria and its Regions. Rapid Appraisal Report, Rural-Urban Governance Arrangements and Planning Instruments*; Federal Institute of Agricultural Economics, Rural and Mountain Research: Vienna, Austria, 2018.
35. Knickel, K.; Kobzeva, M. *Innovation Strategy 2020 for Hessen*; Rhein-Main: Germany, Frankfurt, 2018.
36. Korpa, V.; Siliņa, L.; Šulca, G. Vietējo Ražotāju Produkcijas Īpatsvara Palielināšana Pašvaldības Publiskajos Pārtikas Produktu Iepirkumos: Situācijas Analīze TUKUMA Novadā. SIA Latvijas Lauku konsultāciju un izglītības Centrs. 2015. Available online: http://new.llkc.lv/system/files_force/aktualitates/petijums.pdf (accessed on 26 April 2021).
37. Jones, R.; Goodwin-Hawkins, B.; Woods, M. From territorial cohesion to regional spatial justice: The Wellbeing of Future Generations Act in Wales. *Int. J. Urban Reg. Res.* **2020**, *44*, 894–912. [CrossRef]
38. Sayers, M. The Welsh Doughnut—A Framework for Environmental Sustainability and Social Justice. 2015. Available online: https://oxfamilibrary.openrepository.com/bitstream/10546/346207/7/rr-welsh-doughnut-environmental-sustainability-social-justice-270415-en.pdf (accessed on 26 April 2021).

39. Rovai, M.; Fastelli, L.; Lucchesi, F.; Monacci, F. Integrated urban regeneration: The opportunity of enhancing the open spaces. *Adv. Eng. Forum.* **2014**, *11*, 338–343. [CrossRef]
40. Pina, C. *Regional Spatial Plan for the Territory of the Lisbon Metropolitan Area (PROT-AML)*; Rapid Appraisal Report; Rural-Urban Governance Arrangements and Planning Instruments; CCDR-LVT: Lisbon, Portugal, 2018.
41. Meijers, E.; Van der Wouw, D. Struggles and strategies of rural regions in the age of the urban triumph. *J. Rural Stud.* **2019**, *66*, 21–29. [CrossRef]
42. Rosário Partidário, M. *Municipal Master Plan of Sintra*; ROBUST, Rural-Urban Governance Arrangements and Planning Instruments; Instituto Superior Técnico (IST): Lisbon, Portugal, 2018.
43. Smith, A. *An Inquiry into the Nature and Causes of the Wealth of Nations*; Strahan, W., Cadell, T., Eds.; Librito Mondi: London, UK, 1776.
44. Botsman, R.; Rogers, R. *What's Mine Is Yours. The Rise of Collaborative Consumption*; Harper Business: New York, NY, USA, 2010.

land

MDPI

Article

Classifying New Hybrid Cooperation Models for Short Food-Supply Chains—Providing a Concept for Assessing Sustainability Transformation in the Urban-Rural Nexus

Katrin Martens [1,*], Sebastian Rogga [1], Jana Zscheischler [1,2], Bernd Pölling [3], Andreas Obersteg [4] and Annette Piorr [1]

1 Leibniz Centre for Agricultural Landscape Research, Eberswalder Str. 84, 15374 Müncheberg, Germany; sebastian.rogga@zalf.de (S.R.); jana.zscheischler@uni-vechta.de (J.Z.); apiorr@zalf.de (A.P.)
2 Geography—Faculty II, University of Vechta, Driverstraße 22, 49377 Vechta, Germany
3 Department of Agriculture, Fachhochschule Südwestfalen University of Applied Sciences, Lübecker Ring 2, 59494 Soest, Germany; poelling.bernd@fh-swf.de
4 HafenCity University Hamburg, Henning-Voscherau-Platz 1, 20457 Hamburg, Germany; andreas.obersteg@hcu-hamburg.de
* Correspondence: katrin.martens@zalf.de

Citation: Martens, K.; Rogga, S.; Zscheischler, J.; Pölling, B.; Obersteg, A.; Piorr, A. Classifying New Hybrid Cooperation Models for Short Food-Supply Chains—Providing a Concept for Assessing Sustainability Transformation in the Urban-Rural Nexus. *Land* **2022**, *11*, 582. https://doi.org/10.3390/land11040582

Academic Editors: Stephan Bartke and Sigrun Kabisch

Received: 14 March 2022
Accepted: 11 April 2022
Published: 15 April 2022

Abstract: In response to the negative effects caused by structures of the dominant agricultural system and new market opportunities, increasing food supply structures have re-emerged in the urban-rural context of industrialized countries in recent years. These food supply structures often accompany new forms of hybrid cooperation models, including actors and institutions that have not shared resources previously. They form new alliances for sustainable transformation in the agri-food sector. Simultaneously, discourse has arisen in science and practice about the sustainability potential of such hybrid cooperation, referring to a lack of critical systematization and the necessity for creating an assessment concept. From the latter, one could draw conclusions about the transformative potential of such cooperation models and their potential to serve as blueprints for other regions. In this conceptual paper, a classification approach derived from social enterprise literature is elaborated, extended, and evaluated, to design a classification of new hybrid cooperation models that allow comparisons between regions and are sensitive to their dynamics. We show in an application how the classification approach, considering the dimensions "actors", "resources", and "actions", serves to discover patterns in the development of short food-supply chain practices, identifying individual transition paths and, thus, making statements about their sustainability and challenges.

Keywords: agri-food systems; collective action; cooperation; pooling; urban-rural interlinkages; short food-supply chains; regional food systems; hybrid organizations

1. Introduction

Scientists argue that, exacerbated by the COVID-19 pandemic [1–3], we are at a tipping point of our unsustainable and non-resilient globalized food system [4–6]. The premise of providing sufficient and affordable food to as many people as possible has led to the dominance of global agri-food system structures that are highly concentrated and follow the prevailing paradigm of productivity and efficiency, to the detriment of the environmental, governance and social dimensions of sustainability [7,8]; for example, in terms of global warming [6], there is water and soil pollution [9–12], a decreasing diversity of crop species [13–15], soil erosion [16] and a loss of biodiversity in agricultural systems [17].

Thus, alternative, more sustainable conceptualizations of agri-food systems are increasingly discussed, highlighting the importance of the territorial embeddedness of agri-food systems and the social relations of proximity [18,19]. In this context, short food-supply chains (SFSCs) in the urban-rural nexus [7,20,21] are considered to be one "building block" [22] of the sustainable transformation complex. Many studies have shown that SFSCs contribute to the sustainable transformation of the agri-food supply system [7]. One reason for this is seen in the reconnection of urban consumers to producers in the surrounding areas of the cities [23,24]. This argument is reinforced by Bartke et al. [25], who stress that urban and rural areas are interdependent. Their resources and functions, such as soil, land, freshwater, waste, energy and food supply, are closely intertwined. In addition, the fact that more and more local products and new SFSCs are emerging in Europe [7,26] reflects the increased interest ("turn to quality") of consumers in regional food production and supply [23,27]. This creates a novel opportunity space for the farmers remaining in the urban-rural environment to create new business strategies [28] which has also been recognized and advocated by the European Union (EU), which defines SFSCs as "a supply chain involving a limited number of economic operators, committed to co-operation, local economic development, and close geographical and social relations between producers, processors and consumers" [29].

However, there is a lack of knowledge regarding how new SFSC initiatives emerge in the urban-rural context, and what strategies and organizational models can foster sustainable transformation [30–32]. Georgiadis et al. [33], for example, criticize researchers for ignoring the role of interdependencies between the actors involved in a food system. Doernberg et al. [34] and Ilieva [35] point out that local governments focus primarily on sectoral approaches. For instance, the EU definition of an SFSC ignores the multifunctional and hybrid characteristic of SFSCs as it focuses only on economic operators, ignoring public actors and civil society, which are identified as important actors within SFSCs [22,25,34,36]; this shows that multi-governmental and multifunctional topics, such as the sustainable transformation of urban-rural SFSCs, are still under-researched. While municipalities are beginning to acknowledge that local SFSCs are one field of action that contributes to sustainable transformation [25,37], there are a lack of tools to classify and systematize the projects and activities emerging in their areas, which makes it challenging to contribute in cross-scale cooperation and knowledge-sharing with other governments [22,34,38–40]. This lack of knowledge is seen as one of the reasons why there are not only proponents of strengthening urban-rural linkages through SFSCs [7,41], which highlights the need to provide a concept for assessing SFSCs in the urban-rural context; this can help us to better understand their hybrid nature, resource-pooling strategies, challenges, and potentials, and, thus, their sustainability potential in the long term.

In order to characterize the hybrid nature of SFSCs, we argue that it makes sense to look more closely at their role in the overall economy and their interaction with the market, civil society and public policy [42]; this is because the cooperation of actors, and, thus, the pooling of resources, is considered essential for sustainable transformation [25,43]. This also means that this article's units of analysis are new cooperation models that aim to shorten the food supply chain to drive sustainable transformation. In this context, the new cooperation models are hybrid forms of organization [44], i.e., a new cooperation of at least two actors pursuing the dual mission of financial sustainability and a social purpose [45]. New models of cooperation are about pooling resources to fulfil this dual mission "to the emergence of novel institutional forms that challenge traditional conceptions of economic organizing" [46], cited in [44] (p. 418).

A screening of examples in the literature and practice shows that it is challenging to classify emerging cooperation models in the agri-food sector due to the diversity of actors, strategies, and organizational forms, as they follow the dual mission; thus, they do not follow the usual logic of the market, similar to social enterprises. Social enterprise theory deals with the contextual factors that lead to the creation of social enterprises, the underlying organization dynamics and structures, and how different types generate social impact, mobilize resources, and bring about sustainable social change [47]. To apply the concept of social enterprises to new cooperation models in the agri-food system seems purposeful, and is justified by the explicitness of their objectives; these go beyond the economic dimensions and integrate motivations of environmental, social and political concern [48,49], for example, through action against biodiversity loss, for societal inclusiveness or food sovereignty [21,38,50].

This paper aims to put the aspect of cooperation at the center of the analysis. We propose a classification approach derived from social enterprise literature that allows us to highlight the complexity of the actors, measures, and approaches to sustainable resource use and their impact levels. Building on the increasing pressure to promote sustainable transformation processes in the rural–urban nexus and advancing the emerging scientific discussion on the potential of SFSCs in this context, our concept paper aims to make an important contribution. Therefore, the following Section 2 presents our concept with the dimensions "actors", "resources" and "actions" (ARA). In addition, we evaluated the concept using real-world examples. The evaluation process is presented in Section 3, and in Section 4, the resulting findings are outlined. The discussion in Section 5 focuses on the questions of potential and challenges of the ARA concept.

2. Classifying Urban-Rural Short Food Chain Cooperation Models

Capturing new cooperation models in urban-rural SFSCs requires a classification that maps both established conventional business models and new forms of cooperation, thus, enabling statements about their novelty and distinctive features regarding a transition towards sustainability. To this end, this section presents a classification approach that has been adopted and further developed from the literature on social enterprises. We generally base the classification on three dimensions: firstly, an actor dimension; secondly, a resource dimension; and thirdly, an action dimension. In addition, we adapted the "basic governance triangle", composed of state, market, and civil society, to display the purposes and more straightforward utilization of the classification. Distinguishing key institutional arrangements that may give 'structure' to collective behavior within society is widely used in social sciences [51–54]. The ARA concept covers the following research questions, which will be explained in more detail below in 2.2:

1. Actor dimension: which actors cooperate with which interests?
2. Resource dimension: on what resources is the cooperation built?
3. Action dimension: which steps along the SFSC does the cooperation model represent?

In order to understand the link between social entrepreneurship and urban-rural SFSC cooperation models, Section 2.1 serves to provide background information and implications for our particular context and reflect critically on them. Finally, Section 2.2 presents the three dimensions according to which the cooperation models are to be classified.

2.1. Critical Reflection on the Social Enterprise Classification and Its Use to Classify SFSC Cooperation Models

Within the urban-rural nexus, SFSCs are assumed to contribute to various sustainability dimensions [7]. Although the number of studies in this context is increasing, attention is seldom paid to how these emerging SFSCs are organized and formalized. What is known is that those urban-rural cooperation models, particularly their formalization and organization, do not often fit into the conventional economic landscape represented by the market and the regulatory role of the state. Moreover, their emergent processes and practices are poorly understood. An important issue here is the purpose of these cooperations compared to those in the conventional economic landscape. Similar to what Defourny and Nyssens [42] found in their attempt to systematize social enterprises, new cooperation models in the agri-food system do not often aim primarily to generate profits, but rather, to fulfill one or more social missions, with social enterprises often employing practices such as redistribution or reciprocity [53]. Whether these are also important practices in urban-rural SFSC cooperation models needs to be identified.

Although it has been discussed for a long time that redistribution and reciprocity are essential elements to understand how societies work [55], it was only in the 1980s and 1990s that they were used to conceptualize social enterprises. New entrepreneurial dynamics emerged in Europe at that time, intending to find solutions to structural unemployment, high government budget deficits, the exclusion of certain social groups from the labor market and a lack of adequate public policy programs [56]. In order to capture these activities, Evers and Laville [57] and others labeled these emerging organizations—such as associations, cooperatives and foundations—as private nonprofit organizations, and assigned them to the third sector [53,57]. The latter is the space that accommodates different kinds of nongovernmental and nonprofit institutions [58]. The need to define a new 'action space' for these organizations has also been recognized by policymakers; the Italian government created a legal concept for social cooperatives in 1991, and the EU Commission, in 2017 [59], defined social enterprises as organizations that "combine social objectives with an entrepreneurial spirit" and focus on "achieving broader social, environmental or community objectives" [60].

Social enterprises, thus, emerge when social entrepreneurs establish new cooperation models to apply commercial activities to fulfill their social mission [61]. According to Defourny and Nyssens [62], the social mission can be justified in several ways. Noting that this list is not exhaustive, they identified three possible levels of social mission. The latter exists when, firstly, products or services are produced to address a social problem that other companies or the public sector cannot solve. Secondly, the social mission can be found in how social actors interact, for example, by adopting innovative organizational methods that integrate disadvantaged groups. Thirdly, the social mission can also be fulfilled by addressing broader societal values, such as democracy or sustainable lifestyles [62,63]. These models of cooperation that aim to operate at one or more of these levels vary widely and are, therefore, difficult to classify. This difficulty is also reflected in Defourny and Nyssens' [56] statement when applying their classification approach in Europe. They found that the form and objectives, and the organizational forms chosen, vary between and within countries.

Social enterprise and hybrid organization research arose primarily to find solutions for the social dimension of the sustainability debate, i.e., topics such as labor integration, the financing of social tasks, and education [42,44,59]. Therefore, the proposed classification is also suitable for systematizing new urban-rural cooperation models in SFSCs. While it was primarily social challenges that led to the creation of social enterprises in the past, we argue that today, in early industrialized countries, environmental issues are important drivers alongside new social challenges, such as increasing spatial inequality between urban and rural areas, and unhealthy diets. Thus, cooperation models are now emerging to address these challenges. Defourny and Nyssens [56] agreed that social enterprises serve multiple purposes, such as environmental practices or local development. Indeed, when mapping social missions in social enterprises worldwide, Defourny et al. [53] found that ecology, nutrition, community development, and capacity building are four of the top ten social missions and, thus, have clear links to urban-rural agri-food systems.

Accordingly, social missions do not pursue social goals exclusively, but often follow a sustainability program that links ecological, social and economic objectives to democratic action. These arguments lead to the conclusion that sustainable missions are social missions, and new SFSC cooperation models can be understood as forms of social enterprise. Therefore, new cooperation models targeting the urban-rural SFSCs could also be understood and studied using the social enterprises' classification approach. However, more knowledge about their dynamics is needed at this point. In Section 2.2, we present the different dimensions used to classify social enterprises, which help to specify our research questions.

2.2. *Capturing Cooperation Models Dimensions: The ARA Concept*

The ARA concept focuses on understanding the interplay between actors, resources, and action within urban-rural SFSC cooperation models. These ARA dimensions will be introduced from Section 2.2.1 onward, along with our understanding of urban-rural SFSCs in Section 2.2.3.

2.2.1. Actor Dimension

Identifying important actors within new cooperation models requires linking their actions to specific behaviors and characteristics. Following the classification of Defourny and Nyssens [42] and Defourny and Pestoff [58], we propose focusing on the drivers and interests of actors that also cover the main actors in urban-rural linkages, which is an essential prerequisite for our approach. Learned initially from Gui [64], the three principles of interest—mutual, general, and capital—are distinguished by marking the angles of a triangle (Figures 1 and 3). In this way, it is possible to discuss the relationship between these different motives by focusing on the specific responsibilities and characteristics of the actors [58,65,66]. For simplification, we apply a classification approach, being aware that these ideal actor types are not always reflected in reality [54,62]. Taking the three types of actors—private, state, and community actors—with their motives as a starting point, we describe their "traditional" action space and principles in Table 1a, while Table 1b depicts hybrid cooperation models with enlarged action space, following definitions and empirical descriptions from the literature [42,58,62]. New cooperation models can be mapped in this way. The social enterprise literature argues that individual actors retain their interests and capabilities regardless of their cooperation with other actors in the same space (see Table 1a) or with actors in different spaces (see Table 1b). Thus, a wide variety of hybrid cooperation models can be described theoretically [67].

Table 1. Summary and examples of action spaces and their characteristics.

	Principles/ Mechanism	Main Interest	Cooperation Partners and Examples from the Agri-Food Sector
1a Traditional action spaces			
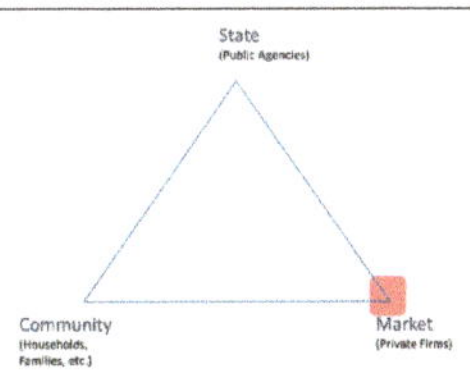 Market Action Space	○ Pricing, contracts, efficiency, (steering) supply-demand relationship, capital interest	○ Maximizing profit	○ Market-space actors ○ Examples: farm businesses, supplier companies, processing, and marketing firms
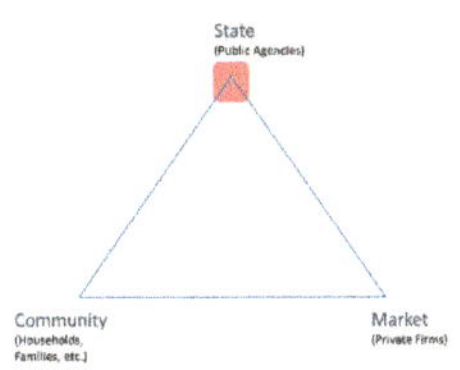 State Action Space	○ Redistribution through, for example, payment rules, policies, subsidies, information, and regulation	○ Providing public goods	○ State-space actors, such as governments on different levels, state employees, research institutions ○ Examples: CAP payments, urban food policies
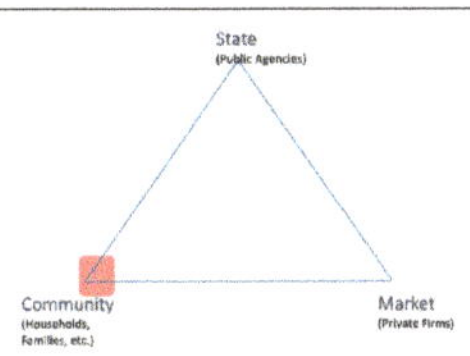 Community Action Space	○ Reciprocity, mutuality, social relationships	○ Circulating, goods and services, fulfilling common interests	○ Community-space actors, such as households, civil society, self-help, and mutual groups ○ Examples: Food Policy Councils

Table 1. *Cont.*

	Principles/ Mechanism	Main Interest	Cooperation Partners and Examples from the Agri-Food Sector
1b Hybrid action spaces			
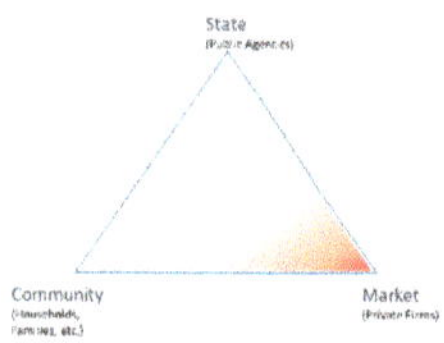 Market-Third Sector Action Space	○ Applying market principles together with redistribution and or reciprocity principles ○ Cooperation models "that operate entirely in the market and seek profits while applying different rules than typical capitalist enterprises" [42] (p. 14)	○ Maximizing profit	○ State-space actors and/or community-space actors ○ Examples: community-supported agriculture
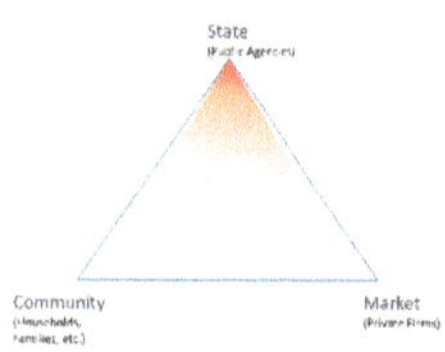 State-Third Sector Action Space	○ Mixing redistribution principles with market principles and/or reciprocity	○ Providing public goods	○ Market-space actors and/or community-space actors ○ Examples: Swedish farmers cooperatives, which were heavily involved in regulating agricultural production, prices, import quotas, and export subsidies before their membership in the EU and the CAP [65]
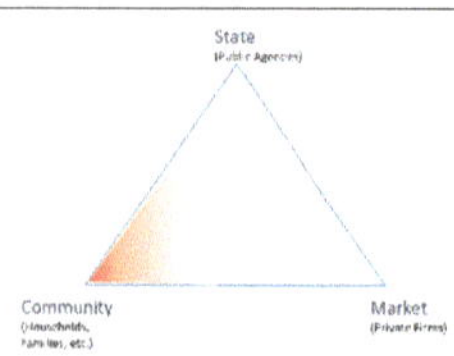 Community-Third Sector Action Space	○ Often characterized as nonprofit ○ Using infrastructure or redistribution capacities of governments and/or market principles	○ Fulfilling common interests	○ Market-space or/and state-space actors ○ Example: agriculture cooperatives

Starting with the market action space, private actors operate primarily according to the market principle. They operate within a convergence between supply and demand and draw their resources from the sale of products and services exchanged through pricing. The relationship between the supplier and the buyer is usually a formal commercial contract. According to Evers and Laville [57], capital interests cooperate with individual business units that focus primarily on profit-making and profit maximization. In most democracies, the state action space is represented either by actors elected to fulfill public interests or people employed by the government. Their task is to put the common good in the foreground with its actions, and to regulate access to public goods. To this end, state actors often use the principle of redistribution. On this basis, decision-making powers are transferred to central authorities, which are responsible for their administration and acting based on established payment rules and targets [57]. The community action space is represented by civil society or the household, and often operates based on the reciprocity principle; this refers to mutuality and is, thus, a fundamental principle of human action. Evers and Laville [57] describe reciprocity as the circulation of goods and services between groups and individuals, which can only take shape if all parties are willing to enter into a social relationship. According to this, reciprocity is an original noncontractual principle of economic activity in which the social connection is more critical than the goods exchanged. Following Gui [64] and Defourny and Nyssens [62], we adopt the view that in this reciprocity, actors aim primarily to satisfy the needs of the members of a community. Accordingly, action is not founded to maximize the return on capital, but to fulfill a general or common interest [64], contribute to the common good, or meet the social demands of specific populations.

These traditional action spaces in Table 1a indicate a separation of action where entrepreneurs only work together within one's action space. However, the literature on social enterprise shows that hybrid cooperation models exist alongside these traditional action spaces [42,58,65]. These activities arise when the market, state, or community services no longer meet the required demands or are no longer accessible to certain groups [58,68]. Martens et al. [54], for example, show multiple cooperatives founded in rural areas due to the increasing lack of public infrastructures, such as freshwater supply or swimming pools. Meanwhile, action outside the market space occurs when the services provided by private actors are not affordable for the people who need these services, for example, in the housing sector [54,58]. Table 1b, therefore, illustrates further action spaces and examples resulting from cooperation models where the market, state and community actors interact with each other, forming hybrid cooperation models. These interactions are not exhaustive, as cooperation models could emerge at any point in the triangle.

2.2.2. Resource Dimension

The question when researching hybrid cooperation models that are founded to apply more sustainable practices is how the accompanying pluralistic objectives can be fulfilled, compared to cooperation models with only one goal. The resource issue plays an important role here. The assumption is that resource pooling is necessary to compensate for the economic disadvantages caused by the rejection of the profit accumulation narrative. Thus, in addition to the actor dimension, another important dimension of Defourny and Nyssens' [62] classification approach is the resource dimension, i.e., the analysis of whether the hybrid cooperation model is built primarily on market revenues, public funds or philanthropic resources. Moreover, the resource dimension can be used to examine which mix of resources is likely to be successful, i.e., can lead to a balance between the social mission of the cooperation model and its long-term sustainability. This also includes the question of how the costs and benefits are shared between the different actors, and shows that the actor and resource dimensions are closely linked [58,62]. Evers [67] argues that academic discourse on this issue lags behind the actions of public authorities, which have long begun to integrate non-state resources into their policies to ensure the sufficient provision of public goods. Even though research on hybrid organizations has increased in recent years, the critique still seems relevant.

As illustrated in Figure 1, the resource types in the classification triangle can be added by dotted lines, indicating the areas where the different resource types are likely to occur; they are dotted because exceptions may occur. Providing examples of hybrid cooperation models of the agri-food system, Defourny and Nyssens [62] explain the boundary between hybrid and market dominance, which converges in the community corner of the triangle, by arguing that this angle is where conservative agricultural cooperatives are located. The latter are member-based organizations, thus, serving a particular community; however, they also operate primarily in the market, drawing their resources predominantly from market action. Another example of hybrid cooperation models situated between the community and the state relying on a mix of nonmarket- and/or market-based resources could be a Food Policy Council initiated by the community but supported by a public actor—for example, by providing infrastructure or funding a coordinating workforce.

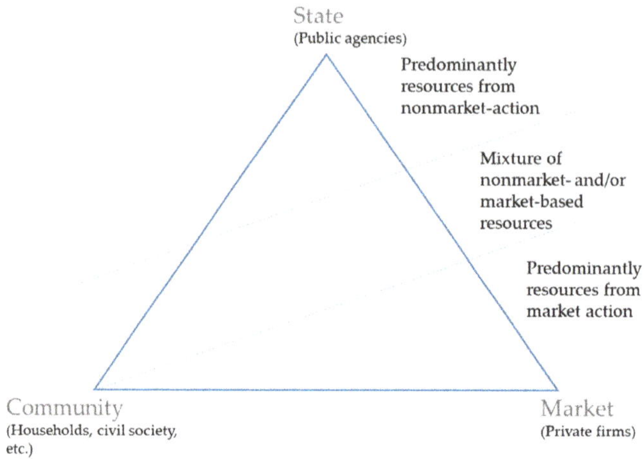

Figure 1. Resource dimension.

2.2.3. Action Dimension

While the actor and resource dimensions are already established in the social enterprise literature, this version of the concept adds the action dimension. It is argued that this fills an essential gap in the literature, as the action dimension allows one to study hybrid cooperation models that emerge in specific sectors; this highlights sector differences, potentials, and challenges due not only to sector specifications, but also to their territorial context. Referring back to the title of this study, it is argued that integrating the action dimension provides a basis for assessing the sustainability transformation potential of hybrid cooperation models.

This study focuses on the agri-food sector within the urban-rural nexus by investigating hybrid cooperation models established to shorten the food supply chains. A lot of studies in recent years have dealt with the definition and importance of SFSCs for sustainable transformation, highlighting the producer-consumer interaction [7,23,69–72]. In this study, we are particularly interested in how hybrid models of cooperation reestablish the links between urban and rural areas and pursue sustainability-oriented goals with their actions in the context of input requirements, such as land availability and seeds, food production, processing, and distribution. We argue that these latter four action fields of SFSCs can be initiated by different actors, namely those presented in the actor dimension in Section 2.2.1. Following this logic, the perspective here is less about the role of the producer or consumer and more about the role of the private, public, and civil society sector; thus, it prompts the following questions: Who provides resources for production? Who produces, processes, and distributes food in the urban-rural nexus? Therefore, it opens new perspectives on the topic. With this approach, we follow Aubry and Kebir [73] and Renting et al. [74], with their more technical understanding of SFSCs, which allows us to examine the new cooperation models according to the four action fields of SFSCs, which are further defined in Table 2. We assume that this will help to investigate the differences between the action fields and to ascertain whether certain areas are more likely to be pooled together than others. It is also possible to illustrate which hybrid cooperation models are created by which actor constellations, and which resource-pooling mechanisms are used.

Table 2. Action fields within urban-rural short food-supply chains (SFSCs).

Icon	Action Fields	Definition
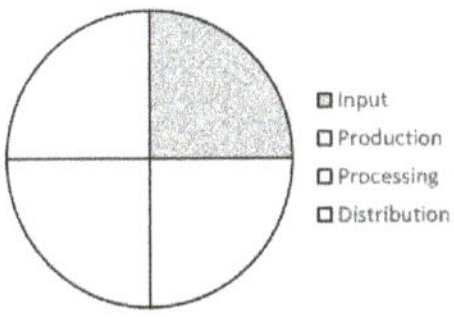	Input	Access to land, knowledge, machinery, and financial resources is essential for implementing new SFSC cooperation models [23]. This action field is fulfilled if the cooperation model contributes purposefully to making land, financial resources, and knowledge accessible for the emergence of SFSCs.
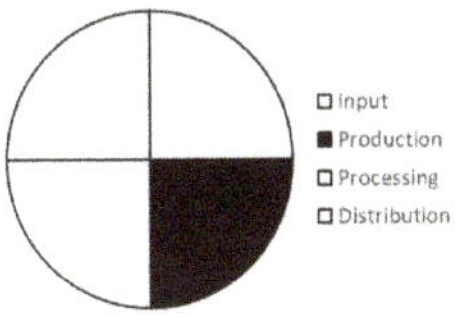	Production	The cooperation model aims at producing local raw products.
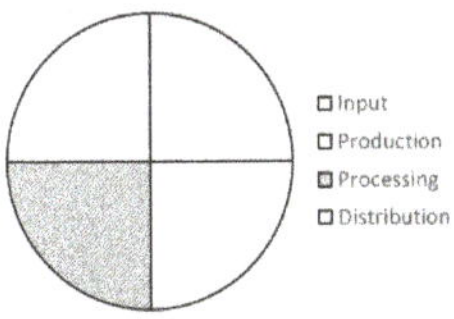	Processing	The cooperation model processes raw products into products of higher/different value.
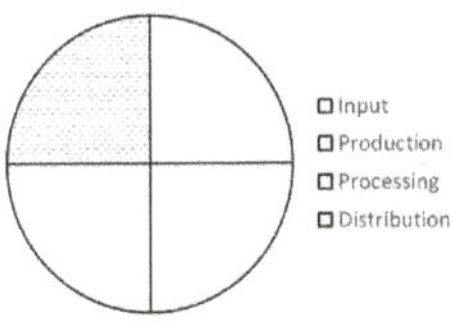	Distribution	The cooperation model supplies or sells products to consumers. Thus, this action field includes different channels of delivering food to the end consumer, such as community catering and farm shops.
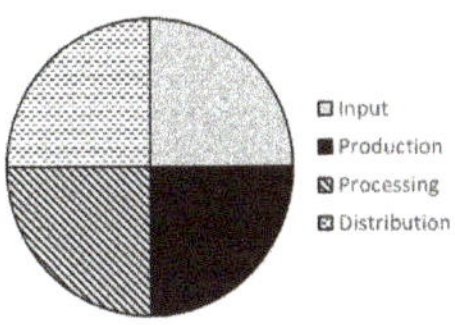	Combinations	All four steps identified can be combined as cooperation models, sometimes taking over not one but several action fields or the whole SFSC.

Following the argument that integrating the action dimension into the analytical concept allows the linking of hybrid cooperation models to their territorial context, other existing definitional approaches for SFSCs are worth considering. The notion of producer and consumer is recognized as valuable for further elaboration of the strong connection between SFSCs and urban-rural linkages. Industrialization, globalization, and agricultural modernization in the Global North terminated the geographical dependency and symbiotic relationship between cities and the local population [75–77]. Farms located in or close to cities and agglomerations exploit the producer–consumer proximity as a locational advantage by establishing SFSCs, which eliminate all or most of the intermediaries of long food supply chains [73,78]. The geographical and organizational proximity of producers and consumers are crucial [73]; while strong geographical proximity can be achieved by establishing urban-rural SFSCs, organizational (solid) proximity is created by direct producer–consumer relationships and channels. Contrarily, weak geographical and organizational proximity are non-regional, and are characterized by indirect and disconnected producer–consumer relationships [19,48,79]. Direct sale is the most direct supply chain connection between producers and consumers. Nevertheless, indirect relationships within urban-rural areas can also be defined as short when only one or very few intermediaries are transparently positioned between producers and consumers.

We argue that this transparency can be ensured by integrating the action dimension and, thus, highlighting actor constellations and resource-pooling mechanisms. An increasing number of publications show that SFSCs target public-, private-, and community-space actors [7], while less is known about how they are connected and organized and, hence, how resource pooling works bests. Ilieva [35], for example, points out that urban areas are "new spaces for food system innovation". At the same time, Cretella [80] and Doernberg et al. [34] showed that urban governmental body areas are increasingly engaging in the local agri-food sector, often in response to self-organized civil society initiatives for fair and healthy food.

Therefore, it is necessary to integrate the action dimension to capture the great diversity of dynamics that can emerge within the urban-rural nexus. It is assumed that SFSCs are rarely mapped holistically from the beginning, but emerge from individual cooperation models whose strategies need to be investigated. The extended version of the system makes it possible to not only analyze the resource mix and entrepreneurs involved in these new initiatives, but also to map the actions addressed by the initiatives in the SFSC and, in a following step, show the differences and reasons behind the discrepancies between regions.

3. Evaluation of the Proposed ARA Concept

This study was carried out as part of the KOPOS research project—a transdisciplinary research project over the period 2020–2025 that focuses on new forms of cooperation and the pooling of resources in the transformation of the agri-food system—and aims to strengthen urban-rural linkages. The project operates closely with science and practice partners, and adopts real-world laboratory design elements [81]. KOPOS draws on real-world issues of societal relevance as starting points of investigation: (1) access and security to arable land for sustainable farming practices in the German region of Berlin Brandenburg, and (2) SFSCs in the German region of Freiburg. KOPOS generally aims to develop and test new cooperation models that tackle the above-mentioned real-world issues in two case study regions, each presenting a real-world laboratory. In addition to the real-world laboratory in Berlin Brandenburg, which deals with access to and the security of land, the practice partners in the Freiburg region are working on the topic of SFSCs. They were subsequently consulted for the evaluation of the proposed ARA concept.

The urban-rural region of Freiburg is scientifically interesting for the investigation of SFSCs. Located in the federal state of Baden Württemberg in Southwestern Germany, it covers the area of the city of Freiburg and the two neighboring districts of Breisgau-Hochschwarzwald and Emmendingen. The project partners chose this spatial definition because it overlaps entirely with the spatial dexterity of existing policy cooperations ("Region Freiburg" and "Biomusterregion Freiburg") and previous studies on regional food supply [82].

The Freiburg region has suitable prerequisites for a substantial regional food supply due to good climatic and geographic conditions, and a large agricultural diversity from arable land, vineyards and grasslands. Furthermore, due to the Biodiversity Strengthening Act (July 2020), the intention is to reach a share of more than 30% of organic farming within the area in Baden Württemberg to be used agriculturally by 2030. The share of organic cultivation in 2020 reached 12.3% [83]. The region is also known as a hotspot of innovative action in the agri-food sector. Some of the farms of the Freiburg region were pioneers in organic farming for Germany as early as the 1950s [84].

The real-world lab in Freiburg involves numerous actors from science and practice. There is coordination between representatives of the city, the district, the regional food council, and non-university and practice-oriented research institutions. In addition, five cooperation models (hereafter, pilots) were selected in the first year of the project through a tendering and selection process. The criteria included: (1) cooperative elements within the pilot; (2) innovativeness of the approach; (3) perspectives for urban-rural linkages;

(4) location and action area within regional borders; (5) relevance to action fields (see Table 2); and (6) sustainability orientation. In addition, we specifically paid attention to the early business/consolidation stages of the pilots. Thus, the latter can be considered innovative forms of potential cooperation models within the SFSC. The five pilots were each supervised and accompanied by a local KOPOS practice partner (aka "mentor") to guarantee two-way communication with the KOPOS project.

In order to discuss the ARA concept from the perspective of not only academic debate but also practitioners, we conducted semi-structured interviews with the five mentors of the five pilots. Due to the COVID-19 situation, the interviews were only partially conducted in person, with four of the five interviews conducted via the digital conference platform Zoom. All interviews were fully recorded and transcribed.

The guiding questions were divided into three parts. First, questions were asked about the mentor's role in the project and their expertise. In addition, questions were asked about the relationship and frequency of interaction with the pilots to ensure that the mentor could make statements about the pilots. The second part focused on the pilots. Mentors were asked to describe what actors, resources, and actions they considered crucial in their pilots. This information was essential to interpret the classification completed in the third part of the interview. In the second part, mentors were also asked what contribution the pilots make to the sustainable transformation process. This information significantly fed into Table 3 in the following section. In the third part, the mentors classified the pilots using the triangle seen in Figure 1. For this purpose, the resources and actors' dimensions of the ARA concept were initially explained to the mentors. The mentors were then asked to mark where they currently saw their pilot and why. Mentors were also asked to discuss where they would like to see the pilot in five years. In doing so, we wanted to see whether the classification tool was understood and if the tool could be used to map the development of hybrid cooperation models. In addition, workshops, project meetings, and websites were consulted for information about the pilots, to conduct the evaluation and determine the strategy plans completed by the pilots.

The pilots are briefly introduced here in order to better follow the presentation of the results. The first pilot aims to bring together farmers, small distillers and beekeepers, and food producers and consumers, by creating a regional brand and a dedicated shop. The project focuses on biodiversity and inclusion, as the shop will employ people with disabilities. The cultivation of old plant and grain varieties by local farmers is intended to give food a more local touch and enhance the ecological value of field cultivation. The second pilot is an association which emerged from a parents' initiative. Utilizing two cooks and many volunteers, the pilot aims to prepare a fresh lunch for the children of the participating day-care centers every day. The focus is on the quality of the food. It should be cooked in a suitable way for children, regional and seasonal, with animal products from organic farming. In order to achieve these goals, contracts are to be concluded with local farmers and cooperations, and entered into with sustainable producers. The third pilot is the emerging food hub in Freiburg. The initiative wants to create places where good food is distributed and sold. Oriented on the principles of the solidarity economy, people throughout the entire region are to be given access to socially and ecologically produced food, and sustainable agriculture is to be ensured. To this end, the network wants to offer advice and networking and the complementary infrastructure of logistics or marketing. The fourth pilot is a spin-off of a larger think tank, which promotes primarily regional social–ecological projects through a public limited company. This new spin-off wants to improve the organic offerings in regional community catering. Accordingly, 100% organic products (mainly) from the region are to be processed and cooked. Different types of kitchens are planned under the roof of the new commercial kitchen. In addition, the latter also offers space for a preserving kitchen to strengthen food craft in the region. Positive synergy effects are to be achieved and food waste avoided by sharing the premises and centralizing work processes (e.g., pooling ordering procedures). The fifth pilot is an organization of various organic farmers who have set themselves the task of building up a previously missing element of SFSCs, namely, a carbon-neutral storage facility to supply the urban population with vegetables produced in the surrounding area, even in winter. A new building constructed according to the latest ecological standards is planned for this purpose.

4. Results

This paper aims to establish an analytical concept for classifying new hybrid cooperation models of SFSCs in urban-rural contexts, which will now be evaluated based on five pilots. Since the pilots were selected from a larger sample according to a set of common criteria, it is argued that they are representative of social enterprises and, thus, organizations that can contribute to the sustainable transformation of a region, as described in Section 2.1. To illustrate what this means in practice, the mentors were asked to describe the added value of their pilots for sustainable transformation, which has been summarized in Table 3. It can be seen that the initiatives often pursue several social missions and, thus, fulfill an important criterion of a social enterprise. In this result section, the dimensions of the ARA concept for all pilots are examined.

Table 3. Identified impact delivered via short food value chain cooperation models (according to the description of the mentors, which was compared with the self-representation of the pilots on the project website).

	Social Dimension	Economic Dimension	Ecologic Dimension	Political Dimension
Pilot 1	○ Farmers see that something is being done in the region ○ Strengthens urban-rural connection ○ Educational concept ○ Inclusive	○ Creation of an own brand for regional products	○ Promote the shortening of supply chains ○ "Not organic farming per se, but still very close to nature"	○ Helps to create knowledge about the legal form chosen, nonprofit limited liability company ○ Role-model function of the mayor, take initiative ○ It is a rural initiative that could exist like this in many other places
Pilot 2	○ Educational offers for children (creating a relationship to food production) ○ Increasing multigenerational learning by bringing together younger and older generations ○ Revitalization of the village center	○ Secure income for farmers ○ Lower transportation costs	○ Promote SFSCs ○ Process mainly organic products ○ Processing of second-choice vegetables (reduce food waste)	○ Reliability of the municipality (security) ○ The idea to extend the project to other municipalities with a similar structure

Table 3. *Cont.*

	Social Dimension	Economic Dimension	Ecologic Dimension	Political Dimension
Pilot 3	○ The initiative connects different associations (one that shows refugees' perspectives in the agri-food sector, another that works with people with disabilities) ○ Initiative tries to find a fair price mechanism so that everyone can afford regional food. ○ Wants to create/increase recognition of producers (e.g., their role as landscape protectors) ○ Strengthening the urban-rural relationship ○ Offer training ○ Raise awareness for children	○ Strengthen small, regional producers	○ Offer seasonal, environmentally friendly cultivated products	○ Integrate politics and authorities in the project, keep politics up-to-date
Pilot 4		○ Promoting regional business with local players	○ There should be 100% organic food, 50% of which should be regional (set incentives for more organic production in the region) ○ Internalization of the ecological costs	○ Contribution to more sustainable public procurement
Pilot 5		○ Creation of missing infrastructure ○ Creation of greater diver-sity/independence ○ Typical marketing structure for ecologically produced food from the region	○ Storage facility built according to ecological standards	○ Contribution to the change in public procurement

4.1. Actor Dimension

In a second step, the pilots were classified by the mentors in the triangle. We applied the mapping to distinguish the current state in Figure 3, and the envisioned state in five years in Figure 2, to grasp the dynamic component of the transformative vision, being part of their initiative concept.

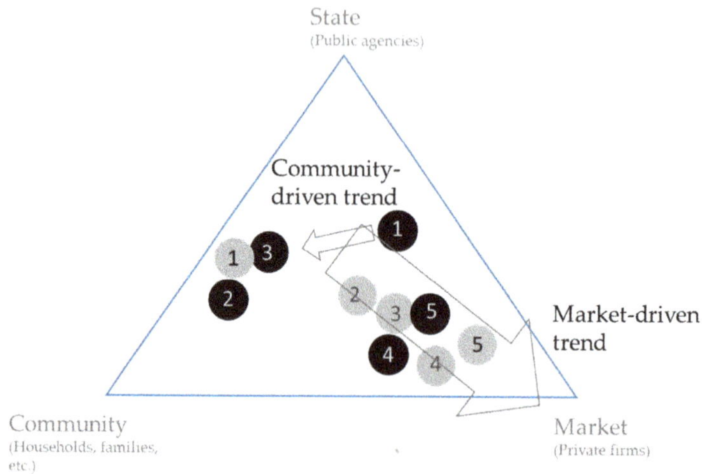

Figure 2. Expected situation in five years, mapped by mentors (current situation—black; situation in five years—grey; arrows show trends identified).

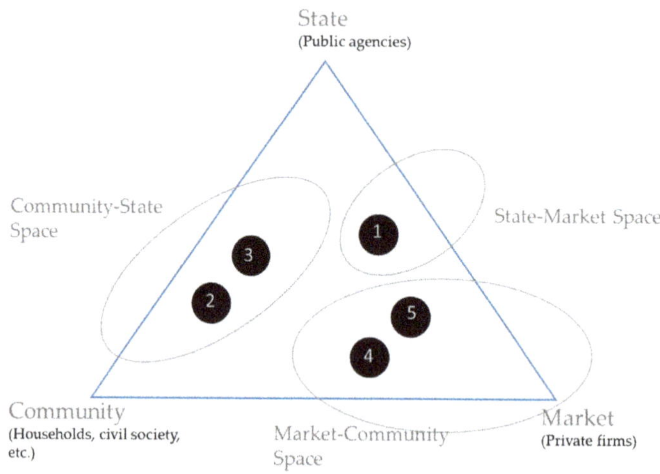

Figure 3. Current situation, mapped by mentors.

Regarding their current status, pilots 4 and 5 are initiated within the market-community space. Pilot 5 is a hybrid of market and community, as the initiative was developed by an existing organic farmers' organization formalized as a limited liability company. Therefore, following the argumentation by Defourny and Nyssens [62], they need to be separated from farmer organizations that chose to be formalized as agricultural cooperatives. This means they do not have such a strong community focus and democratic structure and, thus, need to be classified closer to the market space, as shown in Figure 3. Following the mentor's description, however, pilot 5 also does not operate solely in the market space, as it pursues multiple social and ecological goals, as described in Table 3. Interestingly, pilot 4 is also formalized as a limited liability company (abbr. in German: GmbH), and is classified more as a community space. However, the classification of this actor requires a look at the resource dimension, which will be explained in more detail in Section 4.2. Pilot 2 is located in the community-state space because it was initiated by parents seeking better nutrition for their children. They chose to formalize as an association, which shows that their project is not aimed at maximizing profits but acts in the name of a social mission. Although not an active member in the initiative, the village's mayor is also a member of the association, showing his general support.

Pilot 3 is identified as difficult to fit into the ARA concept, due to its complexity and the fact that it has not yet settled on a formal organizational form. The project, initiated by the Freiburg Food Policy Council, involves numerous actors from all parts of the SFSC and further associations addressing education, all of whom are engaged voluntarily. In this case, again, the classification is linked to the resource dimension of the ARA concept as two members of the coordination team are funded by public money; this ultimately determined its assignment to the community-state space. Finally, pilot 1 is mapped in the state-market space because its primary actors are a mayor and a farmer. Together, they plan a small vendor shop in the community to sell regionally produced products under a new regional brand. This project also wants to cooperate with a charity organization and integrate people with disabilities. This would qualify this pilot to be classified in the core of the third sector, thus, integrating state, community, and market actors. However, since the charity organization is not mentioned as an active partner in this project phase, the mentor did not give so much weight to this actor. Another interesting point about pilot 1 is the choice of business model, namely a nonprofit limited liability company (abbr. in German: gGmbH). This means that the profits of the limited liability company must be used for charitable purposes and, in principle, may not be distributed to the shareholders.

Figure 2 illustrates the indicated position of the five pilots in five years. As not all the initiatives are currently operating, it will be of interest to see how the initiatives' situations might change once the projects have been implemented and are operating fully. Interestingly, two significant trends can be observed. Firstly, most initiatives are mapped more in the market space direction. Among other reasons, this development was justified because all the projects plan sustainable financing through the market. This means they want to finance their future by selling products or charging rent. Pilots 1, 2 and 3 are actively planning to sell products (cooked meals, own-brand products, regional food from the food hubs), pilot 4 is about income from renting its designed commercial kitchen, and pilot 5 is also planning to ensure that food can be stored longer and sold regionally throughout the year.

The other trend is noted only for pilot 1. Here, the mentor's choice was explained by the fact that the initiators plan to integrate the community more in the project in the future. The mayor wants to withdraw from the project in the long term. However, many volunteers have left because of the COVID-19 pandemic and need to be recruited again. The mentor also argued that this project should be financed with public funds to create feasible prices and allow affordable products for everyone. The mentor stated that as long as the globally dominant agricultural system is subsidized, smaller initiatives must also be supported so that they have a chance to build a counterweight to the dominant system.

4.2. Resource Dimension

The results of the mentors' classifications of resource levels are presented next. The respective resources in the tabular summary are not broken down by the pilot so as not to disclose any of the pilots' potentially sensitive data. In addition, the aim of this evaluation is to see whether essential resources can generally be identified based on their classification. As can be seen in Table 4, this is confirmed. It also shows that the initiation of all pilots required the interaction of different actors and their resources, which emphasizes their hybrid character (Figure 4).

Table 4. Identified nonmarket and market resources necessary for initiating short food value chain cooperation models.

Nonmarket Resources		Market Resources
Resources Provided Mostly via Community Actors	**Resources Provided via State Actors**	**Resources Provided via Market Actors**
○ Voluntary workers ○ Donations ○ Human capital spent next to job (people creating flyers for free) ○ Social capital (people have large networks) ○ Altruism ○ Enthusiasm ○ Using publicity of KOPOS to promote themselves	○ Public infrastructure ○ Providing information ○ Public investments in local infrastructure to support initiative ○ Providing land ○ Public financial resources to employ people ○ Social Starter Program (extensions)	○ Providing land ○ Providing financial resources to hire people ○ Existing market activities, rent from housing, selling food collectively

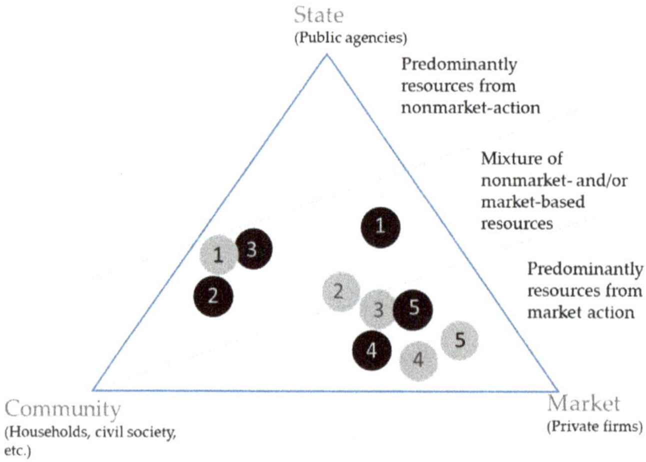

Figure 4. Integrating resource dimension (current situation—black; situation in five years—grey).

The focus is initially on the importance of market resources and illustrates the strategies of hybrid cooperation models. Pilots 4 and 5 rely on preexisting business relationships that probably form the financial basis for the projects. The producer group in pilot 5 already existed and operated in the market. The project to build its storage facility is based on this existing business model and the desire to expand it, while filling an identified gap in regional SFSCs. Pilot 4 was initiated by a non-listed regional stock corporation committed to sustainable business practices. Once the idea for pilot 4 reached a certain maturity level, the project was transferred to another subsidiary. Three positions were funded to implement the project, financed by market revenue from property rentals. The financial basis for the other three initiatives had to be established at the start of the project.

Another interesting finding is that land is provided by different market actors, such as in pilot 1. The project received a building site on the parking lot of a nationally operating supermarket chain, to build a container house for selling its products and processing oilseeds. In addition, the actors of pilot 1 and the supermarket do not see each other as competitors, but want to benefit from each other. The producer organization in pilot 5 is planning its storage site on the land of one of the producers and the main initiator.

The diversity and variety of SFSC cooperation models are also apparent when looking at the role of public funding opportunities, which is shown in Table 4. Three strategies could be identified here. In pilot 1, the local government is directly involved, as the main initiator is also the village's mayor. According to the mentors, this example is crucial, because local politicians and administrations committed to changing the agri-food system are still very rare. Less direct involvement can be found in pilot 2, where the local government is not the primary initiator, but provides the essential infrastructure for the project in the form of the public kitchen. The mayor is also a member of the association. In addition, in pilot 2, it was reported that the project would like to use public funds, but there are no suitable programs at the local, regional, national, or European level that the project could use. This example indicates that the pilots are in their early stages and not yet on the radar of funding agencies, which could change through a classification such as the ARA concept [58]. Pilot 3 illustrates the example wherein two coordination team members are financed by public funds/projects. This proved to be somewhat insufficient or unsustainable, as here, the initiative is dependent on public funds, which are often only paid for short periods of time. Therefore, a lot of time has to be invested in searching and applying for new funds, and continuity is insecure. This problem in social enterprises is often described in the literature [54,57].

Another essential resource mentioned was that of the existing social networks and relationships between public-, private-, and civil-society actors, which were created primarily due to the spatial proximity of the actors. Pilot 3, for example, was founded by the Food Policy Council, which was described as being very recognized and promoted by the region. Pilot 2 builds on a well-connected parents' initiative; the main protagonist, in particular, can draw on good contacts through her workplace in the citizens' council and her involvement in other networks.

4.3. Action Dimension

We added a third dimension to capture the action dimension (Figure 5), to apply the classification approach to specific sectors and the urban-rural nexus. This was identified as necessary when screening the literature about urban-rural linkages and SFSCs. When analyzing SFSCs' potential and challenges, what is talked about from an organizational point of view is often not identified. The pilots used to evaluate the ARA concept demonstrate that none of them cover the whole SFSC at the moment, but they cover different components that might foster urban-rural linkages and sustainable transformation.

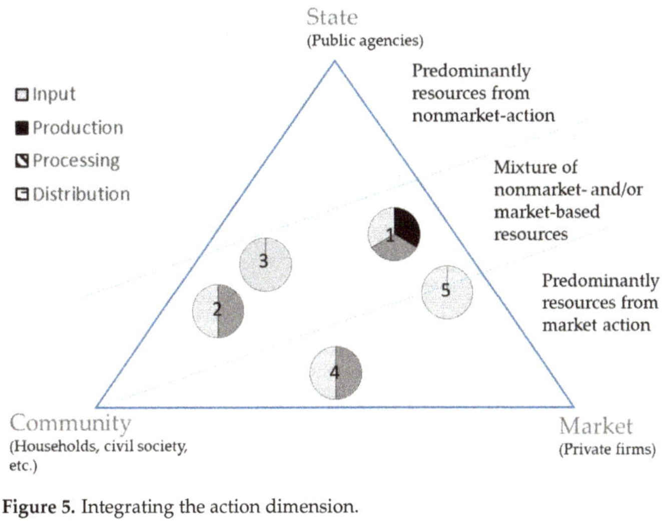

Figure 5. Integrating the action dimension.

Thus, as anticipated when integrating this dimension, the ARA concept allows us to show and analyze the diversity of SFSC value cooperation models, while simultaneously generalizing their activities, which helps to identify patterns. As shown in Figure 5, pilots 2 and 4 map the same parts of the SFSCs; this makes sense, as both initiatives aim to build a catering business. Remarkably, both pilots base their intentions on other resources and actors, hence, building different organizational structures; this shows that the same goals can be achieved with different strategies, depending on the actor capacities and resource availability in the urban-rural nexus.

5. Discussing the Potential and Challenges of the ARA Concept

This article was written to answer the question: How appropriate is the classification concept proposed in the social enterprise literature for capturing new cooperation models in urban-rural contexts in SFSCs? This question was conceptualized by introducing the ARA concept in Section 2, and evaluated by mapping different pilot projects initiated at the urban-rural interface. Furthermore, the application purposes of the ARA concept were evaluated by interviewing selected mentors. This resulted in determination of its potential and challenges, which will be discussed here.

5.1. ARA Concept Potential

In summary, it is clear that the actor dimension of the SFSC cooperation models investigated differ; this allows for the analysis of very different initiatives, and critically reflects the purpose of the ARA concept. In addition, mapping the actors involved within the initiatives showed that none of the cooperation models suit the conservative logic of market and state actors, demonstrating the need for a new classification tool to make them visible, as requested by Spyra et al. [37].

As the evaluation of the results shows, the ARA analytical concept allows the classification of partly very hybridized cooperation models; this shows the multi-governance nature of SFSC cooperation models and their need to be acknowledged by science and governments [58]. This fulfills the aspiration of the classification emerging in the social enterprise literature and shows its usefulness in this context. Furthermore, the ARA concept makes a more realistic classification possible in the context of SFSCs. For example, by integrating the action dimension, SFSC cooperation models could be analyzed regarding their specific action field and, hence, whether cooperation models aim to shorten the food supply chain by focusing on the input, production, processing, and/or distribution of food within the urban-rural nexus. By integrating the actor dimension, it could be shown that SFSC cooperation models emerge from different constellations of actors that do not necessarily originate from the agri-food system. By integrating the resource dimension, information could be gained regarding what resources are necessary and sufficient, for establishing new cooperation models that aim at multiple social missions outside the dominant, competitive global agri-food system. Hence, by including all the ARA dimensions that operate in the urban-rural nexus, a basis can be established that takes the discussion on the effects of SFSCs for sustainable transformation to a new level.

While a lot of discussion on SFSCs and their transformative potential in the producer-consumer nexus seems to be criticized and limited [30–33], we argue that focusing on the organizational level and territorial framing of the urban-rural nexus provides new insights. The latter includes local public, private and civil society actors and their resource highlights: firstly, successful strategies; secondly, the transformation pathways of urban-rural regions; and thirdly, a critical examination of actors' responsibilities to promote sustainable transformation.

Identifying successful strategies: Mapping different cooperation models for shortening the food supply chain according to the ARA principles has shown that the same fields of action, such as public procurement, can be successfully initiated via different strategies based on various resources and actor constellations (see Figure 5). Therefore, an important application of the ARA concept is seen in classifying a lot of successful hybrid cooperation models. As a result, local governments, social entrepreneurs, or civil society can establish appropriate strategies to promote sustainable transformation in the agri-food sector, adapted to their urban-rural context and resource availability.

Identifying sustainable trajectories. The ARA concept allows the identification of development pathways for new cooperation models. This is demonstrated in Section 4.1 through the mapping of the current position of the emerging cooperation models and their potential positions in five years. This potential is of interest to scholars and practitioners seeking to understand how initiating social missions over time leads to functioning hybrid organizations that drive sustainable change. Interesting questions in this area include organizational development topics, such as upscaling sustainable organizations. In addition, the ARA concept makes it possible to examine the trajectories of cooperation models that have been little explored, such as the development of cooperation models in community-state space. These cooperation models are referred to by Rossi and Brunori [85] as "new food governance" and are classified by Doernberg et al. [34] as an essential pillar of current food governance research and policy.

Identifying responsibilities. Sustainable transformation is one of the most complex tasks of our time. A multilevel governance approach is required, including a wide range of actors to enable the transition to a sustainable agri-food system. Although policies are often established at a regional, national, and international level, the implementation of these policies and the fulfillment of objectives is usually the responsibility of local governments and actors. However, how such change is organized and who is responsible for the tasks is often unresolved. Despite the missing guidance, Barling et al. [86] argue that local governments have started to see the local agri-food system as essential for ensuring food security, the local economy, social integration, and environment protection. Large cities, especially, are already beginning to address the issue of food for sustainable transformation [34]. The ARA classification of hybrid cooperation models that link rural and urban actors on this journey shows the possible courses of action and starting points of the responsible government. It provides small farms with new ideas for more sustainable markets. It activates civil society to transform the agri-food system, as already described by Stierand [87]. In addition, Evers [67] highlights the importance of the social enterprise classification, as it helps local governments to acknowledge hybrid cooperation models and their potential. This might lead to new support programs, not just on a local level, but also on a multi-governmental level, showing the transformative potential of new hybrid cooperation models. This point is reflected in pilot 2, where the mentor expressed that they were urgently looking for public funding, but their cooperation models did not suit any taxonomy.

5.2. ARA Concept Challenges

Some challenges also came to light during the evaluation of the ARA concept that should be taken into account in the further development and application of the analytical concept. For example, one challenge in evaluating the ARA concept using the pilot information was that it is sometimes difficult to identify what the actors involved in the cooperation model are doing, what they intend to achieve with their cooperation model in the urban-rural nexus, and what the established hybrid cooperation model is actually performing. To exemplify this, pilot 3 is probably best described, at this time, as a network of many actors already working on or interested in food system change (Figure 5). The mentor pointed out that SFSCs are involved in all parts of the food value chain. However, this does not mean that the output and goals of the network include all the activities of the SFSC, and it is difficult to say, currently, what activities the cooperation model will carry out in the future. To specify, just because farmers are part of the cooperation model, it does not mean that the cooperation model produces anything.

This highlights another challenge that future applicants of the concept need to specify, namely, whether the ARA concept is used to classify emerging hybrid cooperation models or their output dimensions. In our case, all the pilots have been emerging initiatives. Thus, actor and resource pooling are likely, or have already planned, to change when the project is realized; this partly explains the drive towards the market, as people will be employed and products offered (Figure 2). Thus, classifying them provides interesting knowledge about how resource pooling and cooperation emerge in an urban-rural nexus and, therefore, delivers potential points of entrance for local governments and other actors. However, measuring their effects on sustainable transformation at this stage is not possible, but might require already-operating cooperation models. Scholars such as El Ebrashi [47] and Haugh [88] suggest separating the impact of cooperation models from the output, wherein output refers to the direct product and services produced by the hybrid cooperation model. By contrast, impact refers to "sustainable long-term change" [47]. Thus, the classification model may help to map differences in actor constellations throughout the realization of a project and afterward. However, applicants need to define the stage of the project to allow the classification model analysis. Therefore, what can be compared and what cannot needs to be clear in order to compare different cooperation models.

Furthermore, identifying key actors within the project has sometimes been challenging, for different reasons. On the one hand, some initiatives appear to be very complex, such as a conglomerate of various stakeholders not agreeing on a formal business model that would assign board members or people in charge. On the other hand, pilots have one key person in charge of communication most of the time, which sometimes makes it difficult to access the internal working distribution and resource-pooling strategies of other stakeholders. Thus, in order to apply the ARA concept, in-depth case knowledge seems relevant to assigning stakeholders either to community, state or market space.

Finally, in this article, we separated the three ARA dimensions to enhance the paper's structure. However, it has been noted that the resource dimension often defines the actors' positions in the triangle. Thus, it is sometimes difficult to separate these dimensions. This point also includes the question of what kind of cooperation models to classify. As an example, when combining conventional companies with new hybrid cooperation models aimed at independent social missions, it is sometimes difficult to separate them based on the classification approach. Therefore, we suggest that additional parameters should also be further integrated to explore the potential for sustainable transformation. Without knowing the social mission of pilot 5, for example, it could be classified as a pure market actor. However, with the knowledge available, one can see that it is primarily about fulfilling a social mission through market-based practices and resources. This observation is common for social enterprises that aim to increase their social impact and not accumulate private actors' profits [47,62]. Therefore, it is crucial to integrate this knowledge, as well as information about the historical and political context of the urban-rural nexus in which this hybrid cooperation model emerges; Defourny and Pestoff [58] and Evers [67] emphasize that these conditions also influence the transformative potential of the established cooperation model.

6. Conclusions

This article aims to extend and validate a classification approach from the social enterprise literature to classify new collaborative models that aim to shorten the food supply chain in an urban-rural context. The reason for this is the increasing pressure to promote sustainable transformation of the agri-food system and, in particular, the question of how SFSCs can contribute to this in an urban-rural nexus. Based on a literature review, a lack of studies that capture the complexity of SFSCs as super-sectoral and multi-actor entities emerged. In addition, the question of how SFSCs occur and what strategies lead to successful cooperation models has been little highlighted and classified, to serve as blueprints for other governments on the ways in which to foster sustainable transformation.

In order to bring a new perspective to this discussion, we argue that it requires an understanding of the landscape of actors in their territorial context, and of how these actors organize themselves to establish urban-rural cooperation models. There needs to be common ground that provides a basis for evaluating new cooperation models and their contribution to sustainable transformation. This basis can be provided by analyzing and mapping existing initiatives within an urban-rural context by their actor, resource, and action dimensions. The actor and resource dimensions have been used successfully in the social enterprise literature to classify cooperation models aiming to fulfill a social mission. In addition, we added the action dimension, as we argue that according to what sector we look at, different action fields emerge; these are important to know to evaluate their sustainable transformation potential for the urban-rural nexus.

By applying this classification approach, we found that the established ARA concept offers several options for classifying cooperation models in the social economy and, thus, can provide practitioners, researchers and policymakers with a purposeful tool. First of all, the ARA concept demonstrates the diversity and complexity of SFSC cooperation models and is an important tool for making funders, policymakers, and other stakeholders aware of this diversity. Furthermore, the ARA concept allows the analysis of different resource-pooling strategies for establishing cooperation models. This will enable actors to select strategies according to their capabilities and learn from existing examples. In addition, the ARA concept can be used to visualize trajectories by mapping the evolution of emerging cooperation models. This enables actors to question and steer their actions, and to prioritize the social mission of their actions. Finally, the ARA concept helps to identify responsibilities. Governments can set up targeted programs to fill identified gaps, or support cooperation models in the right places by mapping the cooperation models in an urban-rural nexus.

Further research will help us to gain knowledge about the emergence of cooperation models, and address the output and impact dimensions of these cooperation models in terms of sustainable transformation. Therefore, it is essential to acquire in-depth case knowledge about the cooperation models, their actors, and the urban-rural nexus investigated.

Author Contributions: Conceptualization, K.M., S.R. and A.P.; formal analysis, K.M. and S.R.; funding acquisition, J.Z., B.P., A.O. and A.P.; investigation, K.M.; methodology, K.M.; project administration, S.R.; supervision, A.P.; validation, S.R., J.Z., B.P., A.O. and A.P.; visualization, K.M.; writing—original draft, K.M., S.R., B.P. and A.P.; writing—review and editing, K.M., S.R., J.Z., B.P., A.O. and A.P. All authors have read and agreed to the published version of the manuscript.

Funding: This research was funded by the German Federal Ministry of Education and Research (BMBF), grant number 033L221A-J, funding activity urban-rural (Stadt-Land-Plus), project name: "KOPOS: New cooperation for sustainable land use and food supply in urban-rural areas".

Institutional Review Board Statement: Not applicable.

Informed Consent Statement: Informed consent was obtained from all subjects involved in the study.

Data Availability Statement: Some of the data presented in this study are publicly available by visiting the official project website (https://www.kopos-projekt.de/, last accessed on 13 April 2022). More detailed data from the interviews are not publicly available for data protection reasons and to protect the sources.

Conflicts of Interest: The authors declare no conflict of interest.

References

1. Prosser, L.; Thomas Lane, E.; Jones, R. Collaboration for innovative routes to market: COVID-19 and the food system. *Agric. Syst.* **2021**, *188*, 103038. [CrossRef]
2. Rivera-Ferre, M.G.; López-i-Gelats, F.; Ravera, F.; Oteros-Rozas, E.; Di Masso, M.; Binimelis, R.; El Bilali, H. The two-way relationship between food systems and the COVID-19 pandemic: Causes and consequences. *Agric. Syst.* **2021**, *191*, 103134. [CrossRef]
3. Vittuari, M.; Bazzocchi, G.; Blasioli, S.; Cirone, F.; Maggio, A.; Orsini, F.; Penca, J.; Petruzzelli, M.; Specht, K.; Amghar, S.; et al. Envisioning the Future of European Food Systems: Approaches and Research Priorities after COVID-19. *Front. Sustain. Food Syst.* **2021**, *5*, 58. [CrossRef]
4. van Poeck, K.; Östman, L. Learning to find a way out of non-sustainable systems. *Environ. Innov. Soc. Transit.* **2021**, *39*, 155–172. [CrossRef]
5. *Farmers' Cooperatives and Sustainable Food Systems in Europe*; Gonzalez, R.A. (Ed.) Routledge: Abingdon, UK; Oxon, UK; New York, NY, USA, 2018; ISBN 978-1-351-21630-2.

6. Battisti, D.; Naylor, R. Historical Warnings of Future Food Insecurity with Unprecedented Seasonal Heat. *Science* **2009**, *323*, 240–244. [CrossRef]
7. Chiffoleau, Y.; Dourian, T. Sustainable Food Supply Chains: Is Shortening the Answer? A Literature Review for a Research and Innovation Agenda. *Sustainability* **2020**, *12*, 9831. [CrossRef]
8. Willett, W.; Rockström, J.; Loken, B.; Springmann, M.; Lang, T.; Vermeulen, S.; Garnett, T.; Tilman, D.; DeClerck, F.; Wood, A.; et al. Food in the Anthropocene: The EAT–Lancet Commission on healthy diets from sustainable food systems. *Lancet* **2019**, *393*, 447–492. [CrossRef]
9. Corona-López, E.; Román-Gutiérrez, A.D.; Otazo-Sánchez, E.M.; Guzmán-Ortiz, F.A.; Acevedo-Sandoval, O.A. Water-Food Nexus Assessment in Agriculture: A Systematic Review. *Int. J. Environ. Res. Public Health* **2021**, *18*, 4983. [CrossRef]
10. Kurian, M. The water-energy-food nexus. *Environ. Sci. Policy* **2017**, *68*, 97–106. [CrossRef]
11. Lu, Y.; Song, S.; Wang, R.; Liu, Z.; Meng, J.; Sweetman, A.J.; Jenkins, A.; Ferrier, R.C.; Li, H.; Luo, W.; et al. Impacts of soil and water pollution on food safety and health risks in China. *Environ. Int.* **2015**, *77*, 5–15. [CrossRef]
12. Schwarzenbach, R.P.; Egli, T.; Hofstetter, T.B.; von Gunten, U.; Wehrli, B. Global Water Pollution and Human Health. *Annu. Rev. Environ. Resour.* **2010**, *35*, 109–136. [CrossRef]
13. Renard, D.; Tilman, D. National food production stabilized by crop diversity. *Nature* **2019**, *571*, 257–260. [CrossRef] [PubMed]
14. Aguilar, J.; Gramig, G.G.; Hendrickson, J.R.; Archer, D.W.; Forcella, F.; Liebig, M.A. Crop Species Diversity Changes in the United States: 1978-2012. *PLoS ONE* **2015**, *10*, e0136580. [CrossRef] [PubMed]
15. van Dusen, E.; Taylor, J. Missing markets and crop diversity: Evidence from Mexico. *Environ. Dev. Econ.* **2005**, *10*, 513–531. [CrossRef]
16. Pimentel, D.; Burgess, M. Soil Erosion Threatens Food Production. *Agriculture* **2013**, *3*, 443–463. [CrossRef]
17. Poux, X.; Aubert, P.-M. An Agroecological Europe in 2050: Multifunctional Agriculture for Healthy Eating. *Find. Ten Years Agroecol. (TYFA) Model. Exerc.* **2018**, *9*, 18.
18. Lamine, C.; Renting, H.; Rossi, A.; Wiskerke, J.S.C.; Brunori, G. Agri-Food systems and territorial development: Innovations, new dynamics and changing governance mechanisms. In *Farming Systems Research into the 21st Century: The New Dynamic*; Darnhofer, I., Gibbon, D., Dedieu, B., Eds.; Springer: Dordrecht, The Netherlands, 2012; pp. 229–256, ISBN 978-94-007-4502-5.
19. Lamine, C. Sustainability and Resilience in Agrifood Systems: Reconnecting Agriculture, Food and the Environment. *Sociol. Rural.* **2015**, *55*, 41–61. [CrossRef]
20. Haysom, G. Integrating Food Sensitive Planning and Urban Design into Urban Governance Actions. *Urban Forum* **2021**, *32*, 289–310. [CrossRef]
21. Opitz, I.; Berges, R.; Piorr, A.; Krikser, T. Contributing to food security in urban areas: Differences between urban agriculture and peri-urban agriculture in the Global North. *Agric. Hum. Values* **2016**, *33*, 341–358. [CrossRef]
22. Blay-Palmer, A.; Santini, G.; Dubbeling, M.; Renting, H.; Taguchi, M.; Giordano, T. Validating the City Region Food System Approach: Enacting Inclusive, Transformational City Region Food Systems. *Sustainability* **2018**, *10*, 1680. [CrossRef]
23. Opitz, I.; Specht, K.; Piorr, A.; Siebert, R.; Zasada, I. Effects of consumer-producer interactions in alternative food networks on consumers' learning about food and agriculture. *Morav. Geogr. Rep.* **2017**, *25*, 181–191. [CrossRef]
24. Opitz, I.; Zoll, F.; Zasada, I.; Doernberg, A.; Siebert, R.; Piorr, A. Consumer-producer interactions in community-supported agriculture and their relevance for economic stability of the farm—An empirical study using an Analytic Hierarchy Process. *J. Rural Stud.* **2019**, *68*, 22–32. [CrossRef]
25. Bartke, S.; Forster, T.; Githiri, G.; Jering, A.; Kago, J.; Schlimmer, S.; Sietchiping, R. The UN-Habitat Urban-Rural Linkages Guiding Principles: Assessment of the Adoptability to Topical Land Management Challenges in Germany, Kenya and Tanzania. In *International Yearbook of Soil Law and Policy 2019*; Ginzky, H., Dooley, E., Heuser, I.L., Kasimbazi, E., Kibugi, R., Markus, T., Qin, T., Ruppel, O., Eds.; Springer International Publishing: Cham, Switzerland, 2021; pp. 369–398, ISBN 978-3-030-52316-9.
26. Feenstra, G. Local food systems and sustainable communities. *Am. J. Altern. Agric.* **1997**, *12*, 28–36. [CrossRef]
27. Knickel, K.; Brunori, G.; Rand, S.; Proost, J. Towards a Better Conceptual Framework for Innovation Processes in Agriculture and Rural Development: From Linear Models to Systemic Approaches. *J. Agric. Educ. Ext.* **2009**, *15*, 131–146. [CrossRef]
28. Brunori, G. Local food and alternative food networks: A communication perspective. *Anthropol. Food* **2007**. [CrossRef]
29. Regulation (Eu) No 1305/2013 of the European Parliament and of the Council on Support for Rural Development by the European Agricultural Fund for Rural Development (EAFRD) and Repealing Council Regulation (EC) No. 1698/2005. *Off. J. Eur. Union* **2013**, *347*, 487–548.
30. Schmutz, U.; Kneafsey, M.; Sarrouy Kay, C.; Doernberg, A.; Zasada, I. Sustainability impact assessments of different urban short food supply chains: Examples from London, UK. *Renew. Agric. Food Syst.* **2018**, *33*, 518–529. [CrossRef]
31. Benis, K.; Ferrão, P. Potential mitigation of the environmental impacts of food systems through urban and peri-urban agriculture (UPA)—A life cycle assessment approach. *J. Clean. Prod.* **2017**, *140*, 784–795. [CrossRef]
32. Chiffoleau, Y.; Millet-Amrani, S.; Rossi, A.; Rivera-Ferre, M.G.; Merino, P.L. The participatory construction of new economic models in short food supply chains. *J. Rural Stud.* **2019**, *68*, 182–190. [CrossRef]
33. Georgiadis, P.; Vlachos, D.; Iakovou, E. A system dynamics modeling framework for the strategic supply chain management of food chains. *J. Food Eng.* **2005**, *70*, 351–364. [CrossRef]
34. Doernberg, A.; Horn, P.; Zasada, I.; Piorr, A. Urban food policies in German city regions: An overview of key players and policy instruments. *Food Policy* **2019**, *89*, 101782. [CrossRef]

35. Ilieva, R.T. *Urban Food Planning*; Routledge: London, UK, 2016; ISBN 9781317331704.
36. Levkoe, C.Z.; Blay-Palmer, A. Food Counts: Food systems report cards, food sovereignty and the politics of indicators. *Can. Food Stud.* **2018**, *5*, 49–75. [CrossRef]
37. Spyra, M.; La Rosa, D.; Zasada, I.; Sylla, M.; Shkaruba, A. Governance of ecosystem services trade-offs in peri-urban landscapes. *Land Use Policy* **2020**, *95*, 104617. [CrossRef]
38. Binimelis, R.; Rivera-Ferre, M.G.; Tendero, G.; Badal, M.; Heras, M.; Gamboa, G.; Ortega, M. Adapting established instruments to build useful food sovereignty indicators. *Dev. Stud. Res.* **2014**, *1*, 324–339. [CrossRef]
39. Sonnino, R.; Tegoni, C.L.; de Cunto, A. The challenge of systemic food change: Insights from cities. *Cities* **2019**, *85*, 110–116. [CrossRef]
40. Moragues-Faus, A.; Sonnino, R. Re-assembling sustainable food cities: An exploration of translocal governance and its multiple agencies. *Urban Stud.* **2019**, *56*, 778–794. [CrossRef]
41. Edwards-Jones, G.; Milà i Canals, L.; Hounsome, N.; Truninger, M.; Koerber, G.; Hounsome, B.; Cross, P.; York, E.H.; Hospido, A.; Plassmann, K.; et al. Testing the assertion that 'local food is best': The challenges of an evidence-based approach. *Trends Food Sci. Technol.* **2008**, *19*, 265–274. [CrossRef]
42. Defourny, J.; Nyssens, M. The Emes Approach of Social Enterprise in a Comparative Perspective. In *Social Enterprise and the Third Sector*; Routledge: London, UK, 2014.
43. Pestoff, V. Collective Action and the Sustainability of Co-Production. *Public Manag. Rev.* **2014**, *16*, 383–401. [CrossRef]
44. Doherty, B.; Haugh, H.; Lyon, F. Social Enterprises as Hybrid Organizations: A Review and Research Agenda. *Int. J. Manag. Rev.* **2014**, *16*, 417–436. [CrossRef]
45. Hestad, D.; Tàbara, J.D.; Thornton, T.F. The role of sustainability-oriented hybrid organisations in the development of transformative capacities: The case of Barcelona. *Cities* **2021**, *119*, 103365. [CrossRef]
46. Wilson, F.; Post, J.E. Business models for people, planet (& profits): Exploring the phenomena of social business, a market-based approach to social value creation. *Small Bus. Econ.* **2013**, *40*, 715–737. [CrossRef]
47. El Ebrashi, R. Social entrepreneurship theory and sustainable social impact. *Soc. Responsib. J.* **2013**, *9*, 188–209. [CrossRef]
48. Zoll, F.; Specht, K.; Siebert, R. Alternative = transformative? Investigating drivers of transformation in alternative food networks in Germany. *Sociol. Rural.* **2021**. [CrossRef]
49. Zoll, F.; Specht, K.; Opitz, I.; Siebert, R.; Piorr, A.; Zasada, I. Individual choice or collective action? Exploring consumer motives for participating in alternative food networks. *Int. J. Consum. Stud.* **2018**, *42*, 101–110. [CrossRef]
50. Ruiz-Almeida, A.; Rivera-Ferre, M.G. Internationally-based indicators to measure Agri-food systems sustainability using food sovereignty as a conceptual framework. *Food Sec.* **2019**, *11*, 1321–1337. [CrossRef]
51. Creamer, E.; Eadson, W.; van Veelen, B.; Pinker, A.; Tingey, M.; Braunholtz-Speight, T.; Markantoni, M.; Foden, M.; Lacey-Barnacle, M. Community energy: Entanglements of community, state, and private sector. *Geogr. Compass* **2018**, *12*, e12378. [CrossRef]
52. Rhodes, R.A.W. *Understanding Governance: Policy Networks, Governance, Reflexivity and Accountability*; Open University Press: Buckingham, UK, 1997; ISBN 9780335197279.
53. Defourny, J.; Nyssens, M.; Brolis, O. *Mapping and Testing Social Enterprise Models across the World: Evidence from the "International Comparative Social Enterprise Models (ICSEM) Project"*; ICSEM Working Papers No. 50; Belgian Science Policy and EMES: Brussels, Belgium, 2019.
54. Martens, K.; Wolff, A.; Hanisch, M. Understanding social innovation processes in rural areas: Empirical evidence from social enterprises in Germany. *Soc. Enterp. J.* **2020**, 203. [CrossRef]
55. Polanyi, K. *The Great Transformation*; Beacon: Boston, MA, USA, 1944.
56. Defourny, J.; Nyssens, M. Social enterprise in Europe: At the crossroads of market, public policies and third sector. *Policy Soc.* **2010**, *29*, 231–242. [CrossRef]
57. Evers, A.; Laville, J. (Eds.) *The Third Sector in Europe*; Edward Elgar Publishing Limited: Cheltenham, UK; Northampton, UK, 2004; ISBN 1-84376-400-8.
58. Defourny, J.; Pestoff, V. Towards a European Conceptualization of the Third Sector. In *Accountability and Social Accounting for Social and Non-Profit Organizations*; Costa, E., Parker, L.D., Andreaus, M., Eds.; Emerald Group Publishing Limited: Bingley, UK, 2014; pp. 25–87, ISBN 978-1-78441-005-6.
59. Davies, I.; Chambers, L. Integrating hybridity and business model theory in sustainable entrepreneurship. *J. Clean. Prod.* **2018**, *177*, 378–386. [CrossRef]
60. EU. Social Enterprises. Available online: https://ec.europa.eu/growth/sectors/proximity-and-social-economy/social-economy-eu/social-enterprises_en (accessed on 12 April 2022).
61. Kerlin, J.A. Social Enterprise in the United States and Europe: Understanding and Learning from the Differences. *Voluntas* **2006**, *17*, 246–262. [CrossRef]
62. Defourny, J.; Nyssens, M. Fundamentals for an International Typology of Social Enterprise Models. *Voluntas* **2017**, *28*, 2469–2497. [CrossRef]
63. Bauwens, T.; Huybrechts, B.; Dufays, F. Understanding the Diverse Scaling Strategies of Social Enterprises as Hybrid Organizations: The Case of Renewable Energy Cooperatives. *Organ. Environ.* **2020**, *33*, 195–219. [CrossRef]
64. Gui, B. The economic rationale for the "third sector". Nonprofit and other noncapitalistic organizations. *Ann. Public Coop. Econ.* **1991**, *62*, 551–572. [CrossRef]

65. Pestoff, V.A. Third sector and co-operative services—An alternative to privatization. *J. Consum. Policy* **1992**, *15*, 21–45. [CrossRef]
66. Laville, J.; Young, D.; Eynaud, P. (Eds.) *Civil Society, the Third Sector and Social Enterprise: Governance and Democracy*; Routledge Frontiers of Political Economy: Oxford, UK, 2015; ISBN 9781138673502.
67. Evers, A. Part of the welfare mix: The third sector as an intermediate area. *Volunt. Int. J. Volunt. Nonprofit Organ.* **1995**, *6*, 159–182. [CrossRef]
68. Benz, A. (Ed.) *Governance—Regieren in Komplexen Regelsystemen: Eine Einführung*; VS Verlag für Sozialwissenschaften: Wiesbaden, Germany, 2004; ISBN 978-3-531-90171-8.
69. Mundler, P.; Laughrea, S. The contributions of short food supply chains to territorial development: A study of three Quebec territories. *J. Rural Stud.* **2016**, *45*, 218–229. [CrossRef]
70. Ilbery, B.; Maye, D. Alternative (Shorter) Food Supply Chains and Specialist Livestock Products in the Scottish–English Borders. *Environ. Plan. A* **2005**, *37*, 823–844. [CrossRef]
71. Kneafsey, M.; Dowler, E.; Lambie-Mumford, H.; Inman, A.; Collier, R. Consumers and food security: Uncertain or empowered? *J. Rural Stud.* **2013**, *29*, 101–112. [CrossRef]
72. Forssell, S.; Lankoski, L. The sustainability promise of alternative food networks: An examination through "alternative" characteristics. *Agric. Hum. Values* **2015**, *32*, 63–75. [CrossRef]
73. Aubry, C.; Kebir, L. Shortening food supply chains: A means for maintaining agriculture close to urban areas? The case of the French metropolitan area of Paris. *Food Policy* **2013**, *41*, 85–93. [CrossRef]
74. Renting, H.; Marsden, T.K.; Banks, J. Understanding Alternative Food Networks: Exploring the Role of Short Food Supply Chains in Rural Development. *Environ. Plan. A* **2003**, *35*, 393–411. [CrossRef]
75. Thünen, J. *Der Isolirte Staat in Beziehung auf Landwirthschaft und Nationalökonomie: Untersuchungen über den Einfluß, den die Getreidepreise, der Reichthum des Bodens und die Abgaben auf den Ackerbau ausüben*; Perthes: Hamburg, Germany, 1826.
76. Inwood, S.M.; Sharp, J.S. Farm persistence and adaptation at the rural–urban interface: Succession and farm adjustment. *J. Rural Stud.* **2012**, *28*, 107–117. [CrossRef]
77. Akimowicz, M.; Cummings, H.; Landman, K. Green lights in the Greenbelt? A qualitative analysis of farm investment decision-making in peri-urban Southern Ontario. *Land Use Policy* **2016**, *55*, 24–36. [CrossRef]
78. van der Schans, J. Urban agriculture in the Netherlands. *Urban Agric. Mag.* **2010**, *1*, 40–42.
79. Monaco, F.; Zasada, I.; Wascher, D.; Glavan, M.; Pintar, M.; Schmutz, U.; Mazzocchi, C.; Corsi, S.; Sali, G. Food Production and Consumption: City Regions between Localism, Agricultural Land Displacement, and Economic Competitiveness. *Sustainability* **2017**, *9*, 96. [CrossRef]
80. Cretella, A. Urban Food Strategies. Exploring Definitions and Diffusion of European Cities' Latest Policy Trend. In *Metropolitan Ruralities*; Andersson, K., Sjöblom, S., Granberg, L., Ehrström, P., Marsden, T., Eds.; Emerald Group Publishing Limited: Bentley, UK, 2016; pp. 303–323, ISBN 978-1-78560-797-4.
81. Schäpke, N.; Bergmann, M.; Stelzer, F.; Lang, D.J.; Editors, G. Labs in the Real World: Advancing Transdisciplinary Research and Sustainability Transformation: Mapping the Field and Emerging Lines of Inquiry. *GAIA Ecol. Perspect. Sci. Soc.* **2018**, *27*, 8–11. [CrossRef]
82. Moschitz, H.; Oehen, B.; Rossier, R. Regionaler Konsum in Freiburg—Anteil von Lebensmitteln regionalen Ursprungs am Gesamtverbrauch der Stadt Freiburg. 2016. Available online: https://www.freiburg.de/pb/site/Freiburg/get/pa-rams_E138525682/1045847/Regional-Konsum.pdf (accessed on 14 March 2022).
83. Statistisches Landesamt Baden-Württemberg. Online Zugang zu Regionaldaten. Available online: https://www.statistik-bw.de/Landwirtschaft/ (accessed on 14 March 2022).
84. Beckmann, J.; Volz, P.; Vecera, L. *Situationsanalyse. Handlungsfeld "Kurze Wertschöpfungskette" Freiburg. KOPOS Arbeitspapier.* Freiburg i.Br. 2021. Available online: https://www.kopos-projekt.de/sites/default/files/2021-09/Situationsanalyse%20Kurze%20Version%20HF%20Freiburg_210916.pdf (accessed on 14 March 2022).
85. Rossi, A.; Brunori, G. Towards a New Food Governance: Exploring the Development of an Integrated Urban Food Strategy. Pisa. 2015. Available online: https://www.fao.org/fileadmin/templates/ags/docs/MUFN/CALL_FILES_EXPERT_2015/CFP1-18_Full_Paper.pdf (accessed on 17 February 2022).
86. Barling, D.; Lang, T.; Caraher, M. Joined-up Food Policy? The Trials of Governance, Public Policy and the Food System. *Soc. Policy Admin.* **2002**, *36*, 556–574. [CrossRef]
87. Stierand, P. Stadtentwicklung mit dem Gartenspaten: Umrisse einer Stadternährungsplanung. Dortmund. 2012. Available online: https://speiseraeume.de/downloads/SPR-Stadternaehrungsplanung-Stierand.pdf (accessed on 17 February 2022).
88. Haugh, H. Social Enterprise: Beyond Economic Outcomes and Individual Returns. In *Social Entrepreneurship*; Mair, J., Robinson, J., Hockerts, K., Eds.; Palgrave Macmillan UK: London, UK, 2006; pp. 180–205, ISBN 978-1-349-54553-7.

land

MDPI

Article

Addressing Goal Conflicts: New Policy Mixes for Commercial Land Use Management

Hannah Kosow [1,*], Sandra Wassermann [1], Stephan Bartke [2], Paul Goede [3], Detlef Grimski [2], Ines Imbert [4], Till Jenssen [5], Oliver Laukel [6], Matthias Proske [7], Jochen Protzer [8], Kim Philip Schumacher [9], Stefan Siedentop [3,10], Sandra Wagner-Endres [11], Jürgen Wittekind [12] and Karsten Zimmermann [13]

1 ZIRIUS Center for Interdisciplinary Risk and Innovation Studies, University of Stuttgart, 70174 Stuttgart, Germany; sandra.wassermann@zirius.uni-stuttgart.de
2 German Environment Agency, 06844 Dessau-Roßlau, Germany; stephan.bartke@uba.de (S.B.); detlef.grimski@uba.de (D.G.)
3 ILS—Research Institute for Regional and Urban Development & County Government, 48143 Münster, Germany; paul.goede@bezreg-muenster.nrw.de (P.G.); stefan.siedentop@ils-forschung.de (S.S.)
4 European Institute for Energy Research, 76131 Karlsruhe, Germany; ines.imbert2@eifer.uni-karlsruhe.de
5 Ministry of the Environment, Climate Protection and the Energy Sector Baden-Württemberg, 70182 Stuttgart, Germany; till.jenssen@um.bwl.de
6 Chamber of Commerce and Industry, Northern Black Forest, 75173 Pforzheim, Germany; laukel@pforzheim.ihk.de
7 Regional Association Northern Black Forest, 75172 Pforzheim, Germany; proske@rvnsw.de
8 Economic Development Corporation of the Northern Black Forest, 75172 Pforzheim, Germany; protzer@nordschwarzwald.de
9 Institute of Geography, Osnabrück University, 49074 Osnabrück, Germany; kim.schumacher@uni-osnabrueck.de
10 Research Institute for Regional and Urban Development, TU Dortmund University, 44227 Dortmund, Germany
11 German Institute of Urban Affairs, 10969 Berlin, Germany; wagner-endres@difu.de
12 Institut Raum und Energie, 22880 Wedel, Germany; wittekind@raum-energie.de
13 Faculty of Spatial Planning, Department of European Planning Cultures, TU Dortmund University, 44227 Dortmund, Germany; karsten.zimmermann@tu-dortmund.de
* Correspondence: hannah.kosow@zirius.uni-stuttgart.de

Citation: Kosow, H.; Wassermann, S.; Bartke, S.; Goede, P.; Grimski, D.; Imbert, I.; Jenssen, T.; Laukel, O.; Proske, M.; Protzer, J.; et al. Addressing Goal Conflicts: New Policy Mixes for Commercial Land Use Management. *Land* **2022**, *11*, 795. https://doi.org/10.3390/land11060795

Academic Editor: Jacques Teller

Received: 28 April 2022
Accepted: 23 May 2022
Published: 27 May 2022

Abstract: Commercial land use management that focuses on a future-oriented urban and regional development must address multiple goals. Effective policy mixes need to simultaneously (1) improve city-regional and inter-municipal cooperation, (2) reduce land take, and (3) assure the long-term economic development of a region. Using the Northern Black Forest in Germany as a case study, we brought together planning and land use research with public policy analysis. We applied cross-impact balances (CIB) to build and analyze a participatory policy-interaction model. Together with a group of 12 experts, we selected effective individual measures to reach each of the three goals and analyzed their interactions. We then assessed the current policy mix and designed alternative policy mixes. The results demonstrate that current approaches to commercial land use management present internal contradictions and generate only little synergies. Implementing innovative measures on a stand-alone basis runs the risk of not being sufficiently effective. In particular, the current practice of competing for *municipal* marketing and planning of commercial sites has inhibiting effects. We identified alternative policy mixes that achieve all three goals, avoid trade-offs, and generate significant synergy effects. Our results point towards a more coherent and sustainable city-regional (commercial) land-use governance.

Keywords: commercial area; land use governance; city-regional governance; goal conflict; cross-impact balances CIB; policy mix; policy design; policy coherence; urban and regional development

1. Introduction

Reducing land take in urbanized areas and preserving open spaces and conservation land are necessary to preserve the soils with their diverse functions [1]. Land-saving urban development has, therefore, become a guiding principle in several countries around the world, including Germany. However, land is still required for housing and the expansion of infrastructure, as well as for commercial development and renewable energies. At the same time, competition between different land uses is increasing [2]. For decades, the central strategic approach to sustainable land management has been the idea of urban densification and intensification [3], widely recommended by environmental agencies [4,5]. For commercial land management, the redevelopment of brownfield land is a significant concept of growth management and planning [6–8]. Moreover, concepts for multi-story or multifunctional construction in commercial areas also exist [9,10].

In addition to urban densification, inter-municipal or even regional cooperation is recommended by environmental authorities as well as the scientific community with regard to commercial land development. Inter-municipal cooperation is expected to generate economic, social, and ecological benefits [11–13], as well as synergies between municipal partners and moving land-use planning to a more regional level [14–16]. In policy and strategy papers in Germany [10,17], inter-municipal cooperation in land management is often expected to support efficient land use and reduce land take [18]. However, such causality still needs to be empirically proven. Empirically, commercial areas in Germany are still mostly planned and managed in municipal competition [19], and innovative planning processes as well as land- and resource-efficient construction concepts are rarely implemented. In summary, we are confronted with a well-known but persistent issue.

In the following, we briefly outline the reason behind (a) a goal conflict between economic development and reduction in land take and (b) the rarely realized expectation of inter-municipal cooperation to support both goals. First, there is an inherent conflict between economic growth on the one hand and more efficient use of resources, including land, on the other hand. By means of different regulatory and financial control instruments, various environmental policy areas were successful in internalizing the previously externalized negative effects of a growth-oriented economy [20]. However, despite the ideas of a post-growth economy [21] or degrowth [22], companies follow the logic of profit maximization, and the state at all levels is dependent on tax revenues, which particularly applies to local governments in Germany. Therefore, environmental policies are challenged by the need for job and business development and municipal tax revenue generation [23].

Second, inter-municipal, or even regional cooperation in commercial land management, is expected to address this conflict: inter-municipally managed commercial sites are expected to be larger, more professional, and more attractive to businesses than smaller, dispersed, municipal developments. Smaller municipalities, in particular, believe themselves to be in a better position to attract companies and new residents from outside by cooperating in the management of commercial areas and thus realizing larger, more attractive locations [24]. Larger centers often lack suitable land in their jurisdiction. Furthermore, economies of scale are expected through shared infrastructures and development costs [25]. Simultaneously, larger commercial areas operated in inter-municipal cooperation help prevent urban sprawl. Cooperation also counteracts the problem that small municipalities designate and develop too much commercial land, often in poor locations, for which there is no need in reality. In particular, municipalities with poor economic performance tend to develop more commercial areas to attract businesses and industry than the demand requires, thus, wasting land resources [26].

Nevertheless, inter-municipal or even regional cooperation in commercial land management is rare. Reasons include the municipal planning sovereignty in Germany [27] and municipal dependence on business tax revenue [28], leading to competition between municipalities with regard to industrial location [29] and few incentives for cooperation. Although attempts are often made to do so, the cooperation partners are often unable to agree on the legal form to be chosen for the cooperation, voting rights, and cost–benefit distribution that are perceived as fair for all parties involved [19]. Cooperation between cities and their smaller surrounding municipalities is, in addition, hindered by the contrasts in size, interests, and power of the unequal partners [30]. Therefore, cooperation, especially between urban and suburban or rural municipalities, is often conflictive and inhibits sustainable land management [31].

Beginning with the diagnosis of these fundamental tensions and contradictions, our analysis assumes that there are (at least) three fundamental goals that the management of commercial land must address in the future:

- Goal I: Improve inter-municipal, in particular, city-regional, cooperation regarding commercial areas;
- Goal II: Reduce land take for commercial use;
- Goal II: Ensure a future-oriented (i.e., long-term) economic development.

Clearly, this is a simplification, as in different contexts, other goals may be prioritized (e.g., in cases where land is still easily available or in the case of municipalities focusing on housing and tourism). Moreover, even if goals I–III are prioritized, other development goals may also play a role, and this makes the interlinkages, trade-offs, and potential for synergy more complex. Indeed, multiple goals and goal conflicts were recognized as a central challenge for sustainable development. This is currently discussed regarding the 17 Sustainable Development Goals (SDGs)—with 8–12 targets each—of the United Nations Agenda 2030 [32]; see [33] for an overview of SDG interaction research. Policy coherence, i.e., coherence between policies from different sectors and across scales, has been recognized as a key requirement for sustainability transformations [34–37] and formulated in SDG Target 17.14 [32]. A growing body of the literature is using policy coherence for sustainable development (PCSD) as a framework to understand trade-offs and develop new policy methodologies fostering sustainability transformations [37,38]. Striving for synergy and avoiding trade-offs is crucial [39,40]. Land use governance research proposed typologies of common interactions between instruments during different phases of the policy cycle [41]. While goal conflicts have been widely recognized as hindrances to more regional and more sustainable commercial area planning and governance [17], policy combinations regarding commercial areas and their coherence have, to the best of our knowledge, not been considered yet.

Research on policy combinations has been performed in various fields, such as development policy [38], innovation policy [42–44], and mobility policy [45–47]. It analyzes interactions, goal conflicts, temporal logics of (old and newly implemented) strategies, and measures and policy instruments within or between policy fields and between levels of governance. The literature illustrates that consistent bundles of measures are rare. In contrast, a policy patching of different measures is frequently observed, which is often inconsistent in their interactions [48,49]. Public policy analysis provides a rich body of literature on policy mixes [42,50,51], i.e., combinations of policies (tools, instruments, and measures) and their interaction. The main hypothesis that we can draw from the literature is that policies must be considered in their interplay; focusing on individual approaches neglects the interplay of new and old policies as well as policies within and between sectors and on different scales (e.g., municipal, inter-municipal, regional, state). Combinations require to be considered to avoid trade-offs and generate synergies. Regarding commercial areas of the future, lists of individual policies for different governance levels in Germany have been developed [10]; however, we lack a systematic analysis of their interplay and the effects of their combination. Such analysis is required to design approaches for a more coherent (and more sustainable) regional governance of commercial land use. We consider this as the central research gap addressed by this study.

In order to systematically analyze the interplay between policies to reach multiple goals, a new approach was recently developed [52] that uses a qualitative but systematic form of systems analysis, cross-impact balances (CIB) [53]. This approach allows the evaluation of status quo policy-mixes ex post, designing of alternative policy mixes, and their ex ante evaluation regarding their internal consistency and degree of synergy. This approach was developed in the field of water management, and its transferability to other fields was argued [52]. However, this transfer has not been empirically tested, and hence, it is the secondary research gap addressed by this study.

The following sections aim to respond to the two research gaps identified above. We ask the following questions: How to combine policies to improve city-regional cooperation, reduce commercial land take and assure the economic performance of a region at the same time? How do policy mixes appear in which policies interfere with or contradict each other as little as possible in their effectiveness and realize as many synergies as possible? Which policy mixes can achieve the three goals equally and jointly?

Using the case of the Northern Black Forest as a typical example, we brought together environmental, planning, and land use research and public policy analysis. We applied cross-impact balances (CIB) to build and analyze a participatory policy-interaction model. Together with a group of twelve experts from science and practice, we selected effective individual measures on the regional and municipal level to reach each of the three goals (in total, 27 measures) and analyzed their interplay (Section 2). With the help of CIB, we then assessed the current policy mix and designed alternative policy mixes without internal contradictions, as well as their synergy and goal attainment (Section 3). We discussed central contributions, limits, implications for the region, transferability, and avenues for further research (Section 4) and drafted a brief conclusion (Section 5).

This paper aims to address the goal conflicts of commercial land use management. We analyzed the contradictions of the current policy mix and identified effective policy combinations to simultaneously improve city-regional (urban-surrounding) cooperation, reduce commercial land take, and ensure the long-term economic development of a region. Our analysis reveals that the policy mix, which is currently in place on the regional and municipal levels in Germany, depicts internal contradictions and generates only little synergies. Many measures to foster city-regional cooperation and long-term economic development and reduce land take are hindered by the current practice of competing for municipal planning and marketing of commercial areas. Despite this, there are several alternative policy mixes that could achieve all three goals jointly, avoid contradiction, and generate significant synergy effects. Fundamental levers in these alternative mixes involve regional development strategy on commercial land, enforced municipal land policy to strengthen inward development, joint infrastructures in business parks, as well as joint regional commercial area management and marketing. Such policy mixes could overcome goal conflicts, improve city-regional cooperation, reduce land take, and simultaneously improve long-term economic performance.

2. Materials and Methods

2.1. A Case in the Northern Black Forest, Germany

As a case study, our analysis addressed the area of Pforzheim, a regional center of ca. 120,000 inhabitants in South-Western Germany, and its 22 surrounding communities belonging to a county named Enzkreis (with ca. 3000–12,000 inhabitants each). Together, administratively spoken, the center and surroundings form a sub-region (Mittelbereich Pforzheim) of 500 km^2, abbreviated by PFENZ (see Figure 1). Along with two more rural counties (Landkreise), PFENZ is part of the Region's Northern Black Forest. PFENZ, as we argued in the following, is an exemplary case that is typical for several other cases and will prove to be more so in the future. First, the initial situation in PFENZ is challenging and, therefore, particularly instructive [54]. There is not only competition in commercial land planning between the city and surrounding municipalities but also, historically, a rather poor and conflict-laden relationship between the central city and the surrounding municipalities [55]. There is a severe lack of suitable areas for new commercial areas (greenfield development) due to topography and natural and regulatory constraints.

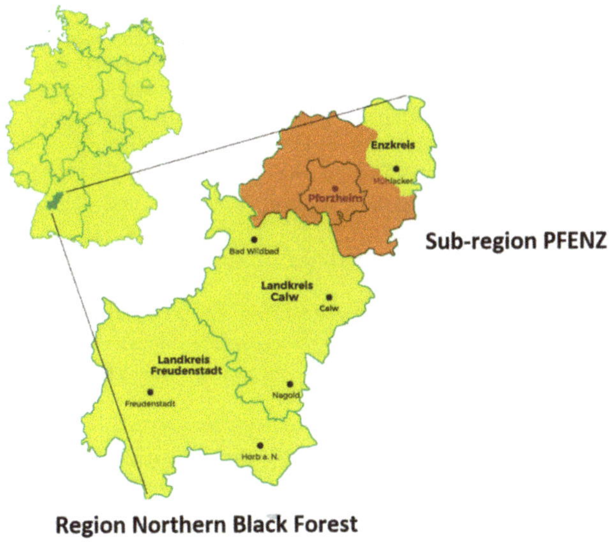

Figure 1. Location of PFENZ (figure adapted with permission from Wirtschaftsförderung Nordschwarzwald GmbH).

In addition, there have been considerable wasteful practices of commercial land development in the past, e.g., by settling single-story logistics centers with large parking spaces on one of the last remaining large commercial areas. Finally, there is a rather difficult economic situation in the center and a heterogeneous economic situation in the surrounding municipalities. Thus, the center perceives a strong need to develop further land for commercial use to assure its economic development but has very few areas left in its own jurisdiction [26,56].

Second, PFENZ can be considered a typical case for many other densely populated and prospering sub-regions in Germany, and despite institutional and regulatory differences, to a certain degree also for city regions with medium-sized core cities in other European countries, such as France, Spain, or Poland. The current situation of policies and measures applied (i.e., the status quo policy-mix, see Section 3.1) is rather typical. The particular challenges regarding land scarcity in the case study area are to anticipate the issues most regions will be confronted with in the future when other land uses compete more intensely with commercial use. This future development is suggested by multiple land-use related political goals of the current Federal Government [57]: (1) reducing the net land take to zero by 2050 (in the Land of Baden-Württemberg, this goal is already to be met in 2035); (2) erecting wind turbines on 2% of the land; and (3) building 400,000 new housing units per year.

2.2. Using Cross-Impact Balances (CIB) for Policy-Interaction Modeling and Policy Design with a Group of Experts

In order to design and assess policies that may lead to future-oriented commercial area management, we used CIB for policy interaction modeling (Section 2.2.1) within an inter- and transdisciplinary workshop series that brought together a group of experts from research and practice (Section 2.2.2).

2.2.1. CIB for Policy Design—Assessing Consistency and Synergy of Policy Mixes

CIB is a qualitative yet semi-formalized form of systems analysis [52]. The principal goal of this method is to increase policy coherence. The method requires identifying system elements and exploring the interrelations found between them. A brief introduction to CIB is given in Supplement S1. Initially, CIB was developed and used to construct future scenarios [58–63]. However, CIB also proved useful for qualitative forms of systems analysis [64,65], and recently, CIB has been transferred to the realm of policy design [52]. The main idea of this new application consists of considering goal conflicts on the level of policies to reach these goals and using the CIB balance algorithm to optimize all goals at the same time.

The approach comprises four steps [52]:

- Step 1: Identify and define central objectives as well as alternative policies to reach the goals;
- Step 2: Assess directed hindering and fostering impacts between policies (pairwise) through expert or stakeholder judgments;
- Step 3: Identify policy mixes with a high level of internal consistency;
- Step 4: Assess policy mixes, e.g., regarding their synergy, goal attainment, or other criteria.

Internal consistency of policies is measured by the CIB balance algorithm, which evaluates the direct and indirect influences of the policies on each other. With the help of the CIB algorithm, all thinkable policy combinations are analyzed, and the consistent mixes are identified. *Consistency of a policy mix* explains whether all sub-goals of a policy mix are present in an optimal state [52], i.e., whether, in a policy mix for each sub-goal, the optimal policy alternative (the one with the highest sum of impact arguments) is selected. Consistent policy mixes represent the Nash optima of the policy-impact network. They avoid major conflicting impacts, i.e., trade-offs, among all policies and maximize all related sub-goals, individually but at the same time. Consistency informs about the inner stability of a policy mix.

Synergy of a policy mix is measured by the sum of positive and negative impacts within each mix, i.e., the sum of interactions or total impact score (TIS) [52]. Synergy explains how well a policy mix combines fostering relations and avoids hindering relations between policies. It must be understood as a relative statement, i.e., policy mix X is more synergetic than policy mix Y. Maximizing synergies in policy mixes allows to benefit from supportive policy interactions and gives information on the overall effectiveness of a policy mix. The mix with the highest synergy (measured by TIS) can be considered the most (overall) effective one.

Regarding the relation between consistency and synergy, it is stated that "CIB solutions imply that each objective is 'choosing' its policy in an attempt to optimize its own synergy gains and the TIS represents the sum of all individual synergy gains" [52] (p. 43). Synergy helps policy makers to decide which mix would be the overall most effective combination (global information). Consistency describes the individual contradictions showed by a mix and helps in indicating the unequal distributions of gains and losses among the goals that also appear in synergetic mixes and can jeopardize the stability of the mix.

In the following, we specified how the four steps of CIB policy-interaction modeling and analysis were implemented.

2.2.2. Online Workshop Series with a Group of Experts from Research and Practice and Model Analysis

We invited a group of twelve experts from research (e.g., planning, political science, economic science) and practice (e.g., the regional planning authority, regional chamber of commerce and industry (CCI), regional economic development agency, Federal Environment Agency, State Ministry of Environment) to a series of online workshops (April to November 2021). We selected four experts with expertise for each of the three domains: inter-municipal cooperation, reduction in land take, and economic development. Four of the twelve experts had, in addition to their domain knowledge, particularly local knowledge about the region.

During the first step, a first online workshop took place during which the three goals, i.e., improving city-regional cooperation, reducing land take, and improving economic performance, were further specified into three sub-goals each (see Table 1). Moreover, key policies were identified to achieve these sub-goals. To be most effective, each of these policies was combined with ancillary measures and preconditions. Finally, the policies currently implemented in PFENZ were identified (status quo) (see Table 1). The selected n = 27 policies (status quo and alternatives) were further defined and detailed following the workshop and circulated among the expert group to assure a joint understanding of all measures (cf. Section 2.1 and Supplement S2).

Table 1. Overview of sub-goals and policies and their initial effectiveness (PFENZ). Status quo policies: target not (fully) achieved. Numerical value in brackets: effectiveness of individual policy with regard to the respective sub-goal. Based on a preliminary survey among the respective 4–5 topic experts, joint validation during workshop 2. Scale: (1) = not effective, (2) = less effective, (3) = rather effective, (4) = effective, (5) = very effective.

Goal	Sub-Goal	Status Quo	Option A	Option B
I Improve city-regional cooperation regarding commercial areas	1 Improve communication and trust	1S Irregular exchange between city and individual surrounding municipalities (3)	1A Informal expert group PFENZ (initiative of the regional planning authority) (4)	1B PFENZ mayors' group (initiative of the municipalities) (4)
	2 Joint development perspective for PFENZ	2S Priority areas for commerce and industry (regional plan) (4)	2A Joint statement of the municipalities in PFENZ on the regional plan (3)	2B Regional development concept commercial areas in PFENZ (4)
	3 Develop commercial areas in inter-municipal cooperation	3S Informal (and not very systematic or transparent) preliminary talks for inter-municipal commercial area development (3)	3A Cooperation agreement for the cooperation between the city and the surrounding municipalities in the development of commercial areas (4)	3B Special-purpose association for joint commercial area development between city and surrounding municipalities (5)
II Reduce land take for commercial purposes	4 Reduction and control of land use for commercial areas	4S Regional plan as a corridor for land use (3)	4A Regional plan (as 4S) + regional land monitoring incl. land development reports to review land policy objectives (4)	4B Regional plan + land-use monitoring (as 4A) PLUS regional quantity target with tradable quotas (5)
	5 Mobilize existing commercial land and buildings	5S Innovative purchase agreements (individual municipalities) (2)	5A Leasehold; interim acquisition (in case of insolvency of companies) (Basic approaches of municipal land policy (3,5)	5B Extended approaches to municipal land policy as interim acquisition, urban development contracts, and deconstruction obligations, etc. (4)
	6 Use of commercial areas in a land-saving and multifunctional way	6S Information on land-saving and multifunctional use (2)	6A Municipal incentives for multi-story and multifunctional construction (especially urban planning competitions, awarding of concepts) (4)	6B Lower land prices in return for multifunctional and public areas (3,5)

Table 1. *Cont.*

Goal	Sub-Goal	Status Quo	Option A	Option B
III Ensure long-term economic development	7 Future-oriented commercial area planning and implementation	7S Commercial areas planned from a municipal perspective (not strategic for PFENZ) (2)	7A Tools for cost–benefit analysis and monitoring of commercial area projects (4,5)	7B Shared infrastructures within commercial areas (4,5)
	8 Securing and creating (sustainable) jobs	8S Regional start-up consulting; cluster strategy of the city (3)	8A Promotion of research and development (R&D) in PFENZ (4)	8B Joint innovative settlement strategy and skilled labor strategy in PFENZ (5)
	9 Improve regional location quality	9S Individual municipal location marketing (heterogeneous, depending on the size and economic power of the municipalities) (2)	9A Gain and communicate a sustainability label for PFENZ (initiative of CCI and business development) (3)	9B Joint (collaborative) commercial area management and location marketing PFENZ (5)

During the second step, we split the expert group by domain of expertise into three groups of four experts. During small-group workshops, we first asked the experts to assess the effectiveness of each of the 27 individual measures to reach its (sub-) goal individually on a five-point scale (1 = ineffective, 5 = very effective). This assessment was discursively validated in the expert group. Second, we asked experts regarding interactions between measures (the status quo and the most effective alternatives). This means that we arranged the 27 measures as lines and rows of a CIB matrix and asked for every pair of measures, "Does measure "x" have an impact on the effectiveness of measure "y"? If yes, is it a fostering or a hindering impact? How strong is the impact?" We coded these impacts using a 7-point scale [53] (−3 = strongly hindering impact, 0 = no impact, +3 = strongly fostering impact). The full matrix can be found in Supplement S3. We also asked for verbal justifications and explanations for these impact assessments and noted and recorded the argumentation as well as the final group agreement for each assessment (see Supplement S4 for an example).

In phase three, we combined the results from all small-group workshops and constructed one joint CIB matrix containing all impact assessments as well as their verbal justifications. With the help of the CIB balance algorithm and the freely available CIB software *ScenarioWizard* we analyzed the matrix. The CIB software is available for free download at www.cross-impact.de (accessed on 1 March 2022). We assessed the consistency and synergy of the status quo (a mix combing only status quo policies) and searched for internally consistent policy mixes, i.e., mixes without contradictions. This means that we searched for combinations of measures that are stable in themselves, i.e., interfere or contradict each other as little as possible in their effectiveness. We sorted these solutions by cluster analysis [66] and diversity sampling [67,68] and selected the most diverse representatives for consistent, i.e., internally stable, and alternative policy mixes for commercial development in the future. An analysis of active and passive sums helped in further understanding our policy interaction model [69].

In phase four, we also analyzed the synergies resulting from the different consistent mixes. For this, we used the CIB total impact score (TIS), summing all positive and negative impacts of each mix. We then asked how well the different mixes reached the three different goals. Subsequently, we developed a new measurement to assess the quality of the different mixes to achieve the three overall goals. *Goal attainment* was calculated by summing the initial effectiveness of the policies with the total impact score TIS (interaction effect) per overall objective. We chose to give the initial effectiveness (rated on a 5-point scale) a double weight, with the argument that one strongly hindering impact from another policy (impact score = −3) should not annulate a policy individually considered rather effective (basic effectiveness score = 3), but the strongly hindering impacts by two other policies (impact sum = −6) might do so. In summary, to calculate goal attainment for each consistent mix, we considered the initial effectiveness of the individual measures to reach each of these goals and summed it with the interaction effects by splitting the total impact score (TIS) by goal. We also calculated correlations between goal achievement. Finally, we presented our results in a third workshop to the expert group and discussed the plausibility of results, gaps, and limits of the analysis, as well as questions on how to implement alternative policy mixes in the region and assess the transferability of the approach and its results.

2.3. Sub-Goals and Individually Effective Measures

Table 1 summarizes the three sub-goals defined by the experts for each of the three overall goals, as well as their most relevant two alternative policies (options) as suggestions on how to reach each sub-goal better than with the current status quo policy (see Table 1). An example of the detailed definitions of sub-goals and policies, including auxiliary measures and conditions for each, is given in Supplement S2. Table 1 also indicates how the experts assessed the effectiveness of the *individual* policies (initial effectiveness). Policies vary in effectiveness when considered individually. Tendentially, status quo policies are considered less effective than alternative options. Initial effectiveness rating seems to be dependent on the clarity and concreteness of the definition of the measure. Examples of rather vague policies include 2A and 6B. Moreover, the critical feasibility—due to legal considerations—might have led to lower assessments of 6B. Experts unfamiliar with the situation in the region tended to rate the status quo policy more positively than the regional experts.

3. Results

In this section, we described why the status quo mix is not an optimal solution (Section 3.1) and how alternative consistent and synergetic policy mixes for commercial area management may appear (Section 3.2). Subsequently, we discussed the policies not seen in any of the consistent mixes and the reason behind their absence (Section 3.3). Finally, we demonstrated how well the different mixes perform regarding the three different overall goals and how these overall goals correlate (Section 3.4).

3.1. The Status Quo Policy-Mix Is Inconsistent and Realizes Little Synergy

The policy-interaction model was first used to check whether individually composed policy mixes (combinations of one policy each per sub-goal) are internally consistent and how well they realize synergies or show conflicts between policies, as well as the reason behind. Analyzing the current status quo policy mix reveals that the nine individual policies have very low consistency values. This means that the respective impact balance of each status quo policy only slightly exceeds or lags behind the impact balance of the best alternative variant. Figure 2 shows that the mix contains several inconsistent policies (colored in orange), i.e., policies that are in contradiction with the other policies. This applies to four out of nine measures (2S, 6S, 8S, 9S).

For three measures (S3, S8, and S5, in gray and italic print in Figure 1), the sum of the influences (impact sum) is negative for all variants (albeit to different degrees). Thus, in the status quo situation, these measures are inhibited in their effect, as their alternatives are. Finally, in the status quo policy mix, two policies, namely the municipal planning (7S) and municipal marketing (9S) of commercial areas, i.e., the municipal competition, strongly hinder the effectiveness of several other measures (red arrows in Figure 1). Most of these other measures are assessed as rather effective in reaching the three overall goals in PFENZ when considered individually: Priority areas for commerce and industry (2S, initial effectiveness value 4), innovative purchase agreements (individual municipalities) (5S, initial effectiveness value 2), informal preliminary talks for inter-municipal commercial area development (3S, initial effectiveness value 3), and the regional start-up consulting and cluster strategy of the city (8S, initial effectiveness value 3) are inhibited in their effectiveness by the *municipal* approaches to planning (7S) and location marketing (9S).

These more innovative status quo approaches are thus blocked from being effective by the traditional municipal competition regarding commercial areas in the status quo mix.

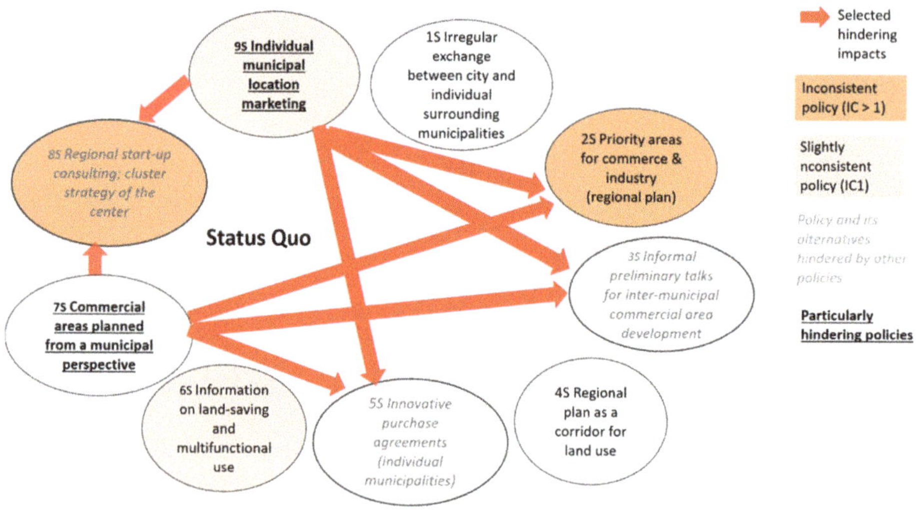

Figure 2. The status quo policy mix in PFENZ: inconsistent policies and hindering effects by municipal approaches to planning and marketing.

The policy-interaction model also helps in discovering the effects of changing individual policies in a mix. In this case, if one replaces the municipal planning (7S) with the planning of joint infrastructures (7B) and the municipal location marketing (9S) with joint area management and location marketing in PFENZ (9B), this leads to fostering relations within the mix (selected impacts represented by green arrows in Figure 3).

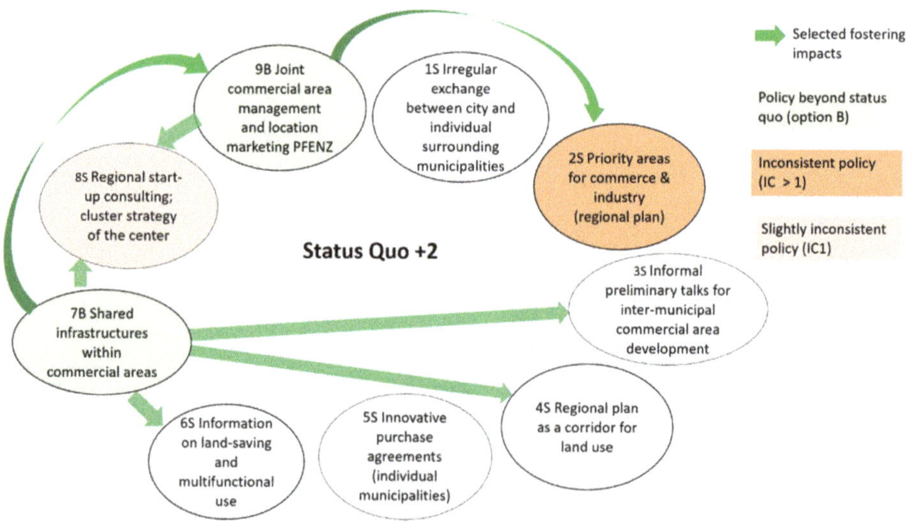

Figure 3. Changing individual policies leads to more mutually fostering relations, but the mix is still inconsistent and not very synergetic (selection of impacts, TIS = 48).

However, this mix is still not completely free of contradictions: The regional policy start-up consulting and cluster strategy of the center (8S) are still slightly inconsistent. The policy of priority areas (2S) remains highly inconsistent, as seven argument points more in favor of its alternative, the regional development concept "commercial areas in PFENZ" exist (2B). In short, it seems insufficient to alter a few status quo policies into more innovative policies, and it is difficult to find mixes "by hand" that are free of contradictions and make optimal use of synergies between policies. Using the CIB balance algorithm to systematically scan the policy-interaction network for overall consistent mixes provides the results presented in the following subsection.

3.2. Consistent Policy Mixes and Their Synergy

From the almost 20,000 theoretically conceivable combinations, 60 combinations, i.e., policy-mixes, are internally consistent. This means that they show no or almost no internal contradictions, implying no inconsistent policies. To account for evaluation uncertainty (e.g., in case an impact was evaluated +1 instead of +2), we also defined those policies as consistent, of which the impact sum is one point lower than the one of their alternatives (CIB inconsistency level 1 or IC1).

Diversity sampling, i.e., choosing mixes that are as diverse as possible and cover the space of possible mixes as broadly as possible, reveals six policy mixes. Cluster analysis [66–68] demonstrates that seven clusters can be identified, with each cluster sharing at least six out of nine policies (see Supplement S5). Behind each of the six diverse mixes, a cluster exists, i.e., a group of similar mixes (with variations), and thus, the diversity sample is a fair representation of the group of n = 60 consistent or stable mixes. To represent the seventh cluster, we added one more mix to the selection, which we briefly present in the following. For an overview, see Table 2.

While the status quo itself is not consistent, interestingly, there is a consistent mix ("Status quo plus R&D") close to the status quo, with eight out of nine status quo policies. In this mix, the regional start-up consulting and cluster strategy of the city (which has a municipal logic) are replaced by R&D funding in PFENZ, which has a regional logic. This new policy seems to stabilize the status quo by assuring the competitiveness of the regions' innovation system by adding a regional perspective that does not contradict or hinder any of the municipal activities. This mix is consistent, meaning that it is internally stable but with very low consistency values per policy and very few synergies (see Table 2 and further interpretation below).

In the mix "First steps", five status quo policies remain. Regarding cooperation (goal I), a regional development concept "commercial areas in PFENZ" (2B) introduces a strategic regional perspective on commercial areas. This policy is found in all internally consistent mixes (except for the mix "Status quo plus R&D"). In this mix, municipalities also increase their influence on commercial land through basic approaches of municipal land policy as leasehold and interim acquisition of commercial real estate (in case of insolvency of companies) (5A). This presents important changes regarding policies for the long-term economic development of the region (goal III) through common/shared infrastructures within commercial areas (7B) and joint commercial area management and location marketing (9B).

The mix "Planning competition" combines a regional development concept (2B) with more extended approaches of municipal land policy as interim acquisition (in the extreme case of expropriation), urban development contracts and deconstruction obligations, etc. (5B), and with municipal incentives for multi-story and multifunctional construction (especially urban planning competitions, awarding of concepts) (6A). This approach is further supported by strong policies to achieve goal III, namely, in addition to shared infrastructures within commercial areas (7B) by a joint innovative settlement strategy and skilled labor strategy in PFENZ (8B) and a sustainability label for PFENZ on the initiative of the chamber of industry and commerce and of the business development agency (9A).

Table 2. Selection of consistent mixes, to be read in columns, status quo white, option A blue , option B green ; unique selling points of each mix within this selection (measures unique to this mix) in **bold print**.

	Main Topic	**Status Quo Plus R&D**	**First Steps**	**Planning Competi-tions**	**Negotiating**	**Enforced Planning**	**Narrow Inter-Municipal and Strategic Cooperation**	
	Diversity	x	x	x	x		x	x
	From Cluster	1	4	3	7	6	5	2
	Variants						Cooperation agreement	Special purpose association
	Consistency	IC1	IC1	IC1	IC1	IC1	IC0	IC1
	Synergy (TIS)	15	68	68	98	92	(a) 84; (b) 97	102
Goal	**Sub-goal**	**Policies**						
I Improve city-regional cooperation on commercial areas	1 Improve communication and trust	1S Irregular exchange between city and individual surrounding municipalities			**1B PFENZ mayors' group (initiative of the municipalities)**	1A Informal expert group PFENZ (initiative of the regional planning authority)	(a) 1S Irregular exchange between city and individual surrounding municipalities (b) 1A Informal expert group PFENZ (initiative of the regional planning authority)	1A Informal expert group PFENZ (initiative of the regional planning authority)
	2 Joint development perspective for PFENZ	**2S Priority areas for commerce and industry (regional plan)**	2B Regional development concept "commercial areas in PFENZ"					
	3 Develop commercial areas in inter-municipal cooperation	3S Informal (and not very systematic or transparent) preliminary talks for inter-municipal commercial area development					**3A Cooperation agreement for the cooperation between the city and the surrounding municipalities in the development of commercial areas**	**3B Special-purpose association for joint commercial area development between city and surrounding municipalities**

Table 2. *Cont.*

	Main Topic	Status Quo Plus R&D	First Steps	Planning Competitions	Negotiating	Enforced Planning	Narrow Inter-Municipal and Strategic Cooperation	
	4 Reduction and control of land use for commercial areas	4S Regional plan as a corridor for land use			**4B Regional plan (as 4S) plus land use monitoring (as 4A) PLUS regional quantity target with tradable quotas**	4A Regional plan (as 4S) plus regional land monitoring incl. land development reports to review land policy objectives	4S Regional plan as a corridor for land use	4A Regional plan (as 4S) plus regional land monitoring incl. land development reports to review land policy objectives
II Reduce land take for commercial purposes in PFENZ	5 Mobilize existing building land and buildings	5S Innovative purchase agreements (individual municipalities)	5A **Leasehold; interim acquisition (in case of insolvency of companies) (*Basic approaches of municipal land policy***	5B Extended approaches to municipal land policy (as interim acquisition, urban development contracts and deconstruction obligations		5S Innovative purchase agreements (individual municipalities)	5B Extended approaches to municipal land policy (as interim acquisition, urban development contracts and deconstruction obligations)	
	6 Use of commercial areas in a land-saving and multifunctional way	6S Information on land-saving and multifunctional use		**6A Municipal incentives for multi-story and multifunctional construction (especially urban planning competitions, awarding of concepts)**	6S Information on land-saving and multifunctional use			
	7 Future-oriented commercial area planning and implementation	**7S Commercial areas planned from a municipal perspective (not strategic for PFENZ)**	7B Shared infrastructures within commercial areas					
III Ensure economic development PFENZ	8 Securing and creating (sustainable) jobs	8A Promotion of research and development (R&D) in PFENZ	**8S Regional start-up consulting; cluster strategy of the city**	8B Joint innovative settlement strategy and skilled labor strategy in PFENZ				
	9 Improve regional location quality	**9S Individual municipal location marketing (heterogeneous, depending on the size and economic power of the municipalities)**	9B Joint business park management and location marketing PFENZ	9A Gain and communicate a sustainability label for PFENZ (initiative of CCI and business development)	9B Joint business park management and location marketing PFENZ			9A Gain and communicate a sustainability label for PFENZ (initiative of CCI and business development)

The mix "Negotiating" combines strong economic policies with different forms of cooperation, namely a mayors' group on the initiative of the municipalities (1B), which supports and is supported by an extension of the regional planning instruments beyond the regional plan in the form of land use monitoring and a regional quantity target with tradable quotas (4B). This planning instrument is consistent only in mixes that also provide an arena for that trade, as reciprocated by the mayors' group.

The Mix "Enforced planning" combines the strong economic policies with an informal expert group PFENZ on the initiative of the regional planning authority (1A) as well as with regional planning instruments beyond the regional plan in the form of land monitoring, including land development reports to review land policy objectives (4B); however, it does not foresee a regional quantity target nor tradeable quota.

Next, we see mixes that can be labeled "Narrow strategic and inter-municipal cooperation". One variant is marked by a "Cooperation agreement for the cooperation between the city and the surrounding municipalities in the development of commercial areas" (3A). This variant is fully consistent (IC0) with both; an irregular exchange between the city and individual surrounding municipalities (1S) or an informal expert group PFENZ (an initiative of the regional planning authority (1A)). The other variant is a mix with a "special-purpose association for joint commercial area development between city and surrounding municipalities" (3B). This mix belongs to a cluster of mixes with a special-purpose association as their "unique selling point".

All 60 consistent mixes (except for the Status Quo plus R&D mix) share the regional development strategy regarding commercial land use (2B) and sharing joint infrastructures in commercial sites (7B). The innovative joint settlement and skilled labor strategy in PFENZ (8B) is almost as dominant as 2B and 7B. This measure for achieving sub-goal 8 secure and create (sustainable) jobs is found in 56 of the 60 mixes, which are further always found to be in combination with 2B and 7B. Another dominant measure is information on land-saving and multifunctional use (6S). It is included in n = 51 of the 60 mixes. Although its initial effectiveness is rather low (2), this informational policy is highly compatible with all other policies and (almost) not hindered by any other policy. These three dominant policies can be considered (quasi) unavoidable for any consistent policy mix going considerably beyond the status quo. However, as our results showed, several different policy mixes are equally consistent.

The consistent mixes diverge regarding their degree of *synergy*, expressed by the Total Impact Scores (TIS), with a range from 15 to 111 points (see Supplement S6). The two mixes close to the status quo clearly illustrate the lowest synergy scores. There are several thousands of inconsistent mixes, which show higher overall synergy values. The two consistent mixes close to the status quo lie at about 3% of the TIS percentile. This phenomenon of low synergy but consistency can be interpreted as a "lock-in effect". The literature states that "consistent mixes can lead to suboptimal global synergy gains, when no objective (or the actors behind it) can pave the way to a globally better solution by its own one-sided policy portfolio" [49] (p. 43). Synergy, i.e., overall effectiveness, of the other mixes is much higher, starting from the 70% TIS percentile. However, higher synergies also vary considerably, with the mixes "Planning competition and "First steps" showing the lowest and mixes from the "Special purpose association" cluster showing the highest synergy values.

3.3. *Policies Absent from Any Consistent Mix*

Three policies from the initially selected list of n = 27 are not consistent with any of the n = 60 mixes and thus, do not appear in any of them. These are the joint statement of the municipalities in PFENZ on the regional plan (2A), lower land prices in return for multifunctional and public land (6B), and finally, tools for cost–benefit consideration and monitoring of commercial land projects (7A). By comparing the column sums (impact sums) of these policies and their alternatives in the matrix (Supplement S3), it can be seen that there are considerably fewer supporting arguments coded for all three vacant policies than for their respective alternatives. This means that these policies have a difficult time from the outset asserting themselves against their alternatives. Albeit they are not seen as being hindered by other policies, experts consider that the other policies do not support them. This results in large differences regarding column sums of alternatives. One part of the explanation for these low impact sums might also be that during impact assessment, experts presented doubts regarding the local feasibility (in the case of 2A) and the legal feasibility (in the case of 6B).

However, for 7A, the initial expert assessment suggested that, when considered individually, tools for cost–benefit analysis and monitoring of commercial area projects would be an active, impactful measure (see Section 2.1). As the policy 7A was assessed as interesting and individually very effective by the experts, as it is not being hindered by other policies, but its alternatives are more strongly promoted, we propose to have a closer look at this policy. In addition, comparing active (line) and passive (row) sums of individual policies (Figure 4) demonstrates that 7A would be an active, impactful measure if only it is promoted and supported by other measures.

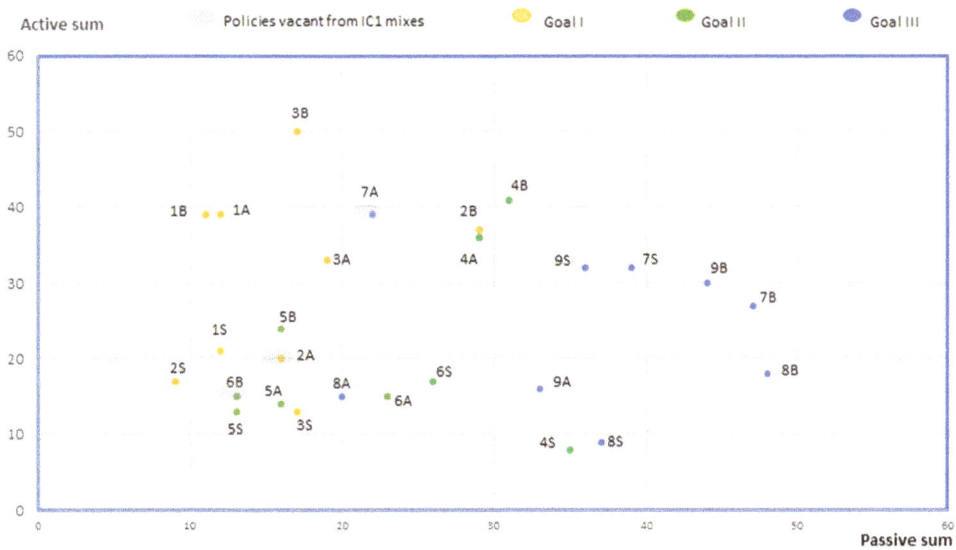

Figure 4. Active and passive sum of all individual policies.

3.4. Goal Attainment by the Consistent Mixes and Correlation between Goals

Therefore, we assume, in the form of a thought experiment, 7A to be set in PFENZ due to further policies outside those considered in our model as legal requirements (federal law) or joint decisions (regional agreement). If we "enforce" 7A within the matrix, additional consistent mixes appear. We found n = 5 solutions that are fully (IC0) and n = 56 further solutions that are almost fully (IC1) consistent. Scanning these solutions for similarity with the mixes identified before and sampling them by diversity shows that tools for cost–benefit analysis and monitoring of commercial area projects (7A) could be included in several types of mixes, such as those of "Narrow strategic and inter-municipal cooperation" as well as "Planning competitions". Jointly, with a regional development concept "commercial areas" in PFENZ (2B), regional land monitoring incl. land development reports to review land policy objectives (4A) and a joint commercial area management and location marketing PFENZ (9B), it could also form another "Enforced planning" mix with five remaining status quo mixes. Nonetheless, these considerations only hold if forces external to our model are strong enough for the tools for cost–benefit analysis and monitoring of commercial area projects (7A) to become an incontrovertible and unavoidable policy for all players in the region.

3.4.1. How Well Do the Different Mixes Achieve the Three Goals?

Figure 5 illustrates how well the different n = 60 mixes achieve the three overall objectives (for the calculation of the goal achievement from initial effectiveness and interaction effects, see Section 2.2.2).

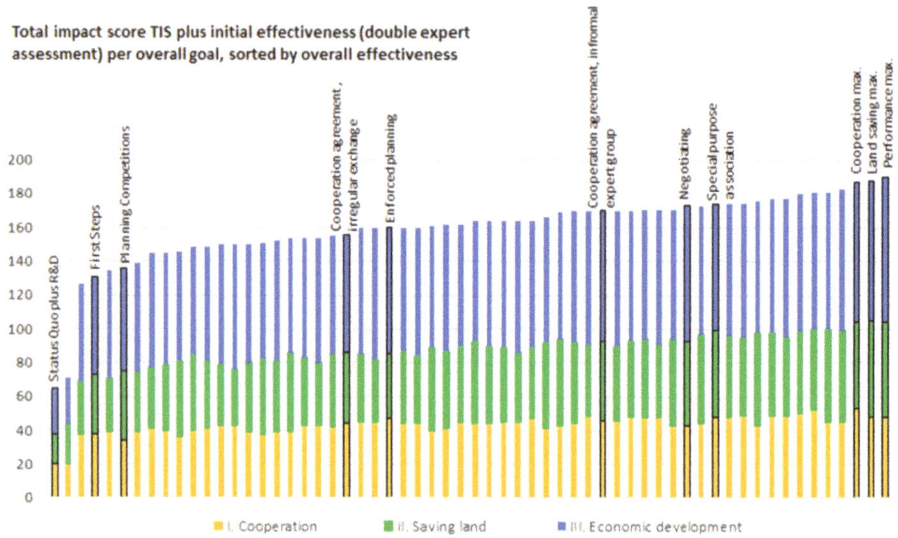

Figure 5. Goal attainment of consistent mixes: How well do the different mixes achieve the three overall goals?

By examining these results, it can be seen that the goal achievement values of the individual mixes for the overall objective of economic development (goal III) demonstrate a significantly higher value than for the other two objectives. There are two reasons for this. First, the nine measures in the matrix to achieve goal III are almost all passive or highly interlinked measures that receive many effects from the other measures. Second, all alternative measures (all "a" and "b" measures) have high initial effectiveness for achieving the sub-goals of goal III. For the interpretation of the results, it should be noted that the measures for goal I, "Improve city-regional cooperation", represent active measures that significantly contribute to the achievement of the other goals. Conversely, however, the measures for achieving the sub-goals of overall goal I benefit less from interaction effects through measures for overall goal II, "primarily buffering or networked measures", and overall goal III, "passive or networked measures". Therefore, to achieve better results, goal I, as well as goal II to a lower degree, would need to be supported by additional framework conditions outside of our analysis, as goal I benefits from fewer "positive side effects" from measures targeting goals II and III.

How well are the various overall goals being achieved by different policy mixes? Overall, the mixes close to the status quo are relatively weak, while the mixes beyond the status quo are much stronger in achieving all three goals; this difference is particularly significant regarding goal I and goal III. The "First steps" mix has nearly twice as high a score on all three goals as the worst-performing "Status quo plus R&D" mix. The mix "Planning competitions" performs better to some degree, especially regarding land saving and economy. However, the other mixes in our selection are significantly stronger in terms of goal achievement. The mix "Enforced planning" has an overall effectiveness of 160 value points, with rather high values for cooperation and economic development but poor results regarding saving land. The mix "Cooperation agreement" in its variant b), the informal expert group on the initiative of the regional planning authority, performs better on all three objectives than its variant a), with an irregular exchange between the city and the surrounding area (+4 index points on objectives I and II; +7 points on objective III). The "Negotiating" mix performs +5 index points better than the "Special purpose association" mix for the economic development objective III, but 5 index points worse for cooperation (goal I).

The mixes that best achieve the overall objectives belong to the "Special purpose association" cluster (see Supplement S5). The mix consisting of "b" measures only and of 6S (on the far right in Figure 4) possesses the best overall goal achievement ("Performance max") and performs best regarding goal III, economic development. A variant of this mix, which includes a sustainability label (9A) instead of joint area management (9B), shows satisfactory performance, especially with respect to objective III (−8 points) and with respect to objective I (−5 points). The mix with the maximum goal attainment regarding cooperation (goal I) ("Cooperation max") is a special-purpose association variant with leasehold (5A) and with area monitoring, including area development reports (4A), but without any regional quantity target and tradable quotas. The mix with the maximum performance in land saving ("Land saving max") corresponds to the mix known as "Performance max" but replaces the mayor's round by the informal expert group of the regional planning authority. In this mix, the assumption appears that experts under the aegis of the regional planning authority possibly place the goal of land saving higher than this is the case in the political considerations of a mayor's round.

3.4.2. How Well Are the Three Overall Goals Achieved Simultaneously in the Consistent Mixes?

Figure 6 shows that the overall objectives in the consistent mixes seem to present a linear relation and positive correlation with each other. In order to support our interpretation of the direction of these correlations, we also considered the active and passive sums of all policies, which inform about their role in the impact network (Figure 4).

First, there is a surprisingly strong positive relationship (r^2 = 0.59) between goal III (ensure a long-term economic development in PFENZ) and goal II (consume less land for commerce in PFENZ) (Figure 6a). This positive correlation seems counter-intuitive at first, as it does not follow the literature's assumption of the classic economy vs. environment conflict. However, the measures formulated for goal III in our policy-interaction model contain a notion of long-term, sustainable economic development that does not exclude land saving. If the (low) values of the two status quo mixes are excluded from the analysis, the correlation between the achievement of both goals in the n = 58 mixes beyond the status quo is still considerable (r^2 = 0.41).

Regarding possible effects of urban-rural cooperation (goal I) on the economy (goal III) and the ecology, namely land use by commerce (goal II), we achieve the following results: The strongest positive correlation (Figure 6b) shows between improving city-regional cooperation in the commercial sector (goal I) and increasing economic performance in PFENZ (goal III) (r^2 = 0.925). Presumably, this is because both goals are strongly linked via regional approaches to commercial land strategy, management, and utilization, as well as synergies between these approaches. Considering that cooperation (goal I) policies in our model are rather active and economic policies are rather passive (Figure 3), improving the city-regional cooperation in the commercial sector (goal I) can be considered to promote long-term economic development (goal III) of a region.

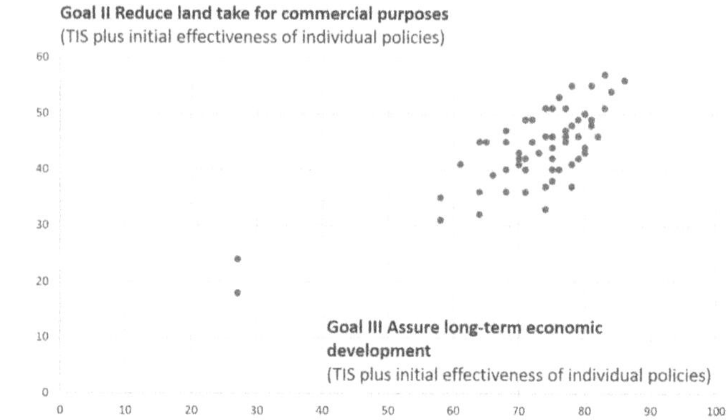

(a)

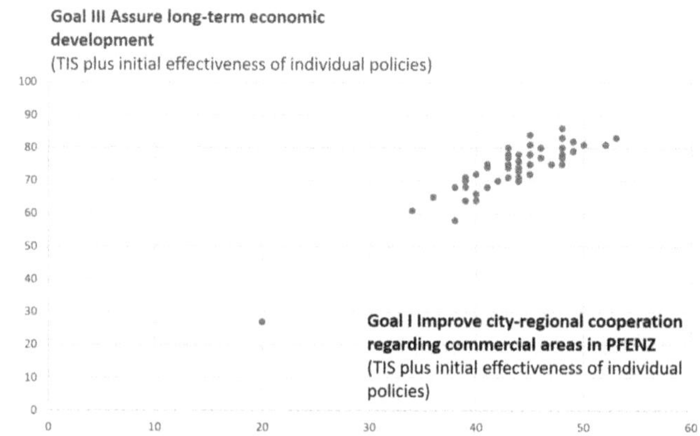

(b)

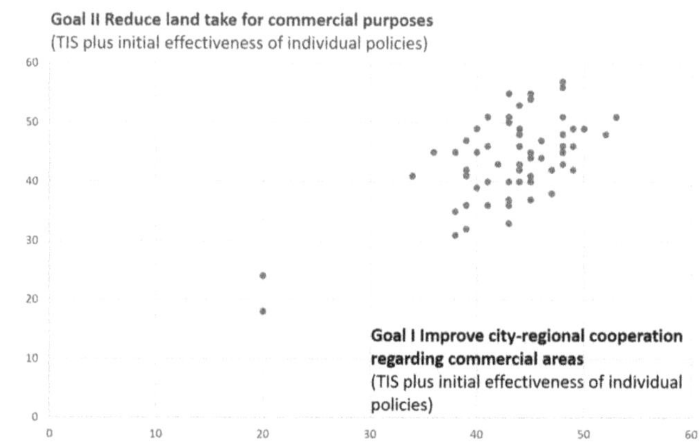

(c)

Figure 6. Relation of the overall objectives to each other in the n = 60 consistent policy mixes (goal attainment calculated by (double) initial effectiveness of the individual measures plus Total Impact Score TIS): (**a**) goal II and goal III; (**b**) goal III and goal I; (**c**) goal II and goal I.

Finally, the correlation between goal I, to improve city-regional cooperation in PFENZ in the commercial sector, and goal II, and take up less land for commercial activities in

PFENZ, is clearly weaker ($r^2 = 0.42$) (Figure 6c). Presumably, the relationship between these two goals is not as immediate and unidirectional as the literature on inter-municipal cooperation and land use saving postulates. Instead, effective land-saving approaches are often largely decided and implemented at the *municipal* level (sub-goal 5 and 6). These policies of (re)densification seem not to contradict the measures to improve inter-municipal or regional cooperation.

4. Discussion

We discussed the central contributions and limits of our study as well as future research avenues regarding policy coherence for a sustainable transformation of commercial land use (Section 4.1) and regarding the use of the method CIB for policy-mix evaluation and design (Section 4.2).

4.1. Towards Policy Coherence for a Sustainable Transformation of Commercial Land Use

Our analysis of the Northern Black Forest region shows that, in order to remain viable in the future, commercial land use management cannot focus unilaterally on short-term economic goals but must, over the long term, realize several goals simultaneously. Today's instruments and measures for regional cooperation, land conservation, and commercial land management depict clear internal contradictions and generate only very few synergies. This is a surprising insight as the German system of regional planning is considered to be comprehensive-integrative and mature in international comparisons [70]. In particular, the current practice of competing for *municipal* planning and marketing of commercial sites has an inhibiting effect on all three goals.

The analysis also presents how effective policy combinations could be designed to improve city-regional cooperation, reduce land take, and simultaneously ensure the long-term economic success of a region. First, introducing single new measures into the existing policy mix will not be sufficient to significantly increase goal attainment. Following or pushing one preferred policy (e.g., leasehold (5A), which is currently prominently discussed in the PFENZ region) is likely to not change the overall governance of commercial land use. Instead, the measures' effect (even with high basic effectiveness given) will become lost in trade-offs through hindering effects by the mutual interrelations of the status quo policies. Instead, new measures must be considered in consistent bundles, which mutually enforce each other. Examples that we found for levers to be set in motion are a regional development strategy regarding commercial land use (2B) and sharing joint infrastructures in commercial areas (7B).

According to our policy-interaction model, changes appear inevitable in all mixes going beyond the status quo; these *basic or inevitable policy changes* are the aforementioned regional development strategy regarding commercial land use (2B) and sharing joint infrastructures in commercial areas (7B) while also implementing a joint, i.e., collaborative, commercial area management (9B), as well as measures to reduce land consumption and strengthen re-densification (see policies to achieve sub-goals 4–6). Including these inevitable changes, *alternative policy mixes,* followed by adding further measures, also achieve the three goals together, avoid contradictions between measures and prove to be more synergetic. The analysis also shows the policies that were assessed to be quite effective individually but do not prevail in any consistent mix. In particular, this is the case regarding "tools for cost-benefit analysis and monitoring of commercial area projects" (7A). This policy, to be part of a consistent policy mix, would require support from further strategies or policies, potentially in the form of mutual agreements. These different mixes—as well as the information on policies that do not find their way in any consistent mix—can be considered valuable information for policy makers on municipal, regional but also superordinate levels. The results, in the form of options, could inform their policy-making processes and are a valuable addition to policy documents considering policies individually [10,17]. With respect to transferability, the diagnosis regarding the (deficient) status quo policy mix seems highly transferable to many other city regions, in which municipal competition and parochial thinking are dominant and suitable commercial land becomes scarcer. Furthermore, the alternative policy mixes identified in this study might be transferable to other densely populated and prosperous regions.

Our results also add to the literature's assumptions that (a) there is a conflict of goals between economic development and land saving, and (b) inter-municipal cooperation can help to achieve these two goals. In our model, mixes with high performance regarding future-oriented economic development (goal III) are simultaneously strong in reducing land take (goal II). This result seems to be highly influenced by the integrated and broad understanding of economic development in this study, which, at least implicitly, integrated social and ecological dimensions in the definition of goal III as well as its sub-goals and the selected measures. Thus, if *sustainable* economic development is the goal, our correlation analysis shows it is not conflicting with the land-saving goal but rather shows synergies. However, such a broad understanding of the practice of local politics requires the inclusion of different administrative and disciplinary perspectives, which is usually not the case [31,71]. However, this is a promising result indicating ways to exit the apparently fundamental conflict of commercial land use between economic and environmental goals and towards one of the rare, win–win solutions in land use planning and management [72].

Our analyses also demonstrated that mixes that performed well in achieving goal I also performed well in achieving goals II and III. Inter-municipal cooperation regarding commercial areas seems to help both to reduce land take [11–13] and to contribute to long-term economic development [24,25]. However, the positive side-effects of policies aiming at cooperation (goal I) and those aiming at the long-term economic development of a region (goal III) seem more direct and clearer than mutually supportive effects between cooperation and saving the use of commercial land (goal II). The positive relation between inter-municipal cooperation (goal I) and reducing land take (goal II) seems far from automatic and might be hampered by a mismatch of scales. Land saving strategies and instruments on the inter-municipal, sub-regional, or even regional level would avoid such a mismatch of scale; however, it would challenge municipal autonomy. The problem of the lack of integration of different administrative and disciplinary perspectives becomes virulent: regional economic development agencies or planning associations often know about the advantages of regional cooperation [71,73]. However, decisions for or against inter-municipal cooperation are made in the municipal councils. On the municipal level, the issue of economic development is still closely tied to commercial tax revenues and considered dominantly from a municipal perspective [29]. The relationship between city-regional cooperation (goal I) and reducing commercial land consumption (goal II) needs further empirical investigation. Finally, our analysis attributes high importance to cooperation policies. However, in order to achieve higher levels of goal attainment, measures to reach goal I would require further support by overarching strategies, potentially on the land or federal level, as subsidies for inter-municipal or regional cooperation that compensate for potential losses of municipal autonomy [74].

A caveat to these results, however, is that they are not based on an empirical analysis of factual land-use decisions. Instead, our model is based on expert judgments, which in turn reflect the assumptions of the literature as well as the knowledge and beliefs of the expert sample. In addition, for reasons of feasibility, we limited our study to the selection of three overarching goals with three sub-goals and considered 27 policies only (the status quo plus two alternative policies per sub goal). This selection clearly is a simplification of the situation, especially regarding the perspective of sustainable development of commercial areas. Sustainable development comprises a multitude of dimensions (see the 17 Sustainable Development Goals SDG [75]). Sustainable commercial land use does not only need to reduce land take, but also reduce the use of other resources (during construction, operation, and reuse of commercial areas). It further comprises social, cultural, ecological, climate-related, and political dimensions that are not considered in our analysis. Future research could perform comprehensive sustainability assessments of the different policy mixes found by our analysis. Possible methodological approaches to comprehensive sustainability assessments within CIB models are pointed out by the literature (e.g., [52,76]).

Our results seem valid for city regions with a comparable policy mix and where comparable (political) main goals apply. For other (sub-)regions (and in different country contexts), which prioritize different goals and have other potential strategies, one could apply the methodology of this study but would need to set up an adapted policy-interaction model. Further research should also differentiate more strongly between the planning, development, and use of small vs. large commercial sites.

Finally, the policy-interaction modeling approach and the results provide new information on policy interactions. This information could be useful in municipal and regional decision-making processes toward more coherent governance of commercial land use. The implementation of more consistent and synergetic policy mixes then remains a political task.

4.2. CIB for Policy Mix Evaluation and Design

With this study, the policy-interaction modeling approach was shown to be transferable to other issues and regions [52]. CIB for policy mix evaluation and design can be used for policy design regarding sustainable land use and regional development. The approach could be especially interesting to analyze increasing land-use conflicts [2], e.g., reducing land take vs. increasing land use for housing, renewable energies, and potentially also for agricultural production.

Regarding more specific methodological issues, our study makes the following contributions. First, our analysis adds to the understanding of the relation between consistency and synergy of policy mixes. Foregoing studies [52] (p. 43) found that "[. . .] in our case [. . .] striving for individual optimization (consistency) can be expected to find reasonable global solutions (synergy)". This case adds to these findings: Some mixes, albeit being consistent (namely the mix combining status quo measures with R&D funding in PFENZ), show very low synergy rates that we interpreted as lock-in effects of the Nash-equilibria (see Section 3.2). The other consistent mixes in our case lay at least in the 70% TIS percentile of the TIS distribution. The best consistent mix with TIS = 111 marks the absolute TIS maximum. Optimizing individual goals thus manages in the best case to reach the cooperative optimum in a self-organized way. However, the individual optimization can exceptionally result in a dead end. While synergy informs us about the policy mix that would be most effective (globally), consistency tells us the policies (even in synergetic mixes) that contradict each other. Thus, resonating with the argument to consider not only synergies but also trade-offs [39], CIB allows us to consider both consistency and synergy and interpret them together. First, this allows for avoiding instabilities, which are indicated by internal inconsistencies within presumably overall *synergetic* mixes. Inconsistencies indicate where individual policies are more strongly hindered than supported in their effectiveness in the interplay with other policies. Second, this allows for avoiding lock-ins that might block internally consistent policy mixes from being synergetic and finally effective. Our conclusion would be that it can be necessary to separate locked-in solutions in fruitful destruction to destroy stability if one can assume justifiably that a better state, in terms of synergy, is achievable.

Second, the validity of expert judgments (impact assessments) required by CIB studies is threatened by subjectivity and bias by nature [60,77,78]. However, we can be confident that the impact assessments in this policy-interaction model go well beyond individual experts' gut feelings, as our study design, confronting the views of local experts with those of external issue experts in small groups during online workshops, fostered the discursive validation of every impact assessment [79,80]. Therefore, it combined the advantage of expert interviews with the advantages of group discussions and supported inter- and transdisciplinary knowledge integration of local expertise and general issue expertise [81].

Third, our study has introduced a new measure into CIB for policy-mix evaluation and policy design, namely the measure of goal attainment (beyond overall effectiveness). Our analysis demonstrated that the initial effectiveness alone is not sufficient to determine the contribution of a policy to goal achievement in the combination of measures, as this effect is moderated by policies' interactions. Therefore, we summed the initial effectiveness with the effects of interrelations between measures into a new measure for goal attainment that is overall differentiated by the goal. We observed that the more abstract or context-dependent the policies, the more difficult to assess their initial effectiveness. Future research could further inquire about the relation between the initial effectiveness of policies and the effects of interrelations on final goal attainment and might propose further approaches. For instance, it would also be possible to consider the initial effectiveness of every policy as a moderating factor within the matrix (see Supplement S8) or to systematically analyze the role of the assumed initial effect on the experts' assessments of policy interrelations.

5. Conclusions

The governance of commercial land use must address multiple goals and design and implement effective policy combinations that improve city-regional cooperation between municipalities, reduce land take, and assure the long-term economic development of a region simultaneously. By using the Northern Black Forest as a case study, we brought together environmental, planning, and land use research as well as public policy analysis. We applied cross-impact balances (CIB) to build and analyze a participatory policy-interaction model, selected effective individual measures to reach each of the three goals, and analyzed their interactions. With the help of CIB, we then assessed the current policy mix and designed alternatives.

Our analysis revealed that current approaches to regional cooperation, reducing land consumption, and commercial land management depict clear internal contradictions and generate very few synergies. In particular, the current practice of competing for municipal planning and marketing of commercial sites has inhibiting effects. Implementing innovative measures on a stand-alone basis runs the risk of not being sufficiently effective. Therefore, to become more effective in reaching fundamental planning objectives, several mutually supportive levers must be moved simultaneously. We identified alternative policy mixes that achieve all three goals, avoid trade-offs, and generate significant synergy effects. Fundamental levers in these alternative mixes are a regional development strategy on commercial land, joint infrastructures in business parks, collaborative regional management, and marketing of commercial areas, and finally, strengthening inward development through enforced municipal land policy. Such policy mixes could overcome goal conflicts and improve city-regional cooperation, reduce land take, and sustain the economic performance of a region at the same time. In summary, the study introduces CIB to analyze goal conflicts, perform policy mix evaluation, and design into land use and planning research. This approach allows for the assessment of consistency, synergy, and goal attainment of different policy mixes, and thus, it is a useful tool for a more coherent and sustainable inter-municipal or even regional (commercial) land-use governance.

Supplementary Materials: The following supporting information can be downloaded at: https://www.mdpi.com/article/10.3390/land11060795/s1, Supplement S1 [82]: CIB in a nutshell; Supplement S2: Example for a detailed definition of a sub-goal and its alternative policies; Supplement S3: CIB Matrix; Supplement S4: Example for impact logics and their verbal justifications; Supplement S5: Cluster analysis; Supplement S6: Total impact score of consistent and inconsistent mixes; Supplement S7: Sampling of mixes by best goal attainment; Supplement S8: Initial effectiveness as a moderating factor within the matrix.

Author Contributions: Conceptualization, H.K. and S.W.; methodology, H.K. and S.W.; data collection, H.K., S.W., S.B., P.G., D.G., I.I., T.J., O.L., M.P., J.P., K.P.S., S.S., S.W.-E., J.W. and K.Z.; formal analysis, H.K. and S.W.; investigation, H.K. and S.W.; data curation, H.K. and S.W.; writing—original draft preparation, H.K., S.W., S.B., P.G., D.G., I.I., T.J., O.L., M.P., J.P., K.P.S., S.S., S.W.-E., J.W. and K.Z.; writing—review and editing, H.K. and S.W.; visualization, H.K. and S.W.; project administration, H.K.; funding acquisition, H.K. and S.W. All authors have read and agreed to the published version of the manuscript.

Funding: This research was funded by BMBF German Federal Ministry of Education and Research, grant number 033L224C.

Institutional Review Board Statement: Not applicable.

Informed Consent Statement: Not applicable.

Data Availability Statement: The full CIB Matrix, including qualitative justifications for all impact assessments, is available through the authors upon reasonable request.

Acknowledgments: We would like to thank our colleagues, Wolfgang Weimer-Jehle and Sigrid Prehofer, who supported us in the design of data collection and data analysis.

Conflicts of Interest: The authors declare no conflict of interest. The funders did not influence the design of the study; the collection, analyses, or interpretation of data; the writing of the manuscript; or the decision to publish the results.

References

1. UBA (Umweltbundesamt). *Umwelt- und Aufenthaltsqualität in Kompakt-Urbanen Und Nutzungsgemischten Stadtstrukturen. Analysen, Fallbeispiele, Handlungsansätze unter Nutzung und Weiterentwicklung des Bebauungs- und Umweltrechts*; UBA: Dessau-Roßlau, Germany, 2018.
2. Sayer, J.; Sunderland, T.; Ghazoul, J.; Pfund, J.-L.; Sheil, D.; Meijaard, E.; Venter, M.; Boedhihartono, A.N.; Day, M.; Garcia, C.; et al. Ten principles for a landscape approach to reconciling agriculture, conservation, and other competing land uses. *Proc. Natl. Acad. Sci. USA* **2013**, *110*, 8349–8356. [CrossRef]
3. Williams, K. Urban intensification policies in England: Problems and contradictions. *Land Use Policy* **1999**, *16*, 167–178. [CrossRef]
4. EEA (European Environmental Agency). Urban Systems. Available online: https://www.eea.europa.eu/soer/2015/europe/urban-systems (accessed on 18 January 2022).
5. EEA (European Environmental Agency). Urban Land Take. Available online: https://www.eea.europa.eu/airs/2018/natural-capital/urban-land-expansion (accessed on 18 January 2022).
6. De Sousa, C. *Sustainable Brownfield Development. Building A Sustainable Future on Sites of Our Polluting Past*; Routledge: London, UK, 2021.
7. Betker, F. Nachhaltigkeit institutionalisieren: Ein neuer Gesellschaftsvertrag für städtische Gewerbe gebiete. *GAIA* **2013**, *22*, 178–186. [CrossRef]
8. Lange, D.; McNeil, S. Clean It and They Will Come? Defining Successful Brownfield Development. *J. Urban Plan. Dev.* **2004**, *130*, 101–108. [CrossRef]
9. Landeshauptstadt Stuttgart. Urban Sandwich. *Steigerung der Flächeneffizienz Durch Stapelung Gewerblicher Nutzungen*. Available online: https://www.stuttgart.de/medien/ibs/Abschlussbericht_Urban_Sandwich_2020.pdf (accessed on 1 April 2022).
10. UBA (Umweltbundesamt). *Instrumente zur Reduzierung der Flächeninanspruchnahme*; UBA: Dessau-Roßlau, Germany, 2018.
11. Akkoyunlu, S. The potential of rural-urban linkages for sustainable development and trade. *Int. J. Sustain. Dev. World Policy* **2015**, *4*, 20–40. [CrossRef]
12. UN-Habitat. United Nations Conference on Housing and Sustainable Urban Development Habitat III: Issue Paper on Urban-Rural Linkages. Available online: https://www.researchgate.net/publication/321168294_Habitat_III_ISSUE_PAPER_10_ON_URBANRURAL_LINKAGES/link/5a12ef53a6fdcc717b522655/download (accessed on 1 March 2022).
13. UN-Habitat. Urban-Rural Linkages: Guiding Principles to Advance Integrated Territorial Development. Nairobi. Available online: https://urbanrurallinkages.files.wordpress.com/2019/04/url-gp.pdf (accessed on 1 March 2022).
14. UBA (Umweltbundesamt). *Rahmenbedingungen und Instrumente für die Gestaltung Nachhaltiger Stadt-Land-Verknüpfungen*; UBA: Dessau-Roßlau, Germany, 2019.
15. UBA (Umweltbundesamt). *Regional Konsolidierte Gewerbeflächenentwicklung (RekonGent)*; UBA: Dessau-Roßlau, Germany, 2019.
16. Difu (Deutsches Institut für Urbanistik). *Region ist Solidarität. Gerechter Ausgleich von Lasetn und Nutzen im Interkommunalen Dialog*; Difu: Berlin, Germany, 2020; Available online: https://repository.difu.de/jspui/bitstream/difu/576260/1/Region_ist_Solidaritaet_300_web.pdf (accessed on 20 February 2022).
17. UBA (Umweltbundesamt). *Innovative Instrumente zum Flächensparen und zur Förderung der Innenentwicklung: Kommunaler Praxistest, Schlussfolgerungen für Akteure in Bund, Ländern, Regionen und Gemeinden & Kommunikation der Ergebnisse*; UBA: Dessau-Roßlau, Germany, 2020.
18. Gust, D.; Kinn, U.; Murschel, B.; Renn, O.; Ruther-Mehlis, A.; Schneider, J. *Verbundprojekt REFINA Regionaler Gewerbeflächenpool Neckar-Alb (REGENA). Abschlussbericht*; Regina: Nürtingen, Germany; Stuttgart, Germany, 2010.
19. Wassermann, S.; Kosow, H.; Kopfmüller, J.; Wagner, U. Interkommunale Kooperation im Gewerbeflächenmanagement. In *Stadtregionales Flächenmanagement*; Henn, S., Zimmermann, T., Braunschweig, B., Eds.; Springer: Berlin/Heidelberg, Germany, 2022. [CrossRef]
20. Sterner, T.; Robinson, E. Selection and design of environmental policy instruments. In *Handbook of Environmental Economics*; Dasgupta, P., Pattanayak, S., Smith, K., Eds.; Elsevier: Amsterdam, The Netherlands, 2018; Volume 4, pp. 231–284.

21. Jackson, T. The Post-growth Challenge: Secular Stagnation, Inequality and the Limits to Growth. *Ecol. Econ.* **2018**, *156*, 236–246. [CrossRef]
22. Reichel, A.; Perey, R. Moving beyond growth in the Anthropocene. *Anthr. Rev.* **2018**, *5*, 242–249. [CrossRef]
23. Westskog, H.; Aarsæther, N.; Hovelsrud, G.K.; Amundsen, H.; West, J.J.; Dale, R.F. The transformative potential of local-level planning and climate policies. Case studies from Norwegian municipalities. *Cogent Soc. Sci.* **2022**, *8*, 2033457. [CrossRef]
24. Wuschansky, B.; König, K. *Interkommunale Gewerbegebiete in Deutschland. Grundlagen und Empfehlungen zur Planung, Förderung und Finanzierung, Organisation, Vermarktung—146 Projektbeschreibungen und Abgeleitete Erkenntnisse*; ILS NRW: Dortmund, Germany, 2011.
25. Niaounakis, T.; Blank, J. Inter-municipal cooperation, economies of scale and cost efficiency: An application of stochastic frontier analysis to Dutch municipal tax departments. *Local Gov. Stud.* **2017**, *43*, 533–554. [CrossRef]
26. Regionalverband Nordschwarzwald. Mangel im Überfluss? Gewerbeflächenstudie der Region Nordschwarzwald. Pforzheim. Available online: http://www.nordschwarzwald-region.de/fileadmin/filemounts/redaktion/Bilder/3_Regionalverband/4_Sitzungen/Beilagen/015_2016/05_PA%2013.04.2016%20FDS/28_M_A%20zu%20Gewerbefl%C3%A4chenstudie.pdf (accessed on 1 March 2022).
27. Seuberlich, M. Interkommunale Zusammenarbeit als Rettungsanker?—Wie die Länder dieses Instrument für ihre Kommunen nutzen. *Dms—Der Mod. Staat Z. Für Public Policy Recht Und Manag.* **2012**, *5*, 105–124. [CrossRef]
28. Ministerium für Finanzen. *Die Kommunen und ihre Einnahmen*; Ministerium für Finanzen: Stuttgart, Germany, 2017.
29. Heinz, W. *Deutsche Städte und Globalisierung. Annäherung An ein Komplexes Thema*; Deutsches Institut für Urbanistik, Difu: Berlin, Germany, 2006.
30. Henn, S.; Zimmermann, T.; Braunschweig, B. *Stadtregionales Flächenmanagement*; Springer: Berlin/Heidelberg, Germany, 2022.
31. Repp, A.; Weith, T. Building Bridges across Sectors and Scales: Exploring Systemic Solutions towards A Sustainable Management of Land—Experiences from 4th Year Status Conference on Research for Sustainable Land Management. *Land* **2015**, *4*, 325–336. [CrossRef]
32. United Nations. Transforming Our World: The 2030 Agenda for Sustainable Development. Resolution Adopted by the General Assembly on 25 September 2015. New York (A/RES/70/1). Available online: https://sustainabledevelopment.un.org/post2015/transformingourworld (accessed on 1 April 2022).
33. Bennich, T.; Weitz, N.; Carlsen, H. Deciphering the scientific literature on SDG interactions: A review and reading guide. *Sci. Total Environ.* **2020**, *728*, 138405. [CrossRef] [PubMed]
34. Nilsson, M.; Zamparutti, T.; Petersen, J.E.; Nykvist, B.; Rudberg, P.; McGuinn, J. Understanding policy coherence: Analytical framework and examples of sector–environment policy interactions in the EU. *Environ. Pol. Gov.* **2012**, *22*, 395–423. [CrossRef]
35. Nilsson, M.; Griggs, D.; Visbeck, M. Map the interactions between Sustainable Development Goals. *Nature* **2016**, *534*, 320–322. [CrossRef] [PubMed]
36. Nilsson, M.; Weitz, N. Governing Trade-Offs and Building Coherence in Policy-Making for the 2030 Agenda. *Politics Gov.* **2019**, *7*, 254–263. [CrossRef]
37. Koff, H.; Challenger, A.; Portillo, I. Guidelines for Operationalizing Policy Coherence for Development (PCD) as a Methodology for the Design and Implementation of Sustainable Development Strategies. *Sustainability* **2020**, *12*, 4055. [CrossRef]
38. Carbone, M.; Keijzer, N. The European Union and Policy Coherence for Development: Reforms, Results, Resistance. *Eur. J. Dev. Res.* **2016**, *28*, 30–43. [CrossRef]
39. Kostetckaia, M.; Hametner, M. How Sustainable Development Goals interlinkages influence European Union countries' progress towards the 2030 Agenda. *Sustain. Dev.* **2022**, *2022*, 1–11. [CrossRef]
40. Weitz, N.; Carlsen, H.; Trimmer, C. SDG Synergies: An Approach for Coherent 2030 Agenda Implementation. SEI Policy Brief May 2019. Available online: https://www.sei.org/wp-content/uploads/2019/04/sei-brief-2019-sdg-synergies-2.pdf (accessed on 4 February 2022).
41. Lambin, E.F.; Meyfroidt, P.; Rueda, X.; Blackman, A.; Börner, J.; Cerutti, P.O.; Dietsch, T.; Jungmann, L.; Lamarque, P.; Lister, J.; et al. Effectiveness and synergies of policy instruments for land use governance in tropical regions. *Glob. Environ. Chang.* **2014**, *28*, 129–140. [CrossRef]
42. Rogge, K.; Reichardt, K. Policy mixes for sustainability transitions: An extended concept and framework for analysis. *Res. Policy* **2016**, *45*, 1620–1635. [CrossRef]
43. Reichardt, K.; Negro, S.O.; Rogge, K.S.; Hekkert, M.P. Analyzing interdependencies between policy mixes and technological innovation systems: The case of offshore wind in Germany. *Technol. Forecast. Soc. Chang.* **2016**, *106*, 11–21. [CrossRef]
44. Kern, F.; Rogge, K.S.; Howlett, M. Policy mixes for sustainability transitions: New approaches and insights through bridging innovation and policy studies. *Res. Policy* **2019**, *48*, 103832. [CrossRef]
45. Feitelson, E. Packaging policies to address environmental concerns. In *Handbook of Transport and the Environment*; Hensher, D.A., Button, K.J., Eds.; Elsevier: Amsterdam, The Netherlands, 2003; pp. 757–769.
46. Givoni, M. Addressing transport policy challenges through Policy-Packaging. *Transp. Res. Part A Policy Pr.* **2014**, *60*, 1–8. [CrossRef]
47. Scheer, D.; Schmidt, M.; Dreyer, M.; Schmieder, L.; Arnold, A. Integrated Policy Package Assessment (IPPA): A Problem Oriented Research Approach for Sustainability Transformations. *Sustainability* **2022**, *14*, 1218. [CrossRef]

48. Kivimaa, P.; Kern, F. Creative destruction or mere niche support? Innovation policy mixes for sustainability transitions. *Res. Policy* **2016**, *45*, 205–217.
49. Kern, F.; Kivimaa, P.; Martiskainen, M. Policy packaging or policy patching? The development of complex energy efficiency policy mixes. *Energy Res. Soc. Sci.* **2017**, *23*, 11–25.
50. Howlett, M.; Rayner, J. Patching Vs Packaging: Complementary Effects, Goodness of Fit, Degrees of Freedom and Intentionality in Policy Portfolio Design. In Proceedings of the ESEE Meetings, Lille, France, 18–21 June 2013.
51. Howlett, M. From the 'old' to the 'new' policy design: Design thinking beyond markets and collaborative governance. *Policy Sci.* **2014**, *47*, 187–207. [CrossRef]
52. Kosow, H.; Weimer-Jehle, W.; Leon, C.; Minn, F. Designing synergetic and sustainable policy mixes—A methodology to address conflictive environmental issues. *Environ. Sci. Policy* **2022**, *130*, 36–46. [CrossRef]
53. Weimer-Jehle, W. Cross-Impact Balances: A System-Theoretical Approach to Cross-Impact Analysis, Technol. *Forecast. Soc. Chang.* **2006**, *73*, 334–361. [CrossRef]
54. Yin, R. *Case Study Research: Design and Methods*, 4th ed.; Sage: Thousand Oaks, CA, USA, 2009.
55. Kosow, H.; Wassermann, S. Conflicts in inter-municipal management of commercial areas—An empirical study using cognitive affective mapping. *Eur. Plan. Stud.* **2022**. *in preparation*.
56. Stadt Pforzheim. Grundsatzentscheidung Gewerbeflächen Pforzheim. Available online: https://www.pforzheim.de/fileadmin/user_upload/bauen/Gewerbefl%C3%A4chenentwicklung/Pr%C3%A4sentation_2020_09_29_Bedarfe_Gewerbefl%C3%A4chen_Pforzheim.pdf (accessed on 3 February 2022).
57. SPD; Bündnis 90; Die Grünen; FDP. Mehr Fortschritt Wagen. Bündnis für Freiheit, Gerechtigkeit und Nachhaltigkeit. Available online: https://www.spd.de/fileadmin/Dokumente/Koalitionsvertrag/Koalitionsvertrag_2021-2025.pdf (accessed on 20 December 2021).
58. Kemp-Benedict, E.; de Jong, W.; Pacheco, P. Forest futures: Linking global paths to local conditions. In *Forest under Pressure—Local Responses to Global Issues. Part IV—Possible Future Pathways*; IUFRO World Series; Katila, P., Galloway, G., de Jong, W., Pacheco, P., Mery, G., Eds.; IUFPRO: Vienna, Austria, 2014; Volume 32, pp. 539–553.
59. Wachsmuth, J. Cross-sectoral integration in regional adaptation to climate change via participatory scenario development. *Clim. Chang.* **2015**, *132*, 387–400. [CrossRef]
60. Weimer-Jehle, W.; Buchgeister, J.; Hauser, W.; Kosow, H.; Naegler, T.; Poganietz, W.-R.; Pregger, T.; Prehofer, S.; von Recklinghausen, A.; Schippl, J.; et al. Context scenarios and their usage for the construction of socio-technical energy scenarios. *Energy* **2016**, *111*, 956–970. [CrossRef]
61. Schneider, M.; Gill, B. Biotechnology versus agroecology—Entrenchments and surprise at a 2030 forecast scenario workshop. *Sci. Public Policy* **2016**, *43*, 74–84. [CrossRef]
62. Schütze, M.; Seidel, J.; Chamorro, A.; León, C. Integrated modelling of a megacity water system—The application of a transdisciplinary approach to the Lima metropolitan area. *J. Hydrol.* **2019**, *573*, 983–993. [CrossRef]
63. Norouzi, N.; Fani, M.; Ziarani, Z.K. The fall of oil age: A scenario planning approach over the last peak oil of human history by 2040. *J. Pet. Sci. Eng.* **2019**, *188*, 106827. [CrossRef]
64. Renn, O.; Jager, A.; Deuschle, J.; Weimer-Jehle, W. A normative-functional concept of sustainability and its indicators. *Int. J. Glob. Environ. Issues* **2009**, *9*, 291. [CrossRef]
65. Weimer-Jehle, W.; Deuschle, J.; Rehaag, R. Familial and societal causes of juvenile obesity—A qualitative model on obesity development and prevention in socially disadvantaged children and adolescents. *J. Public Health* **2012**, *20*, 111–124. [CrossRef]
66. Everitt, B.S.; Landau, S.; Leese, M. *Cluster Analysis*, 4th ed.; Arnold: London, UK, 2001.
67. Carlsen, H.; Eriksson, E.A.; Dreborg, K.H.; Johansson, B.; Bodin, Ö. Systematic exploration of scenario spaces. *Foresight* **2016**, *18*, 59–75. [CrossRef]
68. Carlsen, H.; Lempert, R.; Wikman-Svahn, P.; Schweizer, V. Choosing small sets of policy-relevant scenarios by combining vulnerability and diversity approaches. *Environ. Model. Softw.* **2016**, *84*, 155–164. [CrossRef]
69. Vester, F. *Die Kunst, Vernetzt zu Denken*; DTV: München, Germany, 2002.
70. Nadin, V.; Stead, D. Opening up the Compendium: An Evaluation of International Comparative Planning Research Methodologies. *Eur. Plan. Stud.* **2013**, *21*, 1542–1561. [CrossRef]
71. Weith, T.; Barkmann, T.; Gaasch, N.; Rogga, S.; Strauß, C.; Zscheischler, J. A Knowledge-Based European Perspective on Sustainable Land Management: Conceptual Approach and Overview of Chapters. In *Sustainable Land Management in A European Context. A Co-Design Approach*; Weith, T., Barkmann, T., Gaasch, N., Rogga, S., Strauß, C., Zscheischler, J., Eds.; Springer: Wiesbaden, Germany, 2021; pp. 1–13.
72. Meyfroidt, P.; de Bremond, A.; Ryan, C.M.; zu Emergassen, E.K.H.J. Ten facts about land systems for sustainability. *Proc. Natl. Acad. Sci. USA* **2022**, *119*, e2109217118. [CrossRef] [PubMed]
73. Stemberg, J.; Vogelgesang, M.; Pongratz, P.; Fink, A. *Handbuch Innovative Wirtschaftsförderung. Moderne Konzepte Kommunaler Struktur- und Entwicklungspolitik*; Springer: Wiesbaden, Germany, 2020.
74. MWIDE.NRW (Ministerium für Wirtschaft, Innovation, Digitalisierung und Energie des Landes Nordrhein-Westfalen). Regio NRW. Wirtschaftsflächen. Available online: www.fz-juelich.de/etn/DE/RegioNRW (accessed on 20 August 2021).

75. UN (United, Nations). Resolution Adopted by the General Assembly on 6 July 2017: Work of the Statistical Commission Pertaining to the 2030 Agenda for Sustainable Development. Available online: https://www.un.org/en/development/desa/population/migration/generalassembly/docs/globalcompact/A_RES_71_313.pdf (accessed on 4 March 2022).
76. Kopfmüller, J.; Weimer-Jehle, W.; Naegler, T.; Buchgeister, J.; Bräutigam, K.-R.; Stelzer, V. Integrative Scenario Assessment as a Tool to Support Decisions in Energy Transition. *Energies* **2021**, *14*, 1580. [CrossRef]
77. Weimer-Jehle, W.; Vögele, S.; Hauser, W.; Kosow, H.; Poganietz, W.-P.; Prehofer, S. Socio-technical energy scenarios: State of the art and CIB-based approaches. *Clim. Chang.* **2020**, *162*, 1723–1741. [CrossRef]
78. Schweizer, V. Reflections on cross-impact balances, a systematic method constructing global socio-technical scenarios for climate change research. *Clim. Chang.* **2020**, *162*, 1705–1722. [CrossRef]
79. Baard, P. Knowledge, participation, and the future: Epistemic quality in energy scenario construction. *Energy Res. Soc. Sci.* **2021**, *75*, 102019. [CrossRef]
80. Kurniawan, J. Participatory cross-impact balance analysis. A transdisciplinary approach for developing long-term energy transition scenarios. *Energy Res. Soc. Sci.* **2022**. *submitted*.
81. Prehofer, S.; Kosow, H.; Nägler, T.; Pregger, T.; Vögele, S.; Weimer-Jehle, W. Linking qualitative scenarios with quantitative energy models: Knowledge integration in different methodological designs. *Energy Sustain. Soc.* **2021**, *11*, 25. [CrossRef]
82. Kosow, H. The Best of Both Worlds? An Exploratory Study on Forms and Effects of New Qualitative-Quantitative Scenario Methodologies. Ph.D. Thesis, University of Stuttgart, Stuttgart, Germany, 2016. [CrossRef]

land

MDPI

Article

Expert-Based Maps as a Regional Planning Tool Supporting Nature Conservation and Production-Integrated Compensation—A German Case Study on Biodiversity Offsets

Christian Sponagel [1,*], Andre Raichle [2], Martin Maier [3], Susanne Zhuber-Okrog [4], Ulrike Greifenhagen-Kauffmann [5], Elisabeth Angenendt [1] and Enno Bahrs [1]

1 Department of Farm Management (410b), Institute of Farm Management, University of Hohenheim, Schwerzstraße 44, 70593 Stuttgart, Germany; Elisabeth.Angenendt@Uni-Hohenheim.de (E.A.); bahrs@Uni-Hohenheim.de (E.B.)
2 Raichle-Ecology, Stahlbrunnstraße 20, 73266 Bissingen an der Teck, Germany; a.raichle@raichle-ecology.de
3 Flächenagentur Baden-Württemberg GmbH, Gerhard-Koch-Straße 2, 73760 Ostfildern, Germany; maier@flaechenagentur-bw.de
4 Naturschutzbund Deutschland (NABU) Stuttgart e. V., Charlottenplatz 17, 70173 Stuttgart, Germany; susanne.zhuber-okrog@NABU-Stuttgart.de
5 Landwirtschaft/Grundstücksverkehr, Liegenschaftsamt, Landeshauptstadt Stuttgart, Heustraße 1, 70174 Stuttgart, Germany; Ulrike.Greifenhagen-Kauffmann@stuttgart.de
* Correspondence: Christian.Sponagel@Uni-Hohenheim.de

Citation: Sponagel, C.; Raichle, A.; Maier, M.; Zhuber-Okrog, S.; Greifenhagen-Kauffmann, U.; Angenendt, E.; Bahrs, E. Expert-Based Maps as a Regional Planning Tool Supporting Nature Conservation and Production-Integrated Compensation—A German Case Study on Biodiversity Offsets. *Land* **2021**, *10*, 808. https://doi.org/10.3390/land10080808

Academic Editors: Stephan Bartke and Sigrun Kabisch

Received: 15 July 2021
Accepted: 30 July 2021
Published: 1 August 2021

Abstract: Many countries worldwide have developed guidelines for offsetting impacts on nature and landscape. Suitable locations are the prerequisite for the implementation of these measures, and this might lead to conflicts with agriculture. In addition, comprehensive planning is often lacking and potential added values for nature conservation are not exploited. Concepts such as the so-called production-integrated compensation (PIC) have been introduced to give farmers the opportunity to actively participate in the offsetting process and improve cooperation. However, up to now, PIC has only rarely been put into practice. Against this backdrop, we have developed a regional planning tool for the implementation of PIC in practice. Based on geodata such as soil data, agricultural structure, or natural conditions at the field and landscape level, the general suitability, and specific measure-based recommendations for each plot can be verified with the help of a decision support system. These factors are assessed from both a nature and an agricultural perspective. The goal here is to highlight synergy effects and increase the likelihood of the proposed measures being implemented. Our tool facilitates the integrated planning of biodiversity offsets at regional level. In this way, it can promote the bundling and networking of measures. However, on-site analyses should be undertaken to complement the implementation of measures.

Keywords: biodiversity offsets; offset implementation; production-integrated compensation; nature conservation; landscape planning; agri-environmental policy

1. Introduction

Land taken up by settlement and infrastructure development is one of the major drivers of biodiversity loss [1–3]. As part of their so-called "no-net-loss" policies, many countries around the world have implemented regulations for offsetting impacts on nature and landscapes [4,5]. In Germany, this is anchored in the Nature Conservation Law (*BNatSchG*) in what is known as the "Impact Mitigation Regulation" (IMR) and follows a polluter pays principle [6]. Offsetting is actually the last step in the mitigation hierarchy after avoiding or minimizing impacts on nature and landscapes [7]. The requirement for offsetting is usually linked to various related questions, for instance, where and how offset measures can be conducted in the most efficient and effective manner from an ecological perspective [8,9]. Hence, in addition to the land taken up by impacts, further land is required for the implementation of offset measures [10].

In contrast to strategic planning on a regional level, randomly available sites are often taken up for offsetting [11]. Hence, potential added values for nature conservation generated by pooling and networking measures are often not exploited. In addition, there may be land use conflicts with agriculture as farmland is frequently used for biodiversity offsets [10,12]. Despite this, agriculture in particular could be a potential and important partner in nature conservation, as it occupies the largest area of land in Germany [13]. It is therefore an important stakeholder in biodiversity offsetting [14–17]. Concepts such as the so-called production-integrated compensation (PIC) have been introduced to give farmers the chance to actively participate in the offsetting process, and to improve cooperation between nature conservation and agriculture [18–20]. In general, there is no one single definition of PIC. However, PIC can be broadly defined as management or maintenance measures as set out in Article 15 (3) *BNatSchG* on agricultural and forestry land with continued agricultural and forestry use. It leads to a permanent enhancement of nature or landscapes. However, it is not always possible to clearly distinguish PIC measures from other offset measures on agricultural land in individual cases. In addition to the type of measure, individual farm conditions play a role here. For example, the creation of a meadow orchard or the conversion of arable land into grassland may well be a PIC measure for an individual farm if it fits into the farm concept. For other farms, however, this would not constitute a PIC measure as they would not be able to derive a monetary return from the land.

In general, one goal of PIC is to minimise the loss of agricultural productivity and to preserve agricultural land use [18]. In particular, measures that only take up a small area of a parcel of land and yet lead to an upgrading of the entire area would count as PIC. Therefore, the following conditions should be met from our perspective, according to Mössner [21], to address PIC from an agricultural angle:

I. The measures are implemented in a consensus with agriculture.
II. There is still a monetary return from agricultural land through production.
III. Both forms of "land sparing" and "land sharing" are possible, i.e., spatial separation between extensification and intensification or extensification and production on the same area (e.g., flower strips and extensive cultivation of cereals).

Especially in metropolitan areas, the multifunctionality of agriculture takes on a special significance [22]. In addition to the production of high-quality food, the emphasis is also on the cultivation of cultural landscapes. It is precisely here that PIC could also contribute directly to the diversity of nature and landscapes as a recreational area for the population in a conurbation [23]. Agriculture also recognises the need of these groups especially in these peri-urban areas. Consequently, PIC could also generate added value in the context of social recognition [24]. From a political point of view, the topic of biodiversity and agriculture is currently very much on the agenda and is shaped at EU level by the EU Biodiversity Strategy [25] or the EU Farm-to-Fork Strategy [26]. The main objectives include strengthening protected areas and reducing the use of plant protection products. The political demands, therefore, also present a challenge for many farms from an economic point of view [27]. At this point, PIC measures could, of course, also offer economic options that enable synergies to be tapped into. By means of the guaranteed maintenance of the measures, nature conservation could also benefit from PIC [28]. According to a study by Rabenschlag et al. [29] in Baden-Württemberg, there are often many deficits in the implementation of biodiversity offsets at the present time.

However, production-integrated compensation measures are rarely implemented in practice [21]. As far as agriculture is concerned, PIC measures are particularly in demand and could be expanded in the future [30]. There is often a lack of information and on-site communication, as well as a general lack of overarching planning [30]. In the field of nature conservation and ecosystem services, expert-based assessment approaches using geodata have been used to map the potential for specific ecosystem services [31] or green infrastructure planning [32], for example. Against this backdrop, we have developed a regional planning map as a tool for promoting PIC in the Stuttgart Region in the German state of Baden-Württemberg based on expert knowledge. In this context, we primarily proceeded from the hypothesis that there may be synergy effects between nature conservation objectives and agricultural interests. This is based on previous studies which have shown that farmers are indeed willing to implement voluntary measures under certain conditions [16,19].

Based on available and suitable geodata for soil conditions, agricultural structure, protected areas, etc., various PIC options have been evaluated at the field level to derive specific recommendations for offset measures. In this context, we focus on measures on arable land and grassland. Both agricultural and nature conservation perspectives were taken into account in order to elucidate the effectiveness and likelihood of implementation. In addition, potential synergy effects have been identified to promote the networking and bundling of measures. This means that our regional map for PIC could be a relevant aid for decision-makers at the municipal level when planning offset measures and could serve as a basis for constructive discussions with farmers. Hence, measures that appear to make sense for both agriculture and nature conservation can be selected.

2. Characterisation of the Study Area

The Stuttgart Region is one of the strongest economic regions in Germany with a high volume of land taken up for settlement and transport infrastructure development [33,34]. It accounts for approximately 10% of the total area of Baden-Württemberg. Moreover, 16% of the total land use for settlement and transport infrastructure in Baden-Württemberg was earmarked here between 2000 and 2016 [35]. Therefore, offsetting is an important factor in this region.

In total, the utilised arable land (UAA) in the Stuttgart Region amounts to approximately 133,766 ha according to the 2019 dataset of the Integrated Administration and Control System (IACS), which was provided by the Baden-Württemberg Ministry of Rural Affairs and Consumer Protection. About 54% of this is arable land (ARA) and 38% grassland. The remaining area consists of vineyards and permanent crops. There are major spatial disparities in the region's agricultural structure. Whereas the districts Böblingen and Ludwigsburg are dominated by arable farming, the eastern districts Rems-Murr-Kreis and Göppingen have a high proportion of grassland. Especially in the urban district of Stuttgart, specialty crops, such as fruits and vegetables, are cultivated on more than 10% of the arable land (Figure 1).

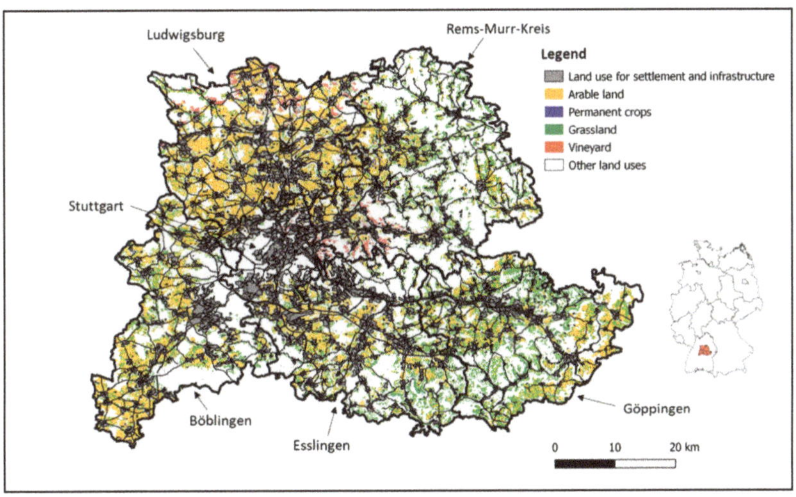

Figure 1. Land use in the Stuttgart Region based on IACS data and BKG [36] and Verband Region Stuttgart [37].

3. Material and Methods

3.1. General Procedure and Input Data for the Regional Map

The site-specific selection of measures plays a major role in the successful implementation of PIC measures [23]. Therefore, both nature conservation and agricultural concerns must be taken into account. For example, a measure may be appropriate and prudent from a nature conservation point of view on the one hand, but acceptance by agriculture may be low on the other. As a result, the likelihood of implementing the measure may be limited. Therefore, in the following, the interfaces between nature conservation and agricultural concerns in the compensation process are examined on the regional level. Figure 2 gives an overview of the entire approach and process that was used to draw up the regional map for the implementation of PIC measures on agricultural land. The aim is to derive recommendations for PIC measures at the plot level. To this end, the immediate surroundings of a plot (field level) and the wider spatial environment (landscape level) are taken into account. In this context, plots mean the arable and grassland plots taken from the IACS dataset. Beside the IACS dataset also agricultural statistical data was used regarding the agricultural perspective. This data is based on the so-called Agrarstrukturerhebung in Germany. This is a nationwide and regularly conducted survey of farms on farm structure, land use, and livestock production, etc. [38].

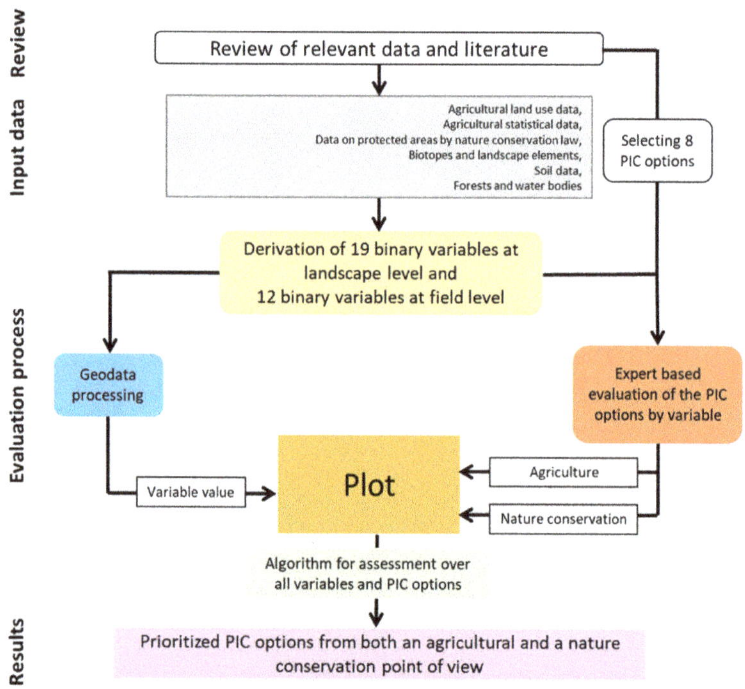

Figure 2. Overview of the entire approach and process of the development of a regional map for PIC measures.

3.2. PIC Options Considered for the Regional Map

Basically, there are numerous proposed measures for PIC that needed to be discussed at the outset [18,23,39,40]. Based on the literature review carried out in this context, 14 potential and typical PIC measures were initially identified, and a description was provided for the model (Table 1).

Table 1. Overview of the selected PIC measures and their description.

PIC Measure	Description
Temporary Greening	Temporary greening is carried out, for instance, of tramlines using a flower mixture in the tramline width of usually 2.50 m to 3.00 m. Fertilisers may be used on these tramlines but not pesticides.
Rotating Perennial Flower Strips	A system of annual/perennial flower strips is developed to ensure that sufficiently developed flower strips are always available. Before a flowering strip is removed, a new flowering strip is planted with sufficient lead time to ensure that the habitat characteristics of the flowering strips are uninterrupted. If, due to crop rotation, it is not possible to establish new flowering strips with annual flowering strips in advance, individual flowering strips must be left for 1.5 to 2 years. The continuous presence of flowering strips enables them to serve as a refuge for several species in the field.
Permanent Flower Strips and Areas	Permanent flowering strips or areas are established on arable land, each of which makes up only a small part of the arable field. This can be done along paths and field edges, for example, where the flower strips are established as green bands. Regular re-establishment takes place (e.g., every 3 to 5 years). In order to regulate undesirable plant species, annual planting can also be undertaken from time to time as an exceptional measure.
Fallow Land	This is arable fallow land with self-vegetation and annual mowing, but no removal. In addition, no fertilisers or pesticides are used. Fallowing is done in rotation with a standing time of 3–5 years on one area.
Extensive Used Arable Land	A field is only cultivated extensively on a permanent basis. Pesticides and synthetic fertilisers are no longer used. Reduced mechanical weed control and organic fertilisation are possible. Seeds are sown with wider or double seed row spacing, for instance 25–30 cm. This encourages the growth of wild herbs on the field. This measure is carried out in rotation in accordance with the crop rotation.
Skylark Windows	In the cereal crop, about 5–6 lark windows per ha are created, each at least 25 m^2 in size. The usual use of fertilisers and pesticides is maintained.
Infield Nature Protection Spots	In an arable field, individual sites with low yield but high conservation potential (e.g., wet sites, small water bodies, dry knolls) are removed from agricultural land use.
Stubble Fallow	After the grain harvest, the stubble is left on the field at a height of about 20 cm until at least February. Fertilisation and synthetic chemical pesticides are used during the growing season of the arable crop. In the period from harvest to February there is no fertilisation, no use of plant protection products and no mechanical weed control.

Table 1. *Cont.*

PIC Measure	Description
Unharvested Strips of Cereals	A portion of a crop is not harvested, but is left on the field until at least February. It serves as a food base or refuge for specific species such as field hamsters. No synthetic fertilisers or pesticides are used in this area.
Reduced Tillage	Only greatly reduced tillage is used, i.e., usually no-till and little tillage before and after harvest.
Conversion of Arable Land into Grassland	Arable land is converted into extensively used grassland. The use of synthetic fertilisers and pesticides is avoided. There is at least one annual mowing with removal.
Extensification of Grassland	A permanent grassland area is only used extensively, i.e., no fertilisers or pesticides are used. In addition, strips of old grass are left when mowing.
Uncut Hay Meadow Strips	In meadows, individual strips are left out during the first mowing. These strips of old grass, about 6 m wide, are then mown from mid-June at the earliest.

Some measures do show similarities with regard to the site conditions. The measures have, therefore, been grouped into a total of eight PIC options for evaluation in the map. In this context, all measures relating to the extensification of arable land or grassland use were combined in each case. Table 2 gives an overview of how the individual measure were grouped in the options.

Table 2. Grouping of the measures in PIC options.

PIC Measure	PIC Option
Greening	Greening
Annual flower strips	Flower strips
Permanent flower strips	Permanent flower strips
Fallow land	Fallow land
Extensively used arable land	Extensive arable farming
Skylark windows	Extensive arable farming
Infield nature protection spots	Extensive arable farming
Stubble fallow	Extensive arable farming
Unharvested strips of cereals	Extensive arable farming
Reduced tillage	Reduced tillage
Conversion of arable land into grassland	Grassland
Extensification of grassland	Extensification of grassland
Uncut hay meadow strips	Extensification of grassland

3.3. *Description of Data Input and Processing*

Based on a review of relevant and available data, 19 variables were derived at the landscape, i.e., the municipal level, and 12 variables at the field level (Tables 3 and 4). From the perspective of nature conservation, special attention was paid to protected areas under nature conservation law and elements of the biotope networks that may be especially relevant for PIC [23]. From an agricultural perspective, the agricultural structure, for instance the share of grassland or arable land in the utilised agricultural area per municipality or the share of specific crops, such as specialty crops, that also reflect agricultural opportunity costs, for example, were some of the criteria considered [41].

The variables were then treated as binary variables, i.e., a specific characteristic is present or not. Certain limits were set for each variable on the municipal or field level based on expert knowledge. The definition of these limits at municipal level was based on the typical values in the Stuttgart Region. The selected values are predominantly above the median for all municipalities. For the proportion of grassland at the municipal level, this limit was set at 50%, for example. If a municipality exceeds this proportion, the characteristic of a high proportion of grassland in the municipality is met at the landscape level.

Table 3. Variables used to create the PIC map at the landscape, i.e., municipal level.

Variables	Source of the Data
Share of grassland (>50%)	IACS dataset 2019
Share of specialty crops on arable land (>5%)	IACS dataset 2019
Share of cereals in the crop rotation (>60%)	IACS dataset 2019
Share of root crops in the crop rotation (>10%)	IACS dataset 2019
Share of cattle (>1 GV */ha grassland +forage crops)	IACS dataset 2019; Statistisches Landesamt [42]
Share of dairy cows (>0.5 GV/ha grassland + forage crops)	IACS dataset 2019; Statistisches Landesamt [42]
Share of AEMs ** on arable land (>10%)	IACS dataset 2019
Share of AEMs on grassland (>20%)	IACS dataset 2019
Share of organic farming on arable land (>10%)	IACS dataset 2019
Share of organic farming on grassland (>10%)	IACS dataset 2019
Share of arable land in nature conservation areas (>0.5%)	IACS dataset 2019; BfN [43]
Share of grassland in nature conservation areas (>2%)	IACS dataset 2019; BfN [43]
Share of arable land in Special Protection Areas (>10%)	IACS dataset 2019; BfN [44]
Share of grassland in Special Protection Areas (>20%)	IACS dataset 2019; BfN [44]
Share of arable land in FFH *** areas (>2%)	IACS dataset 2019; BfN [45]
Share of grassland in FFH areas (>10%)	IACS dataset 2019; BfN [45]
Share of arable land in landscape protection areas (>25%)	IACS dataset 2019; BfN [46]
Share of grassland in landscape protection areas (>40%)	IACS dataset 2019; BfN [46]
Share of grassland in legally protected biotopes (>0.5%)	IACS dataset 2019

* livestock unit (1 GV equals 500 kg live weight); ** Agri-environmental measures; *** Flora-Fauna-Habitat areas as defined in the European Union's Habitats Directive (92/43/EEC).

Table 4 gives an overview of the selected variables at field level. As is the case of the landscape level, a characteristic of the plot, such as high soil quality, is either present or absent. This was defined by the intersection of the plot polygons and the respective geodata layer, for instance the soil map using ArcGIS [47]. Some variables at field level were defined by the distance in meters between the respective plot and forest, biotopes, landscape elements, etc. Plots within areas protected under nature conservation law were identified, i.e., a distance of zero meters. For this purpose, the distances between the polygons of the plots and the selected geodata features were analysed using the "Near" Tool in ArcGIS [47]. The outlines of the plots were taken from the Integrated IACS 2019 dataset.

Table 4. Binary variables used to create the PIC map at field level.

Variables at Field Level	Description of the Data		
	Source	Attribute	Attribute Level
Forest ($\leq$20 m distance)	ALKIS [48]	GISELANAME	Forest
Water bodies ($\leq$20 m distance)	LUBW [49]	Stream network	
Biotopes ($\leq$20 m distance)	IACS dataset 2019	IACS code	924 *, 925 *
Woody structures ($\leq$20 m distance)	ALKIS [48]	GISELANAME	Field copse, field boundary, grove, hedge
Wetland areas ($\leq$20 m distance)	ALKIS [48]	GISELANAME	Pond, wetland, grassland, wetland;
	LUBW [50]	Core area	
Dry areas ($\leq$20 m distance)	LUBW [51]	Core area	
Soil quality (low)	LGRB [52]	Soil fertility	1.0, 1.5
Soil quality (high)	LGRB [52]	Soil fertility	3.0, 3.5, 4.0
Within nature conservation area (yes)	BfN [43]	BfN_ID	

Table 4. *Cont.*

Variables at Field Level	Description of the Data		
	Source	Attribute	Attribute Level
Within Natura 2000 network (yes)	BfN [45]; BfN [44]	BfN_ID	
Groundwater and backwater up to 1 m depth (yes)	LGRB [52]	GRUSTAWR	
Risk of soil erosion (yes)	LGRB [53]	Mean long-term soil erosion in t/ha/year	>2

* Agriculturally used and unused biotopes.

3.4. *Assessment of the PIC Options According to the Variables*

For each field or grassland plot, 31 binary variables are available at the municipal and at the plot level. The respective eight PIC options were evaluated at the municipal and field level in an assessment matrix. It was then decided, from a nature conservation and agricultural perspective, whether a PIC measure is desirable or rather not desirable for a certain expression of a variable. For this assessment, an expert group consisting of three people each from the field of nature conservation and agriculture was formed. The experts in the field of nature conservation were recruited from both local stakeholders (non-governmental organisations) and people who are active as experts in the field of planning offset measures in the Stuttgart Region having specific local knowledge. The agricultural experts came from academia and local administration, i.e., they also have specific local knowledge.

Hence, 248 (8 × 31) combinations of PIC options and variables were evaluated twice from the agricultural and nature conservation point of view, i.e., 496 decisions in total (Table S1 in the supplementary material). The evaluation was made in each case in four integer levels from +1 (desirable) to −2 (rather unsuitable), which is shown schematically in Figure 3. This is designed to anticipate that an unfavourable expression with regard to just one variable can already markedly affect the suitability of the location. The assessment was based on expert knowledge as well as on specific literature studies [19,23,40,54]. For example, due consideration was given to the fact that a measure such as extensification of grassland may have an impact on feed value and that this option may not be optimal in municipalities with a high volume of dairy farming [55].

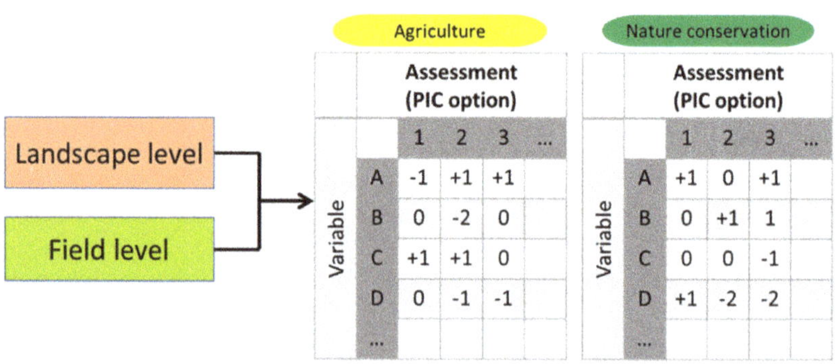

Figure 3. Schematic representation of the valuation of the PIC options in the assessment matrix.

3.5. *Plot-Specific Evaluation of the PIC Options*

The above-mentioned evaluations of the PIC options by variable were then used in an algorithm to generate the map in R [56]. In the algorithm, the value (1;0) of all 31 variables was recorded for each plot, and the evaluation result was calculated for each PIC option on each plot as the sum of the products of the value, of the variable and the evaluation in the assessment matrix, according to the example in Table 5.

Table 5. Example calculation of the evaluation of a measure (permanent flower strips from a nature conservation point of view) on a specific field in the model using the field-level variables.

Variable	Presence of the Feature (Binary Variable)	Evaluation in the Matrix (Integer Variable)	Evaluation of the Measure on the Plot (Integer Variable)
Forest (≤20 m distance)	0 (no)	−1	0
Water bodies (≤20 m distance)	1 (yes)	−1	−1
Biotopes (≤20 m distance)	1 (yes)	+1	+1
Woody structures (≤20 m distance)	1 (yes)	+1	+1
Wetland areas (≤20 m distance)	0 (no)	+1	0
Dry areas (≤20 m distance)	0 (no)	+1	0
Soil quality (low)	0 (no)	0	0
Soil quality (high)	1 (yes)	+1	+1
Within nature conservation area (yes)	0 (no)	+1	0
Within Natura 2000 network (yes)	1 (yes)	+1	+1
Groundwater and backwater up to 1 m depth (yes)	0 (no)	−2	0
Risk of soil erosion (yes)	0 (no)	0	0
Sum			+3

In addition to the assessment described above, the current level of intensification is also important for the PIC option extensification of grassland. Based on the information in the IACS dataset, certain grassland areas were excluded from this option because they are most likely already in extensive use. Consequently, grassland areas with agri-environmental measures, orchard meadows, sheep pastures, set-aside permanent grassland areas, or biotopes with grassland use were excluded. Hence, the option was set at zero in the total evaluation of the plot.

3.6. Generation of the Regional PIC Map

Based on the total evaluation, according to Section 3.5, each PIC option was then given two ratings on each plot from an agricultural and nature conservation point of view, for example, from zero to 16 for permanent flower strips from a nature conservation perspective. Since we focused on a regional map, one aim was to select the most suitable plots for the PIC options in the region. Therefore, for each PIC option, the plots with assessment values above the 75% quantile in relation to the entire region were filtered (e.g., scoring > 5 for permanent flower strips). As the assessment values are integer numbers, it is possible that more than 25% of the plots were considered in this way. The remaining plots were then attributed the value "0" for the specific PIC option. This ensures that plots in the region, that are particularly suitable for a PIC option, are represented in the map. Afterwards, the PIC option with the highest score per plot was selected. In the case of the attribution of zero for all PIC options, no option is recommended in the map. In the evaluation, of course, several options were sometimes close to each other. Therefore, to encourage a compromise between agriculture and nature conservation and allow a margin for potential on-site coordination, almost equivalent options should be proposed in some cases. However, the proposed options should not be too general. In this context, Table 6 analyses the range between the best, second best PIC option etc. (1. Max, 2. Max …) with mean, median, and the 25% quantile. For this analysis, all plots were considered where at least one option had a higher rating than zero. Based on this, a delta of 3 was selected as the cut-off value for recommending an option.

Table 6. Consideration of the delta between options using the example of the nature conservation perspective, restricted to plots where at least one option is valued higher than zero (1. Max > 0).

Range	Mean Difference	Median	Q25-Quantil
1. Max and 2. Max	4.3	1	0
1. Max and 3. Max	5.1	4	1
1. Max and 4. Max	5.9	5	3
1. Max and 5. Max	7.8	8	6
1. Max and 6. Max	8.1	8	6
1. Max and Min	8.1	8	6

The process described above was carried out from the point of view of both nature conservation and agriculture (Figure 4). Finally, there might be one, several, or even no suggestions for PIC on a specific plot for both sides. The plots and proposed options could be displayed in a GIS-map. Based on the two-sided recommendations for PIC options, it was now possible to identify the overlaps between the agricultural and nature conservation perspectives.

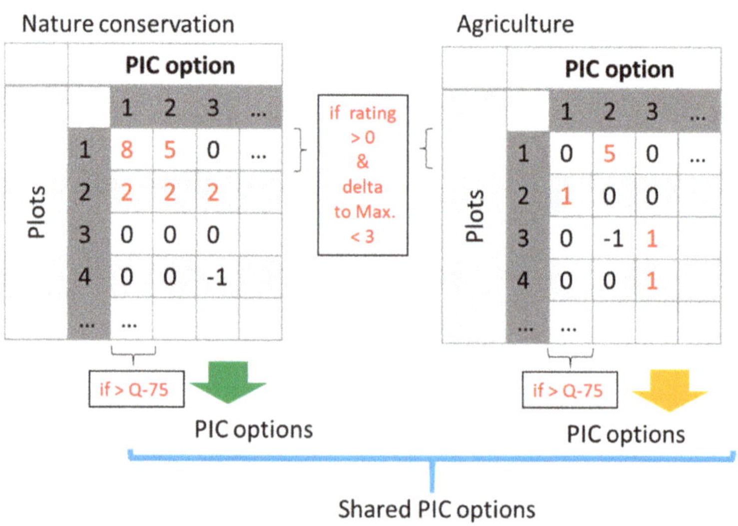

Figure 4. Derivation of plot-specific PIC option recommendations from a nature conservation and agricultural point of view based on the plot-specific overall assessment.

4. Results

PIC Map for the Stuttgart Region

From the perspective of nature conservation, approximately 93% of the arable land and 69% of grassland in the Stuttgart Region seem to be suitable for implementation of at least one of the considered PIC options (Table 7). Especially in the urban district of Stuttgart, almost 100% of arable land and grassland would be suitable for PIC.

Table 7. Proportion of arable land and grassland in % and in ha that is suitable for the implementation of PIC from the perspective of nature conservation.

Urban/Rural District	Land with at Least One Suggested Option from the Perspective of Nature Conservation			
	Proportion in %		In ha	
	Arable Land	Grassland	Arable Land	Grassland
Böblingen	78.3	58.2	11,584	4282
Esslingen	96.8	58.2	9375	5539
Göppingen	95.6	75.6	11,479	11,804
Ludwigsburg	96.0	74.7	22,516	3912
Rems-Murr-Kreis	97.0	71.1	10,825	8824
Stuttgart	100	95.6	1330	656
Stuttgart Region	92.7	68.9	67,110	35,016

Figure 5 shows the PIC map from the nature conservation perspective. The option with the highest rating, or no option, is presented for each plot in the map. Should two measures have an equally high rating, they were presented according to a fixed ranking, ranging from permanent flower strips, flower strips, grassland, extensive used arable land, fallow land, reduced tillage, to temporary greening. The map clearly shows the described agricultural structural features of the region. Hence, for instance in the district of Ludwigsburg with more than 75% utilised arable land (UAA), the extensification of arable land is the major recommended option. In addition, the share of specialty crops there is rather low, and it is mainly cereals that are grown. In the centre of the region, the implementation of permanent flower strips is particularly recommended. The ecological success of flower strips is rather independent of soil quality, and therefore more suitable in areas with very high soil quality, for instance, in Stuttgart. The districts of Göppingen und Rems-Murr-Kreis are characterised by a high proportion of grassland. Consequently, extensification of grassland is also a frequently suggested option in these areas.

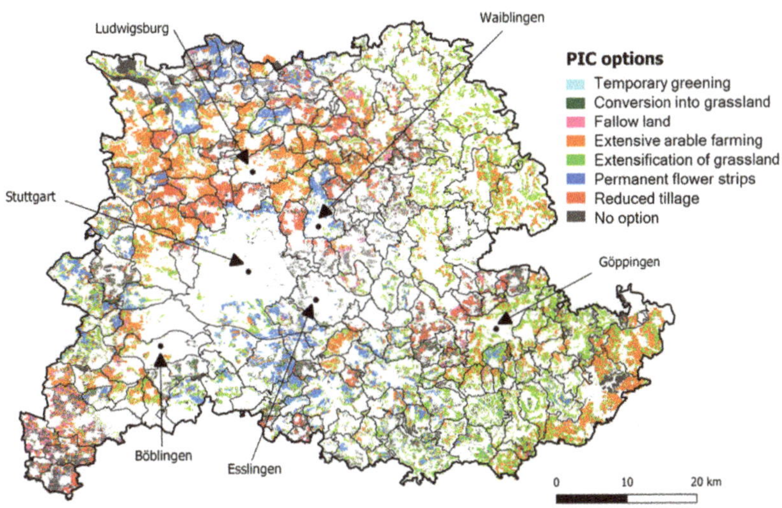

Figure 5. PIC map from the nature conservation perspective showing the priority option based on IACS data and BKG [36].

Generally speaking, PIC is often less recommended from the perspective of agriculture than from the nature conservation perspective. Furthermore, there are also clear regional disparities. For example, from an agricultural point of view, only about 25% of the areas in Stuttgart would be suitable for PIC. This constitutes a considerable discrepancy vis-a-vis the nature conservation assessment (Table 8).

Table 8. Proportion of arable land and grassland in % and in ha that is suitable for the implementation of PIC from the perspective of agriculture.

Urban/Rural District	Land with at Least One Suggested Option from the Perspective of Agriculture			
	Proportion in %		In ha	
	Arable Land	Grassland	Arable Land	Grassland
Böblingen	79.2	50.4	11,718	3709
Esslingen	76.3	50.3	7392	4796
Göppingen	86.1	47.8	10,347	7459
Ludwigsburg	82.9	46.7	19,445	2447
Rems-Murr-Kreis	82.8	42.4	9247	5257
Stuttgart	24.2	95.6	321	656
Stuttgart Region	80.7	47.8	58,470	24,325

In addition, the suggested options may differ between the agricultural and nature conservation perspective. Especially from the point of view of agriculture, the option of temporary greening is seen as a preferred option according to the model (Figure 6). There are also spatial differences within the region. In the western part of the region, temporary greening is often the preferred option, while in the centre and eastern part of the region, reduced tillage and extensification of grassland are often recommended.

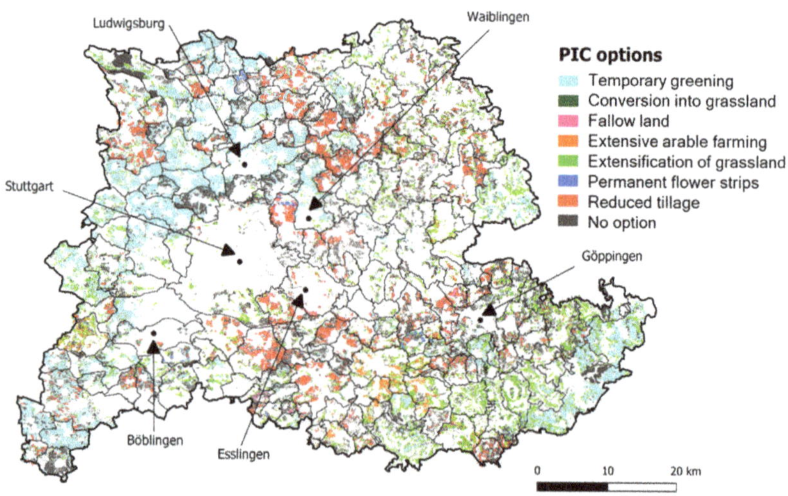

Figure 6. PIC map from the agricultural perspective illustrating the priority option based on the IACS data and BKG [36].

However, there are also overlaps between the nature conservation and agricultural assessments. This applies to about 13% of arable land and 43% of grassland. In this context, there are clear spatial disparities between the districts. Whereas in Esslingen, more than 23% of arable land would be suitable for the implementation of PIC, but in Stuttgart, this figure is only 0.1% of the arable land. Nevertheless, there seems to be a potential for PIC measures on grassland in the district of Stuttgart (Table 9).

Table 9. Proportion of arable land and grassland in % and in ha that is suitable for the implementation of PIC from the perspective of both nature conservation and agriculture.

Urban/Rural District	Land with at Least One Shared Suggested Option from the Perspective of Agriculture and Nature Conservation			
	Proportion in %		In ha	
	Arable Land	Grassland	Arable Land	Grassland
Böblingen	12.3	43.6	1818	3209
Esslingen	23.8	46.7	2309	4449
Göppingen	15.5	45.5	1857	7112
Ludwigsburg	6.8	43.0	1605	2256
Rems-Murr-Kreis	18	34.4	2005	4275
Stuttgart	0.1	95.6	9	656
Stuttgart Region	13.3	43.2	9602	21,956

Figure 7 shows these overlaps, i.e., land plots with a potential for PIC from both the nature conservation and agricultural perspective on arable land and grassland. Accordingly, the potential for PIC seems to be higher, in terms of area, in the east of the region than in the western districts.

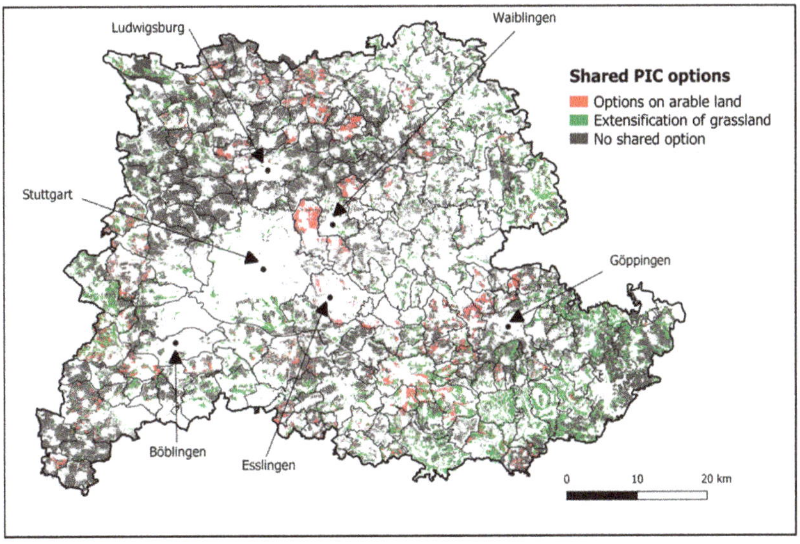

Figure 7. PIC map from both the nature conservation and agricultural perspective based on the IACS data and BKG [36].

On average, 0.23 PIC options per plot are recommended from both a nature conservation and an agricultural perspective. As described in Section 3.6, for each PIC option the plots were selected where the score was above the 75% quantile related to the region. Using the 60% quantile, there are, on average, 0.25 shared PIC options per area. At the 90% quantile, the average number of shared PIC options drops to 0.20 per plot. This sensitivity results specifically from the perspective of the nature conservation assessment since, in the context of the agricultural assessment, Q-60, Q-70, and Q-90 are identical with the exception of temporary greening.

5. Discussion

5.1. Case-Study Application Using the Example of Filderstadt for the Plausibility Check

To prove plausibility, the regional map was applied to individual landscape sections in the urban area of Filderstadt as part of the Stuttgart Region (Figure 8). In the process, the map was assessed by an expert using various aerial photographs and specific site knowledge. Overall, the map yielded plausible recommendations for PIC measures from both a nature conservation and an agricultural perspective. Nevertheless, minor adjustments are worth considering. For example, from a nature conservation perspective, the "conversion to grassland" measure might generally be the first choice over flower strips, especially as a buffer to water bodies (Figure 6). In addition, not all grassland plots with extensive land use seem to be covered, as the data are based on the IACS. Therefore, not all arable and grassland plots are considered, for instance, private land for which no direct payments have been applied for. Furthermore, it might also make sense to recommend the conversion of arable land to grassland in the vicinity of FFH meadows (e.g., 50 m) from a nature conservation perspective. In summary, the conclusion is positive, although minor adjustments are worth considering. Grassland in the floodplain is typical for central Europe. The locations in valleys are often too damp for arable farming or they flood annually, for instance [57–59]. Grassland protects the waterbodies against the entry of phosphate, nitrate, or pesticides. In addition, grassland can prevent soil erosion, serve as a habitat for numerous endangered species, and is important for the biotope network, for instance [60–62].

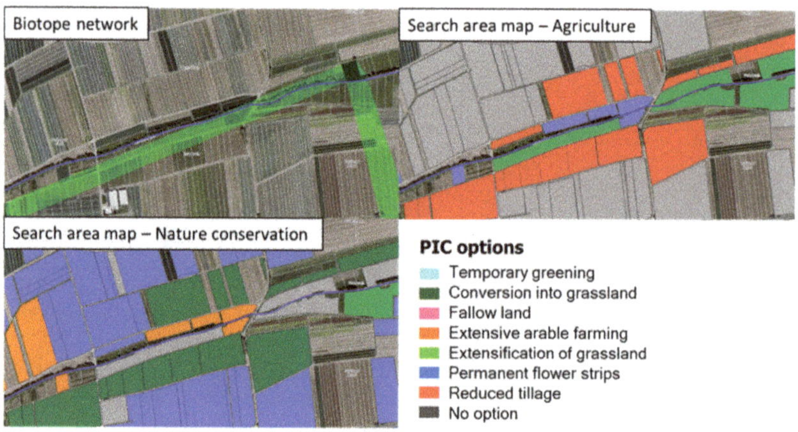

Figure 8. Application of the regional map using the example of an agricultural landscape segment in Filderstadt based on the IACS data and LUBW [63].

5.2. Discussion of the Approach

Based on the hypothesis that there may well be synergies between agricultural interests and conservation objectives in the planning of compensation measures, we produced a regional map for production-integrated compensation measures. We found that, in terms of suitable areas and measures, there is indeed a relevant overlap. This can, therefore, confirm our hypothesis. According to our model, there is significant implementation potential for PIC on more than 10% of arable land and more than 40% of grassland from a nature conservation and agricultural perspective. However, we also identified conflicting interests between nature conservation and agriculture. On the one hand, the PIC option of temporary greening permits good integration into crop rotation, for instance by greening tramlines. It, therefore, offers a high degree of flexibility from an agricultural point of view. On the other hand, this option is viewed critically from a nature conservation perspective as it can have a trapping effect on insects. In addition, we also found spatial disparities within the region. From an agricultural point of view, PIC measures may be less attractive in the centre of the region, due to a high proportion of specialty crops in the crop rotation with correspondingly high gross margins and high soil quality. This supports the findings of Sponagel et al. [41]. Furthermore, the municipalities in the district of Ludwigsburg have an average share of more than 10% of root crops in the crop rotation, which is markedly above the average for the region and may help explain the regional disparities. In this context, the distribution of protected areas under nature conservation law is also relevant, which differs between the districts. In Ludwigsburg, for example, in the average of all municipalities, almost 5% of arable land are located in FFH areas, whereas, in the Rems-Murr-Kreis, this is below 1%. All in all, there are many potential indicators that influence the spatially differentiated recommendation of PIC-options. In addition, the extensification of grassland might often be a less favourable option due to the reduced amount and quality of fodder [64]. In the context of PIC implementation in practice, it should also be mentioned that the costs of PIC measures may differ in the end due to different land prices. In general, measures require legal certainty, for instance, in the form of a land register entry, which can go hand in hand with a loss of market value of the land plot [65]. Especially in the centre of the Stuttgart Region, land prices for arable land are comparatively high, which might be challenging for the implementation of measures, especially from an agricultural perspective [41,66].

We used relevant, available geospatial data to evaluate the PIC options. In particular, protected areas under nature conservation law, biotope networks and landscape elements were included in the analysis from a nature conservation perspective. From an agricultural perspective, one focus was on the proportions of individual crop types in the crop rotation, such as root crops or cereals, soil quality as an indicator of the yield capacity of the plots, and data on livestock density. The crop sequence influences, for example, the option of arable extensification, which is less recommended from an agricultural point of view if the share of root crops is high [54]. In addition to variables at the landscape level, variables at the field level were also considered. For example, from an agricultural perspective, the proximity of a parcel to a forest edge, water body, or other landscape feature may be associated with management constraints and may predispose the location for PIC implementation. There may be restrictions on the use of pesticides and fertilisers, for example. With the arrival of the Act amending the Nature Conservation Act (*NatSchG*) and the Act concerning agriculture and landscape management (*LLG*) from 22 July 2020 in Baden-Württemberg, future restrictions on the use of plant protection products in protected areas were also adopted. In this context, avoiding the use of chemical pesticides can result in yield losses of about 30% for winter wheat, for example [67]. Therefore, future potential farming restrictions may lower the revenue of agricultural production. In addition, protected areas under nature conservation law often already have a negative impact on land prices [65,68]. In this respect, it can be assumed that the obstacle to implementation of PIC from an agricultural point of view is lower in these cases than in other areas with high productivity and suitability for special crops, etc. In addition, marginal sites with low yield capacity can be generally regarded as appropriate from an agricultural perspective. From the point of view of nature conservation, the strengthening of biotope networks and protected areas is also an important concern. Furthermore, from a nature conservation point of view, lean sites are often suitable for the establishment of fields with wild herbs or species-rich grassland [69]. This overlap of interests is thus also reflected in the model. In the future, for example, the management plans for Natura 2000 sites could be integrated into the model that set development targets for specific areas. In this way, further information related to the specific plot could be integrated.

In addition, it might be worthwhile to also address interests such as regional food supply with regard to the implementation of PIC measures in future studies. On the one hand, measures such as the extensification of arable land might enable further agricultural land use. On the other, they may be linked to rather high yield losses and respective low contributions to food supply [70]. Consequently, the relationship between the improvement in terms of nature conservation and land occupation is also relevant. In Baden-Württemberg, biodiversity offsets for habitat banking are assessed in eco credits, for example [71].

Some simplifications had to be made when selecting the PIC measures and bundling them into eight options. In addition, there are measures such as short-rotation plantations that were not taken into account because they could not be adequately assessed on the basis of the available data and are also under discussion [72]. Furthermore, there are definitely relevant criteria that could not be taken into account due to poor data availability. When planning measures, the possible presence of rare arable wild herbs, for example, should be checked on site. In addition, the presence of certain species such as partridges (*Perdix perdix*) should be examined on site. In this case, measures such as flower strips in particular would be an appropriate option, although this may not be apparent from the model [73].

Although our regional map is designed for intervention regulation under nature conservation law, there are, of course, parallels to agri-environmental measures or similarities with measures such as flower strips. Biotope connectivity, in particular, is an important political goal that is defined in section 22 *NatSchG* and can be supported by our model. In Baden-Württemberg, in particular, the biotope network is to be expanded to 15% of open land by 2030, and refugial areas are to be created on 10% of agricultural land. Hence, cooperation with agriculture will probably be necessary. Consequently, the activities of individual farmers should be coordinated at the landscape level in order to ensure the delivery of certain ecosystem services [74]. Ecological networks, particularly on a regional level are, therefore, essential for the preservation of biodiversity [75].

Furthermore, this study could not take into account acceptance of the measures. Some farmers may also reject PIC measures outright. In this context, the required compensation payment for the measures may, of course, vary as well [19]. The strength of our model is, therefore, that it can identify the options that have the highest probability of being implemented on the land.

As our study refers specifically to agricultural land, integration into other models or extension is possible at any time. This also allows for adaptation to possibly deviating local conditions. In this way, our model also ties in with other studies that deal with modelling the ecological effectiveness of biodiversity offsets at the landscape level [76,77].

5.3. Feedback from a Stakeholder Workshop

At a workshop on production-integrated compensation within the project RAMONA, as part of the BMBF funding activity Stadt-Land-Plus, the developed PIC map was presented in January 2021 to more than 25 participants from the agricultural and nature conservation administration, relevant ministries, the agricultural professional representation in Baden-Württemberg, as well as cultural land foundations from various other federal states. One question that was raised had to do with whether such a map could contribute to the improved establishment of PIC measures in practice. Of the 21 respondents, 96% answered this in the affirmative, illustrating the need and potential opportunities for a PIC map.

Nevertheless, in the context of the application of such a regional PIC map, essential criteria were mentioned by the participants from agriculture. Firstly, a regional map must not lead to a top-down change in agricultural land use. Secondly, agriculture must always be able to participate in the decision-making process. On-site communication, therefore, plays a central role.

Furthermore, it was mentioned that such a regional map could be helpful especially for municipalities that own a lot of agricultural land in order to come up with reasonable solutions for all the parties involved. In this respect, the map could also serve as a basis for discussing the implementation of PIC. Ultimately, of course, the current land use and the species and habitat endowment on site must be assessed and taken into account. In addition, other existing nature conservation objectives such as the biotope network and existing measures in the surrounding area must be considered. This also includes any land uses in the surrounding area that could have a negative impact on the success of the offset measure.

5.4. Provision of the PIC Map in a Web Application

The generated regional map is to be made available by the regional planning association, Verband Region Stuttgart, in a web application. This is intended for municipalities in particular. Within the application, suggestions from an agricultural and nature conservation point of view can be retrieved for each plot in the region. Furthermore, it should be possible to display suitable plots for specific PIC options or plots with an overlap between proposals from an agricultural and nature conservation perspective.

6. Conclusions

Geodata were used to create a regional map for production-integrated compensation measures that will be made available for practical use. We found that there may be significant potential for PIC in the Stuttgart Region from an agricultural and nature conservation perspective. The map can thus provide an initial overview when planning and designing biodiversity offsets on farmland. On a regional level, structures for networking and bundling measures can be derived. This can improve the ecological effectiveness of biodiversity offsets. All in all, the final decision on the type of offset measure to be implemented can and must always be made on site. However, in the dialogue between the intervening party, for instance a municipality and a farmer, the map can serve as a foundation for a constructive discussion to determine acceptable solutions for both nature conservation and agriculture. Our regional map can also be integrated into other areas of nature conservation planning such as the biotope network, whereby synergy effects with nature conservation measures, on areas outside agricultural use, can be derived.

A wide range of usable geodata are available, especially for the Stuttgart Region, which could be used to the advantage of this study. In general, the approach requires environmental data such as protected areas under nature conservation law, data on the agricultural structure and land use as well as suitable soil data. Within Germany and Europe, it can be assumed that similar data are available for other regions, for instance, soil maps such as the Corine European soil database [78] or data on agricultural land use. Accordingly, there should be good transferability to individual regions outside the Stuttgart Region with comparable data availability. For this purpose, our chosen approach is generally very flexible, i.e., further variables can be added or it can be adapted for other regions. If the approach is to be transferred to other regions, it should be noted that the selection of the PIC options, the definition of the variables such as distances to water bodies etc. and, at least, the assessment of the options were based on expert knowledge with regard to the rather small-scale agricultural structures in Baden-Württemberg [79]. For regions with less structurally rich agricultural landscapes and larger field structures such as eastern Germany [80], for example, the preselected PIC options should possibly be adjusted. For example, short rotation coppice was not selected for the Stuttgart Region because of the structure of the agricultural landscape. However, it may be a potential PIC option for other regions [72]. Based on our experience, we recommend that new expert groups be set up to roll out the approach to other regions. Ideally, the groups should be made up of equal numbers of people from the field of nature conservation and the field of agriculture. These individuals should be familiar with the specific local conditions, i.e., local stakeholders such as nature conservation authorities and farmers' associations, would be the predestined interlocutors in this case. In addition, a transdisciplinary approach, with cooperation between science and practice, could be advantageous in order to represent different perspectives. From our point of view, it should also be borne in mind that an intensive exchange within an expert group may be necessary over a period of several months.

Supplementary Materials: The following are available online at https://www.mdpi.com/article/10.3390/land10080808/s1, Table S1: Assessment matrix.

Author Contributions: Conceptualization, C.S.; Methodology, C.S., M.M. and A.R.; Software, C.S.; Validation, A.R., M.M., U.G.-K., S.Z.-O. and E.A.; Formal Analysis, C.S.; Investigation, C.S.; Resources, C.S. and E.A.; Data Curation, C.S.; Writing—Original Draft Preparation, C.S.; Writing—Review & Editing, E.B., E.A., M.M., A.R., U.G.-K., S.Z.-O. and C.S; Visualization, C.S. and A.R.; Supervision, E.B.; Project Administration, E.A.; Funding Acquisition, E.B. All authors have read and agreed to the published version of the manuscript.

Funding: This research was funded by the German Federal Ministry of Education and Research (BMBF) within the RAMONA project (grant ID: 033L201B) as part of the funding activity Stadt-Land-Plus.

Conflicts of Interest: The authors declare no conflict of interest.

References

1. Fletcher, R.J., Jr.; Didham, R.K.; Banks-Leite, C.; Barlow, J.; Ewers, R.M.; Rosindell, J.; Holt, R.D.; Gonzalez, A.; Pardini, R.; Damschen, E.I.; et al. Is habitat fragmentation good for biodiversity? *Biol. Conserv.* **2018**, *226*, 9–15. [CrossRef]
2. IPBES. *Summary for Policymakers of the Global Assessment Report on Biodiversity and Ecosystem Services of the Intergovernmental Science-Policy Platform on Biodiversity and Ecosystem Services*; Díaz, S., Settele, J., S, E.S.B.E., Ngo, H.T., Guèze, M., Agard, J., Arneth, A., Balvanera, P., Brauman, K.A., Butchart, S.H.M., Eds.; IPBES Secretariat: Bonn, Germany, 2019; 56p.
3. Laurance, W.F.; Peletier-Jellema, A.; Greenen, B.; Koster, H.; Verweij, P.; van Dijck, P.; Lovejoy, T.E.; Schleicher, J.; van Kuijk, M. Reducing the global environmental impacts of rapid infrastructure expansion. *Curr. Biol.* **2015**, *25*, 259–262. [CrossRef] [PubMed]
4. GIBOP. Global inventory of biodiversity offset policies (GIPOP). In *International Union for Conservation of Nature: The Biodiversity Consultancy*; Durrel Institute of Conservation & Ecology: Canterbury, UK, 2019.
5. Bull, J.W.; Strange, N. The global extent of biodiversity offset implementation under no net loss policies. *Nat. Sustain.* **2018**, *1*, 790–798. [CrossRef]
6. Albrecht, J.; Schumacher, J.; Wende, W. The German Impact-Mitigation Regulation. *Environ. Policy Law* **2014**, *44*, 317–333.
7. Arlidge, W.N.S.; Bull, J.W.; Addison, P.F.E.; Burggass, M.J.; Gianuca, D.; Gorham, T.M.; Jacob, C.; Shumway, N.; Sinclair, S.P.; Watson, J.E.M.; et al. A global mitigation hierarchy for nature conservation. *BioScience* **2018**, *68*, 336–347. [CrossRef]
8. Kiesecker, J.M.; Copeland, H.; Pocewicz, A.; McKenney, B. Development by design: Blending landscape-level planning with the mitigation hierarchy. *Front. Ecol. Environ.* **2010**, *8*, 261–266. [CrossRef]
9. Gelcich, S.; Vargas, C.; Carreras, M.J.; Castilla, J.C.; Donlan, C.J. Achieving biodiversity benefits with offsets: Research gaps, challenges, and needs. *Ambio* **2017**, 184–189. [CrossRef] [PubMed]
10. Tietz, A.; Bathke, M.; Osterburg, B. Art und Ausmaß der Inanspruchnahme landwirtschaftlicher Flächen für außerlandwirtschaftliche Zwecke und Ausgleichsmaßnahmen. In *Thuenen Working Papers*; Johann Heinrich von Thuenen-Institut (vTI), Federal Research Institute for Rural Areas, Forestry and Fisheries: Braunschweig, Germany, 2012; p. 137038.
11. Wende, W.; Tucker, G.-M.; Quétier, F. *Biodiversity Offsets: European Perspectives on No Net Loss of Biodiversity and Ecosystem Services*; Springer International Publisher: Cham, Switzerland, 2018; ISBN 9783319725819.
12. Le Coent, P.; Préget, R.; Thoyer, S. Compensating Environmental Losses Versus Creating Environmental Gains: Implications for Biodiversity Offsets. *Ecol. Econ.* **2017**, *142*, 120–129. [CrossRef]
13. Umweltbundesamt. Struktur der Flächennutzung. Available online: https://www.umweltbundesamt.de/daten/flaeche-boden-land-oekosysteme/flaeche/struktur-der-flaechennutzung#die-wichtigsten-flachennutzungen (accessed on 10 January 2021).
14. Koh, N.S.; Hahn, T.; Boonstra, W.J. How much of a market is involved in a biodiversity offset? A typology of biodiversity offset policies. *J. Environ. Manag.* **2019**, *232*, 679–691. [CrossRef]
15. Taherzadeh, O.; Howley, P. No net loss of what, for whom?: Stakeholder perspectives to Biodiversity Offsetting in England. *Environ. Dev. Sustain.* **2018**, *20*, 1807–1830. [CrossRef]
16. Vaissière, A.C.; Tardieu, L.; Quétier, F.; Roussel, S. Preferences for biodiversity offset contracts on arable land: A choice experiment study with farmers. *Eur. Rev. Agric. Econ.* **2018**, *45*, 553–582. [CrossRef]
17. Busse, M.; Heiterpriem, N.; Siebert, R. The Acceptability of Land Pools for the Sustainable Revalorisation of Wetland Meadows in the Spreewald Region, Germany. *Sustainability* **2019**, *11*, 4056. [CrossRef]
18. Druckenbrod, C.; Beckmann, V. Production-Integrated Compensation in Environmental Offsets—A Review of a German Offset Practice. *Sustainability* **2018**, *10*, 4061. [CrossRef]
19. Czybulka, D.; Hampicke, U.; Litterski, B. (Eds.) *Produktionsintegrierte Kompensation: Rechtliche Möglichkeiten, Akzeptanz, Effizienz Und Naturschutzgerechte Nutzung*; Erich Schmidt: Berlin, Germany, 2012; ISBN 3503138323.
20. Druckenbrod, C.; van Elsen, T.; Hampicke, U. Produktionsintegrierte Kompensation: Umsetzungsbeispiele und Akzeptanz: Ackerwildkrautschutz mit Hilfe der Eingriffsregelung sowie Ergebnisse einer Befragung von Akteuren. *Nat. Landsch.* **2011**, *43*, 111–116.
21. Mössner, R. Produktionsintegrierte Kompensationsmaßnahmen in der Landwirtschaft. *Landinfo* **2019**, 16–20.
22. Gullino, P.; Battisti, L.; Larchner, F. Linking Multifunctionality and Sustainability for Valuing Peri-Urban Farming: A Case Study in the Turin Metropolitan Area (Italy). *Sustainability* **2018**, *10*, 1625. [CrossRef]

23. Etterer, F.; Fritzsch, S.; Lau, M. Arbeitshilfe Produktionsintegrierte Kompensation: Empfehlungen für die Praxis AUS Dem Forschungsvorhaben Stadt PARTHE Land. Available online: https://www.researchgate.net/profile/Florian_Etterer/publication/341540667_Arbeitshilfe_Produktionsintegrierte_Kompensation/links/5ec62ef6458515626cbbfecb/Arbeitshilfe-Produktionsintegrierte-Kompensation.pdf (accessed on 28 January 2021).
24. Zasada, I. Multifunctional peri-urban agriculture—A review of societal demands and the provision of goods and services by farming. *Land Use Policy* **2011**, *28*, 639–648. [CrossRef]
25. European Commission. Communication from the Commission to the European Parliament, the Council, the European Economic and Social Committee and the Committee of the Regions: EU Biodiversity Strategy for 2030 Bringing Nature Back into Our Lives. Available online: https://eur-lex.europa.eu/legal-content/EN/TXT/?qid=1590574123338&uri=CELEX:52020DC0380 (accessed on 11 October 2020).
26. European Commission. Communication from the Commission to the European Parliemant, the Council, the European Economic and Social Committee and the Committee of the Regions: A Farm to Fork Strategy for a Fair, Healthy and Environmentally-Friendly Food System. Available online: https://eur-lex.europa.eu/resource.html?uri=cellar:ea0f9f73-9ab2-11ea-9d2d-01aa75ed71a1.0001.02/DOC_1&format=PDF (accessed on 2 July 2021).
27. Malorgio, G.; Marangon, F. Agricultural business economics: The challenge of sustainability. *Agric. Food Econ.* **2021**, *9*, 6. [CrossRef]
28. Etterer, F.; Voges, A.C.; Fritzsch, S.; Pietsch, M. Tagungsbeitrag: Produktionsintegrierte Kompensation (PIK) im Großraum Leipzig: Mehr Vielfalt in Agrarlandschaften III Am 13. September 2016 in Magdeburg. Available online: https://stadtpartheland.de/wp-content/uploads/2017/07/MehrVielfaltAgrarlandsch-III_Tagungsbeitrag_161206final.pdf (accessed on 5 February 2021).
29. Rabenschlag, J.; Schoof, N.; Schumacher, J.; Reif, A. Evaluation der Umsetzung baurechtlicher Ausgleichsmaßnahmen: Am Fallbeispiel Schönberg bei Freiburg. *Nat. Landsch.* **2019**, *51*, 434–442.
30. PAN. Evaluation der Ökokonto-Verordnung Baden-Württemberg. Available online: https://www.lubw.baden-wuerttemberg.de/documents/10184/389779/Evaluation_OekokontoVO_Endbericht.pdf/2aef1af9-d532-420e-bd37-747eb24270c2 (accessed on 6 October 2020).
31. Müller, F.; Bicking, S.; Ahrendt, K.; Kinh Bac, D.; Blindow, I.; Fürst, C.; Haase, P.; Kruse, M.; Kruse, T.; Ma, L.; et al. Assessing ecosystem service potentials to evaluate terrestrial, coastal and marine ecosystem types in Northern Germany–An expert-based matrix approach. *Ecol. Indic.* **2020**, *112*, 106116. [CrossRef]
32. Kopperoinen, L.; Itkonen, P.; Niemelä, J. Using expert knowledge in combining green infrastructure and ecosystem services in land use planning: An insight into a new place-based methodology. *Landscape Ecol.* **2014**, *29*, 1361–1375. [CrossRef]
33. IREUS. Entwicklung der Ländlichen Räume in Baden-Württemberg. Available online: https://lel.landwirtschaft-bw.de/pb/site/pbs-bw-new/get/documents/MLR.LEL/PB5Documents/alr/06_Veranstaltungen_2020/pdf_Vortr%C3%A4ge_PM/2019_11_11_IREUS2.0_Abschlussbericht.pdf (accessed on 19 January 2021).
34. Dispan, J.; Koch, A.; König, T. Strukturbericht Region Stuttgart. 2019. Available online: https://www.region-stuttgart.org/strukturbericht/ (accessed on 19 January 2021).
35. LUBW. Bodenzustandsbericht Region Stuttgart. Landesanstalt für Umwelt Baden-Württemberg: Karlsruhe. Available online: https://pd.lubw.de/35192 (accessed on 21 December 2020).
36. BKG. NUTS Regions. Available online: https://gdz.bkg.bund.de/index.php/default/open-data.html?p=2 (accessed on 25 March 2020).
37. Verband Region Stuttgart. GIS-Daten aus dem Biotop-Informations- und Managementsystem (BIMS). Available online: http://webgis.region-stuttgart.org/Web/BIMS/ (accessed on 28 July 2021).
38. Statistisches Bundesamt. Land- und Forstwirtschaft, Fischerei Methodische Grundlagen der Agrarstrukturerhebung. 2016. Available online: https://www.destatis.de/DE/Themen/Branchen-Unternehmen/Landwirtschaft-Forstwirtschaft-Fischerei/Landwirtschaftliche-Betriebe/Publikationen/Downloads-Landwirtschaftliche-Betriebe/grundlagen-strukturerhebung-2032605169004.pdf?__blob=publicationFile (accessed on 28 July 2021).
39. Lischka, A.; Rode, M. Umsetzung von Blühstreifen als produktionsintegrierte Kompensationsmaßnahme (PIK). *Umw. Raun* **2018**, *9*, 307–322.
40. Gödeke, K.; Schwabe, M.; Bärwolff, M.; Marschall, K.; Hering, T.; Degner, J.; Hochberg, H.; Maier, U.; Druckenbrod, C. *Produktionsintegrierte Kompensation (PIK): Maßnahmenvorschläge*; Thüringer Landesanstalt für Landwirtschaft: Jena, Germany, 2014.
41. Sponagel, C.; Back, H.; Angenendt, E.; Bahrs, E. Development of Supply Curves for Biodiversity Offsets on Agricultural Land—A Case Study from the Stuttgart Region. *Ger. J. Agric. Econ.* **2021**, *70*, 70–83. [CrossRef]
42. Statistisches Landesamt. Rinderbestand und -Haltungen (HIT-Auswertung). Available online: https://www.statistik-bw.de/Landwirtschaft/Viehwirtschaft/05035050.tab?R=LA (accessed on 18 November 2020).
43. BfN. Naturschutzgebiete Deutschlands: nsg_2017.shp. Available online: https://www.bfn.de (accessed on 29 April 2020).
44. BfN. Vogelschutzgebiete Deutschlands (SPA): spa_de_2018.shp. Available online: https://www.bfn.de (accessed on 29 April 2020).
45. BfN. FFH-Gebiete Deutschlands: ffh_de_2018.shp. Available online: https://www.bfn.de (accessed on 29 April 2020).
46. BfN. Landschaftsschutzgebiete Deutschlands: lsg_2017.shp. Available online: https://www.bfn.de (accessed on 29 April 2020).
47. ESRI. *ArcGIS Desktop: Release 10.5.1: Copyright © 1995–2017*; Environmental Systems Research Institute: Redlands, CA, USA, 2016.

48. ALKIS. Liegenschaftskataster. Available online: https://www.lgl-bw.de/unsere-themen/Produkte/Geodaten/Daten-des-Liegenschaftskataster/ALKIS/ (accessed on 4 February 2021).
49. LUBW. Fließgewässernetz: Daten- und Kartendienst der LUBW. Available online: https://udo.lubw.baden-wuerttemberg.de/public/pages/url/show.xhtml?url=https%3A%2F%2Frips-dienste.lubw.baden-wuerttemberg.de%2Frips%2Fripsservices%2Fapps%2Fgeodatenexport%2Fudo%2Fdownload.aspx%3Fid%3D413 (accessed on 18 November 2020).
50. LUBW. Biotopverbund Feuchte Standorte: Daten- und Kartendienst der LUBW. Available online: https://udo.lubw.baden-wuerttemberg.de/public/api/navigatorurl/show?globalId=naturLand.Biotopverbund.nais%3A%24SYSTEM%7BCADENZAWEB_GEODOWNLOAD%7Did%3D600 (accessed on 31 July 2021).
51. LUBW. Biotopverbund Trockene Standorte: Daten- und Kartendienst der LUBW. Available online: https://udo.lubw.baden-wuerttemberg.de/public/api/navigatorurl/show?globalId=naturLand.Biotopverbund.nais%3A%24SYSTEM%7BCADENZAWEB_GEODOWNLOAD%7Did%3D600 (accessed on 31 July 2021).
52. LGRB. GeoFachdaten BW—Bodenkunde: BK 50. Available online: https://produkte.lgrb-bw.de/catalog/list/?wm_group_id=11113&p=geowakJJg56JWEXCmuURWPlQSviDEUQ09 (accessed on 18 November 2020).
53. LGRB. Daten zur Bodenerosion in Baden-Württemberg: Stand Juli 2015. Available online: https://produkte.lgrb-bw.de/catalog/list/?wm_group_id=11321&p=geozUKeaX6ZyzM7iDytSxI8bS37bagxwD (accessed on 18 November 2020).
54. Meyer, S.; Leuschner, C. (Eds.) *100 Äcker Für Die Vielfalt: Initiativen zur Förderung der Ackerwildkrautflora in Deutschland*; Universitätsverlag Göttingen: Göttingen, Germany, 2015; ISBN 9783863951849.
55. Hochberg, H.; Zopf, D. Untersuchungsergebnisse zur standort- und bestandesdifferenzierten Grünlandbewirtschaftung in Thüringen. In Proceedings of the 58. Jahrestag. Arb. Grünland Futterb. Geselschaft Pflanzenbauwiss, Arnstadt, Germany, 28–30 August 2014; pp. 31–40.
56. R Core Team. R: A Language and Environment for Statistical Computing. R Foundation for Statistical Computing. Available online: https://www.R-project.org/ (accessed on 1 March 2019).
57. Zerbe, S.; Wiegleb, G. *Renaturierung Von Ökosystemen in Mitteleuropa*; Springer: Berlin/Heidelberg, Germany, 2009; ISBN 978-3-662-48516-3.
58. Ellenberg, H.; Leuschner, C. *Vegetation Mitteleuropas mit den Alpen*; Ulmer: Stuttgart, Germany, 2010; ISBN 9783825281045.
59. Dierschke, H.; Briemle, G. *Kulturgrasland: Wiesen, Weiden und Verwandte Staudenfluren; 20 Tabellen*; Ulmer: Stuttgart, Germany, 2002; ISBN 9783800156412.
60. Jedicke, E. *Biotopverbund: Grundlagen und Massnahmen Einer Neuen Naturschutzstrategie; 32 Tabellen*; 2., überarb. und erw. Aufl.; Ulmer: Stuttgart, Germany, 1994; ISBN 3800133245.
61. Colditz, G. *Auen, Moore, Feuchtwiesen*; Birkhäuser: Basel, Switzerland, 1994; ISBN 9783764350192.
62. DWA-Arbeitsgruppe HW-4.3 "Dezentraler Hochwasserschutz". *Dezentrale Maßnahmen zur Hochwasserminderung; DWA-Regelwerk*; DWA Deutsche Vereinigung für Wasserwirtschaft, Abwasser und Abfall: Hennef, Germany, 2015; ISBN 978-3-88721-262-9.
63. LUBW. Daten aus dem Umweltinformationssystem (UIS) der LUBW Landesanstalt für Umwelt Baden-Württemberg. Available online: https://www.lubw.baden-wuerttemberg.de/umweltinformationssystem (accessed on 25 June 2021).
64. Seither, M.; Engel, S.; King, K.; Elsäßer, M. FFH-Mähwiesen Grundlagen-Bewirtschaftung-Wiederherstellung. Available online: https://fortbildung-lazbw.lgl-bw.de/lazbw/webbasys/download/Shop/2018_GL_lazbw_FFH_Maehwiesen_Grundlagen.pdf (accessed on 25 May 2021).
65. Mährlein, A.; Jaborg, G. Wertminderung landwirtschaftlicher Nutzflächen durch Naturschutzmaßnahmen. Eine Bestandsaufnahme mit den Ergebnissen der HLBS-Expertenbefragung. *Agrarbetrieb (AgrB)* **2015**, *3*, 60–64.
66. Sponagel, C.; Angenendt, E.; Piepho, H.-P.; Bahrs, E. Farmers' preferences for nature conservation compensation measures with a focus on eco-accounts according to the German Nature Conservation Act. *Land Use Policy* **2021**, *104*, 105378. [CrossRef]
67. Schwarz, J.; Klocke, B.; Wagner, C.; Krengel, S. Untersuchungen zum notwendigen Maß bei der Anwendung von Pflanzenschutzmitteln in Winterweizen in den Jahren 2004 bis 2016. *Gesunde Pflanz.* **2018**, *70*, 119–127. [CrossRef]
68. Isermeyer, F.; Nieberg, H.; Banse, M. *Auswirkungen aktueller Politikstrategien (Green Deal, Farm-to-Fork, Biodiversitätsstrategie 2030; Aktionsprogramm Insektenschutz) auf Land- und Forstwirtschaft sowie Fischerei*; Thünen-Institut, Bundesforschungsinstitut für Ländliche Räume, Wald und Fischerei: Braunschweig, Germany, 2020.
69. Gilhaus, K.; Boch, S.; Fischer, M.; Hölzel, N.; Kleinebecker, T.; Prati, D.; Rupprecht, D.; Schmitt, B.; Klaus, V.H. Grassland management in Germany: Effects on plant diversity and vegetation composition. *Tuexenia* **2017**, *37*, 379–397.
70. Geisbauer, C.; Hampicke, U. *Ökonomie Schutzwürdiger Ackerflächen: Was Kostet der Schutz von Ackerwildkräutern?* Institut für Dauerhaft Umweltgerechte Entwicklung von Naturräumen der Erde (DUENE) e.V. and Georg-August-Universität Göttingen, Albrecht-von-Haller-Institut für Pflanzenwissenschaften, Abteilung Ökologie und Ökosystemforschung: Greifswald, Germany, 2012.
71. Mazza, L.; Schiller, J. *CASE STUDY REPORT: The Use of Eco-Accounts in Baden-Württemberg to Implement the German Impact Mitigation Regulation: A Tool to Meet EU's No-Net-Loss Requirement?: A Case Study Report Prepared by IEEP with Funding from the Invaluable and OPERAs Projects*; Institute for European Environmental Policy (IEEP): London, UK, 2014.
72. Zehlius-Eckert, W. Moderne Agroforstsysteme als Option für die produktionsintegrierte Kompensation (PIK)—Potenzial, aktuelle Situation und Verbesserungsvorschläge: In: Bäume in der Land(wirt)schaft—von der Theorie in die Praxis. Tagungsband. Mit Beiträgen des 5. Forums Agroforstsysteme 30.11. bis 01.12.2016 in Senftenberg (OT Brieske). Available online: https://opus4.kobv.de/opus4-btu/files/4148/Tagungsband_5_Forum_Agroforstsysteme__PDFA.pdf#page=25 (accessed on 6 July 2021).

73. Buner, F.; Jenny, M.; Zbinden, N.; Naef-Daenzer, B. Ecologically enhanced areas—A key habitat structure for re-introduced grey partridges Perdix perdix. *Biol. Conserv.* **2005**, *124*, 373–381. [CrossRef]
74. Goldmann, R.L.; Thompson, B.; Daily, G.C. Institutional incentives for managing the landscape: Inducing cooperation for the production of ecosystem services. *Ecol. Econ.* **2007**, *64*, 333–343. [CrossRef]
75. Samway, M.J.; Bazelet, C.S.; Pryke, J.S. Provision of ecosystem services by large scale corridors and ecological networks. *Biodivers. Conserv.* **2010**, *19*, 2949–2962. [CrossRef]
76. Tarabon, S.; Dutoit, T.; Isselin-Nondedeu, F. Pooling biodiversity offsets to improve habitat connectivity and species conservation. *J. Environ. Manag.* **2021**, *277*, 111425. [CrossRef] [PubMed]
77. Moilanen, A.; Kujala, H.; Mikkonen, N. A practical method for evaluating spatial biodiversity offset scenarios based on spatial conservation prioritization outputs. *Methods Ecol. Evol.* **2020**, *11*, 794–803. [CrossRef]
78. European Environment Agency. Corine European Soil Database. Available online: https://www.eea.europa.eu/data-and-maps/data/soil-type (accessed on 2 July 2021).
79. Klüter, H.; Bastian, U. Gegenwärtige Strukturen und Entwicklungstendenzen in der Brandenburger Landwirtschaft im Ländervergleich.: Endbericht. Stand Anfang Mai. 2012. Available online: https://www.agrarbuendnis.de/fileadmin/DAten_AB/Projekt_Bodenmarkt/Material/2012_Gutachten_Prof._Klueter_Entwicklungstendenzen_brd._Landwirtschaft_03.07.2012_gesamt-1.pdf (accessed on 6 July 2021).
80. Batáry, P.; Gallé, R.; Riesch, F.; Fischer, C.; Dormann, C.F.; Mußhoff, O.; Császár, P.; Fusaro, S.; Gayer, C.; Happe, A.-K.; et al. The former Iron Curtain still drives biodiversity-profit trade-offs in German agriculture. *Nat. Ecol. Evol.* **2017**, *1*, 1279–1284. [CrossRef] [PubMed]

land

MDPI

Article

The Rhenish Coal-Mining Area—Assessing the Transformational Talents and Challenges of a Region in Fundamental Structural Change

Stefan Greiving [1,*], Dietwald Gruehn [1] and Christa Reicher [2]

1 Department of Spatial Planning, TU Dortmund University, 44227 Dortmund, Germany; dietwald.gruehn@tu-dortmund.de
2 Chair of Urban Design and Institute for Urban Design and European Urbanism, RWTH Aachen University, 52062 Aachen, Germany; reicher@isl.rwth-aachen.de
* Correspondence: stefan.greiving@tu-dortmund.de

Abstract: This paper addresses the extensive structural changes of the Rhenish coal-mining area in Germany. Coal mining was and still is a relevant economic activity throughout Europe and is the focus of many political and societal debates, as well as research activities in the Rhenish coal-mining area. The project DAZWISCHEN followed the concept of evidence-based planning and therefore identified, by means of a GIS-based analysis, the structural changes within the Rhenish coal-mining area for the conflicting thematic clusters for settlement development and open space. Moreover, we investigated the complex multi-level governance that the region is characterized by. The results suggest an increased pressure on blue and green infrastructure by new urban development, especially in the northern part of the Rhenish coal-mining area. On the other hand, the southern part of the Rhenish coal-mining area will be more likely to undergo a process towards an increase in green infrastructure. Thus, the future development of the whole area is segregated in two different development trends in the north and south parts. The complex governance structure in the Rhenish coal-mining area requires an in-depth view of the ongoing working processes in the development of ideas and visions for regional mission statements of different planning areas, levels, and network partners in a real-life laboratory.

Keywords: structural change; coal mining; evidence-based planning; real-life laboratory; regional development

Citation: Greiving, S.; Gruehn, D.; Reicher, C. The Rhenish Coal-Mining Area—Assessing the Transformational Talents and Challenges of a Region in Fundamental Structural Change. *Land* **2022**, *11*, 826. https://doi.org/10.3390/land11060826

Academic Editors: Stephan Bartke and Sigrun Kabisch

Received: 27 April 2022
Accepted: 23 May 2022
Published: 31 May 2022

1. Introduction

Coal mining was and still is a relevant economic activity throughout Europe. Due to its economic importance, the severe local environmental effects, and the given global climatic importance of fossil fuels such as coal, coal mining is the focus of many political and societal debates [1–4]. These debates focus in particular on the phasing out of coal mining that causes a couple of economic, social, and environmental structural changes. Structural change is defined as "a complex, intertwined phenomenon, not only because economic growth brings about complementary changes in various aspects of the economy, such as the sector compositions of output and employment, the organization of industry, the financial system, income and wealth distribution, demography, political institutions, and even the society's value system, but also because these changes can in turn affect the growth processes" [5] (p. 2).

In 2015, 128 coal mines operated in 12 European member states, with a total annual production capacity of 498 million tons. Germany is the largest producer (184 million tons annually), followed by Poland (135 million), Greece, and the Czech Republic (46 million tons each) [6].

The coal-mining industry and coal as a fossil energy source as a whole are significant contributors to global greenhouse gas emissions. In 2018, EU ETS emissions from EU member states' stationary installations had already fallen by 29% since 2005—primarily due to a decrease in the use of hard coal and lignite fuels [7], but still accounted for 16% of gross inland energy consumption in the EU and 24% of the power generation mix [6].

According to ESPON [8] (p. 5), Europe's remaining coal-mining regions will be confronted with a substantial job loss till 2030. The Rhenish coal-mining area, which is located between Aachen and Cologne in the German federal state of North Rhine-Westphalia, is characterized by a substantial job loss of about 50–85% of all existing jobs in this sector (see Figure 1).

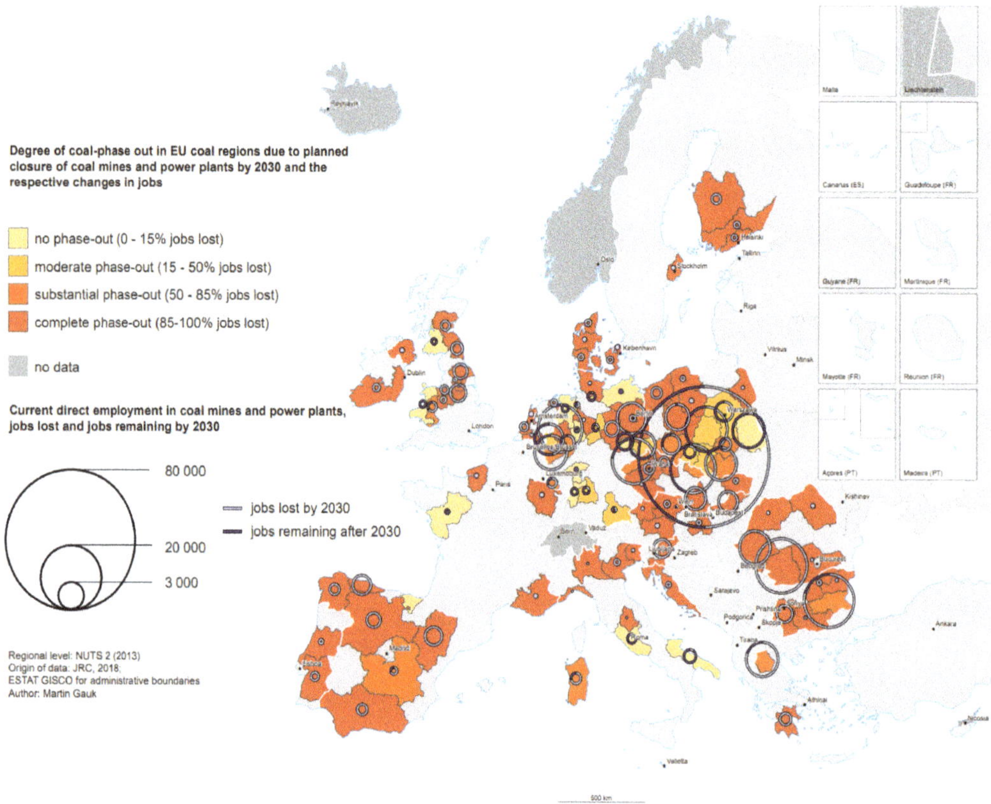

Figure 1. Estimated job losses in coal phase-out regions by 2030. Reprinted with permission from [8] (p. 5).

The European Commission wants to support territories facing serious socio-economic challenges deriving from the transition process towards a climate-neutral economy in the Union, as propagated by the European Green Deal [9] (p. 6). For this purpose, the so-called "Just Transition Fund (JTF)" is to be the first of three pillars that constitute the "Just Transition Mechanism (JTM)". The JTF consists of the three pillars of economic revitalization, social support, and land restoration [8].

However, the coal phase-out pathways differ considerably between the coal-producing European member states [2,10–12]. First, the temporal perspective: while the UK agreed on a coal phase-out by 2024, in Germany, the coal phase-out is determined by law till 2038 at the latest, and in Poland, representatives of the national government and the workers' unions signed a treaty in 2021 that defines 2049 as the phase-out date. Second, the transition strategy: while land reclamation in Germany's surface mining areas is predominantly focusing on an acceptable environmental state, other countries, such as the Czech Republic and Poland, pay more attention to an economically speaking productive use of the former mining area for industrial development and energy production.

Although there are a couple of similar or even more severely affected regions in Europe than the Rhenish coal-mining area, the magnitude of the structural change cannot be solely determined by economic indicators such as production capacity or the number of jobs. The specific feature of this region is its surface mining areas—the largest in Europe—that are characterized by severe changes in the landscape and related forced relocations of human settlements, as well as serious impacts on the regional aquifer and other ecosystem services. In consequence, the still-ongoing mining are activities confronted with serious political protests [13].

Figure 2 shows the location of the several surface lignite-mining areas in the Rhenish coal-mining region.

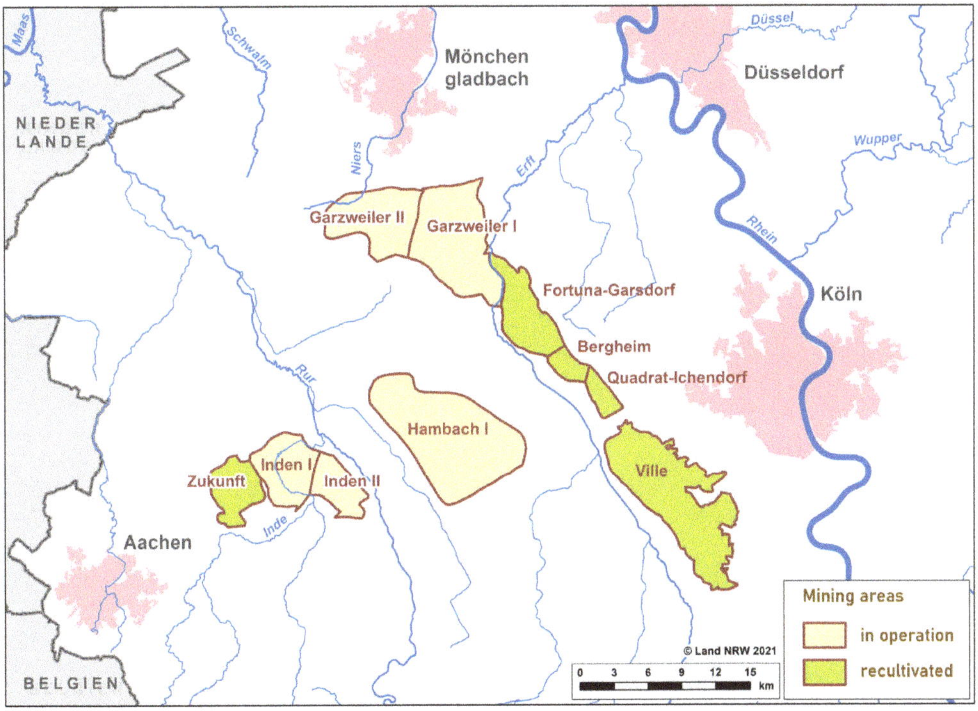

Figure 2. The Rhenish coal mines. Adapted with permission from [14].

Land reclamation is an integral part of any mining project. In Germany, the so-called "Rahmenbetriebsplan" (a general operating plan) in accordance with Art. 52 § 2a Federal Mining Act (Bundesberggesetz) must already consider the necessary land reclamation in the aftermath of the mining activities.

The reclamation of a mine site can involve a number of activities that aim at returning the land and watercourses to an acceptable environmental state and productive use. These activities typically begin with clean-up actions to remove or isolate contaminants, the collection and treatment of any contaminated mine, preventing negative effects on streams and groundwater, the physical stabilization of landforms and structures (mine shafts, tailings, etc.), and the restoration of topsoil [6,15] or, specifically in the case of large-scaled surface mines, the creation of artificial lakes (either by groundwater infiltration or controlled flooding) and the stimulation of related touristic activities. Some other, mostly underground mine sites have been reconverted into museums or destined for other cultural activities (such as the World Heritage Site "Zeche Zollverein" in Essen, Germany) [16].

The energy policy in Germany in the past 20–25 years was subjected to several changes, partly supporting renewable energies, partly re-establishing traditional energy forms. While the German government (first Schröder cabinet) in 2000 officially announced the nuclear phase-out by supporting renewable energy at the same time, in 2005, the first Merkel cabinet re-established nuclear power as a bridging technology, at the cost of renewable energies [17–19]. After the Fukushima Daiichi nuclear disaster in March 2011, the German government (second Merkel cabinet) decided again for a nuclear phase-out until 2022 because of the risk potential linked to nuclear technology [19]. Consequentially, it was decided to use black coal and lignite as a bridging technology until this energy source could be replaced by renewables after a prospective transition time of several decades [17].

In 2016, Germany signed the Paris Agreement under the United Nations Framework Convention on Climate Change, which aims at the reduction of greenhouse gas emissions as a climate protection strategy so that global warming will be limited to a maximum of 2 degrees centigrade. Although the results of this agreement were partly criticized because of the acceptance of global warming up to 1.5 degrees centigrade, the missing of legally binding emission targets as well as specifics on financial support, and no liability provision linked to financial compensation [20], the agreement undoubtedly had a strong impact on politics in Germany in the following years. Not only the Fridays for Future movement, but also statements of unions, churches, environmental and conservational associations, and others, gave evidence of a public awareness change, which in turn triggered the German government (fourth Merkel cabinet) to accelerate the energy transition from carbon-based energy sources towards renewable energies. This process was supported by scientists who argued that the existent German climate protection policy targets were just appropriate to limit global warming on a 2 °C level, while the achievement of more ambitious goals, such as a 1.5 °C limit, would need quicker and more effective measures [21]. As a result of these discussions, in 2018, the German government founded an expert commission on growth, structural change, and employment, which was expected to elaborate a roadmap for a future coal phase-out and energy transition in Germany, based on the necessities and findings of the Paris agreement, as well as on the claims of scientists, environmentalists, and other relevant socio-economic groups [22]. Specific agreements of the commission's work are:

- Stepwise shut down of 84 coal-fired power plants;
- Finalization of electricity generation based on coal until 2038;
- Financial compensation for the operators of the coal-fired power plants in the amount of more than 4 billion EUR;
- Specific measures in the energy sector;
- Development of perspectives for existing, new, and future-proof workplaces;
- Evaluation and monitoring of the initiated measures.

The above-mentioned measures in the energy sector comprise a broad range of different activities, such as specific measures for:

- Climate protection;
- The energy market, electricity tariffs for industry, business, and private end-use customers;
- The security of energy supplies;
- The power grid, accumulators, sector coupling, and innovation potential;
- The added value and employment;
- Open-pit mining and secure maintenance of the pits.

The above-enumerated development of perspectives for existing, new, and future-proof work places includes a consideration of impacts, structural policy effects, and visions of the mining regions' future, explicitly mentioning the Rhenish coal-mining area, which is supplemented by principles of the structural development policy, measures accompanying the structural change, and institutional incorporations [23]. Finally, the annex of the commission's final report contains a detailed lists of a broad range of projects for each mining region, covering the issues of the promotion of economic development, promotion of infrastructure and services for the public, promotion of research and development and science and innovation, legal options to experimentalize new issues, and living labs, as well as regulatory and other measures.

The coal phase-out in Germany was regulated by law in 2020 (Kohleausstiegsgesetz, Kohleverstromungsbeendigungsgesetz) [24], accompanied by critical votes from environmentalists and Green party members, as well as the Fridays for Future movement [13,25]. On the other hand, the German government is entitled to accelerate the process of coal phase-out after a mid-term evaluation and monitoring [24]. Fired by a decision of the Federal Constitutional Court in 2021, the German government was forced to exacerbate and to specify the German Climate Protection Law, especially concerning the period until CO_2 neutrality would be ensured [26].

Currently, the share of mining, energy, and water management in the overall gross value-added in the Rhenish mining region is 9.0 percent, which is around three times as high as in North Rhine-Westphalia (3.7 percent) or Germany as a whole (3.0 percent). The number of people employed directly in lignite mining is in long-term decline. It reached a peak at around 26,000 in the Rhenish mining region at the end of the 1950s. Between 1990 and 2016, employment, excluding power plant workers, fell from around 15,300 persons to around 8400. The production volume developed cyclically, but declined overall, from a maximum of around 108 million tons in 1992 to around 90 million tons in 2016 [27]. The general development perspectives in the Rhenish mining area are currently good. Growth in value-added, as well as the number of employees and inhabitants, has been above average since 2000. At indicators of regional location quality, the Rhenish mining area performs better than the state of North Rhine-Westphalia or Germany as a whole. Particularly with regard to factors that are important for a knowledge-based economy, such as the workforce qualification, R&D, broadband infrastructure, universities, and research institutes in the surrounding area, the Rhenish mining area is characterized by several positive development factors. However, important interconnection effects between the lignite industry and the downstream industrial companies are endangered by the transition process, and their competitiveness is severely impacted by rising electricity prices [27].

It is expected that the end of the mining activities and, as a direct consequence, the shutdown of the thermal power plants powered by lignite coal, will result in a loss of value-added and employment. In total, 14,400 jobs are expected to be lost by 2039. More than two-thirds of this loss of jobs will take place by 2030 due to the rapid closure of a large part of the power plants. However, the 14.8 billion euros of public funding that will flow into the Rhenish mining region by 2038 could generate considerable innovation and growth effects. It is expected that this public funding will provide a private investment impetus of about 22.2 billion euros. Over the funding period, these investments will result in an average of just under 27,000 additional jobs compared with 2021 (trend scenario)—much more than the expected job losses [28].

Against this background, our paper wants to respond to the following research questions: What are the key challenges of structural change? What are the key physical and institutional characteristics of the Rhenish coal-mining region? What are the key development perspectives of the region?

For this purpose, a GIS-based analysis of key physical and spatial characteristics was implemented. The results of this analysis are presented in Section 3.1. Institutional characteristics and the governance setting were studied by means of a desk-top analysis of existing planning and policy documents (see Section 3.2), followed by key elements of the transformation strategy (Section 3.3).

2. Materials and Methods

The research project DAZWISCHEN ("in between") was part of the funding activity "Urban-Rural" (Stadt-Land-Plus) that is administered by the German Federal Ministry of

Education and Research (https://www.zukunftsstadt-stadtlandplus.de/at-a-glance-in-english.html, accessed on 21 November 2021). The funding activity follows the transdisciplinary real-life laboratory approach [29]. Thus, universities and other research institutions collaborate with local and regional practice partners from public administration and civil society. In particular, the funding activity addresses the growing gap between the living environments of urban and rural areas—in dynamic as well as shrinking regions. While the demand for the resource land is intensifying in flourishing regions, shrinking regions face a growing inequality of living conditions. The ultimate goal of the funding activity is the development of innovative solutions for the shared sustainable development of urban, peri-urban, and rural areas, such as a sustainable regional circular economy and the improvement of shared information and data foundations to support the decision-making of actors involved in sustainable land management.

DAZWISCHEN was coordinated by the Department of Spatial Planning at the TU Dortmund University. Further contributing research partners were the Institute for Urban Design at the RWTH University of Aachen and the Institute for Applied Social Sciences. The consortium was complemented by the Zukunftsagentur Rheinisches Revier (regional development agency of the Rhenish coal-mining area), the county of Euskirchen, and the city of Düren as practice partners. They represent the multi-level governance system of the Rhenish mining area.

DAZWISCHEN in particular addressed the extensive structural changes of the Rhenish coal-mining area. The aforementioned German coal phase-out and energy turnaround leads to specific challenges and opportunities for sustainable village, neighborhood, open space, and urban development in the DAZWISCHEN region. In addition to structural change, exogenous developments, especially from the east adjacent Cologne-Bonn metropolitan area, have an impact on the Rhenish mining area.

Furthermore, the project determined the effects of these changes on the functional interdependencies of the whole area, with surrounding large cities and intraregional interdependencies. For consistent data management, a multi-hierarchical spatial information system is to be developed.

Open source data were used for the GIS analyses (Figures 3–8), as shown by Table 1:

Table 1. Overview on data sources.

	Figure 3. Land use and land distribution in the Rhenish mining area	Figure 4. Settlement density in the Rhenish mining area at municipality level	Figure 5. Physical regions within the Rhenish coal-mining area	Figure 6. Habitat networks in the Rhenish coal-mining area	Figure 7. Political and association-based bodies in the Rhenish coal-mining area	Figure 8. Transformation map
Description	The map shows the current land use distribution of the three areas under consideration: the Rhenish Coal-Mining Area, the County of Euskirchen and the City of Duren	The map shows the number of inhabitants per km^2 for all settlement and mobility areas per municipality	The map shows the location and spatial delineation of main natural units in North Rhine-Westphalia. Geomorphological, geological, hydrological, and pedological factors are taken into account for the division into spatial units with largely uniform natural features	The map shows the location and spatial delineation of habitat networks of special and outstanding signifi-cance. The biotope network is intended to permanently safeguard the populations of wild animals and plants including their habitats, biotopes and biotic commu-nities as well as to preserve, restore and develop ecological interrelationships	The map shows the political borders and the regions under various associations	The map shows the different characteristics, talents and gifts that need to be taken into account for spatial development.

Table 1. *Cont.*

Data Source	IT.NRW	IÖR Monitor, Leibniz-Institut für ökologische Raumentwick-lung	LANUV NRW (Landschafts informations-sammlung LINFOS)	LANUV NRW (Landschafts informations-sammlung LINFOS)	IRR: Zukunftsagentur Rheinisches Revier, 2018 Köln-Bonn e.V.	own Figure
Dataset	Land area by type of actual use	B02DT_Nordrhein-Westfalen_KR_2018	Main physical natural regions	Habitat networks	-	Future potential land-uses
From	31 December 2018	per municipality 2015–2018	28 September 2016	2 August 2016	2018	2021
Accessed	19 May 2020	7 June 2020	19 May 2020	19 May 2020	12 December 2021	12 February 2022
Link	https://www.landesdatenbank.nrw.de/ldbnrw/online/	https://www.ioer-monitor.de	https://www.opengeodata.nrw.de/produkte/umwelt_klima/naturschutz/linfos/	https://www.opengeodata.nrw.de/produkte/umwelt_klima/naturschutz/linfos/	https://www.region-koeln-bonn.de/uploads/media/Flyer_RKB_web.pdf	-
Com-ments	Settlement areas including settlement open spaces, residential building area, industrial and commercial area, incl. mining operation, area of mixed use, area of special functional character, sports, leisure and recreation area incl. green space and cemetery	Settlement areas including settlement open spaces	Map based on vectors from: Basis DLM Digitales Basis-Landschaftsmodell	Map based on vectors from: Basis DLM Digitales Basis-Landschaftsmodell	Map based on vectors from: Basis DLM Digitales Basis-Landschaftsmodell	Map based on vectors from: Basis DLM Digitales Basis-Landschafts-modell, Open Street Map

3. Results

3.1. The Physical and Spatial Characteristics of the Rhenish Coal-Mining Area

The following chapter addresses the physical and spatial characteristics of the Rhenish coal-mining area. The far-reaching spatial restructuring processes that have been taking place in the region for decades in the course of lignite use and subsequent recultivation are leading to major landscape and spatial changes. The Rhenish mining area has a variety of settlement structures and landscapes. In addition to the large cities of Mönchengladbach, Aachen, and Neuss, as well as medium-sized towns such as Düren, there are numerous municipalities with centuries-old village structures. Such rural structures stand in stark contrast to new housing developments, which for years have been increasing in number and size along the edges of the agglomerations of the before-mentioned larger cities, as well as the bordering corridor along the Rhine River, where the cities of Düsseldorf, Cologne, and Bonn are situated. In terms of landscape, the diversity in the region, with its large contiguous agricultural areas and important nature conservation areas with essential climatic effects, is no less.

The large open-cast mines with their barrier effect have prevented a coherent development of the area for decades, which means that the decision to phase out the coal mining in the near future entails major potential in terms of settlement and landscape structure. Beyond structural change, other external developments, such as settlement pressure from the mentioned Rhine corridor, have an impact on the Rhenish mining area. Despite the predominantly small-scale settlement structure, the region between Aachen, Mönchengladbach, Düsseldorf, Cologne, and Bonn is an "in-between area" characterized by the agglomeration periphery, which is coming under strong pressure from the population growth in the metropolises. The region as a whole is expected to continue to have a high demand for development areas for housing, commerce, and industry, which the metropolises alone will not be able to meet. At the moment, the Rhenish mining area is losing large areas due to the migrating open-cast mines, which are being prepared for reuse with a time lag, so the land disposition necessary for shaping structural change is being restricted. In addition, land competitions arise that can only be solved across the different spatial levels and must take future trends—such as climate change or mobility change—into account.

The Rhenish mining area is also characterized by socio-demographic change. The region is currently faced simultaneously with growth impulses from the metropolises (Aachen and the Rhine corridor) and shrinking processes in the peripheral parts of the region (especially in the Eifel). However, the increasing ageing and heterogenization of society must also be addressed, with these changes having a region-wide impact on requirements for mobility, forms of housing, infrastructures, open spaces, and, last but not least, the guarantee of services of general interest. This includes, for example, a sustainable realignment of transport infrastructures or opportunities for a balance as well as an improvement in living conditions and the quality of services of general interest between the metropolitan core areas on the one hand and the peripheral parts of the region on the other.

The land distribution within the Rhenish mining area, shown in Figure 3, shows a settlement- and mobility-percentage of 25%. If we look at exemplary sub-areas, the diversity in the settlement structure already mentioned becomes particularly clear. The municipalities of Jüchen, Hürth, Wesseling, and Alsdorf, which border the cities of Mönchengladbach, Cologne, and Aachen, have a percentage of settlement and mobility areas of over 50%, while municipalities such as Hellenthal or Dahlem in the rural south have values of less than 10%.

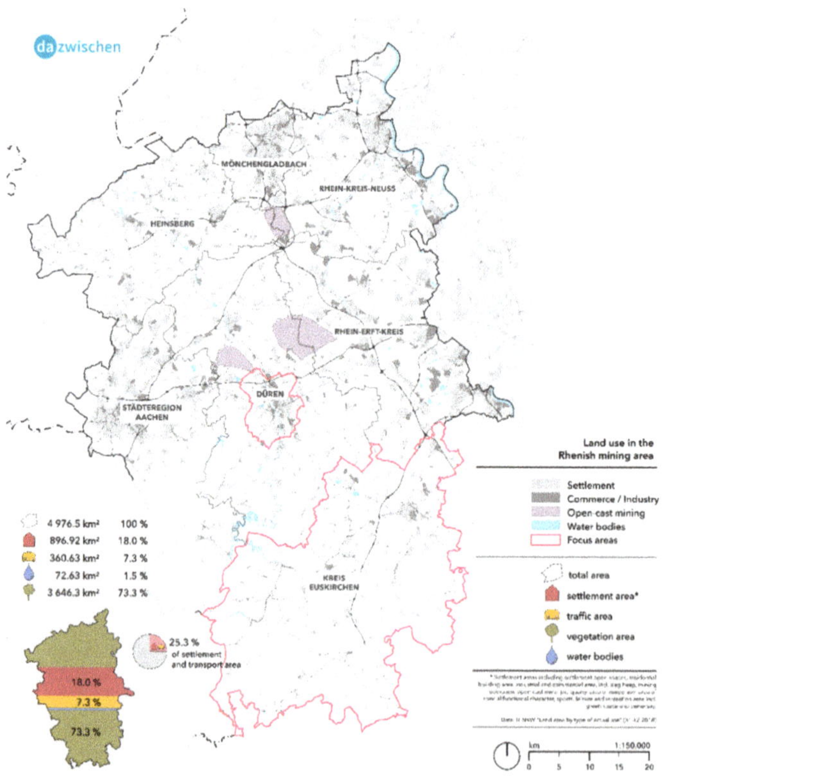

Figure 3. Land use and land distribution in the Rhenish mining area. Source: own Figure.

A similar pattern emerged with the analysis of the settlement density in the Rhenish mining area in Figure 4, which shows the number of inhabitants per km^2 of settlement and mobility area (incl. settlement open spaces) at the municipality level. The densely populated agglomerations around Aachen, Mönchengladbach, and along the Rhine corridor contrast with the rural area of the northern Eifel, thus creating a clear north–south divide within the region.

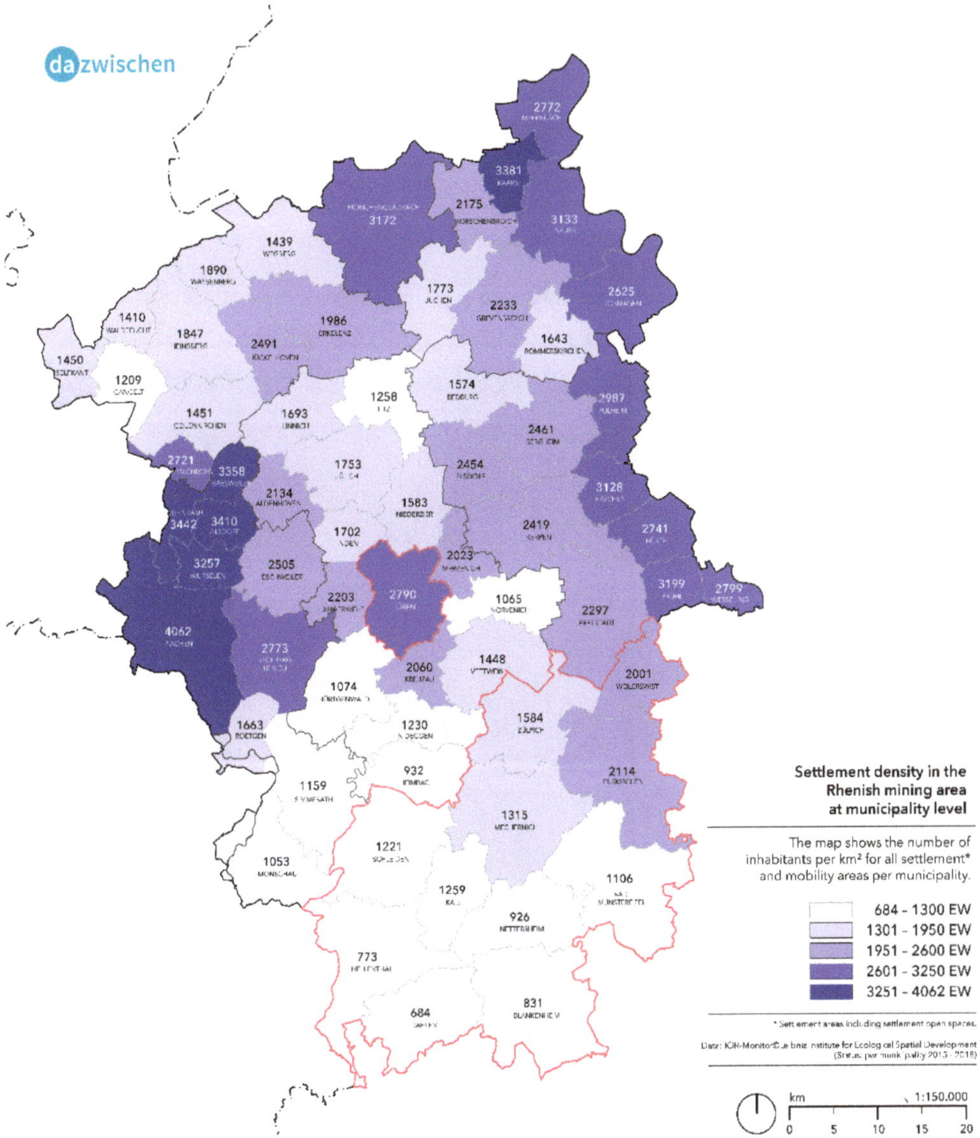

Figure 4. Settlement density in the Rhenish mining area at municipality level. Source: own Figure.

When considering the development of settlement areas, the data from 2016 onwards are only comparable with the previous years to a limited extent due to the changeover from the automated land register ("automatisiertes Liegenschaftsbuch", ALB) to the automated land register information system ("automatisierete Liegenschaftskataster-Informationssystem", ALKIS). However, the period under consideration, 1995 to 2015, showed a trend towards a steady increase in buildings and open spaces, recreational areas, and transport areas. It is only the agricultural land that constantly decreased.

The Rhenish coal-mining area can be roughly characterized by two different landscape types: the average mountain landscape in the southwest with high annual precipitation rates (≥800 mm/a), and the lowlands in the northeast with a mean precipitation of about 650 mm/a, including flat loess soil areas as well as Holocene zones adjacent to the Rhine River and its tributaries (Figure 5).

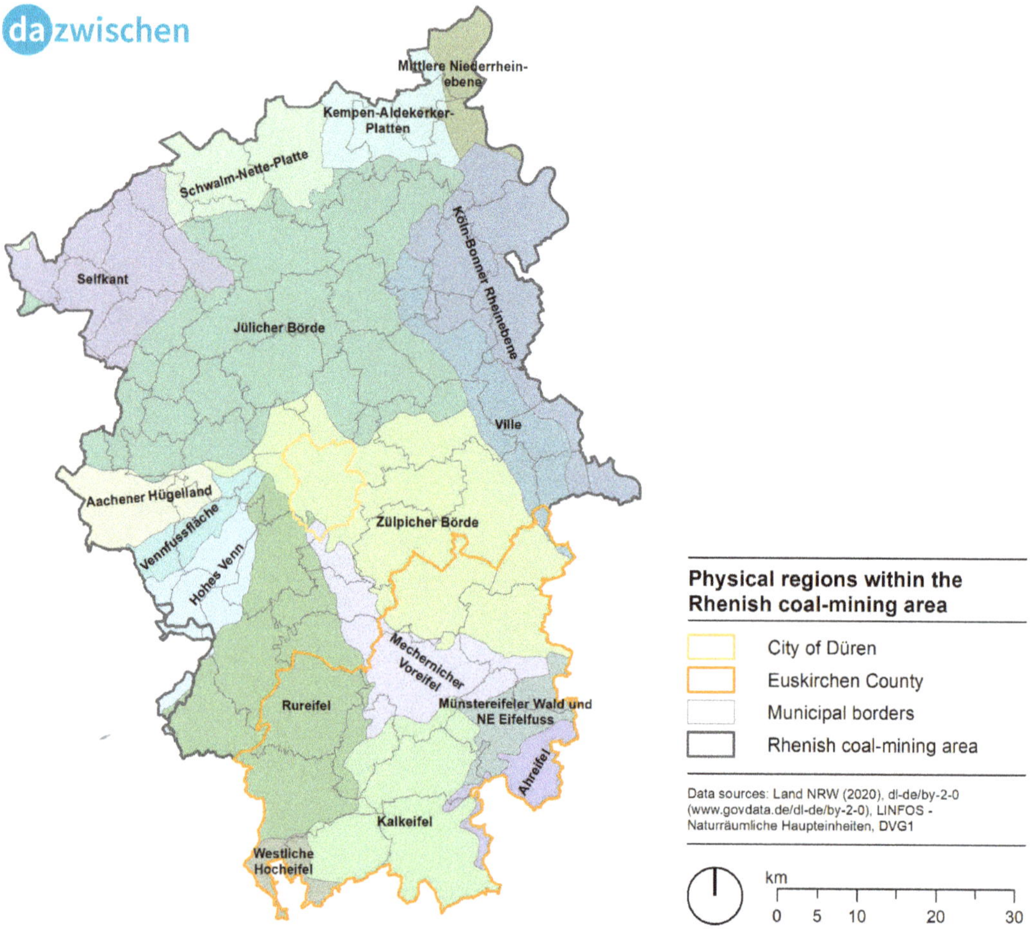

Figure 5. Physical regions within the Rhenish coal-mining area. Source: own Figure.

While the lowlands in the northeast comprise "Jülicher Börde", "Zülpicher Börde", and all other landscapes north from the two mentioned landscapes, the average mountain landscapes are located south from "Jülicher Börde" and "Zülpicher Börde". Because of the high fertility of the loess soils, the Börde landscapes, at least since Roman times, have been, to a high extent, used as arable land. The soil quality in the hilly average mountain landscape is much lower, leading to a higher degree of grassland and forests, with the consequence of more diversified habitats and a rich, attractive, and partly protected landscape, which can be used for recreational purposes, as well (Figure 6).

Urban development is most dynamic around the cities of Cologne (Köln-Bonner Rheinebene) and Aachen (Aachener Hügelland). As pointed out in Figure 2, the coal mines are mainly located in the center of the northern part of the Rhenish coal-mining area (between Cologne and Aachen), mainly in the above-mentioned Börde landscapes, thus creating a lot of land-use conflicts between preserving the given agricultural functions and settlement demands in this area.

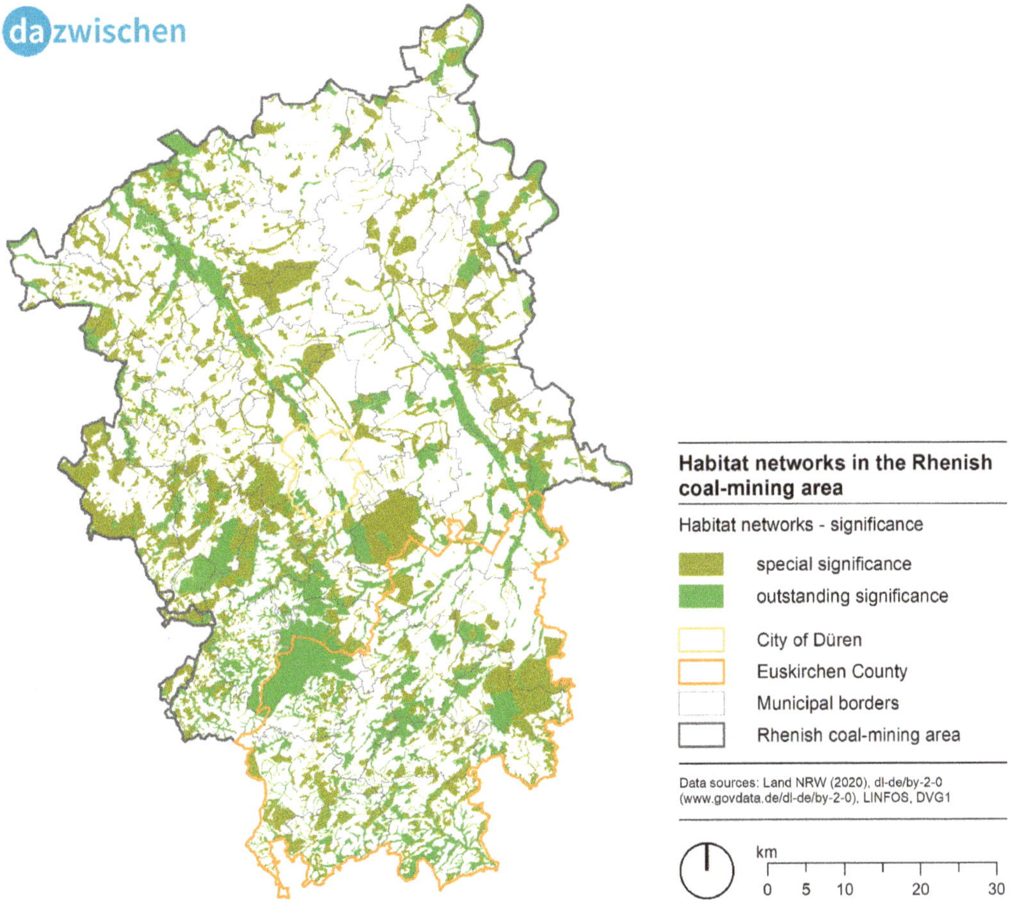

Figure 6. Habitat networks in the Rhenish coal-mining area. Source: own Figure.

3.2. *The Governance Structure of the Rhenish Coal-Mining Area*

In order to counteract this structural change in the Rhenish lignite region in a preventive manner, the "Innovationsregion Rheinisches Revier GmbH" (Innovation Region Rhenish Mining Region, IRR) was founded in 2014, which was subsequently transferred to the "Zukunftsagentur Rheinisches Revier" (Rhenish Mining Region Agency for the Future, ZRR) as a joint platform and trailblazer for the transformation process. The state of North Rhine-Westphalia (NRW) has given the ZRR the task of managing the process of structural change. In order to develop the post-mining landscapes, the former open-cast mining areas are to be reconnected with the surrounding space and thus open up innovative and sustainable development perspectives. The Rhenish mining area is to be transformed into a sustainable mobility, industry, and energy region. This transformation process is to be developed by the "Revierknoten" (a territory node) in cooperation with the regional planning authorities, the municipalities in the region. With the corresponding specifications in regional and local legally binding land-use planning, the development perspectives can be taken into account in the final mining plans still to be approved under mining law, so that municipal planning concepts can be realized [30]. The "Revierknoten" (a territory node), with their respective chairmen, accompany and shape the future issues. In a first step, the district node chairmen and their teams played a key role in drafting the economic and structural program and thus provided the guidelines for the operational implementation of structural change.

The Economic and Structural Program for the Rhenish Mining Area 1.0 (WSP 1.0) was published in December 2019 as a first draft of a vision for the future of the Rhenish mining area for the period after lignite. Responsible for the Economic and Structural Program 1.0 was the ZRR, whose supervisory board and shareholders' meeting adopted it. The program was drawn up in just a few weeks—between September and November 2019—in close consultation with the people in the region and on the basis of insights gained at specialist conferences and workshops. It was handed over to the North Rhine-Westphalian state government as part of the precinct conference in December 2019.

The WSP 1.0 addressed the future fields of "Energy and Industry", "Resources and Agribusiness", "Innovation and Education", and "Space and Infrastructure", which were rounded off by a perspective outlook at an International Building and Technology Exhibition.

The objective of the program is to set out the approach and relevant content so that the funding provided is used with the greatest possible impact. In this sense, the WSP represents the content-related basis for future project and site developments, as well as funding decisions, and is reflected, discussed, and updated accordingly in a broad-based participation process with the people, actors, and institutions in the Rhenish mining area. In an intensive participation process, common guidelines, the so-called "Revier-Charta", were drawn up, which define the guard rails for the further process design.

In June 2021, the updated version of the WSP 1.1 was presented to the public and will henceforth serve as the basis for the content-related orientation of the transformation process. On the way to a mission statement for the Rhenish mining area, guidelines were defined that must guide the shaping of the region:

- A livable region that combines attractiveness with value creation;
- A high-performance region that remains competitive within Europe;
- An innovative region that exploits the future potential of its companies and research institutions;
- A climate-neutral region that is consistently oriented toward a circular economy, resource efficiency, and climate neutrality;
- A networked region that generates new value-added networks through infrastructure, new technologies, and the interlinking of players.

The territory nodes are oriented to these guidelines, in particular the "Revierknoten Raum", which has taken on two central functions: to establish a space laboratory as a knowledge repository for spatial development and to develop an informal spatial strategy with the Rhenish Revier 2038+ (a platform which manages the spatial strategy).

The research project DAZWISCHEN was closely linked to the spatial strategy process in terms of structure and content. Thus, analysis results from the DAZWISCHEN project have been used as a basis for the spatial strategy and—vice versa—the project was able to benefit from initial scenarios and events of the "Revierknoten Raum".

This policy decision and the associated approach are due to the recognition that informal planning processes are becoming increasingly relevant in making land-use designation decisions and in gaining the acceptance of local people. The relevance of an informal set of plans is strengthened by the fact that two formal regional plans for the Düsseldorf and Cologne administrative districts are relevant for the Rhenish mining area. Informal planning can overcome this demarcation in jurisdiction.

The creation of the informal planning strategy has a major impact on the decision makers in formal planning already in the process. These are actively involved in the discourse. The spatial strategy is a technical contribution ("Fachbeitrag") to the formal regional plan. Moreover, the informal planning strategy addresses different time horizons, beyond the temporal validity of the regional plans.

The Spatial Strategy 2038+ is an informal planning process carried out by territorial nodes on behalf of the Ministry of Economic Affairs, Innovation, Digitalization, and Energy of the Federal State of NRW. The process of developing the Spatial Strategy 2038+ itself is structured in different phases:

- Phase 0: inventory, analysis, and preparation of planning works;
- Phase 1: development of spatial images in alternatives;
- Phase 2: deepening of content and spatial guidelines;
- Phase 3: participation and consolidation in an integrated spatial image.

The political representatives, as well as the planning experts in the municipal administrations, are intensively involved in the preparation process.

After alternative development concepts have been presented to a broad public for discussion, a process of participation by the various municipalities, initiatives, and associations follows in order to arrive at a synthesis that is supported as widely as possible and represents a consensus for spatial development in the Rhenish mining area.

The expert findings are fed into the formal planning processes of the regional plans of the districts of Cologne and Düsseldorf. As a technical contribution, the spatial strategy supports the preparation and updating of the two regional plans that contain binding designations on land-use for the Rhenish mining region.

In the Rhenish mining region (see Figure 7), the Cologne district government, which is responsible for the Heinsberg County, the Düren County, the Euskirchen County, the Rhine–Erft County, and the Aachen city regions, and the Düsseldorf district government, which includes the city of Mönchengladbach and the Rhine county of Neuss, are two administrative units with different focuses and processes. The respective regional plans are in different processing phases and have different methods for calculating the land requirements both for the ASB (priority zones for general settlement purposes) and for the GIB areas (priority zones for commercial and industrial purposes). For instance, although both district governments rely on population projections up to the year 2040 when calculating residential land requirements, they already follow different approaches in dealing with replacement requirements and fluctuation reserves. The subsequent distribution to the municipalities is also carried out with different weightings and parameters.

In addition to the district governments, the Rhenish mining area is home to various interest groups, including the Köln-Bonn e.V. (Club) and the association (Zweckverband) of the Aachen region; each covers only parts of the area. Countless economic development agencies, associations, chambers, and political bodies with their own interests make the region a complex conglomeration of interests at different planning levels.

One example is the decisions on new land-use designations, in which the respective district government sets out the binding but rather abstract requirements in the form of the regional plan. It is ultimately up to the political decisions within each municipality to (re-)activate reserve areas or to designate new areas in the local land-use plan. The counties, with their overarching perspective, have little influence on such local developments and can only intervene locally on a small scale to a limited extent with their (informal) development concepts (Kreisentwicklungskonzept).

A look at the different scale levels and planning instruments shows that an informal planning strategy tailored to the challenges is becoming increasingly important. These include information bases and concepts, communicative and cooperative approaches, and formats for planning concepts such as the Spatial Strategy 2038+ and the spatial visions for the Rhenish mining area, developed within the research project DAZWISCHEN.

Figure 7. Political and association-based bodies in the Rhenish coal-mining area. Source: own Figure.

3.3. *Intended Open-Pit Reuse in the Rhenish Coal-Mining Area*

The transformation map (Figure 8) of the Rhenish Revier, which is a result of the still-ongoing regional development process, organized by the "Revierknoten Raum", shows that the region has different characteristics, talents, and gifts that need to be taken into account for spatial development. One main focus is on extending and upgrading the regional transport network and the follow-up uses for the surface mining areas.

However, the reuse of open pits in the Rhenish coal-mining area at least theoretically contributes to a higher degree of freedom concerning the allocation of future land use, because the mining areas can be replaced in the long term by other land-use types. The public discussion mainly focuses on the development of new recreation areas, including a vital blue and green infrastructure, consisting, among others, of woods, meadows, and lakes (see Figure 8). The filling of the lakes will take until about 2090 and depends on the amount of available water, which may decline according to climate change. The "transformation areas" (marked in purple), currently used by thermal power plants, powered by lignite coal, are the designated incubators for new economic activities. In addition, the valid state development plan already contains two large development areas that can be used for large-scale industrial projects.

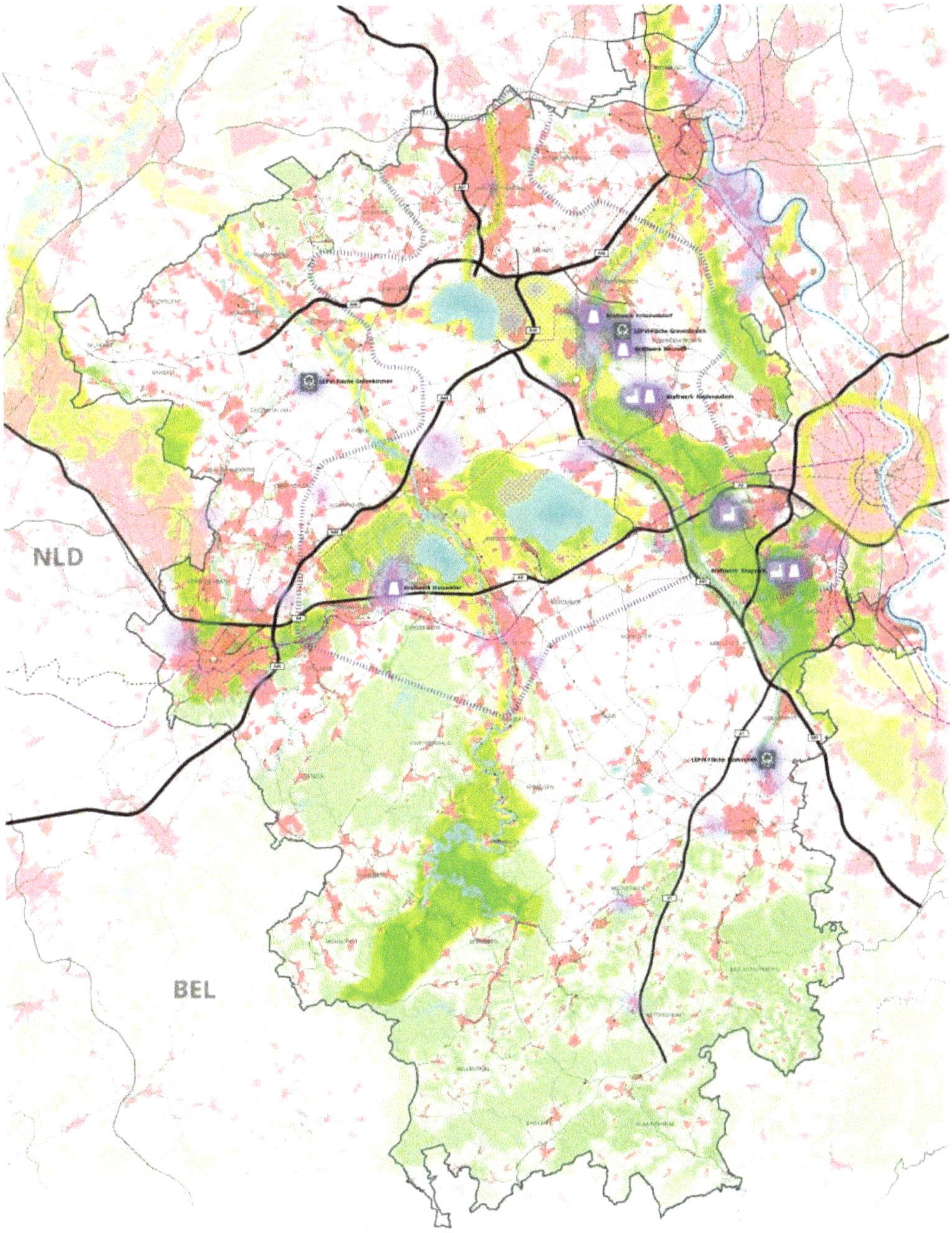

Figure 8. *Cont.*

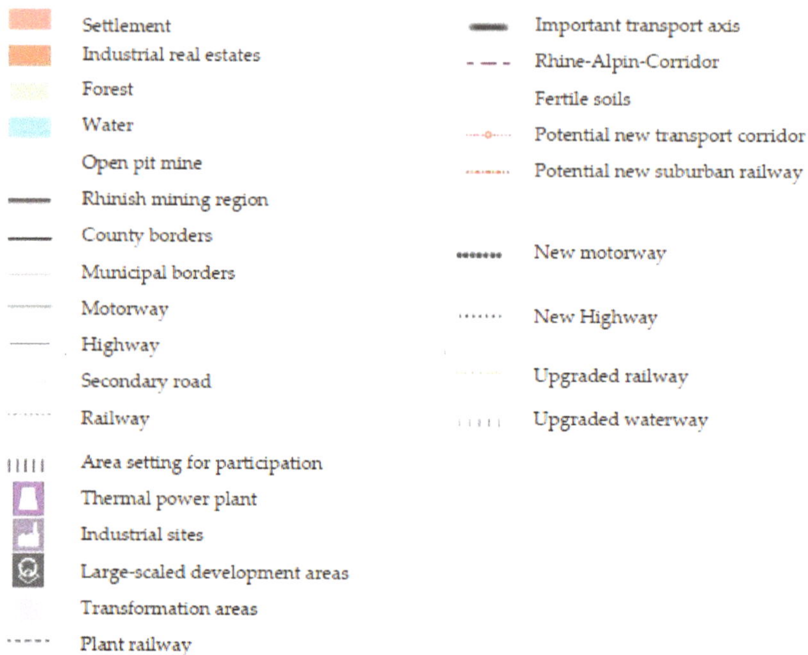

Figure 8. Structure plan: transformation map. Source: own Figure.

4. Discussion

This paper presents the complex spatial structural challenges facing coal-mining regions in general and the Rhenish coal-mining region in particular. This gives rise to a number of further research needs.

The complexity of the governance structure in the Rhenish coal-mining area raises the question of how discourse and decision-making processes take place in the context of creating development strategies and why which interests prevail [31,32]. This requires an in-depth view of the ongoing working processes in the development of ideas and visions for regional mission statements of different planning areas, levels, and network partners in a real-life laboratory, as described by Section 3.2 [29]. Informal planning comprises procedures and instruments of spatial planning that are not legally formalized, standardized, and directly legally binding and at the same time characterized by high flexibility, adaptability, and openness [33].

For this purpose, a multi-stage workshop process seems to be appropriate. In a first step, the results of the analysis, which were developed together with the county of Euskirchen and city of Düren, could be presented to a broader professional public in order to validate them. At the same time, the previous regional and local planning objectives should be reviewed and compared with the results of the analysis. The goal is to identify possible inconsistencies between the structural challenges on the one hand and the political goals on the other. The final step of interaction could then work with the political bodies to adapt the political goals in such a way that they better meet the identified structural challenges.

In this paper, we touched the issue of settlement development pathways and related planning processes. Nevertheless, it seems necessary to take a closer look at the inherent uncertainties, methodological assumptions for land-use modelling, and regional planning approach involved, which is characterized by the so-called counterbalance principle of local and regional interests [34].

As already discussed by the introduction to this paper, the Rhenish coal-mining area is characterized by surface mining activities, which are related with serious impacts on the environment and its ecosystem services [1]. Our trend scenarios suggest an increased pressure on blue and green infrastructure by new urban development, especially in the northern part of the Rhenish coal-mining area, where such landscape elements traditionally have been rare. On the other hand, the southern part of the Rhenish coal-mining area, which is characterized by a high degree of natural structures, such as woods, meadows, and other specific habitats, will be more likely to undergo a process towards an increase in green infrastructure. Thus, the future development of the whole area is segregated in two different development trends in the north and south parts. This process is accompanied by the loss of fertile soils, especially in the northern part of the coal-mining area, which will be a future limitation for conventional as well as organic agriculture practiced close to densely populated areas. Necessarily, this these trends go in line with the deterioration of several ecosystem services. These current and foreseeable future impacts need to be investigated in greater detail [35].

The structural changes in the Rhenish coal-mining areas are influenced by global climate change and its regional and local effects. These effects should be estimated by means of scenario pathway-based parallel modelling of socio-demographic and climatic changes for a reference period and different future scenarios [36,37]. At the same time, these expected climate impacts should be taken into account when deriving spatial development strategies [38,39].

Settlement development has always been closely interlinked with mobility issues [40]. Therefore, scholars argue for an integrated land-use and transport planning approach [41,42], which requires a sound evidence basis of the current and future mobility patterns in the area under investigation. Furthermore, the spatial development strategy should be designed in such a way that the use of public transport is optimized—for example, by allocating further settlement areas close to network nodes of the public transport system.

The structural changes in the region are not only caused by the phase-out of mining activities, but are also related to land-use and demographic changes and, as mentioned above, also influenced by climate and mobility changes. However, a further key determinant is the provision of services of general interest. They are, on the one hand, indispensably needed for maintaining equal living conditions that are guaranteed by the German constitution law (the principle of equality of Art. 3 in conjunction with Art. 20 § 1 on the welfare state principle), but also related to territorial cohesion, which is a key policy concept of the European Union. It aims at the reduction of disparities between the levels of development of the various regions and the backwardness of the least favored regions [43]. The existence and quality of service provision, on the other hand, is an important framework condition for land-use modelling and land-use policies (see, for instance, the central-place theory) [44]. These aspects require a sound knowledge of the spatial distribution of services of general interest.

Any spatial development should be based on a sound evidence basis [45,46], and furthermore, require the European Strategic Environmental Directive and the Environmental Impact Assessment Directive to engage in a continuous monitoring of unforeseen environmental impacts [47,48]. This context should be addressed by a spatial information system that is to be designed to enable the regional and local actors to determine structural changes and their foreseeable impacts, as well as to continuously monitor the effects of their interventions.

5. Conclusions

The Rhenish coal-mining region is a region under fundamental structural changes. The current economic profile will fundamentally change, but the inherent strengths and changes due to the enormous public funding as well as the political pressure on the energy turnaround outweigh the given risks. Our contribution revealed the different spatial talents of the region and its parts. Nonetheless, there are foreseeable conflicts between its socio-economic development perspectives and the restoration (in the northern part) and preservation of ecosystem services (primarily in the southern part). Moreover, the highly fertile loess soils need to be protected. The way to sustainable regional development is through a broad participatory process and an open debate about the given conflicts of interest, which also require further investigations on various research topics, as outlined in our discussion section.

Author Contributions: Conceptualization, S.G.; methodology, S.G.; validation, S.G.; formal analysis, S.G.; investigation, S.G.; resources, S.G., D.G. and C.R.; data curation, D.G. and C.R.; writing—original draft preparation, S.G., D.G. and C.R.; writing—review and editing, S.G., D.G. and C.R.; visualization, S.G., D.G. and C.R.; supervision, S.G.; project administration, S.G.; funding acquisition, S.G., D.G. and C.R. All authors have read and agreed to the published version of the manuscript.

Funding: This research was funded by the German Federal Ministry of Education and Research under Grant 033L217.

Institutional Review Board Statement: Not applicable.

Informed Consent Statement: Not applicable.

Data Availability Statement: The data presented in this study are available on request from the corresponding author.

Conflicts of Interest: The authors declare no conflict of interest.

References

1. Haigh, M. Surface mining and the environment in Europe. *Int. J. Surf. Min. Reclam. Environ.* **2008**, *91*, 104. [CrossRef]
2. Harfst, J.; Wirth, P. Structural change in former mining regions—Problems, potentials, and capacities in multi-level-governance systems. *Procedia—Soc. Behav. Sci.* **2011**, *14*, 167–176. [CrossRef]
3. Harrahill, K.; Douglas, O. Framework Development for "Just Transition" in Coal Producing Jurisdictions. *Energy Policy* **2019**, *134*, 110990. [CrossRef]
4. Skoczkowski, T.; Bielecki, S.; Kochański, M.; Korczak, K. Climate-Change Induced Uncertainties, Risks and Opportunities for the Coal-Based Region of Silesia: Stakeholders' Perspectives. *Environ. Innov. Soc. Transit.* **2020**, *35*, 460–481. [CrossRef]
5. Matsuyama, K. Structural change. In *The New Palgrave Dictionary of Economics*, 2nd ed.; Durlauf, S.N., Blume, L.E., Eds.; Palgrave Macmillan: Basingstoke, UK, 2008; ISBN 9780333786765.
6. Alves Dias, P.; Kanellopoulos, K.; Medarac, H.; Kapetaki, Z.; Miranda Barbosa, E.; Shortall, R.; Czako, V.; Telsnig, T.; Vazquez Hernandez, C.; Lacal Arantegui, R.; et al. *EU Coal Regions: Opportunities and Challenges Ahea*; EUR 29292 EN; Publications Office of the European Union: Luxembourg, 2018; ISBN 978-92-79-89884-6/978-92-79-89883-9. [CrossRef]
7. European Environment Agency. *Trends and Projections in Europe 2019. Tracking Progress towards Europe's Climate and Energy Targets*; European Environment Agency: Copenhagen, Denmark, 2019; ISBN 978-92-9480-103-6. [CrossRef]
8. ESPON ETGC. Structural Change in Coal Phase-Out Regions. 2020. Policy Brief. Available online: https://www.espon.eu/sites/default/files/attachments/Policy%20Brief%20structural%20change%20in%20coal%20phase-out%20regions.pdf (accessed on 21 November 2021).
9. European Commission. The European Green Deal. Brussels, 11.12.2019 COM(2019) 640 Final. Available online: https://eur-lex.europa.eu/resource.html?uri=cellar:b828d165-1c22-11ea-8c1f-01aa75ed71a1.0002.02/DOC_1&format=PDF (accessed on 9 May 2022).
10. Harfts, J. Utilizing the past: Valorising post-mining potential in Central Europe. *Extr. Ind. Soc.* **2015**, *2*, 217–224.
11. Brauers, H.; Oei, P.-Y.; Walk, P. Comparing coal phase-out pathways: The United Kingdom's and Germany's diverging transitions. *Environ. Innov. Soc. Transit.* **2020**, *37*, 238–253. [CrossRef]
12. Schulz, S.; Schwartzkopf, J. *European Lignite-Mining Regions in Transition. Challenges in the Czech Republic and Germany*; Prague Office of the Heinrich-Böll-Stiftung and Deutsche Umwelthilfe in Cooperation with EG3 and Glopolis: Prague, Czech Republic, 2018; ISBN 978-80-88289-04-3.
13. Available online: https://www.ende-gelaende.org/en (accessed on 29 March 2022).
14. Available online: https://www.lanuv.nrw.de/umwelt/wasser/grundwasser/folgen-des-braunkohleabbaus/das-braunkohlerevier (accessed on 29 March 2022).
15. Oei, P.-Y.; Hermann, H.; Herpich, P.; Holtemöller, O.; Lünenbürge, B.; Schult, C. Coal phase-out in Germany—Implications and policies for affected regions. *Energy* **2020**, *196*, 117004. [CrossRef]
16. Available online: https://whc.unesco.org/en/list/975 (accessed on 21 November 2021).
17. Kramm, L. The German Nuclear Phase-Out After Fukushima: A Peculiar Path or an Example for Others? *Renew. Energy Law Policy Rev.* **2012**, *3*, 251–262.

18. Bruninx, K.; Madzharov, D.; Delarue, E.; D'haeseleer, W. Impact of the German nuclear phase-out on Europe's electricity generation—A comprehensive study. *Energy Policy* **2013**, *60*, 251–261. [CrossRef]
19. McCauley, D.; Brown, A.; Rehner, R.; Heffron, R.; van de Graaf, S. Energy justice and policy change: An historical political analysis of the German nuclear phase-out. *Appl. Energy* **2018**, *228*, 317–323. [CrossRef]
20. Gruehn, D. Germany Goes Green—Innovations towards a Sustainable Regional Development. *World Technopolis Rev.* **2013**, *1*, 230–239. [CrossRef]
21. Clémençon, R. The Two Sides of the Paris Climate Agreement: Dismal Failure or Historic Breakthrough? *J. Environ. Dev.* **2016**, *25*, 3–24. [CrossRef]
22. Kumar, S.; Madlener, R. Energy systems and COP21 Paris climate agreement targets in Germany: An integrated modeling approach. In Proceedings of the 7th International Energy and Sustainabililty Connference (IESC) 2018, Cologne, Germany, 17–18 May 2018. [CrossRef]
23. Federal Ministry for Economic Affairs and Energy (BMWi). Commission on Growth, Structural Change and Employment Final Report. Berlin 2019. Available online: https://www.bmwk.de/Redaktion/EN/Publikationen/commission-on-growth-structural-change-and-employment.pdf?__blob=publicationFile&v=3 (accessed on 9 May 2022).
24. Bundesregierung. *Gesetz zur Reduzierung und zur Beendigung der Kohleverstromung und zur Änderung Weiterer Gesetze (Kohleausstiegsgesetz)*; Bundesgesetzblatt Jahrgang: Berlin, Germany, 2021; 2020 Teil I Nr. 37.
25. FAZ, Frankfurter Allgemeine Zeitung 3.7.2020. Bis Spätestens 2038. Bundestag Beschließt Kohleausstieg. Available online: https://www.faz.net/aktuell/wirtschaft/bundestag-beschliesst-kohleausstieg-in-deutschland-16844265.html (accessed on 21 November 2021).
26. BVerfG, Bundesverfassungsgericht 2021. Verfassungsbeschwerden Gegen das Klimaschutzgesetz Teilweise Erfolgreich. Pressemitteilung Nr. 31/2021 Vom 29. April 2021. Available online: https://www.bundesverfassungsgericht.de/SharedDocs/Pressemitteilungen/DE/2021/bvg21-031.html (accessed on 21 November 2021).
27. Bertenrath, R.; Bähr, C.; Kleissner, A.; Schaefer, T. IW-Gutachten Folgenabschätzung Klimaschutzplan und Strukturwandel in den Braunkohleregionen. 2018. Available online: https://www.iwkoeln.de/fileadmin/user_upload/Studien/Gutachten/PDF/2018/Gutachten_Endbericht_DEBRIV_final.pdf (accessed on 9 May 2022).
28. Kempermann, H.; Ewald, J.; Fritsch, M.; Kestermann, C.; Okos, T.; Zink, B. Wertschöpfungs- und Beschäftigungseffekte der Strukturförderung im Rheinischen Revier. Available online: https://www.wirtschaft.nrw/sites/default/files/asset/document/20211209_impact_der_strukturfoerderung_im_rheinischen_revier_final_1.pdf (accessed on 10 May 2022).
29. Sieber, R.; Faulenbach, L.; Fuchs, M.; Gülleken, L. The challenges of co-research in labs in real-world contexts: Empirical findings from four labs in the context of urban climate-change research. *Town Plan. Rev.* **2022**, *93*, 139–163. [CrossRef]
30. Ministerium für Wirtschaft, Innovation, Digitalisierung und Energie des Landes Nordrhein-Westfalen (MWIDE). Leitentscheidung 2021: Neue Perspektiven für das Rheinische Braunkohlerevier Kohleausstieg Entschlossen Vorantreiben, Tagebaue Verkleinern, CO2 noch Stärker Reduzieren. 2021. Available online: https://www.wirtschaft.nrw/sites/default/files/asset/document/leitentscheidung_2021_0.pdf (accessed on 9 May 2022).
31. Kooimann, J. *Governing as Governance*; SAGE Publications: London, UK, 2003. [CrossRef]
32. Shevah, D. Negotiating Urban Space: Initiatives and Innovations in Spatial Governance. *Planext—Next Gener. Plan.* **2017**, *5*, 6–13. [CrossRef]
33. Danielzyk, R.; Sondermann, M. Informelle Planung. In *Handwörterbuch der Stadt- und Raumentwicklung*; ARL—Akademie für Raumentwicklung: Hannover, Germany, 2018; pp. 963–974.
34. Koomen, E.; Stillwell, J. Modelling Land-Use Change. In *Modelling Land-Use Change*; Koomen, E., Stillwell, J., Bakema, A., Scholten, H.J., Eds.; The GeoJournal Library; Springer: Dordrecht, The Netherlands, 2007; Volume 90. [CrossRef]
35. Klopfer, F.; Greiving, S.; Gruehn, D. Creating an evidence base for managing structural change in the Rhineland Mining Area—Evaluating open space applying a comprehensive set of indicators. *J. Environ. Plan. Manag.* **2022**, 1–18. [CrossRef]
36. Van Ruijven, B.; Levy, M.A.; Agrawal, A.; Biermann, F.; Birkmann, J.; Carter, T.R. Enhancing the relevance of Shared Socioeconomic Pathways for climate change impacts, adaptation and vulnerability research. *Clim. Chang.* **2014**, *122*, 481–494. [CrossRef]
37. Greiving, S.; Arens, S.; Becker, D.; Fleischhauer, M.; Hurth, F. Improving the assessment of potential and actual impacts of climate change and extreme events through a parallel modelling of climatic and societal changes at different scales. *J. Extreme Events* **2017**, *4*, 18500033. [CrossRef]
38. Aguiar, F.C.; Bentz, B.; Silva, J.M.N.; Fonseca, A.L.; Swart, R.; Santos, F.D.; Penha-Lopes, G. Adaptation to climate change at local level in Europe: An overview. *Environ. Sci. Policy* **2018**, *86*, 38–63. [CrossRef]
39. Reckien, D.; Salvia, M.; Heidrich, O.; Church, J.M.; Pietrapertosa, F.; Gregorio-Hurtado, S.; D'Alonzo, V.; Foley, A.; Simoes, S.G.; Krkoška, E.; et al. How are cities planning to respond to climate change? Assessment of local climate plans from 885 cities in the EU-28. *J. Clean. Prod.* **2018**, *191*, 207–219. [CrossRef]
40. UN Habitat. Planning and Design for Sustainable Urban Mobility; Global Report on Human Settlements. 2013. Available online: https://www.eltis.org/sites/default/files/tool/planning_design_sus_mob.pdf (accessed on 5 October 2021).
41. Geerlings, H.; Stead, D. The integration of land use planning, transport and environment in European policy and research. *Transp. Policy* **2003**, *10*, 187–196. [CrossRef]
42. Hrelja, R. Integrating transport and land-use planning? How steering cultures in local authorities affect implementation of integrated public transport and land-use planning. *Transp. Res. Part A Policy Pract.* **2015**, *74*, 1–13. [CrossRef]

43. European Commission. My Region, My Europe, Our Future; Seventh Report on Economic, Social and Territorial Cohesion. 2017. Available online: https://ec.europa.eu/regional_policy/sources/docoffic/official/reports/cohesion7/7cr.pdf (accessed on 21 November 2021).
44. Christaller, W. *Die Zentralen Orte in Süddeutschland. Eine ökonomisch-Geographische Untersuchung über die Gesetzmäßigkeit der Verbreitung und Entwicklung der Siedlungen mit Städtischen Funktionen*; Darmstadt Wissenschaftliche Buchgesellschaft: Darmstadt, Germany, 1968.
45. Faludi, A.; Waterhout, B. Introducing Evidence-Based Planning. *disP* **2006**, *165*, 4–13. [CrossRef]
46. Davoudi, S. Evidence-Based Planning—Rhetoric and Reality. *disP* **2006**, *165*, 14–24.
47. European Commission. Directive 2001/42/EC of the European Parliament and of the Council of 27 June 2001 on the Assessment of the Effects of Certain Plans and Programmes on the Environment. Available online: https://eur-lex.europa.eu/legal-content/EN/TXT/PDF/?uri=CELEX:32001L0042&from=EN (accessed on 27 February 2021).
48. European Commission. Directive, 2014/52/EU of the European Parliament and of the Council of 16 April, 2014 Amending Directive, 2011/92/EU on the Assessment of the Effects of Certain Public and Private Projects on the Environment. Available online: https://eur-lex.europa.eu/legal-content/EN/TXT/PDF/?uri=CELEX:32014L0052&from=EN (accessed on 27 February 2021).

land

Article

Climate Impact Assessment for Sustainable Structural Change in the Rhenish Lignite Mining Region

Dennis Becker *, Felix J. Othmer and Stefan Greiving

Department of Spatial Planning, TU Dortmund University, 44227 Dortmund, Germany;
felix.othmer@tu-dortmund.de (F.J.O.); stefan.greiving@tu-dortmund.de (S.G.)
* Correspondence: dennis3.becker@tu-dortmund.de; Tel.: +49-231-755-2433

Abstract: The Rhenish lignite mining region is facing enormous structural changes due to the withdrawal from opencast mining. The current planning of the regional transformation process, however, has so far only insufficiently considered the local impacts of climate change and thus also existing needs for action as well as opportunities in the context of sustainable regional development. However, the integrated consideration of these challenges is probably one of the biggest strategic challenges for this region. The aim of our study was therefore to apply a method for the integrated analysis of structural and climatic changes in the context of a climate impact assessment. We follow a parallel spatial modelling approach and use scenario corridors to describe the bandwidth of potential future conditions. The results clearly show the influence of other drivers such as changes in population and land use, and thus the adaptation options within the context of a sustainable transformation process. Structural changes should be considered in the context of climate impact analyses, as well as climate changes in the management of structural change.

Keywords: climate change; structural change; climate impact assessment; scenario corridors; heat stress

Citation: Becker, D.; Othmer, F.J.; Greiving, S. Climate Impact Assessment for Sustainable Structural Change in the Rhenish Lignite Mining Region. *Land* **2022**, *11*, 957. https://doi.org/10.3390/land11070957

Academic Editors: Stephan Bartke and Sigrun Kabisch

Received: 26 April 2022
Accepted: 20 June 2022
Published: 21 June 2022

1. Introduction

Global climate change is one of the greatest challenges of our time. The changes in the global climate are measurable, have been proven in many ways, and have become more evident than ever in recent years. The assessment reports of the Intergovernmental Panel on Climate Change (IPCC) between the years 2007 and 2014 already make clear the ongoing climate change and the associated diverse impacts [1,2]. The new Sixth Assessment Report arrives at an even more comprehensive picture. New and enhanced climate models and simulations, new analyses, and methods improve the understanding of the human influence on climate variables. Worldwide, the increasing impacts of global climate change are present and noticeable. Weather extremes, such as heat waves, heavy rainfall events, storms, or floods illustrate the problem [3].

Impacts of climate and weather-related extremes depend on the magnitude of climate hazards in combination with locational exposure and vulnerability. In Europe, regions and cities are exposed to different climate and weather-related hazards depending on their geographical location and topography, whereas their physical and socio-economic characteristics influence their vulnerability [4]. The main hazards and impacts of climate change, both current and projected, vary across European regions and cities. In the continental region, where Germany is mainly located, increasing heat extremes are a primary hazard. Together with reduced summer precipitation, they can increase drought risk, risks of forest fires, health risks, and energy demand during the summer. Climate change is also projected to change the river floods regimes in winter and spring because of seasonal precipitation changes [4–7].

Metropolitan regions and urban areas are particularly affected by the negative consequences of climate change [1,2,8]. This is where assets, sensitive facilities and vulnerable groups of people are concentrated, so that a considerable potential for damage can unfold because of climatic changes. Furthermore, climatic changes are interrelated with the type and degree of building and land use as well as the existing urban and natural structure and socio-economic conditions [4,9]. In any case, there are interrelationships between climate change and other processes/megatrends such as demographic change, land use change or structural change [10]. The spatial and temporal convergence of those megatrends can unfold different impact relationships and challenges that need to be managed in an integrated approach. Climate change has regional and local impacts on social, ecological, and economic systems. To deal with these challenges, it is necessary to have knowledge about the possible impacts, their causes and the related synergies and trade-offs. An integrated approach to climate adaptation strategies and measures can be promoted through appropriate instruments and structures of policy governance and coordination. Spatial planning in particular offers considerable opportunities for integrated approaches to dealing with climate change, strengthening resilience, reducing emissions, and promoting sustainable development [1,4,9].

The challenges of climate change do not occur detached from other challenges such as structural change. Climate change and structural change are interrelated, as Matsuyama's definition of the phenomenon of 'structural change' illustrates: "It is a complex, intertwined phenomenon, not only because economic growth brings about complementary changes in various aspects of the economy, such as the sector compositions of output and employment, the organization of industry, the financial system, income and wealth distribution, demography, political institutions, and even the society's value system, but also because these changes can in turn affect the growth processes" [11] (p. 2). On the one hand, climate change is driving these changes as Bardt shows in his work: "Climate change, climate protection and climate policy must be regarded as drivers of structural change" [12] (p. 56). Both climate change adaptation and climate protection can lead to structural changes. Companies and their infrastructures must adapt to the climate and changing weather conditions. As a result, some of the previous products, production methods, locations and markets will be questioned. At the same time, adaptation can create opportunities for new markets and products [12,13]. On the other hand, structural change can influence climate change and its impacts. Moving the economy from a climate deconstructive, mainly carbon-based economy to a sustainable alternative economy can have positive effects on climate mitigation. The transition from a land-intensive economy to a knowledge-based and service economy can also reveal huge spatial potentials that can be used sustainably. For countries in Europe such as Germany, Greece, Poland, or the Czech Republic, these effects mainly impact the traditional lignite mining areas [14]. Thus, the energy transition and the associated structural change in the German lignite mining areas in the Rhineland, Lusatia, and Central Germany, which are still heavily dependent on the old energy industry based on fossil fuels, also require an integrated view of the challenges described so that a sustainable transformation process can be designed [15,16]. To achieve this sustainability in managing structural change, it is essential that the challenges and impacts of climate change are considered for the areas affected by structural change.

The following explanations and results refer to the project DAZWISCHEN (Future-oriented structural change in the Rhenish lignite mining area). DAZWISCHEN is funded by the Federal Ministry of Education and Research (BMBF) within the framework of the Stadt-Land-Plus funding programme (German BMBF Funding Activity Urban-Rural). From 2020–2023, the project examines the structural changes in the region of the Rhenish lignite mining area and their local characteristics, both currently and in the future. In particular, the interrelationships between structural change, settlement and open space development, the mobility transition, the securing of services of general interest and, finally, impacts of climate change, will be examined. The project considers the spatial effects and changes on the functional interdependencies of the entire region with the surrounding cities (Aachen, Bonn and Cologne). The aim is to develop action strategies for sustainable land management in cooperation with the municipalities, political actors, and stakeholders in the region.

As described, a transformation process in the region should consider the challenges of both structural change and climatic change. An integrated consideration of the local impacts and resulting needs for action and opportunities should be incorporated into sustainable regional development. For this purpose, there is a lack of sound evidence based on the challenges of climate change in the region. Our contribution is therefore guided by the following research questions:

1. What contribution can a climate impact analysis make to the successful management of structural change?
2. What is the importance of considering scenario corridors for structural changes such as land use and population change for valid climate impact analyses?

2. Study Area—The Rhenish Lignite Mining Region

The Rhenish lignite mining region is in the west of Germany in the federal state of North Rhine-Westphalia (NRW), Germany's largest opencast lignite mining area and one of the largest in Europe. The region is shaped by a few larger cities, medium-sized centres, small towns, and numerous villages (See Figure 1). Despite the rather small-scale settlement structure the region is not rural at all, but an area formed by the fringe of conurbations, which is coming under pressure from the population growth in the cities (Aachen, Cologne and Bonn). The spatial structural changes in the Rhenish mining region due to surface lignite mining development led to major challenges and opportunities for sustainable village, neighbourhood, open space, and urban development [17]. As a result of the long-standing coal and lignite mining, many energy-intensive industries have settled in the region, such as the chemical, paper and plastics industries, aluminium, and copper production, as well as the processing and food industries [16,17]. The upcoming structural change requires that the jobs lost due to the end of lignite mining be compensated as the demand for high-quality housing and new commercial and industrial areas continues. This is creating land competition that can only be sustainably resolved across the different spatial levels when future trends are considered. The strategic orientation of the Rhenish mining region towards a pioneering role for the initiation of the resource turnaround in Germany offers great opportunities for jobs and value creation in addition to the realisation of ecological sustainability goals [17]. To govern these challenges, the regional development agency of the Rhenish lignite mining area (Zukunftsagentur Rheinisches Revier, ZRR) has been commissioned which comprises the territorial units of the counties of Düren, Euskirchen, Heinsberg, the Rhein-Erft-Kreis and the Rhein-Kreis-Neuss, the Städteregion Aachen and the City of Mönchengladbach (see Figure 1).

The challenges of structural change do not occur separately from other challenges within the Rhenish mining region. Thus, a demographic change is taking place which is characterised by a simultaneity of growth impulses radiating from the metropolis of Cologne and shrinkage in the peripheral parts of the region in the Eifel as well as an overall aging society. This not only creates growing problems for the provision of services of general interest, but also demands on space and infrastructures.

The Rhenish lignite mining area has several locational advantages compared to other lignite mining regions. With Aachen and Mönchengladbach, two major centres belong to the region and the proximity to the centres of the neighbouring Rhineland (Bonn, Cologne, and Düsseldorf). Its location within the border triangle with the Netherlands and Belgium also offer opportunities for the region. The infrastructure, supply and transport connections are good, but the upcoming transformation process must also include the new challenges in terms of infrastructure and be adapted accordingly. The region can build on its strong economic structure. In addition to the energy industry and the energy-intensive industries, this includes, for example, companies in the areas of resource efficiency, mobility and logistics, the digital economy and agriculture. There are numerous starting points for creating new innovative, climate-neutral, and resilient structures. These include, for example, the re-use of power plant sites, the creation of new multifunctional landscapes and/or lakescapes, new model neighbourhoods, new value creation in the bioeconomy, digitalisation, suitable transport infrastructures, and the expansion of climate-neutral mobility in rural areas. These developments could help to better connect the region with the surrounding countryside, to overcome distances more easily, to better link urban and rural qualities and, finally, to create climate-adapted and resilient structures for the future [18].

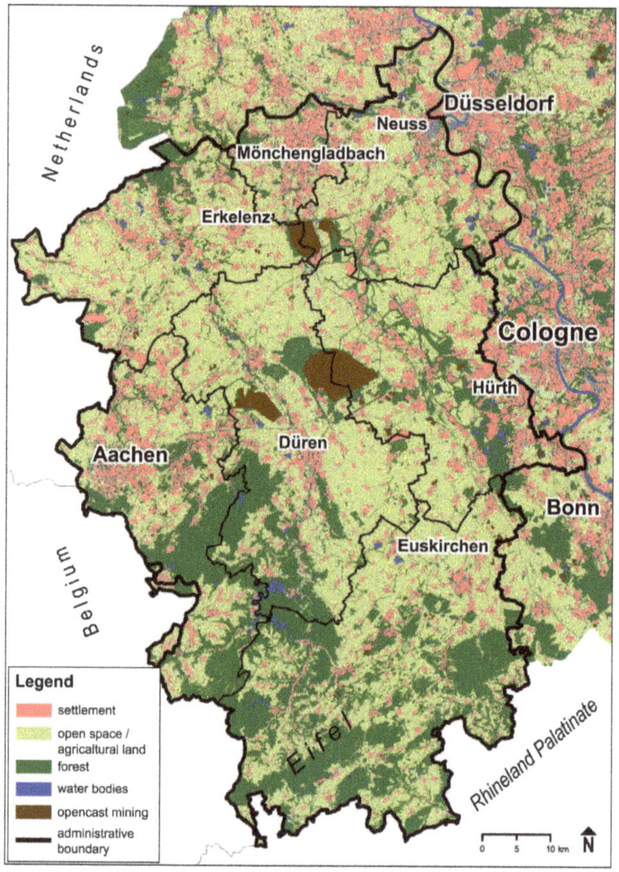

Figure 1. Overview of the Rhenish lignite mining region and its land use. Own figure based on data licence Germany—Attribution—Version 2.0.

The landscape and open spaces of the Rhenish mining region each have a different effect on the regional climate conditions and on the urban climate. The extensive agricultural land acts as cold air production areas and can have a positive effect on the climate of the neighbouring settlements. According to the LANUV climate information system and the explanatory technical report [19], cold air outflows from the Eifel towards the Rur River and from the Jülich-Zülpicher-Börde towards the Erft River are of regional importance. The forest areas of the region act as fresh air production areas. During the day, the forests are cooler than their surroundings and can serve as climatic recreation areas on hot days (days with a maximum air temperature of $\geq$30 °C). Water areas have a similar function with an overall balancing effect, cooling during the day and releasing heat at night. The numerous smaller and larger watercourses act as cold air channels, which supply fresh and cold air to the settlements and thus contribute to the climatic balance in the settlements. The larger settlement areas in towns and cities are to be regarded as heat islands, because buildings and sealed areas heat up more than vegetated areas, store heat and release it again at night. During climate change, a strong increase in hot days and tropical nights (with a minimum air temperature $\geq 20\ ^\circ C$) is expected, which will further promote inner-city heating and affect previously unaffected settlement areas. Against this background, cold and fresh air production areas gain new importance [19,20]. Beside the projected rise in temperature, the region is also affected by other climatic impacts and extremes which pose major challenges for the region, as tragically demonstrated by the heavy rainfall in July 2021, which led to catastrophic flooding [21]. The rise in temperature will also lead to a higher risk of forest fires and droughts in the Rhineland region and thus increase the need for sustainable use of water resources. This will also affect the major rivers such as the Rhine and the Erft and lead to increasing low water problems and to changed flood regimes [22,23]. The region has not only contributed and still contributes with an annual production volume of 90 million tons of lignite coal to climate change for more than a century [24], but is already and will be affected by the impacts of climate change.

Considering all these challenges in an integrated way is probably one of the greatest strategic challenges for the region. Against the current backdrop of change processes in the region, there is a window of opportunity to address this. Therefore, we follow a concept to promote an integrated approach to climatic and structural change, which is addressed in the present climate impact assessment and its underlying methodology.

3. Methodology

In the following section, the underlying methodology of the climate impact analyses carried out is described. For this purpose, it is necessary to first introduce the essential terminology of the method and to emphasize the role of scenarios as well as the parallel modelling of climate and socioeconomic variables.

3.1. Climate Impact Assessments (CIAs)

Climate change adaptation actions need a reliable data basis. Climate impact assessments (CIA) are therefore important prerequisites for the development of adaptation strategies: they identify which regions or sectors are particularly affected by climate change and where adaptation is especially needed [9,25]. To cope with the challenges of climate change, it is important to agree on and define the methodological framework, the key terminology, and the outcome of the assessment which is dealing with the possible impacts [9,25]. Corresponding studies show that there are still existing incongruities in vulnerability concepts, analyses, and assessments, both on a terminological and on a conceptual level [26,27]. This concerns, for example, the reference to place or scales of analysis, key components of vulnerability, involvement of stakeholders or, for example, dealing with uncertainty [26]. This is also linked with the evolution of assessments dealing with climate impacts. Within the evolution (see [27–29]) the approach and its definitions have been changed from a vulnerability approach [1,30] to a risk-based approach as currently used by the IPCC [2,3].

The framework of the CIA used within this article follows the IPCC's understanding of vulnerability and is based on the German Federal Environment Agency's (FEA) Guidelines for Climate Impact and Vulnerability Assessments [25] and visualised in Figure 2.

In contrast to the IPCC's current risk approach, a vulnerability approach for the present assessment was chosen. This was particularly done because, compared to the risk approach of the IPCC (2014), the application of the vulnerability understanding of the IPCC (2007) avoids the problem of having to determine probabilities of occurrence for climatic and socio-economic developments [29]. In comparison to the framework of the FEA, the result or output of the CIA is also not the vulnerability but the climate impact, as the adaptive capacity was not considered in the framework of the CIA. This is due to two reasons: First, the identification and especially the projection of adaptive capacity is very complex, and the projection already assumes actions that must first be identified and politically chosen. Second, spatial planning, as a primary actor in climate adaptation, is an essential component of adaptive capacity on its own since the spatial distribution of assets and values by land-use plans also has a direct influence on the (sub-)components 'exposure' and 'sensitivity'.

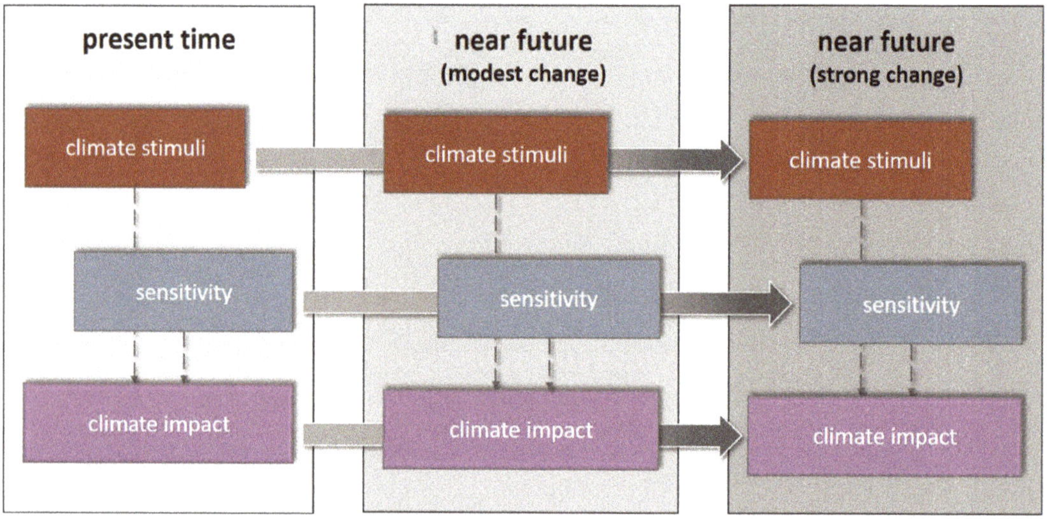

Figure 2. Methodological framework for the CIA. Own figure based on UBA 2015.

The climate impact results from the intersection of climate stimuli and sensitivity (Figure 2). The climate stimuli indicate the expression of a certain climatic parameter at a specific time at a defined location. For example, the average number of hot days per year per grid cell or aggregated on spatial divisions. Sensitivity describes the degree to which a social, ecological or economic system reacts to a specific climate stimulus. This depends on the characteristics of a system as well as its exposure to climate stimuli. Compared to the framework used in the guideline of the FEA there is no distinction between exposure and sensitivity, since both are combined within the sensitivity component.

Two principles are of key relevance in our methodological approach. First, projections are not only used for climate change, but also for the socio-economic changes that have a significant influence on sensitivities. Second, for both components, a future state is not assumed, but rather the possibility range for the future is shown by creating two scenarios and combining them into a scenario corridor of possible futures. These principles are explained in more detail in the following subsections.

3.2. The Parallel Modelling of Climate Stimuli and Sensitivity

The CIA approach follows the so-called 'Parallel Modelling Approach' [10] which emphasises the importance of considering dynamic vulnerabilities or sensitivities. Many current CIAs assume that the climate is changing, but that the vulnerability and sensitivity of the system remains static. The result of these assessments cannot be a reliable basis for adaptation measures since, in addition to climate change as a driver of altered climate stimuli, non-climatic drivers (e.g., land use change, demographic change) can also have an impact on sensitivities, exposures and thus on vulnerabilities, and can significantly influence them [2,29–31]. Therefore, it is unavoidable to take the dynamics of the sensitivity into account. Considering the status quo for the future can lead to maladaptation and failed planning directions [32]. Trends such as demographic change with its facets of 'ageing', 'shrinking', and 'heterogenization' illustrate very clearly why it is particularly relevant to consider socio-economic changes in the context of climatic changes. Some studies assume that the change in non-climatic factors in the future may even have a greater influence on climate impacts than climate change itself [2,33]. The Report on Germany's Vulnerability to Climate Change in 2015 pointed out quite appropriately that every projection of sensitivity, no matter how uncertain, increases the confidence of the statements on future climate impacts, since they must be wrong if they are based on the status quo of the socio-economic systems [34,35], using the example of the German city of Ludwigsburg, that cities and urban planning in general can influence the demographic structure and therefore future vulnerability [35]. This parallel modelling of climate and socio-economic factors was also published with examples from Germany in [10]. The mentioned examples also demonstrate quite clearly why it is important to look at the socio-demographic future in the context of climate change on different levels. There may be interactions between the development of climatic parameters and socio-demographic variables, which should be considered to adapt successfully. Ref. [36] also illustrates well that demographic factors, and in particular their change over time, have a significant influence on the management of natural hazards and thus on the challenges of climate [36].

The integrated consideration of socio-economic and climatic changes within the framework of a CIA not only leads to an improved knowledge base and thus more well-founded results on which measures can be based, but also to the possibility of considering interactions between the different developments in terms of synergies and conflicts. For example, the ageing of the population leads to an increase in the proportion of particularly sensitive population groups, who are much more sensitive to heat stress, thus increasing vulnerability [37]. As [38] stated, "[…] ignoring society aging may lead to underestimated climate change risks" [38] (p. 1).

3.3. The Role of Scenarios for CIAs

Statements about future climate are associated with great uncertainties, as the results are based on model assumptions that can always change. For example, ref. [39] note that in the report of Working Group 1 of the 5th IPCC Assessment Report 2013, the word uncertain and its derivatives (e.g., uncertainty) occur more than 2200 times, which corresponds to an average of 1.5 times per printed page [39].

The fact that climate projections are given in ranges that refer to the representative concentration pathway scenarios (RCPs) lead directly to the assumption that there is no single future state of the climate. Rather, it is common to map the climatic futures using a range of projections, or even to convert them into possible future scenarios that follow an RCP scenario but choose different percentiles. However, as already shown by the parallel modelling approach, it is not only necessary to consider the climate in the future, but also the future characteristics of the system that could be affected by it. Consequently, several possible futures are needed, which are transferred into scenarios along with the climate projections. The consideration of scenarios, especially for non-climatic factors, is significant in the literature and demonstrated not at least by the example of the IPCC's Shared Socioeconomic Pathways (SSPS) [33,40–43]. The climate projections as output of climate ensembles can be used for scenario-building, since good scenarios include projections [44]. This is especially the case for socio-economic forecasts, which are known not to provide any punctual information. Again, the use of bandwidths of potential future conditions can promote acceptance and a more targeted selection of interventions. The familiar example of 'no-regret' measures is particularly beneficial if they are oriented towards the upper edge of the corridor, i.e., towards the worst case.

In addition to taking socio-economic scenarios into account, it is important to merge them with the climatic scenarios and, on this basis, to indicate a scenario corridor of possible future developments. On this basis, a decision can be made on which edge of the scenario corridor the adaptation options should be oriented. Ref. [45] has accurately stated that in a scenario-based approach, the objectives of the analysis and the selection of the scenarios are first normatively set to then determine the factual situation. Based on this analysis, a planner can then exercise the assessment prerogative by deciding whether the planning specifications should be based on the upper edge of the possibility space ('worst case') or on a moderate development [45]. In probabilistic risk calculation, a frequency-magnitude function is usually derived from time series, which are often based on observations from the past. In this respect, approaches to managing the consequences of today's climate can be described as 'decisions under risk' (according to [46]). However, the reaction to a possible future climate change, which can be determined in its concrete spatio-temporal manifestation and not via probability statements, cannot necessarily be grasped in this way. So, the term 'decisions under deep uncertainty' (according to [46]) should be used in the context of a future climate change [29,47,48].

4. Data

The data preparation and analysis were carried out in the ArcGIS programme, to provide the results in descriptive maps as a further working and decision-making basis for the planning processes of local actors and stakeholders.

4.1. Overview of Indicators Used

Since the impacts of climate change affect almost all sectors of society, a set of indicators was developed to cover the diversity of these impacts. The aim was not a comprehensive but an actor-oriented approach, adapted to local challenges. First, a pre-selection of indicators was made based on an internal project discourse with special consideration of the requirements and needs of the joint project partners. In addition, it was possible to refer to the existing climate change prevention strategy of the neighbouring Cologne/Bonn region, which has already been applied as a practical aid regionally since 2019 [49]. Finally, two workshops were held with selected experts from relevant fields as well as federal and state institutions.

The aforementioned climate impact assessment was applied to the Rhenish lignite mining area with a total of eight indicators, which can be classified into three different fields of action: 'living', 'infrastructures', and 'open spaces'. Table 1 provides an overview of the applied indicators, the assigned field of action, and the proxy indicators used for the components' climate stimuli and sensitivity. All fields of action are directly related to the challenges of structural change and the associated transformation process in the region.

For each indicator, the current data basis was first collected and then coordinated and completed with the actors and stakeholders involved. This included a review of the existing data to determine whether forecasts and projections were available and whether the data was available for both the present and the future, to enable the creation of scenarios. The premise here was the use of existing and freely available data sets for the climate stimuli and sensitivity components. In the following, the data basis used for the time-consistent determination of the climate impact and its components as well as the underlying calculations are explained using the example indicator for 'heat stress of the population'. The selected indicator shows hot spots at the municipal level of the Rhenish lignite mining area in which a comparatively high heat stress for the population currently exists and is to be expected in the future by linking the climate stimulus of the hot days with a weighted sensitivity value of the inhabitants.

4.2. Climate Stimuli

In order to determine the climate stimulus, data from the State Office for Environmental and Consumer Protection of the Federal State of North Rhine-Westphalia (LANUV NRW) were used. There, data from the EURO-CORDEX-project [50], which are available to the German Weather Service (DWD), are processed and made available on request. On this basis, data sets could be used for both the present and the future, including RCP scenarios. From this data set, the number of hot days for the period 1971–2000 could be used as representative of the current situation. It should be noted that the selection of the period 1971–2000 for the present is since the climate data for the future always refer to this reference period at the time of data acquisition. For the future, values from the RCP

8.5 ensemble results were used. The advantage of the climate ensemble results lies in the fact that they are available in different bandwidths or the division into percentiles. For the future, a range of possible climatic changes could be estimated by selecting the 15th and 85th percentiles. The time period of the climate data for the future refers to 2021–2050 in relation to 1971–2000. As a result, the ensemble outcomes for the individual indicators for the 15th percentile could be interpreted as a 'moderate change' scenario and for the 85th percentile as a 'strong change' scenario.

Table 1. Overview of indicators, their field of action and the used sub-indicators for climate stimuli and sensitivity within the CIA for the Rhenish lignite-mining area.

Fields of Action	Indicator	Climate Stimuli	Sensitivity
Living	Heat stress of population [1]	Average number of hot days per year	Population sensitivity as the result of the number and relative percentages of the population aggregated as the result of a sensitivity score per municipality
Infrastructures	Impacts of flash floods on the population	Flash flood potential as a result of the combination of terrain slope and average precipitation sums	Population sensitivity as the result of the number and relative percentages of the population aggregated as the result of a sensitivity score per municipality
	Impact of floods on residential areas	Extension of floodplains	Absolute extent and relative share of residential and mixed-use areas per municipality
	Heat impact on social infrastructures	Average number of hot days per year	Absolute extent and relative share of social infrastructure areas per municipality
	Impacts of flash floods on transport infrastructure	Flash flood potential as a result of the combination of terrain slope and average precipitation sum	Absolute extent and relative share of transport infrastructures per municipality combined with the total commuters per year
	Impact of floods on transport infrastructure	Extension of floodplains	Absolute extent and relative share of transport infrastructures per municipality combined with the total commuters per year
Open spaces	Drought stress potential of agriculture	Average precipitation in the cropping season as well as the soil water available to plants	Absolute extent and relative share of agricultural land including grassland per municipality
	Drought stress potential of forests	Average precipitation in the growing season as well as the soil water available to plants	Absolute extent and relative share of forests including wooded areas per municipality weighted according to tree species

[1] This indicator is described in the following as an example in terms of methodology and content, and the corresponding results are presented and discussed. For all indicators, the underlying methodology, data used, calculation rules and sources are available in the Supplementary Material.

Initially, the absolute number of hot days was aggregated at grid levels (in the resolution of 1 × 1 km) using an area-weighted approach at the municipality level. An aggregated value of hot days at the municipality level could then be determined for both the present and the two future scenarios. The data were normalised using a min-max normalisation to transfer the values to a scale between 0 and 1 and thus prepare them for linking with the sensitivity. Since the socioeconomic data in this resolution were not available at the raster cell level but at the community level, all data sets had to be transferred to a common scale at municipality level.

On this basis, it is already possible to estimate where in the Rhenish lignite mining area a high degree of this climate stimulus can already be expected today and in the future. To relate the temporal development to the respective expression, a normalisation over the temporal level was carried out for each of the individual components. By normalising across all temporal levels and thus the entire value range, the results of all time periods are transformed to a scale between 0 and 1. This allows the expression of each time period to be put in relation to the others, and thus the temporal changes can be clearly visualised and communicated. Normalisation across all time levels was carried out for all components of the respective indicators.

4.3. Sensitivity

The appropriate representation of sensitivity remains a major challenge in climate impact research. Key questions that must therefore be answered within the framework of a climate impact assessment concern the extent to which each resource/receptor is sensitive to which stimuli and at what point in time. These questions were answered for each indicator within the framework of the assessment presented here. Various population groups are known to respond differently to climate stimuli. When considering heat-related climate impacts, age is often used as a key factor for determining the sensitivity [51–53]. For example, older or very young people are more sensitive to heat stress due to their physiological characteristics [54]. However, the importance of socio-economic status or living and working conditions in general for heat vulnerability, morbidity and mortality should not be negated by our selection (e.g., [55]).

In the context of determining the sensitivity of the population to heat stress, it was possible to access age-related population data from the statistical office of the federal state of North Rhine-Westphalia. The database provides population data at the level of age years both for the base year 2018 and a population projection for the year 2040 and thus can be used to create own cohorts for each time slice. For the present assessment, the assumption is made that both older and younger people are particularly sensitive to heat stress. Therefore, the total number of inhabitants was transferred into cohorts, which enabled a differentiated representation of sensitivity to be drawn. When spatially calculating population data, not only the absolute number of inhabitants should be used, but this should also be combined with the relative distribution. An exclusive consideration of both absolute and relative statements can lead to significant distortion. Therefore, the calculation also included proportional values of inhabitants per hectare of residential area. The cohorts are weighted for age-related heat sensitivity based on studies of heat stress mortality rates [53,56–58] according to the following formula:

$$sensheatstress(0-5\ years \times 0.25) + (6-64\ years \times 0.05) + (65-79\ years \times 0.25) + (>80\ years \times 0.45))$$

According to this formula, the classic focus is not only on the very young or very old population, but the entire population is considered differently according to its age structure. A special feature, for example, is the explicit consideration of the age group of 6–64-year-olds, who, according to recent surveys, feel particularly affected by heat stress [59,60]. As a result, more detailed statements on the sensitivity to heat stress can be visualised at the level of the municipalities, which consider the differentiated sensitivity of certain age cohorts as well as their absolute and relative distribution within residential areas.

For the socio-demographic future, it was not possible to refer directly to generally available projection data. The development of the population at municipal level and differentiated according to different age cohorts based on the population forecast of the State Office for Information and Technology (IT.NRW) could be used [61]. Since the population projection only provides information on the total population for the target year 2040, a differentiated evaluation was not carried out due to the lack of data. Instead, this data was used to generate assumptions about future land-use development. This was prepared under two different assumptions, so that the future land demand resulting from the population development could be determined quantitatively. The first scenario 'internal development' assumes that 50% of the land demand determined from the population development would be covered within existing built-up areas, and the remaining 50% would be allocated in newly designated building areas. The second scenario 'external development' assumes that 75% of the demand would be met by new construction and only 25% by existing structures.

The scenario of increased internal development is classified as climatically 'strong change' and the scenario of increased external development as climatically 'moderate change' since a densification of already built-up areas is considered as trigger for urban heat stress. In general, our classification is not to be understood in a judgmental way, as an increased external development must always be considered qualitatively from a climatic point of view (e.g., the obstruction of cold air corridors or reduced retention capacity of the peri-urban areas). As a result, a scenario corridor was created for the sensitivity that considers not only the absolute number of population developments but also their change in age structure and the resulting demand of residential areas. In order to merge the sensitivity component with the climate stimulus component, the calculated values were also normalised across all time levels and scenarios.

4.4. Climate Impact

The climate impact was calculated by linking the normalised values for the sensitivity with the climate stimuli. The linkage was carried out multiplicatively, as we follow the assumption that no climate impact can be assumed if one of the two parameters equals zero. The scenarios for the climate stimuli as well as the sensitivity were then merged for the different indicators so that statements on the climate impact could be made for both the assumptions of a strong and a more moderate change.

5. Results

The results of the climate impact analysis follow the triad of Climate Stimuli, Sensitivity and Climate Impact already described under Data and Methodology.

5.1. Climate Stimuli and Scenarios

The intensity of hot days is differentiated in the Rhenish lignite mining area. For example, the current hot days are much more frequent in the eastern part of the region along the Rhine River than in the more predominantly rural parts of the Eifel. This east-west and north-south gradient will continue to develop in the future, both under the assumptions of a moderate and a strong change. Especially when comparing the status quo with the moderate change, it is obvious that even under the assumptions of a less severe change, there will be a significant increase in hot days in the parts of the region that are already stressed today. This is made very clear by the characteristics under the assumptions of a strong change. In the overall view, it should also be emphasized that the impact along the Rhine, for example, is already greater today than it will be in the southern Eifel region under the assumptions of a strong change (see Figure 3).

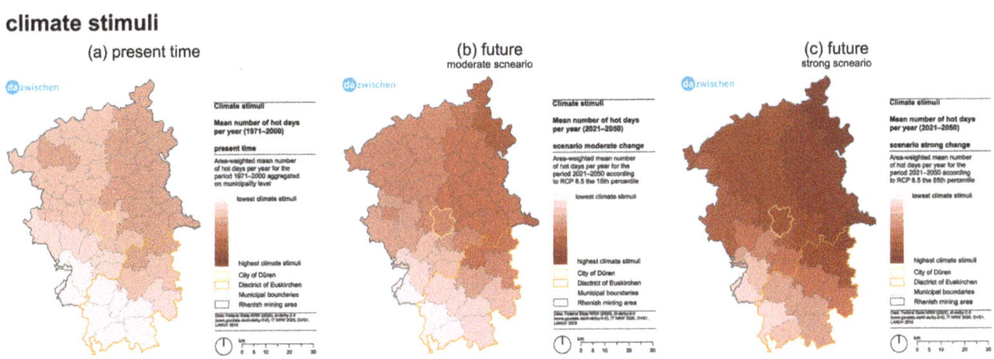

Figure 3. Results for the climate stimuli of the indicator "Heat stress of population": (**a**) Indicator expression for the present time (**b**) Indicator expression for the future under a moderate scenario (**c**) Indicator expression for the future under a strong scenario.

5.2. Sensitivity

The sensitivity component shows a rather differentiated picture of the region. In the present as well as in the future, the densely populated cities of Aachen and Mönchengladbach show the highest sensitivity. The municipalities of the conurbation fringe along the Rhine River are also already significantly more sensitive to the heat stress of the population compared to, for example, the municipalities of the County of Euskirchen except for the city of Euskirchen itself. Viewed over time, two specific effects lead to an increase in sensitivity in parts of the region. On the one hand, demographic change and the associated ageing processes of the population leads to an increased sensitivity to heat stress overall. On the other hand, the projected population increase, especially along the conurbation fringe, also leads to an increased sensitivity. The municipalities along the Rhine can be observed to have a higher sensitivity to heat stress, both in the present and in the future, than, for example, municipalities in the County of Euskirchen (see Figure 4).

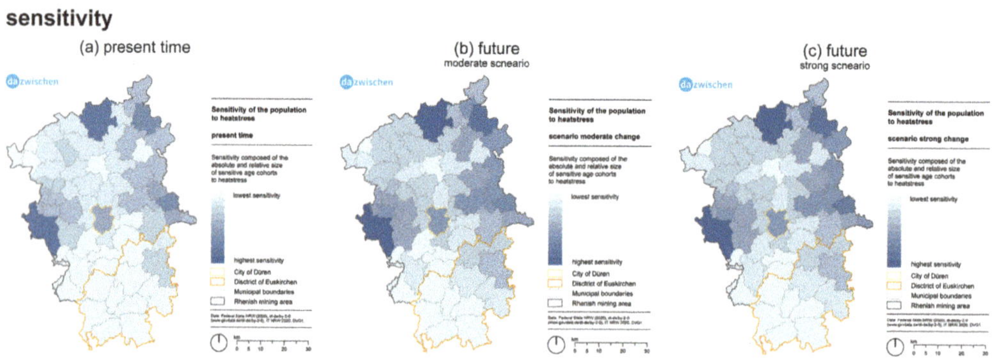

Figure 4. Results for the sensitivity of the indicator "Heat stress of population": (**a**) Indicator expression for the present time (**b**) Indicator expression for the future under a moderate scenario (**c**) Indicator expression for the future under a strong scenario.

5.3. Climate Impact

The figures of the climate impact complete the image of the differentiated impact in the Rhenish lignite mining area that was already drawn in the description of the stimuli and sensitivity components. At the present time, the impacts are comparatively low; only the larger cities such as Aachen and Mönchengladbach, as well as other large and medium-sized cities along the conurbation fringe such as Neuss and Hürth already have a high impact today. Assuming moderate change in the future, this image will further differentiate. The densely populated cities continue to be strongly affected and some municipalities in the northern part of the region will also be more affected. The City of Düren is also strongly affected in this future scenario. Under the assumptions of a strong change, the north-south and east-west gradients will become even more visible. The most pronounced change across all time periods is expected in the cities of Neuss and Mönchengladbach, followed by Aachen, Hürth and Düren. In contrast, the County of Euskirchen is comparatively less affected due to its rather sparsely populated and rural structures, which barely change over time. Here it should be emphasised that some areas are currently more affected today than others will be in the future scenarios. Consequently, a corresponding need for adaptation and action can already be derived for the present (see Figure 5).

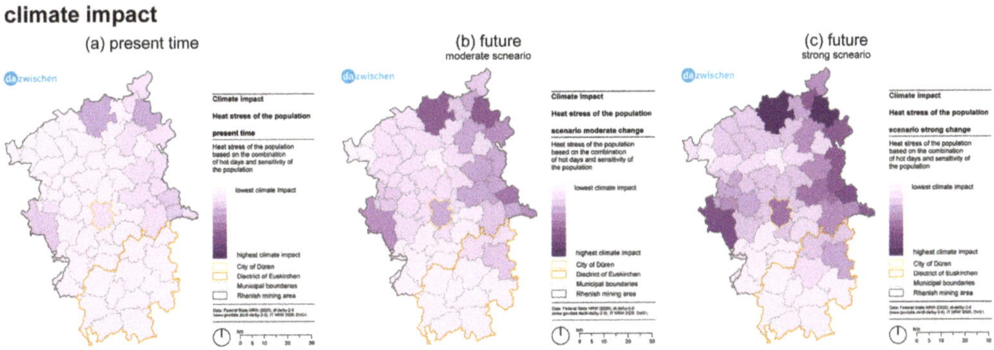

Figure 5. Results for the climate impact of the indicator "Heat stress of population": (**a**) Indicator expression for the present time (**b**) Indicator expression for the future under a moderate scenario (**c**) Indicator expression for the future under a strong scenario.

The results of our study show the clear influence of sensitivity to global warming and increase in heat days on the corresponding climate impact. In terms of negative effects, this also applies to other sectors such as agriculture, forestry, and tourism. Positive effects of climate change on summer tourism or viticulture, for example, are also possible. These results can be used by spatial planning in the management of structural change, particularly to differentiate between action areas and to develop and implement spatially specific adaptation measures accordingly, especially for hotspots.

6. Discussion

Any spatial development strategy should be based on sound evidence [62]. This is also a truth for local adaptation processes [52,63,64]. In addition, the relevant European framework directives require continuous monitoring of unforeseen environmental impacts [65,66]. Structural change is an interdisciplinary topic in the practice of regional and municipal planning that requires the coordination and cooperation of the economy, society, administration, politics, and science. The inclusion and close cooperation of all actors involved makes it possible to integrate the expertise of different disciplines in the transformation process and to achieve balanced goals and strategies for sustainable land management [67]. The main addressee apart from the DAZWISCHEN project itself is the ZRR which currently prepares a regional development strategy for the entire region [68]. For this purpose, the results of the CIA qualify the analytical basis for the planning teams currently developing regional spatial visions on behalf of the ZRR. The authors had the opportunity to present their findings to the planning teams and to discuss fields of application. These give more than a mere consideration of areas that are to be kept free of development due to their climatic significance. One field of application is also the differentiated handling of densification from the urban perspective, both in terms of health impacts due to overheating and the need to adapt to heavy precipitation events.

A comprehensive understanding of the relationship between economic, environmental, and social conditions is essential for sustainable structural change. This includes proper knowledge on the interactions between different aspects of structural change and environmental impacts, such as climate change [12,69]. Against the background of long planning periods, the persistence of built infrastructures as well as the upcoming complex challenges, structural change must prepare for the expected consequences and impacts of climate change at an early stage and take these into account in the planning and transformation process. The presented paper proves the importance of a parallel modelling approach for identifying the potential range of future conditions and thereby contributes to a better understanding of the interconnected climatic and societal changes. The way climate impacts are communicated must also be tailored to the needs and experiences of stakeholders [70].

Considering the diverse land capacities that arise from the abandonment of opencast mining sites, the question must not only be asked about the role of these areas for climate protection, but also how these areas can be used to adapt to local climate impacts. The design of new landscape parks and the planned conversion with large water bodies as lakescapes are often promoted as a promising way of enhancing the value of the area and is seen as a motor for positive structural change. However, the development of such a landscape must be planned and implemented against the background of climatic changes. As the results of the climate impact analysis for the Rhenish mining region show, there will be a significant increase in heat stress in the future and thus also a change in the water balance. On the one hand, this raises the question of where the water to fill the new lakes is to come from, when in summer low water levels are already increasingly being observed in the catchment areas. On the other hand, questions arise about ensuring the quality and quantity of the existing and planned water bodies.

Nevertheless, in view of climate change and the associated shifts, the climatic functions of open spaces and water bodies are of great importance. The protection and strengthening of exchange pathways (e.g., significant biotopes, fresh/cold air corridors) and the provision and design of recreation and climatic compensation areas around the heat-affected settlement areas appear to be particularly important. There is a need for action, on the one hand, to secure existing climate-relevant areas and structures, and, on the other hand, to consistently take climate aspects into account in all future land developments. This shows the importance of climate impact analysis in a structural change. In addition to major regional land-related developments, it is also a matter of shaping structural change in the neighbouring and surrounding municipalities in a sustainable way. For example, the temporary loss of jobs and the migration of energy-intensive production and supplier companies for the mining industry are major challenges but also offer opportunities. This is because structural change creates areas of possibilities that can meet the challenges of climate change.

The ecosystem services, which are mainly provided by the rural south of the region, serve particularly to make the urban climate of the urban core areas bearable by securing cold air volume flows. Finally, the protection of cold air production areas and air guiding paths restricts the development of settlements in rural areas to maintain good living conditions in urban areas. In this context, a balancing of burdens and benefits between rural and urban areas should be discussed in the context of a regional development strategy.

7. Conclusions

Regarding research question one, our contribution has demonstrated the relevance of a climate impact analysis for the management of structural change. The interactions between the foreseeable as well the desired land-use changes and the regional climate are evident and are to be seen as an important framework condition for any development strategy. That is why the upcoming informal spatial strategy of the ZRR considers climate change adaptation as an important element and is going to explore the implications of different development pathways.

Using an example indicator on heat stress of the population, we have addressed research question two and shown how important is the consideration of scenario corridors for structural changes such as land use and population change for a consistent climate impact analysis. This fundamental insight is not put in question by the shortcomings of the old and outdated time slice of 1971–2000 as a reference period. Other appropriate data were simply not available at the time of data acquisition. Nonetheless, these scenario corridors enable decision-makers not only to select a certain scenario pathway, but also to take decisions based on the precautionary principle that considers a worst-case scenario.

However, the chosen procedure of making values on different scales comparable by means of a normalization is very suitable for identifying intraregional differences. This evidence base can be used to identify priority areas for adaptation measures within the region and subsequent funding priorities. Moreover, it supports decision-makers to identify eco-system services which are crucial for stabilizing the urban climate.

At the same time, the normalization allows only relative statements of the municipalities among each other. This is also true for all indicator-based approaches in general, as the findings are not easily transferable to other spatial scales. Thus, for intra-municipal decisions, for instance, in urban land use planning, a higher-resolution investigation is required that focuses on the absolute climate impacts and presents the results on a grid cell basis.

The approach we used in this study is in principle transferable to other regions. The fact that we have limited our analysis to free and freely available data sets increases the transferability. In addition, the detailed documentation of the method as well as the supplementary material allow a step-by-step reproduction of the analysis and its results. Individual improvements could already be identified during the analysis. In order to represent the heat stress indicator more adequately, the number of tropical nights should be taken into account. Even though the probability that a hot day is followed by a tropical night can be considered high, the additional consideration of the proxy indicator leads to an improved data basis. Furthermore, the consideration of an additional population projection can also improve the estimation of future sensitivity and thus the scenarios. Based on this, a moderate and stronger development of the population could also be assumed and then transferred into a differentiated demand for land.

Supplementary Materials: The following supporting information can be downloaded at: https://www.mdpi.com/article/10.3390/land11070957/s1, Factsheets on Indicators of the Climate Impact Assessment.

Author Contributions: Conceptualization, D.B. and F.J.O.; methodology, D.B., F.J.O. and S.G.; software, D.B. and F.J.O.; validation, D.B. and F.J.O.; formal analysis, D.B. and F.J.O.; investigation, D.B. and F.J.O.; resources, D.B. and F.J.O.; data curation, D.B. and F.J.O.; writing—original draft preparation, D.B.; writing—review and editing, D.B. and F.J.O.; visualization, D.B. and F.J.O.; supervision, D.B. and F.J.O.; project administration, D.B., F.J.O. and S.G.; funding acquisition, S.G. All authors have read and agreed to the published version of the manuscript.

Funding: This research is funded from 2020–2023 by the Federal Ministry of Education and Research (BMBF) within the framework of the Stadt-Land-Plus funding programme under the grant 033L217 (German BMBF Funding Activity Urban-Rural).

Institutional Review Board Statement: Not applicable.

Informed Consent Statement: Not applicable.

Data Availability Statement: The data presented in this study are available on request from the corresponding author.

Acknowledgments: This paper is based on results of the project DAZWISCHEN (Future-oriented structural change in the Rhenish lignite mining area).

Conflicts of Interest: The authors declare no conflict of interest.

References

1. Intergovernmental Panel on Climate Change. Climate Change 2007: Synthesis Report. In *Contribution of Working Groups I, II and III to the Fourth Assessment Report of the Intergovernmental Panel on Climate Change*; Core Writing Team, Pachauri, R.K., Reisinger, A., Eds.; IPCC: Geneva, Switzerland, 2007.
2. Intergovernmental Panel on Climate Change. Climate Change 2014: Impacts, Adaptation, and Vulnerability. In *Part A: Global and Sectoral Aspects. Contribution of Working Group II to the Fifth Assessment Report of the Intergovernmental Panel on Climate Change*; Field, C.B., Barros, V.R., Dokken, D.J., Mach, K.J., Mastrandrea, M.D., Bilir, T.E., Chatterjee, M., Ebi, K.L., Estrada, Y.O., Genova, R.C., et al., Eds.; Cambridge University Press: Cambridge, UK; New York, NY, USA, 2014.
3. Intergovernmental Panel on Climate Change. Climate Change 2021: The Physical Science Basis. In *Contribution of Working Group I to the Sixth Assessment Report of the Intergovernmental Panel on Climate Change*; Masson-Delmotte, V., Zhai, P., Pirani, A., Connors, S.L., Péan, C., Berger, S., Caud, N., Chen, Y., Goldfarb, L., Gomis, M.I., et al., Eds.; Cambridge University Press: Cambridge, UK; New York, NY, USA, 2021; *in press*.

4. European Environment Agency. *Urban Adaptation in Europe: How Cities and Towns Respond to Climate Change*; EEA Report 12/2020; Publications Office of the European Union: Luxembourg, 2020. [CrossRef]
5. European Environment Agency. *Climate Change, Impacts and Vulnerability in Europe 2016: An Indicator-Based Report*; EEA Report 01/2017; Publications Office of the European Union: Luxembourg, 2017. [CrossRef]
6. Blöschl, G.; Hall, J.; Parajka, J.; Perdigão, R.A.P.; Merz, B.; Arheimer, B.; Aronica, G.T.; Bilibashi, A.; Bonacci, O.; Borga, M.; et al. Changing Climate Shifts Timing of European Floods. *Science* **2017**, *357*, 588–590. [CrossRef]
7. Blöschl, G.; Hall, J.; Viglione, A.; Perdigão, R.A.P.; Parajka, J.; Merz, B.; Lun, D.; Arheimer, B.; Aronica, G.T.; Bilibashi, A.; et al. Changing Climate Both Increases and Decreases European River Floods. *Nature* **2019**, *573*, 108–111. [CrossRef] [PubMed]
8. Bubeck, P.; Klimmer, L.; Albrecht, J. Klimaanpassung in der rechtlichen Rahmensetzung des Bundes und Auswirkungen auf die Praxis im Raumordnungs-, Städtebau- und Wasserrecht. *Nat. Recht* **2016**, *38*, 297–307. [CrossRef]
9. European Environment Agency. *National Climate Change Vulnerability and Risk Assessments in Europe, 2018*; EEA Report 01/2018; Publications Office of the European Union: Luxembourg, 2018. [CrossRef]
10. Greiving, S.; Arens, S.; Becker, D.; Fleischhauer, M.; Hurth, F. Improving the Assessment of Potential and Actual Impacts of Climate Change and Extreme Events Through a Parallel Modeling of Climatic and Societal Changes at Different Scales. *J. Extrem. Events* **2017**, *04*, 1850003. [CrossRef]
11. Matsuyama, K. Structural change. In *The New Palgrave Dictionary of Economics*, 2nd ed.; Durlauf, S.N., Blume, L.E., Eds.; Palgrave Macmillan: Basingstoke, UK, 2008; ISBN 9780333786765.
12. Bardt, H. Klima- und Strukturwandel: Chancen und Risiken der deutschen Industrie. In *IW-Analysen*; Inst. der Dt. Wirtschaft Köln Medien GmbH: Köln, Germany, 2011; p. 69.
13. Bardt, H.; Hüther, M. Angebotsorientierte Umweltpolitik: Positionsbestimmung und Perspektiven. In *IW-Positionen*; Deutscher Instituts-Verl: Köln, Germany, 2006; p. 21.
14. Skoczkowski, T.; Bielecki, S.; Kochański, M.; Katarzyna, K. Climate-change induced uncertainties, risks and opportunities for the coal-based region of Silesia: Stakeholders' perspectives. *Environ. Innov. Soc. Transit.* **2020**, *35*, 460–481. [CrossRef]
15. Haigh, M.J. Surface mining and the environment in Europe. *Int. J. Surf. Min. Reclam. Environ.* **1993**, *7*, 91–104. [CrossRef]
16. Schulz, S.; Schwartzkopf, J. European Lignite-Mining Regions in Transition. In *Challenges in the Czech Republic and Germany*; Prague Office of the Heinrich-Böll-Stiftung and Deutsche Umwelthilfe in cooperation with EG3 and Glopolis: Prague, Czech Republic, 2018; ISBN 978-80-88289-04-3.
17. Zukunftsagentur Rheinisches Revier GmbH (ZRR). *Wirtschafts- und Strukturprogramm 1.1 für das Rheinisches Zukunftsrevier*; ZRR: Jülich, Germany, 2021.
18. Bundesministerium für Wirtschaft und Energie (BMWi). Kommission. In *Wachstum, Strukturwandel und Beschäftigung. Abschlussbericht*; BMWi: Berlin, Germany, 2019.
19. Landesamt für Natur, Umwelt und Verbraucherschutz Nordrhein-Westfalen (LANUV). *Klimaanalyse Nordrhein-Westfalen. LANUV-Fachbericht 86*; LANUV: Recklinghausen, Germany, 2018; ISSN 1864-3930.
20. Zukunftsagentur Rheinisches Revier GmbH (ZRR). *Regionales Freiraumsystem Rheinisches Revier 1.0. Erfassung und Konzeption*; ZRR: Jülich, Germany, 2019.
21. Dietze, M.; Ozturk, U. A flood of disaster response challenges. *Science* **2021**, *373*, 1317–1318. [CrossRef] [PubMed]
22. Umweltbundesamt (UBA). *Climate Impact and Risk Assessment 2021 for Germany*; Climate Change 27/2021; Umweltbundesamt: Dessau-Roßlau, Germany, 2021.
23. Brasseur, Guy, Daniela Jacob, and Susanne Schuck-Zöller. *Klimawandel in Deutschland: Entwicklung, Folgen, Risiken und Perspektiven*; Springer Spektrum: Berlin, Germany, 2017.
24. Bertenrath, R.; Bähr, C.; Kleissner, A.; Schaefer, T. *IW-Gutachten Folgenabschätzung Klimaschutzplan und Strukturwandel in den Braunkohleregionen*; Auftraggeber: Köln, Germany; Deutscher Braunkohlen-Industrie-Verein e.V. (DEBRIV): Köln, Germany, 2018.
25. Buth, M.; Kahlenborn, W.; Greiving, S.; Fleischhauer, M.; Zebisch, M.; Schneiderbauer, S.; Schauser, I. *Guidelines for Climate Impact and Vulnerability Assessments: Recommendations of the Interministerial Working Group on Adaptation to Climate Change of the German Federal Government*; Umweltbundesamt: Dessau-Roßlau, Germany, 2017; ISSN 2363-832X.
26. Soares, M.B.; Gagnon, A.S.; Doherty, R.M. Conceptual elements of climate change vulnerability assessments: A review. *Int. J. Clim. Chang. Strateg. Manag.* **2012**, *4*, S.6–S.35. [CrossRef]
27. Klopfer, F.; Westerholt, R.; Gruehn, D. Conceptual Frameworks for Assessing Climate Change Effects on Urban Areas: A Scoping Review. *Sustainability* **2021**, *13*, 10794. [CrossRef]
28. Füssel, H.-M.; Klein, R.J.T. Climate Change Vulnerability Assessments: An Evolution of Conceptual Thinking. *Clim. Chang.* **2006**, *75*, 301–329. [CrossRef]
29. Birkmann, J.; Greiving, S.; Serdeczny, O. Das Assessment von Vulnerabilitäten, Risiken und Unsicherheiten. In *Klimawandel in Deutschland: Entwicklung, Folgen, Risiken und Perspektiven*; Brasseur, G., Jacob, D., Schuck-Zöller, S., Eds.; Springer Spektrum: Berlin, Germany, 2017; Volume 75, pp. 267–276. [CrossRef]
30. Intergovernmental Panel on Climate Change. Managing the Risks of Extreme Events and Disasters to Advance Climate Change Adaptation. In *A Special Report of Working Groups I and II of the Intergovernmental Panel on Climate Change*; Field, C.B., Barros, V., Stocker, T.F., Qin, D., Dokken, D.J., Ebi, K.L., Mastrandrea, M.D., Mach, K.J., Plattner, G.-K., Allen, S.K., et al., Eds.; Cambridge University Press: Cambridge, UK; New York, NY, USA, 2012.

31. Birkmann, J.; Schanze, J.; Müller, P.; Stock, M. *Anpassung an den Klimawandel Durch Räumliche Planung: Grundlagen, Strategien, Instrumente*; E-Paper der ARL Nr. 13; Akademie für Raumforschung und Landesplanung (ARL): Hannover, Germany, 2012.
32. Jurgilevich, A.; Räsänen, A.; Juhola, S. Assessing the dynamics of urban vulnerability to climate change: Case of Helsinki, Finland. *Environ. Sci. Policy* **2021**, *125*, 32–43. [CrossRef]
33. Birkmann, J.; Cutter, S.L.; Rothman, D.S.; Welle, T.; Garschagen, M.; van Ruijven, B.; O'Neill, B.; Preston, B.L.; Kienberger, S.; Cardona, O.D.; et al. Scenarios for vulnerability: Opportunities and constraints in the context of climate change and disaster risk. *Clim. Chang.* **2015**, *133*, 53–68. [CrossRef]
34. Umweltbundesamt (UBA). *Vulnerabilität Deutschlands Gegenüber dem Klimawandel*; Climate Change 24/2015; Umweltbundesamt: Dessau-Roßlau, Germany, 2015; ISSN 1862-4359.
35. Birkmann, J.; Sauter, H.; Garschagen, M.; Fleischhauer, M.; Puntub, W.; Klose, C.; Burkhardt, A.; Göttsche, F.; Laranjeira, K.; Müller, J.; et al. New methods for local vulnerability scenarios to heat stress to inform urban planning—Case study City of Ludwigsburg/Germany. *Clim. Chang.* **2021**, *165*, 37. [CrossRef]
36. Clar, C. How demographic developments determine the management of hydrometeorological hazard risks in rural communities: The linkages between demographic and natural hazards research. *WIREs Water* **2019**, *6*, 9. [CrossRef]
37. Haq, G.; Gutman, G. Climate gerontology: Meeting the challenge of population ageing and climate change. *Z. Für Gerontol. Geriatr.* **2014**, *6*, 462–467. [CrossRef] [PubMed]
38. Lee, J.Y.; Kim, H. Projection of Future Temperature-Related Mortality Due to Climate and Demographic Changes. *Environ. Int.* **2016**, *94*, 489–494. [CrossRef] [PubMed]
39. Lewandowsky, S.; Ballard, T.; Pancost, R.D. Uncertainty as Knowledge. *Philos. Trans. R. Soc. A* **2015**, *373*, 20140462. [CrossRef] [PubMed]
40. Kriegler, E.; O'Neill, B.C.; Hallegatte, S.; Kram, T.; Lempert, R.J.; Moss, R.H.; Wilbanks, T. The need for and use of socio-economic scenarios for climate change analysis: A new approach based on shared socio-economic pathways. *Glob. Environ. Chang.* **2012**, *22*, 807–822. [CrossRef]
41. O'Neill, B.C.; Kriegler, E.; Riahi, K.; Ebi, K.L.; Hallegatte, S.; Carter, T.R.; Mathur, R.; van Vuuren, D.P. A new scenario framework for climate change research: The concept of shared socioeconomic pathways. *Clim. Chang.* **2014**, *122*, 387–400. [CrossRef]
42. O'Neill, B.C.; Kriegler, E.; Ebi, K.L.; Kemp-Benedict, E.; Riahi, K.; Rothman, D.S.; van Ruijven, B.J.; van Vuuren, D.P.; Birkmann, J.; Kok, K.; et al. The roads ahead: Narratives for shared socioeconomic pathways describing world futures in the 21st century. *Glob. Environ. Chang.* **2017**, *42*, 169–180. [CrossRef]
43. van Ruijven, B.J.; Levy, M.A.; Agrawal, A.; Biermann, F.; Birkmann, J.; Carter, T.R.; Ebi, K.L.; Garschagen, M.; Jones, B.; Jones, R.; et al. Enhancing the relevance of Shared Socioeconomic Pathways for climate change impacts, adaptation and vulnerability research. *Clim. Chang.* **2014**, *122*, 481–494. [CrossRef]
44. Glenn, J.C.; Gordon, T.J. *Futures Research Methodology: Version 3.0*; The Millennium Project: Washington, DC, USA, 2009.
45. Greiving, S. Analyse- und Bewertungskonzepte für Risiken im Vergleich. *Inf. Zur Raumentwickl.* **2019**, *46*, 62–73.
46. Laux, H. *Entscheidungstheorie*, 7th ed.; Springer-Lehrbuch: Berlin/Heidelberg, Germany, 2007; ISBN 978-3-540-71161-2.
47. van Asselt, M.B.A. The complex significance of uncertainty in a risk era: Logics, manners and strategies in use. *Int. J. Risk Assess. Manag.* **2005**, *5*, 125–158. [CrossRef]
48. Walker, W.; Haasnoot, M.; Kwakkel, J. Adapt or Perish: A Review of Planning Approaches for Adaptation under Deep Uncertainty. *Sustainability* **2013**, *5*, 955–979. [CrossRef]
49. Region Köln/Bonn e.V. *Klimawandelvorsorgestrategie für die Region*; Praxishilfe: Köln/Bonn, Germany, 2019.
50. Jacob, D.; Petersen, J.; Eggert, B.; Alias, A.; Christensen, O.B.; Bouwer, L.M.; Braun, A.; Colette, A.; Déqué, M.; Georgievski, G.; et al. EURO-CORDEX: New high-resolution climate change projections for European impact research. *Reg. Environ. Chang.* **2014**, *14*, 563–578. [CrossRef]
51. Apreda, C.; D'Ambrosio, V.; di Martino, F. A climate vulnerability and impact assessment model for complex urban systems. *Environ. Sci. Policy* **2019**, *93*, 11–26. [CrossRef]
52. Donner, J.; Müller, J.M.; Köppel, J. Urban Heat: Towards Adapted German Cities? *J. Environ. Assess. Policy Manag.* **2015**, *17*, 1550020. [CrossRef]
53. Fouillet, A.; Rey, G.; Laurent, F.; Pavillon, G.; Bellec, S.; Guihenneuc-Jouyaux, C.; Clavel, J.; Jougla, E.; Hémon, D. Excess Mortality Related to the August 2003 Heat Wave in France. *Int. Arch. Occup. Environ. Health* **2006**, *80*, 16–24. [CrossRef] [PubMed]
54. Kenney, W.L.; Craighead, D.H.; Alexander, L.M. Heat waves, aging, and human cardiovascular health. *Med. Sci. Sports Exerc.* **2014**, *46*, 1891–1899. [CrossRef] [PubMed]
55. Reid, C.E.; O'Neill, M.S.; Gronlund, C.J.; Brines, S.J.; Brown, D.G.; Diez-Roux, A.V.; Schwartz, J. Mapping Community Determinants of Heat Vulnerability. *Environ. Health Perspect.* **2009**, *117*, 1730–1736. [CrossRef] [PubMed]
56. D'Ippoliti, D.; Michelozzi, P.; Marino, C.; de'Donato, F.; Menne, B.; Katsouyanni, K.; Kirchmayer, U.; Analitis, A.; Medina-Ramón, M.; Paldy, A.; et al. The Impact of Heat Waves on Mortality in 9 European Cities: Results from the EuroHEAT Project. *Environ. Health* **2010**, *9*, 37. [CrossRef]
57. an der Heiden, M.; Muthers, S.; Niemann, H.; Buchholz, U.; Grabenhenrich, L.; Matzarakis, A. Schätzung hitzebedingter Todesfälle in Deutschland zwischen 2001 und 2015. *Bundesgesundheitsblatt Gesundh. Gesundh.* **2019**, *62*, 571–579. [CrossRef] [PubMed]

58. an der Heiden, M.; Muthers, S.; Niemann, H.; Buchholz, U.; Grabenhenrich, L.; Matzarakis, A. Heat-Related Mortality: An Analysis of the Impact of Heatwaves in Germany Between 1992 and 2017. *Dtsch. Ärzteblatt Int.* **2020**, *117*, 603–609. [CrossRef]
59. Baldin, M.-L.; Sinnig, H. Hitzeresiliente Städte und Quartiere: Sichtweisen von Bewohnern und Experten zur urbanen Transformation in Erfurt und Dresden. *Verb. Bundesverb. Wohn. Stadtentwickl. (Vhw)* **2019**, *5*, 244–250.
60. Sandholz, S.; Sett, D.; Greco, A.; Wannewitz, M.; Garschagen, M. Rethinking urban heat stress: Assessing risk and adaptation options across socioeconomic groups in Bonn, Germany. *Urban Clim.* **2021**, *37*, 100857. [CrossRef]
61. Landesdatenbank NRW. Gemeindemodellrechnung-Basis-2018 bis 2040 nach Altersjahren und Geschlecht-Kreisangehörige Gemeinden -Stichtag. Landesdatenbank NRW Website. 2021. Available online: https://www.landesdatenbank.nrw.de/ldbnrw//online?operation=table&code=12422-02ir&bypass=true&levelindex=1&levelid=1637924177620#abreadcrumb (accessed on 7 May 2021).
62. Davoudi, S. Evidence-Based Planning. *Disp Plan. Rev.* **2006**, *42*, 14–24. [CrossRef]
63. Biesbroek, G.R.; Klostermann, J.E.M.; Termeer, C.J.A.M.; Kabat, P. On the nature of barriers to climate change adaptation. *Reg. Environ. Chang.* **2013**, *13*, 1119–1129. [CrossRef]
64. Reckien, D.; Salvia, M.; Heidrich, O.; Church, J.M.; Pietrapertosa, F.; de Gregorio-Hurtado, S.; D'Alonzo, V.; Foley, A.; Simoes, S.; Lorencová, E.K.; et al. How are cities planning to respond to climate change? Assessment of local climate plans from 885 cities in the EU-28. *J. Clean. Prod.* **2018**, *191*, 207–219. [CrossRef]
65. European Commission. Directive 2001/42/EC of the European Parliament and of the Council of 27 June 2001 on the assessment of the effects of certain plans and programmes on the environment. *Off. J. Eur. Union* **2001**, L197/30. Available online: https://eur-lex.europa.eu/legal-content/EN/TXT/PDF/?uri=CELEX:32001L0042&from=EN (accessed on 20 May 2022).
66. European Commission. Directive 2014/52/EU of the European Parliament and of the Council of 16 April 2014 amending Directive 2011/92/EU on the assessment of the effects of certain public and private projects on the environment. *Off. J. Eur. Union* **2014**, L124/1. Available online: https://eur-lex.europa.eu/legal-content/EN/TXT/PDF/?uri=CELEX:32014L0052&from=EN (accessed on 20 May 2022).
67. Molenveld, A.; van Buuren, A.; Ellen, G.-J. Governance of climate adaptation, which mode? An exploration of stakeholder viewpoints on how to organize adaptation. *Clim. Chang.* **2020**, *162*, 233–254. [CrossRef]
68. Website Raum-Strategie. Available online: https://www.raum-strategie.de (accessed on 20 May 2022).
69. Ciarli, T.; Savona, M. Modelling the Evolution of Economic Structure and Climate Change: A Review. *Ecol. Econ.* **2019**, *158*, 51–64. [CrossRef]
70. Mabon, L. Making climate information services accessible to communities: What can we learn from environmental risk communication research? *Urban Clim.* **2020**, *31*, 100537. [CrossRef] [PubMed]

land

Article

Service of General Interest in the Rhenish Coal-Mining Area in Context of Structural Change

Saskia Gall-Roehrig [1,*], Steffen Rosier [2], Katharina Riss [1] and Stefan Greiving [3]

1 County Euskirchen, 53879 Euskirchen, Germany; katharina.riss@kreis-euskirchen.de
2 Community Düren, 52349 Düren, Germany; s.rosier@dueren.de
3 Department of Spatial Planning, TU Dortmund University, 44227 Dortmund, Germany; stefan.greiving@tu-dortmund.de
* Correspondence: saskia.gall-roehrig@kreis-euskirchen.de

Abstract: Territorial cohesion is of main interest in the EU and its member state Germany. It should lead to a reduction of disparities between different stats of development and is closely related to the accessibility to Service of General (SGI) in different areas. The investigated area in this article is characterized by the end of coal fired power generation by the Coal-Phase-out act. In this article the accessibility to SGI is discussed. Additionally, the accessibility to specific SGI is investigated with a special view to the southern rural parts of the Rhenish coal-mining area. It is outlined that the provision to SGIs is high near to agglomerations and decreases in rural parts. This can also be observed for broadband expansion, while the access to internet allows a higher accessibility to SGIs for different analysed indicators. This research is used to establish a regional development strategy to maintain equal living conditions for the entire region.

Keywords: Service of General interest; Rhenish coal-mining area; structural change; equal living conditions

Citation: Gall-Roehrig, S.; Rosier, S.; Riss, K.; Greiving, S. Service of General Interest in the Rhenish Coal-Mining Area in Context of Structural Change. *Land* **2022**, *11*, 956. https://doi.org/10.3390/land11070956

Academic Editor: Alexandru-Ionuţ Petrişor

Received: 29 April 2022
Accepted: 14 June 2022
Published: 21 June 2022

1. Introduction

The Legal and Political Frameworks for SGI in the European Union and Its Member State Germany

Territorial cohesion is a key policy concept of the European Union. It aims at the reduction of disparities between the levels of development of the various regions and the backwardness of the least favoured regions [1] and is therefore closely related to the provision of SGI that are imperative for economic and social well-being [2,3].

SGIs "are services that public authorities of the Member States classify as being of general interest and, therefore, subject to specific public service obligations (PSO). The term covers both economic activities (...) and non-economic services" [4] (p. 3).

Moreover, SGI should not be limited to the quality of and the access to physical infrastructure facilities only, but encompass [5]:

1. Basic physiological needs (eating, drinking, sleeping, living);
2. Security needs (order, safety, health, protection against natural hazards);
3. Social needs (education, cohesion, accessibility).

A subcomponent of SGI includes the so-called "Services of general economic interest" (SGEI). This type of services encompasses "economic activities which deliver outcomes in the overall public good that would not be supplied (or would be supplied under different conditions in terms of quality, safety, affordability, equal treatment or universal access) by the market without public intervention." [4] (p. 3).

SGEI are granted a special legal position in the internal market in the Treaty on the Functioning of the European Union (TFEU). It constitutes with Art. 14 TFEU that there are areas of nonsovereign service provision that lie beyond the market-based governance system. As these services are of general interest and important for the territorial and social cohesion of the people, the member states have to ensure that these services "operate on the basis of principles and conditions, particularly economic and financial conditions, which enable them to fulfil their missions."

Therefore, SGEI are only conditionally subject to the rules of the internal market as laid down in Art. 101-109 TFEU. Art. 102 § 2 clarifies: "Undertakings entrusted with the operation of services of general economic interest or having the character [...] shall be subject to the rules contained in the Treaties, in particular to the rules on competition, in so far as the application of such rules does not obstruct the performance in law or in fact, of the particular tasks assigned to them."

Beyond the legal basis of SGI, the European Union assigns SGI a special role in European cohesion policy. The Urban Agenda for the EU considers the "Provision of adequate public SGI" under "Priority Themes and cross-cutting issues of the Urban Agenda for the EU" [6] (p. 8). The new Territorial Agenda 2030 as adopted at the Informal Meeting of Ministers responsible for Spatial Planning and Territorial Development and/or Territorial Cohesion on 1 December 2020 points at increasing imbalances and inequalities and identifies a need for action among other fields in regard to SGI: "The accessibility, proximity, affordability and quality of public services is important to quality of life and business development. Needs for greater cost-effectiveness and efficient public management can lead to the withdrawal or clustering of services in certain locations. At the same time, expectations of availability and quality increase" [7] (p. 8).

On the national level in Germany, the legislative framework for SGI is based on the Constitution Law (Grundgesetz, GG) in accordance with the Federal Regional Planning Act (Raumordnungsgesetz, ROG). The provision of SGI can be derived from the principle of the welfare state (Art. 20 § 1 GG) and the principle of equality of Art. 3 GG. Art. 2 § 1 GG (freedom of development) comes also into play, since it imposes an obligation on the federal and federal state governments to reduce regional disparities in individual living conditions or at least not to exacerbate them, since freedom of development presupposes equality of opportunity. Equal provision of SGI should give all citizens equal opportunities to develop their personalities.

A key instrument for equal-living conditions are central-place systems [8]. Thus, the special role of spatial planning to ensure the provision of public services should be emphasized (Art. 1 ROG points at ensuring equal living conditions in all regions). Thereby, the 41st Standing Conference of Ministers responsible for Spatial Planning adopted in Berlin on 9 March 2016 the "Concepts and Strategies for Spatial Development in Germany" [9].

Concept 2 is "Ensure the provision of public services". It identifies four priority actions:

- Consistently apply the central places system in order to minimize distances to service facilities for citizens and maximize the attractiveness of the locations for operators
- Develop cooperative systems among municipalities
- Ensure the supply of sparsely populated rural areas with SGI
- Ensure accessibility by maintaining and improving the quality of public transport services

Concept 2 contains a visual overview on the current provision of SGI (see Figure 1). The concept points at present unequal living conditions and identifies related need for action due to demographic reasons.

The county of Euskirchen, shown by the red cycle, belongs partly to areas with limited accessibility to SGIs and endangered accessibility to public transport.

Moreover, the German Federal Ministry of the Interior, Building and Community launched in 2019 a strategy on securing equal living conditions all over the country (BMI 2019). This strategy underlines the importance of SGI as key approach for securing equal living conditions (own translation): "Germany needs good and accessible SGI so that all people everywhere have good prospects for life and development. Reliable education, care, cultural and leisure services are of the utmost relevance for the decision: leave-or stay?" [10] (p. 23).

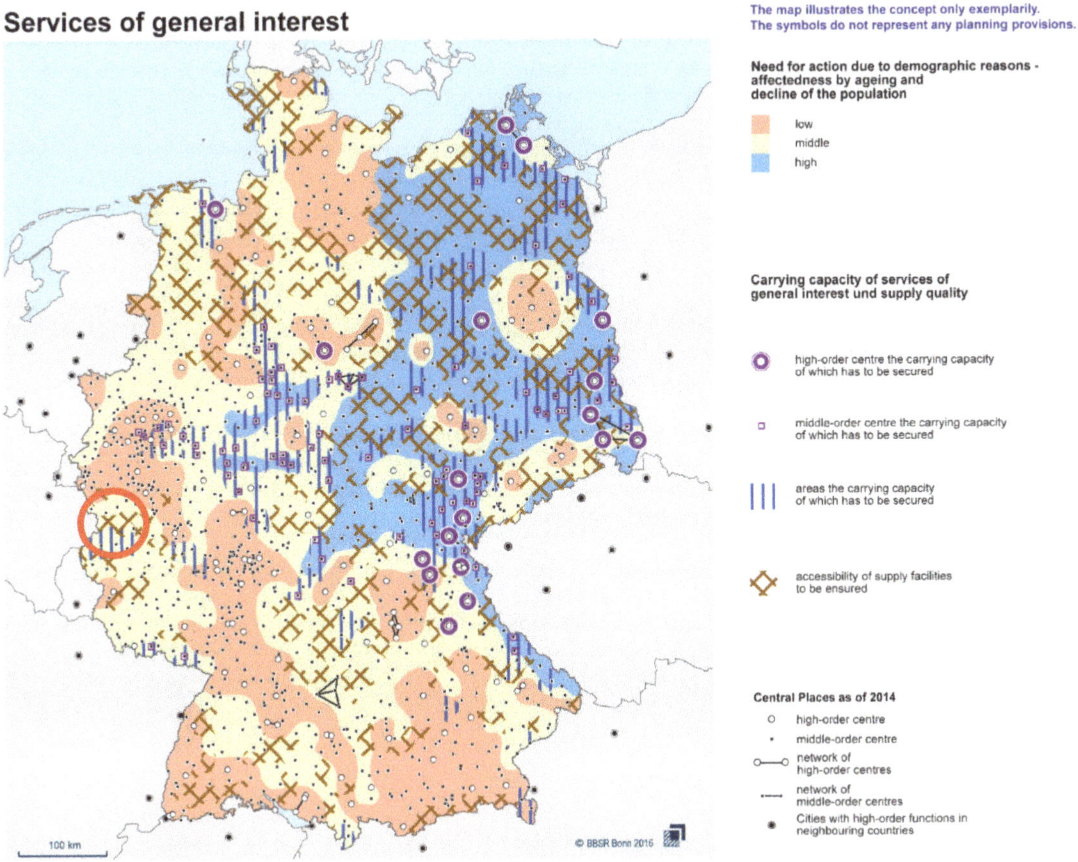

Figure 1. SGI in Germany [9] (p. 20). Reprinted with permission from Ref. [9]. 2016, BBSR.

In 2020, the German Federal Government established an equality check to be applied for all legislative acts on the federal level. The government intends to prove whether and how federal legislative projects affect the equivalence of people's living conditions, for example, reduce, do not affect/consolidate, or reinforce unequal living conditions. For this purpose, among other indicators, the provision of SGI matters. Legislative projects must therefore be examined for their effects, particularly in the areas of local supply; health and care; participation of people with disabilities; reconciliation of family, care and work; childcare and education; security (police, fire department, disaster control) and administrative services [11].

The provision of SGI has a specific local dimension. Art. 28 § 2 GG guarantees the inviolability of local self-government: "The municipalities must be guaranteed the right to regulate all matters of the local community within the framework of the law under their own responsibility." These matters of the local community also encompass the independent provision of local SGI such as health care and education. The role of local communities is also laid down by Art. 8 § 1 Municipal Code of the Federal State of North-Rhine Westphalia of which the Rhenish mining area is a part of: "Municipalities shall create, within the limits of their capacity, the public facilities necessary for the economic, social and cultural care of their inhabitants" (own translation). That is why we focus in this paper on the quality-of-service provision of SGI in a local context by the example of the county of Euskirchen. The results were derived from the research project DAZWISCHEN (future-oriented structural change in the Rhenish lignite mining area). The project examined the structural changes in the region of the Rhenish lignite mining area and their local characteristics, both currently and in the future. In particular, the interrelationships between structural change, settlement and open space development, the mobility transition, the securing of SGI and, finally, impacts of climate change were studied.

This paper focuses on the spatial distribution of SGI. The related research questions are as follows:

1. How are SGI clustered and spatially distributed in the Rhenish coal-mining area?
2. Do all settlements in the Rhenish coal-mining area have equal access to SGI?
3. How can the provision of and access to SGI be improved in sparsely populated areas in order to reduce regional disparities in individual living conditions?

These questions are answered to quantify and, more important, qualify the present SGI in order to identify areas which are, in the context of the territorial cohesion, currently left behind. By acquiring this knowledge this paper aims to contribute towards a strategic policy regarding the access and the quality of SGI in rural areas of the Rhenish coal-mining area, with a special spotlight on the county of Euskirchen. The last questions imply and recommend specific actions to overcome existing inequalities.

Section 1 explains the legal and political framework of SGI within the European Union and Germany specifically and aims to highlight the necessity to achieve equal living standards on a countrywide scale. Afterwards the Materials used to quantify the status of SGI are presented. In addition, Section 2 introduces the scientific method that was used. In Section 3 the achieved results illustrate the current state of SGI within the Rhenish coal-mining area, replenished with an investigation regarding the broadband development. The following discussion compares the results with existing investigations and explains the differences between the outcomes. The conclusion at the end of the paper connects the research with ongoing strategic efforts in the county of Euskirchen to establish a regional development policy addressing existing issues.

2. Materials and Methods

The method used to display the status quo of the accessibility to SGI within the Rhenish coal-mining area is based on so called "kernel-density" estimates. Kernel density estimates have been used since the late 1950s [12,13], as an alternative to representing point distributions in histograms [14]. Previous studies have already used Grids for the detection of intra-community centrality or location clusters [15]. The results depend on the selected cell size (bandwidth), and also on the location of the grid cell network laid over the study area. Compared to this method, the kernel density approach has a clear advantage, because single kernel features can be "superimposed" on each point feature and thus not only the exact position of the point in space is weighted but all spatial units that lie within the range/bandwidth selected for the kernel.

The kernel density approach works in such a way that a "kernel" (density function) is created for each feature to be investigated. Raster cells close to the feature are weighted higher, and the weighting decreases with increasing distance decreases up to the outer bandwidth. The overlapping density functions are summed to produce a kernel density estimate (see Figure 2).

A kernel function is particularly useful for phenomena which have a spatial character such as market areas, a provider such as a hospital or school offers services for. Then, for each grid cell, the kernel density value of the kernels within range is calculated by addition of the overlapping kernels, whereby smaller distances between the points are reflected in increased values.

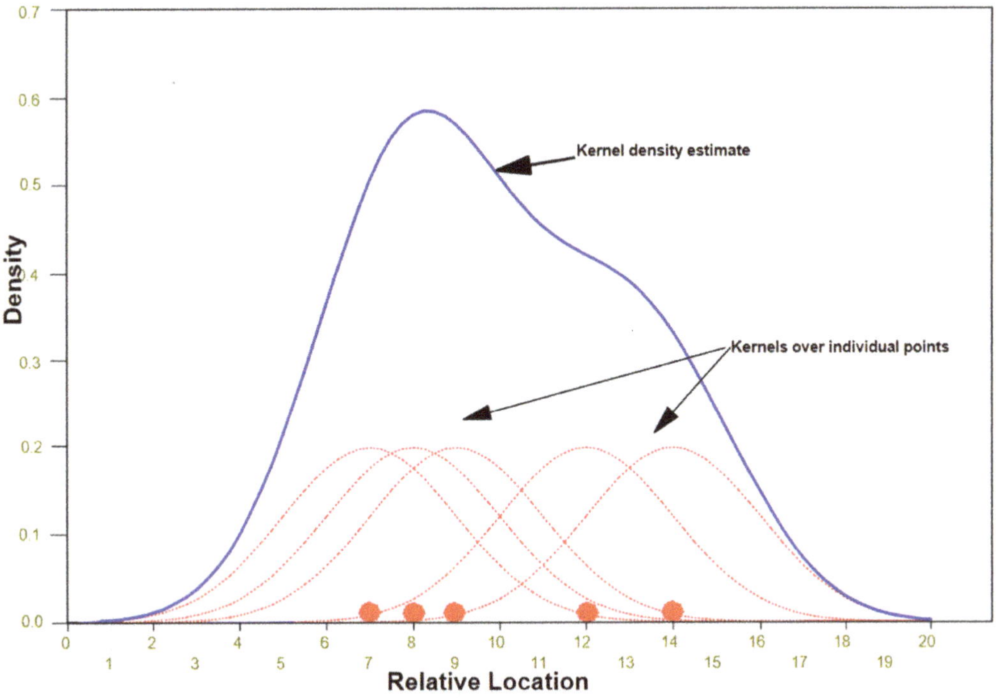

Figure 2. Kernel-density function [14]. Reprinted with permission from Levine [14]. 2019, Levine.

The determined core densities are influenced much more strongly by the bandwidth used than by the shape of the kernel (core function). In this study the bandwidth (100 m × 100 m) was primarily chosen by visual inspection of preresults. In addition, several different bandwidths were discussed. Because the investigation area covers rural areas (such as the county of Euskirchen) same as dense areas, choosing a higher bandwidth could result in a higher quantity of extreme values.

The analysis performed in this case (as explained below) in that respect is not a sole kernel-density estimation., but a combination of two methods. The calculation of the distance is based on the nearest neighbour method while the chosen distance categories imply a kernel-density function. For example, a single cell without any facility types within its expansion is able to achieve high values if the cell is near an accommodation of facilities in adjacent cells. In this specific case the nonexisting facilities within the cell do not account towards the cells rating.

This analysis is based on data concerning several address-based geocoded Points of Interest provided by infas 360, Institut für angewandte Sozialwissenschaften GmbH (project partner in DAZWISCHEN). Including SGI of the health care sector, education, retail food sales, banking and post offices. Prior analyses dataset was checked, and duplicates of single facilities were removed.

To analyse the distribution of SGI in the Rhenish coal-mining area and their accessibility, a method is used which allows a quantification of the gravity of SGI. For this purpose, the zASB were chosen. zASBs are a main component of the regional plan issued by the district government of Cologne and embed in the principle 6.2-1 of the State Development Plan of North-Rhine Westphalia (LEP NRW) [16]. Principle 6.2-1 states that "the expansion of settlements and their development should focus on those areas where a spatial accumulation of SGI (provided by state and private sector) is available" [16] (p. 57).

The area under consideration (Rhenish coal-mining area) is overlaid by a vector raster with a specific cell size of 100 × 100 m using ArcGIS 15. This results in a grid layer with over 500,000 single cells. Due to the topography, not all grid cells have the same size; cells alongside the borders of North-Rhine Westphalia have a nonrectangular expansion. This guarantees that no space within the investigation area is left out of the analysis.

Based on the "zASB" evaluation of the district government of Cologne, a list of infrastructures was chosen. The list includes the following facility types (provided by infas 360):

The chosen types of facilities are important to evaluate the accessibility of SGI, mostly based on the studies of Borchardt and Gericke [17,18]. Each SGI within the Rhenish coal-mining area listed in Table 1 is georeferenced in ArcGIS 15. Some SGI analysed by Borchardt and Gericke are not included such as stops of public transport, because they do not contribute to the gravity of SGI, but their accessibility which is analysed by QGIS.

Table 1. Facility types and used distance categories. Source: own table.

Facility Type	Distance [17]	Distance Category
Physician/medical specialist Dentists Pharmacies Library Post office Hair stylist Food sales	1250 m 1250 m 1250 m 1250 m 1250 m 600 m 500–750 m	up to 500 m (3 points) 501–1500 m (2 points) 1501–3000 m (1 point) >3000 m (0 points)
Hospitals Retirement homes	6000 m (12 km radius) 300–400 m	up to 2000 m (3 points) 2001–5000 m (2 points) 5001–8000 m (1 point) >8000 m (0 points)
Kindergarten	500–600 m	up to 500 m (3 points) 501–1000 m (2 points) 1001–2000 m (1 point) >2000 m (0 points)
Bank Branch/ATM	Not specified	up to 1000 m (3 points) 1001–2500 m (2 points) 2501–3000 m (1 point) >3000 m (0 points)
Schools (all types)	500–4000 m	up to 1000 m (3 points) 1001–2500 m (2 points) 2501–5000 m (1 point) >5000 m (0 points)

Opposite to the zASB analysis by the district government of Cologne, the following facilities are not analysed: sports facilities and public baths were not included in the data provided by infas. Furthermore, types of schools were not differentiated for the same reason mentioned above, and the provided data only contained a single category for all types of schools.

To group the accessibility of different SGI, categories of distances are built and rated from 0 to 3 points (Table 1). The shortest distance of the centroids of each grid cell to each particular SGI is calculated using open software QGIS plugin feature NNJoin (QGIS Version 3.16.10, NNJOin Version 3.1.3 by Håvard Tveite, NMBU). Resulting in a separate layer for each facility including a table with shortest distances of each grid cell centroid to different SGI (Table 1).

This process results in 12 separate grid-based layers. Distances to each SGI are classified into 4 categories (from 0 to 3), where "0 points" represents the furthest distance of grid centroids towards SGI, and "3 points" represents the closest (see Table 1). Categories distribution of four different SGI is displayed in Figure 3. To evaluate the provision of all analysed SGI for each grid cell, points of each grid cell for different SGI were summed up, resulting in maximal 36 points and minimal 0 points. The values are classified into 6 equally distributed classes with values from 0–5, 6–11, 12–17, 18–23, 24–29, 30–36 using Excel 2016. The final values for each grid cell were rejoined into the input grid, to proceed with the final symbolization. A colour design was chosen from red (high score) to blue (low score). The resulting map was joined with districts of the Rhenish coal-mining area to display the accessibility of SGIs by all settlement areas (Figure 4).

In addition, the status quo of broadband development (dataset provided by infas 360 GmbH) was taken into account. Therefore, the availability of different connection quality was investigated. The accessibility ranges from 30 Mbit/s over 100 Mbit/s, 400 Mbit/s to 1000 Mbit/s.

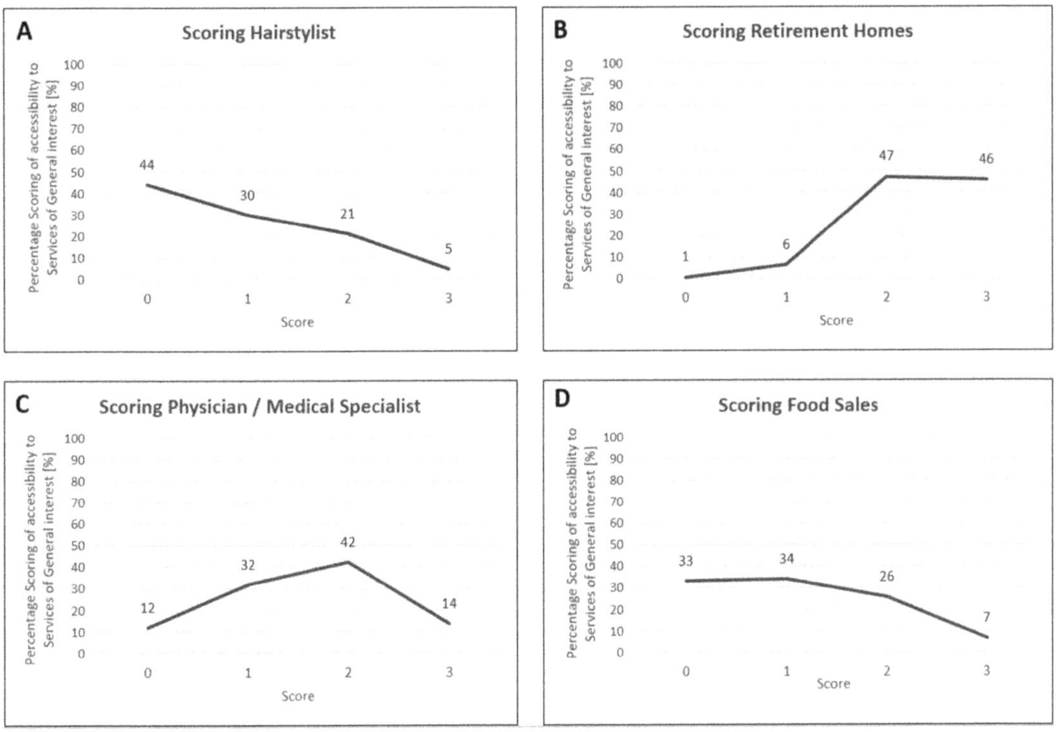

Figure 3. Percentage Scoring of Accessibility to SGI. Source: own figure. (**A**) Scoring of Accessibility to hairstylist within the Rhenish coal-mining area. (**B**) Scoring of Accessibility to Retirement Homes within the Rhenish coal-mining area. (**C**) Scoring of Accessibility to Physicians/Medical Specialist within the Rhenish coal-mining area. (**D**) Scoring of Accessibility to Food sales within the Rhenish coal-mining area.

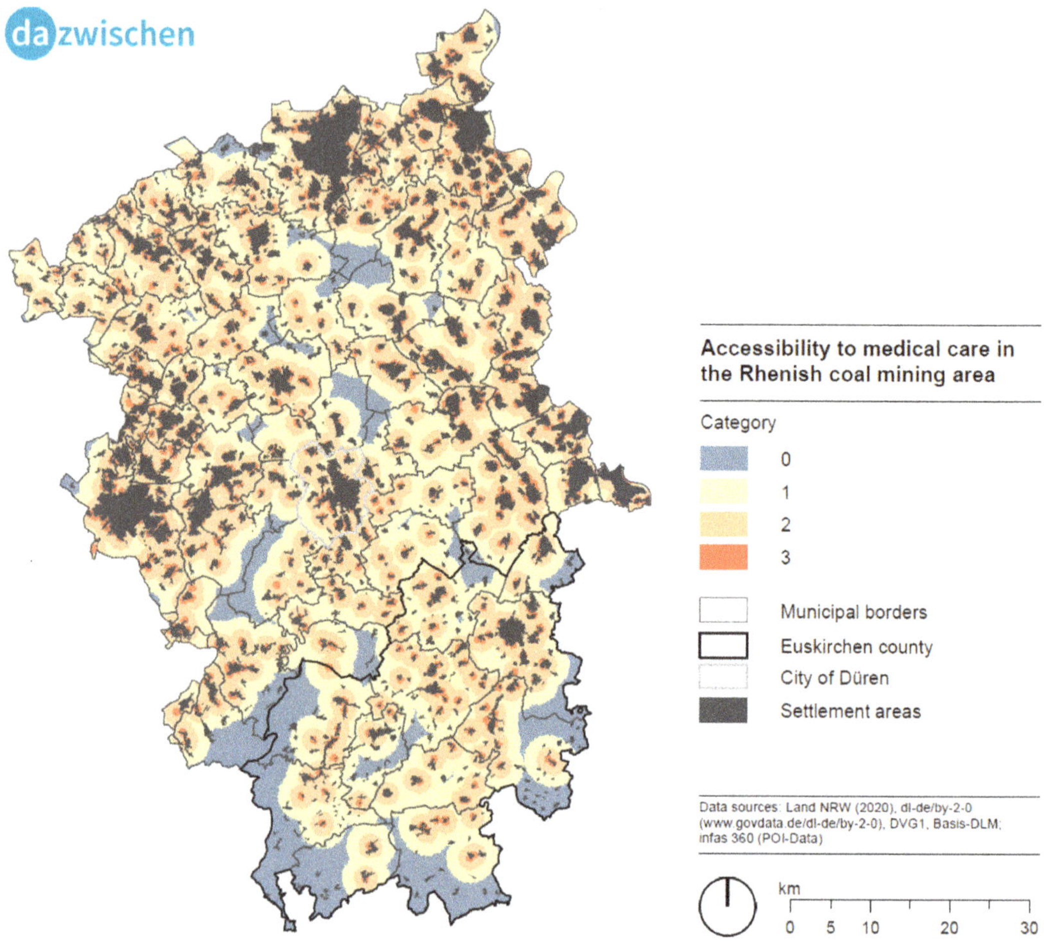

Figure 4. Accessibility to medical care. Source: own figure.

3. Results

The spatial distribution of the different SGI is quite different. For example, the SGI hair stylist shows the highest amount with 44 % of grid cells with the category 4 (longest distance) and only 5% in category 1 (closest distance). On the other hand, retirement homes throughout the Rhenish coal-mining area are up to 46% close to the grid cells centroid, while only 1% are considered to be in category 4 (0 points), which is the lowest amount of all SGI. Of course this can be explained with the distance categories used for the analysis. While hair stylists within <500 m distance to a grid cell centroid are considered to be near, for retirement homes a distance of <2000 m counts as near. On the other hand, the analysis was not only based on settled areas resulting in a wide range of values, too (Figure 3).

The accessibility to food supply and medical care based on the calculated categories shows that the distribution of accessibility to food supply in the Rhenish coal-mining area is in category 1–2 (3–2 points) near to agglomerations while rural parts often belong to category 3–4 (0–1 Points). The same can be observed for the accessibility to medical care with a lower expression (Figures 4 and 5).

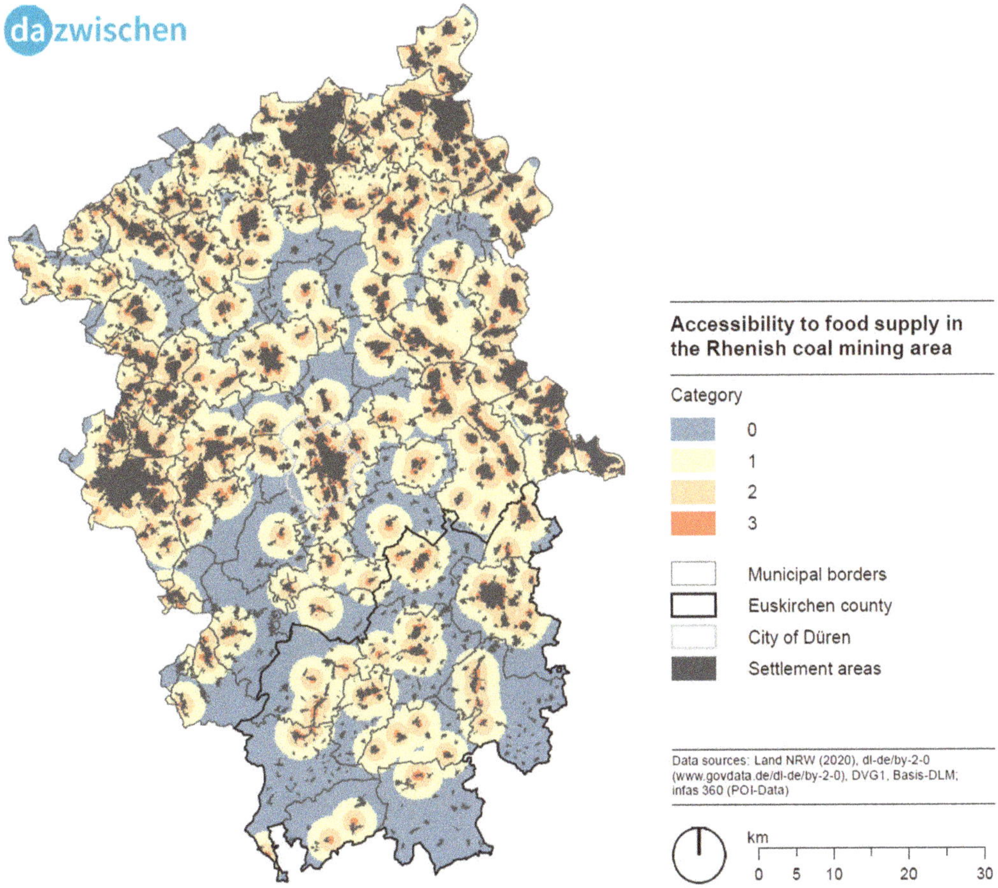

Figure 5. Accessibility to food supply. Source: own figure.

Therefore, it is important to classify the calculated values and display them in the context of the settled areas within the Rhenish coal-mining area. Figure 6 shows that rural areas are more likely in classes 1 or 2 (0–5 points, 6–11 points) while areas next to central districts are more likely in classes 3–6 (12–17 points, 18–23 points, 24–29 points, 30–36 points) (Figure 3). The northern as well as eastern and western parts of the Rhenish coal-mining area show areas next to large cities, which mostly belongs to class 4–6 (18–23 points, 24–29 points, 30–36 points) of accessibility to SGI. The southern part of the investigated area shows more rural parts, which displays a class range from 1–3 (0–5 points, 6–11 points, 12–17 points). In this area, higher classes of accessibility to SGI are more frequently located in centralized settlements next to tiny villages (Figure 6). Above all parts of municipality Hellenthal, Dahlem, Blankenheim and Bad Münstereifel in the southern part of the Rhenish coal-mining area show accessibility classes of 1–2, more frequently than the other parts of this area.

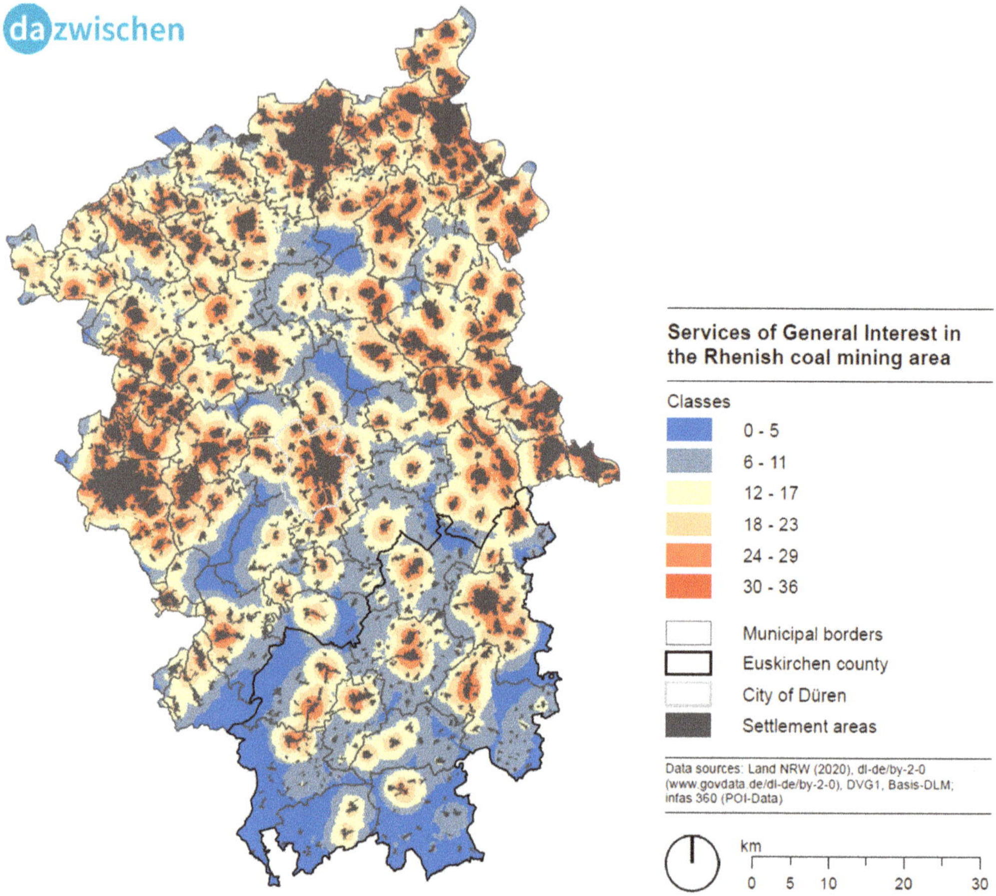

Figure 6. Analysis of accessibility to SGI in the Rhenish coal-mining area. Source: own figure.

Broadband

In the Rhenish coal-mining area the northern parts as well as the western regions alongside the Dutch and Belgium borders and the Rhein-Kreis-Neuss District show a high coverage regarding access to 1000 Mbit/s. From these regions towards the City of Düren, which is located in the centre of the Rhenish coal-mining area, the coverage rate drops significantly. The same can be witnessed around the City Euskirchen and the City Schleiden. A lot of the regions with poor access to fibre optic line are currently used by the extractive industry. The broadband quality in the investigation area (by district) is listed below in descending order:

Rhein-Kreis-Neuss (74%), Stadt Mönchengladbach (68%, the only City in this enumeration), Städteregion Aachen (66%), Kreis-Heinsberg (59%), Rhein-Erft-Kreis (42%), Kreis Euskirchen (41%), Kreis Düren (39%). The average cover rate is 55%.

The county of Euskirchen can be separated into two regions regarding the broadband coverage, showing that in the northwestern parts of the county the coverage is way higher than in the southeastern parts. In general, 41% of all households are covered with a bandwidth of 1000 Mbit/s, and only 18% reach a bandwidth of less than 100 Mbit/s. While in the municipalities of Weilerswist, Dahlem, Stadt Schleiden and the city Euskirchen more than 50% of all households could have access to a fibre optic line, communities such as Blankenheim or Nettersheim reach a coverage of less than 20%.

The broadband coverage with a bandwidth of 1000 Mbit/s in the City of Düren is available for 63% of all households. In the centre of the city and the neighbourhoods Rölsdorf, Derichsweiler, Gürzenich, Arnoldsweiler and Grüngürtel more than 75% are covered with fibre optic line. For the northern and the southwestern parts of the city less than 10% of all households achieve this quality.

4. Discussion

The accessibility to SGIs in rural areas of the Rhenish coal-mining area is limited compared to areas near to agglomerations (Figure 5). Nevertheless, more people move to suburban areas for different reasons. One reason is price development in urban areas [19] and availability of affordable housing. Another reason is the establishment of mobile working opportunities during the COVID-19 pandemic, which resulted in a decoupling of place of residence and working place. Additional preference of living close to nature is an important factor leading people to move into suburban areas. This preference is invigorated during the lockdown periods during the course of the COVID-19 pandemic [20].

In this context it seems to be that the inhabitant's attitude towards a duration of accessibility of a SGI is not that important to some parts of population; other values are more taken into account.

The accessibility to different SGIs varies from region to region. Two indicators are discussed in more detail with a view to the southern rural part of the investigated area, especially the investigated subspace of the county of Euskirchen. Access to food supply in rural and suburban areas is lower than in urban areas (Figure 5). However, rural areas in the southern part of the investigated area are often equipped with directly marketed farm products that are not included in this analysis. This is also important in view of petrol stations, which also often provide SGI in terms of food supply or post offices. They are mostly reachable in <15 min by car in the complete investigated area but have a limited assortment and higher prices than local food suppliers. Additionally, they are often only reachable by car, which limits the accessibility for people without one [21]. Mobile shops and home delivery services are also not included as well as specialized food sales, such as bakeries, butcher or cheese factories, in the analysis, which may lead to lower categories for this indicator.

This assumption is also supported by a comparison of the accessibility of food sales to physicians/medical specialists. The availability of medical support is in categories 3 and 4, higher than the availability of food sales in the background of the same distance categorisation (Table 1, Figures 3–6). The well-structured accessibility to medical care in the Rhenish coal-mining area is in accordance with an analysis of the Association of Statutory Health Insurance Physicians, which emphasize a well-accessible primary health care in comparison to the nationwide average in Germany [22].

Nonetheless, the availability of physicians and medical specialists in southern part is lower compared to the suburban and urban areas around the investigated area (Figure 4). That means the time to reach medical care is higher in this part and is often linked to the usage of a car and longer time to arrange an appointment with a medical specialist. A times, doctors' offices do not find a successor after retirement resulting in a sparseness of offices in rural parts. On average 28.9% of physicians in NRW are aged over 60 years, which may lead to the same development to other parts of the Rhenish coal-mining area as well [22].

Another effect is that medicine graduates like to work part-time, which leads to three follow-ups to two retired physicians being needed [23]. This situation has been tightened in the aftermath devastating flood disaster in July 2021. Centrally located settlement areas were destroyed and were no longer able to provide accessibility to different SGI including critical infrastructures. This is the case for public transport, sales, medical care and power supply. The highest affectedness is found in the county of Euskirchen within the Rhenish coal-mining area, which also already displays the lowest classes of accessibility to SGIs in ordinary times. Nontheless, Küpper and Steinführer point out that across all settlement and spatial types, improvements and deteriorations are perceived with regard to the provision of public services close to the place of residence [24].

Digital platforms become more important in this context. In the course of digitalisation, the availability of different SGIs by such platforms is improved. The services provided by these platforms improve the availability of SGIs in rural areas, as well. These effects can be recognized for local supply with goods but also for basic medical care and education, as well as for social interactions. As these digital services were not investigated in this analysis in detail, the broadband expansion in rural parts of the Rhenish coal-mining area is an important indicator. Against the background, rural areas display a lower class of accessibility to SGIs; the allocation to a well-developed broadband access can be an essential factor to provide a basic supply [25]. Digital platforms can be one way to improve the availability to SGIs in rural parts, which shows the importance of a broadband expansion.

For a valid statement more investigation is needed, because the demographic trend also needs to be included. Older segments of population are often not that familiar with digital platforms and internet applications. This might be induced by missing knowledge of how to use these services but also by a low confidence to digitalization and usage of digital services.

Even though the method used to analyse the structure of SGI within the Rhenish coal-mining area was inspired by the methodological approach, the district government of Cologne follows for the identification of suitable future settlement areas that are characterised by a spatially clustered profile of SGI (in German "zentralörtlich bedeutsame allgemeine Siedlungsbereiche", zASB).

First, the amount of SGI in investigation. Where the district government of Cologne used 16 indicators, the authors of this article selected 12. As already mentioned in Section 2, sports facilities, public baths and stops of public transport are not taken into account, but retirement homes are part of the analysis. Accessibility of retirement homes is important in the context of demographic change above all in the rural parts of the Rhenish coal-mining area, where, for example, the County Euskirchen belongs to, but also in the context of the demographic development of the whole area in future. In addition, there is no distinction between different types of school forms, such as elementary schools, high schools and professional schools. The distances to schools are rated with the same categorisation. The same procedure applies to physician and medical specialists, as explained in Section 2.

Second, it is important to acknowledge that the investigated area does not focus on zASB [26]. Despite the different spatial foci, there are some similarities to be mentioned. The Rhenish coal-mining area consists of the counties Heinsberg, city region Aachen, Düren, Euskirchen, Rhein-Erft and Rhein-Kreis Neuss. Rhein-Kreis Neuss is not part of the government district of Cologne. In addition, the City of Cologne, Rheinisch-Bergischer Kreis, Oberbergischer Kreis and Rhein-Sieg Kreis are not a part of the Rhenish coal-mining area. While these parts (except for Cologne) do have some rural parts to some extent, the City of Cologne (with the highest and densest population) does not count into the analysis. Cologne with its high-ranking supply utilities generates distant values far off from those achievable in rural areas. This can be seen alongside the eastern border of the Rhein-Erft-Kreis (Figure 6).

The Rhenisch coal-mining area includes various parts of two government districts, the district government of Cologne and of Dusseldorf. Both districts have identified zASBs, as it is mandatory according to Principle 6.2-1 State Development Plan of North-Rhine Westphalia (LEP NRW) [16].

What distinguishes both methods further are the chosen indicators and measurement values. For example, the district government of Cologne takes hospitals and gymnasiums into account, while these facilities are not part of the survey in Dusseldorf. On the other hand, Dusseldorf lists travel agencies, citizens' offices and sports grounds, opposite of Cologne [27].

As expected, the validity of the chosen method is limited. One of the main limitations is caused by the borders of the investigation area. All SGI considered by the analysis are located within the Rhinish coal-mining area. Facilities outside of this area are not taken into account, which may lead to lower classes of accessibility to SGI in areas close to the borders. Citizens in these areas are not bound to accomplish their needs within these boarders. If the next (regarding the distance) supermarket is located on the other side of the border, the values may not show the reality of facilities available to inhabitants. The same issue can be found alongside the borders towards the Netherlands and Belgium. In these areas in particular, the distances to certain facilities are not the main driver for choosing an SGI or a SGEI since economic reasons such as lower prices or certain goods are considered to be more important to people than the travel time.

As explained, the decision if a certain area is well supplied or not is based on different concepts. The categories cannot be transferred to any other region in Germany because the analysis and evaluation is not standardized. For example, in the low mountain range "Spessart" (Bavaria, Hesse) the standard for accessibility to a supermarket (full supply) is 15 min by car. In the county of Coburg on the other hand, the benchmark for 60% of all inhabitants in a single city quarter is set to a 5 min drive [28] for the same facility type. For some parts of the Rhinish coal-mining area, above all for the rural areas, a different definition for accessibility maybe more specific targeted.

The provided data sets only contained facilities within the investigation area. Therefore, neither the positive nor the negative side effects of cross-border interactions to fulfil the daily needs of SGIs have impacts on the results.

However longer distances to reach SGIs in rural areas, of course, also induce that nearly two-thirds of all routes are covered by using a car [29]. This is also supported by disadvantageous station times in rural areas [30].

Although the county of Euskirchen is not directly affected by the lignite mining phase-out according to the definition of the EEFA, since there are no workplaces directly related to brown coal production or workplaces of RWE AG (coal mine operator and main energy supplier in western Germany) in the district, the occupation of energy-intensive companies with 16% of the employees subject to social security contributions nevertheless harbours an indirect impact [31,32]. Therefore, the loss of direct or indirect jobs will have an adverse effect on purchasing behaviour in the county of Euskirchen. A decrease in these expenditures may lead to a further decrease in jobs, which currently still benefit from the income-induced effects (e.g., retail trade) [32]. The provision services of general interest in the county of Euskirchen may therefore also be affected by the phase-out of lignite mining and, as a consequence, deteriorate.

Based on the analysis carried out along with the inclusion of various concepts of the county of Euskirchen, such as the climate protection concept, the climate change adaptation strategy and the sustainability strategy, the DAZWISCHEN project provides the opportunity to create a strategy specifically customised for the region. This strategy offers the framework for a positive development of the county of Euskirchen with regard to the provision of SGI within the structural change. Reitzenstein et al. point out that achievements for one region are hardly transferrable to other (coal) regions due to very distinct characteristics within a country [33]. Only building on the strength and assets of the region at hand can provide a successful structural change process [33]. This is based on structural differences such as the number of inhabitants, density or the amount energy-intensive companies and the industrial workforce. Therefore, the established strategy and associated projects are due to the governance-based approach suitable for a future orientated and sustainable handling of lignite mining phase out based structural change [34]. According to Art. 72 § 2 GG, a contribution to the preservation of equal living conditions is to be made by strengthening the individual responsibility for the development of regions. Based on Kötter et al. and Kersten et al., the inclusion of citizens in "strategies with cooperative approaches and private initiatives" can therefore contribute an essential part of future settlement development [35,36]. The developed strategy is suitable to improve the provision of SGIs by the identification of low accessibility to different services in rural parts on the one hand and development of governance based projects with relation to public services on the other hand.

5. Conclusions

The results we presented in this paper will be used for setting up a regional development strategy. One guidance principle is the maintenance of equal living conditions in the entire region and the reduction of intraregional disparities in individual living conditions. This requires a sound accessibility of SGI which is partly questioned in some southern parts of the region. In this context, reference has been made to the aforementioned four priority actions of the "Concepts and Strategies for Spatial Development in Germany":

- Consistent application of central system: the methodological approach the district government of Cologne follows for the identification the settlement areas stipulates the allocation of future settlement areas in the vicinity of infrastructure clusters.
- Develop cooperative systems among municipalities
- Ensure the supply of sparsely populated rural areas with SGI
- Ensure accessibility by maintaining and improving the quality of public transport services

The central-location system of the district government is pursued by the strategy, since its focus is primarily intended to strengthen the provision of public services in the respective municipalities. Due to the fact that the strategy was developed in a participatory approach involving the different stakeholders, the system can be considered as a cooperative system among municipalities.

Sparsely populated areas within the county are considered within the strategy along with corresponding projects. Food vending machines are an example of this approach aiming to improve local supply and thus the provision of public services in rural areas.

In order to maintain and improve the quality of public transportation, projects such as the electrification of certain rail roads as well as a revised, tighter scheduling of trains are already being planned in order to improve the linkage between the individual towns and villages in the country where possible. The expansion of on-demand buses or car-sharing possibilities within sparsely populated regions are also being implemented.

As shown above, a regional development strategy can lead to a governance-based system between municipalities, which enables the region to identify sparsely supplied regions and can prevent a further decline of SGI in rural areas through specific future-orientated projects. Addressing disadvantageous situations on a countywide level using a cooperative policy and strategic approach can therefore provide an important keystone to encounter the problems of the structural change the county has to face in the near future.

Further research needs are related to the potential, partly deeply uncertain changes in the quality and accessibility due to demographic change, digitalisation and transport. Changes in location and quality of service provisions which are mostly privately operated cannot be precisely predicted. Transportation is considerably influenced by political decisions and related funding programs, but also marked-based factors such as energy prices. In consequence, scenario corridors seem to be appropriate for an identification of a bandwidth of potential future evolving conditions.

Author Contributions: Conceptualization, S.R. and S.G.-R.; Methodology, S.R. and S.G.-R.; Validation, S.R., S.G.-R. and S.G..; Formal Analysis, S.G.-R. and S.R.; Investigation, S.R. and S.G.-R.; Resources, County Euskirchen, infas and LLPD; Data Curation, S.G.-R. and S.R.; Writing—Original Draft Preparation, S.G.-R., S.R. and S.G., Writing—Review and Editing, S.G. and K.R.; Visualization, S.G.-R., County Euskirchen; Supervision, S.G.; Project Administration, S.G.; Funding Acquisition, S.G. All authors have read and agreed to the published version of the manuscript.

Funding: This research was funded by the Federal Ministry of Education and Research (BMBF) within the framework of the Stadt-Land-Plus funding programme under the grant agreement number 033L217A/B/C/E/F (German BMBF Funding Activity Urban-Rural).

Institutional Review Board Statement: Not applicable.

Informed Consent Statement: Not applicable.

Data Availability Statement: The data presented in this study are available on request from thecorresponding author.

Conflicts of Interest: The authors declare no conflict of interest. The funders had no role in the design of the study; in the collection, analyses, or interpretation of data; in the writing of the manuscript; nor in the decision to publish the results.

References

1. European Commission. My Region, My Europe, Our Future. Seventh Report on Economic, Social and Territorial Cohesion. 2017. Available online: https://ec.europa.eu/regional_policy/sources/docoffic/official/reports/cohesion7/7cr.pdf (accessed on 28 April 2022).
2. Davoudi, S. Understanding Territorial Cohesion. *Plan. Pract. Res.* **2005**, *20*, 433–441. [CrossRef]
3. Medeiros, E. Territorial Cohesion: An EU concept. *Eur. J. Spat. Dev.* **2016**, *60*, 1–30. Available online: https://www.researchgate.net/publication/302580113_Territorial_Cohesion_An_EU_concept (accessed on 28 April 2022).
4. European Commission. *A Quality Framework for Services of General Interest in Europe. Communication from the Commission to the European Parliament, the Council, the European Economic and Social Committee and the Committee of the Regions*; European Commission: Brussels, Belgium, 2011; p. 3.
5. Akademie für Raumforschung und Landesplanung (ARL). Daseinsvorsorge und Gleichwertige Lebensverhältnisse neu Denken–Perspektiven und Handlungsfelder. Hannover. Positionspapier aus der ARL 108. 2016. Available online: http://nbn-resolving.de/urn:nbn:de:0156-01086 (accessed on 28 April 2022).
6. Dutch Presidency. Urban Agenda for the EU Pact of Amsterdam. Agreed at the Informal Meeting of EU Ministers Responsible for Urban Matters on 30 May 2016 in Amsterdam, The Netherlands. 2016, p. 8. Available online: https://ec.europa.eu/regional_policy/sources/policy/themes/urban-development/agenda/pact-of-amsterdam.pdf (accessed on 28 April 2022).
7. Bundesministerium des Inneren und für Heimat. Territorial Agenda 2030. A Future for All Places. Adopted at Informal Meeting of Ministers Responsible for Spatial Planning and Territorial Development and/or Territorial Cohesion 1 December 2020, Germany. 2020, p. 8. Available online: https://territorialagenda.eu/wp-content/uploads/TA2030_jun2021_en.pdf?_cldee=dGlpdS5tYWRhbEBlbGwuZWU%3D&recipientid=contact-3239f1f44373e51180d6005056a00920-42045b934c25478f9856e15b17dec456&esid=b1cb8571-0a2c-ec11-8118-005056a043ea (accessed on 28 April 2022).
8. Greiving, S.; Flex, F.; Terfrüchte, T. Vergleichende Untersuchung der Zentrale-Orte-Konzepte in den Ländern und Empfehlungen zu ihrer Weiterentwicklung. *Raumforschung und Raumordnung—Spat. Res. Plan.* **2015**, *73*, 285–297. [CrossRef]
9. Bundesinstitut für Bau-, Stadt- und Raumforschung im Bundesamt für Bauwesen und Raumordnung (BBSR). *Concepts and Strategies for Spatial Development in Germany. Decision of the 41st Standing Conference of Ministers responsible for Spatial Planning in Berlin 09 March 2016*; Secretariat of the Standing Conference of Ministers Responsible for Spatial Planning Federal Ministry of Transport and Digital Infrastructure (BMVI): Berlin, Germany, 2016. Available online: https://www.bbsr.bund.de/BBSR/DE/themen/_alt/Raumentwicklung/RaumentwicklungDeutschland/Leitbilder/leitbildbroschuere-englisch.pdf?__blob=publicationFile&v=1 (accessed on 28 April 2022).
10. Bundesministerium des Innern, für Bau und Heimat. Unser Plan für Deutschland. Gleichwertige Lebensverhältnisse überall. Schlussfolgerungen von Bundesminister Horst Seehofer als Vorsitzendem sowie Bundesministerin Julia Klöckner und Bundesministerin Dr. Franziska Giffey als Co-Vorsitzenden zur Arbeit der Kommission "Gleichwertige Lebensverhältnisse". Berlin, Germany, 2019; p. 23. Available online: https://www.bmi.bund.de/SharedDocs/downloads/DE/veroeffentlichungen/themen/heimat-integration/gleichwertige-lebensverhaeltnisse/unser-plan-fuer-deutschland-langversion-kom-gl.pdf?__blob=publicationFile&v=4 (accessed on 28 April 2022).
11. Bundesministerium des Inneren und für Heimat. Leitfaden zur Durchführung des „Gleichwertigkeits-Checks" (GL-Check) bei Gesetzesvorhaben des Bundes. 2020, p. 3. Available online: https://www.bmi.bund.de/SharedDocs/downloads/DE/veroeffentlichungen/themen/heimat-integration/gleichwertige-lebensverhaeltnisse/gleichwertigkeits-check.pdf?__blob=publicationFile&v=1 (accessed on 28 April 2022).
12. Parsen, E. On the estimation of a probability density and mode. *Ann. Math. Stat.* **1962**, *33*, 1065–1076. [CrossRef]
13. Gatrell, A. Spatial point pattern analysis and its application in geographical epidemiology. *Trans. Inst. Br. Geogr.* **1996**, *21*, 256–274. [CrossRef]
14. Levine, N. CrimeStat: A Spatial Statistics Program for the Analysis of Crime Incident Locations (v 4.02). Houston (TX), Washington (DC). 2019. Available online: https://nij.ojp.gov/sites/g/files/xyckuh171/files/media/document/CrimeStat%2520IV%2520Chapter%252010.pdf (accessed on 28 April 2022).
15. Boustedt, O.; Lellau, W. Zentrale Standorte zur Versorgung der Bevölkerung in Hamburg 1961 *HiZ*, 1970, H. Sonderheft 1.
16. Ministerium für Wirtschaft, Innovation, Digitalisierung und Energie des Landes Nordrhein-Westfalen. Landesentwicklungsplan Nordrhein-Westfalen (LEP NRW). 2020. Available online: https://landesplanung.nrw.de/system/files/media/document/file/20201104_druckversion_lep.pdf (accessed on 28 April 2022).
17. Gerike, R. Wie Kann das Leitbild Nachhaltiger Verkehrsentwicklung Konkretisiert Werden?—Ableitung Grundlegender Aufgabenbereiche. Ph.D. Dissertation, Technische Universität Dresden, Dresden, Germany, 12 April 2005. Available online: https://tud.qucosa.de/api/qucosa%3A24597/attachment/ATT-0/ (accessed on 28 April 2022).

18. Borchardt, K. *Der Flächenbedarf der Siedlung und ihre Folgekosten*; Institut für Städtebau und Wohnungswesen der Deutschen Akademie für Städtebau und Landesplanung: München, Germany, 1974.
19. IT NRW. Verbraucherpreisindex für NRW – Dezember 2021, Artikel-Nr. M123 2021 12, Kennziffer M I–m 12/21. Verbraucherpreisindex für Nordrhein-Westfalen Dezember 2021 (nrw.de). Available online: https://www.it.nrw/statistik/eckdaten/verbraucherpreisindex-fuer-nrw-1891 (accessed on 28 April 2022).
20. Dolls, M.; Mehles, J. Wie beeinflusst die Corona-Pandemie die Wohnortpräferenzen?—Evidenz aus einer großangelegten Umfrage in Deutschland. *ifo Schnelld.* **2021**, *74*, 27–31. Available online: https://www.ifo.de/DocDL/sd-2021-08-dolls-suburbanisierung_1.pdf (accessed on 28 April 2022).
21. Neumeier, S. Street patrol station shops as an alternative to missing local food suppliers—Contribution to the German discourse on 'Daseinsvorsorge': A German view. *Appl. Geogr.* **2015**, *60*, 150–164. [CrossRef]
22. Kassenärztliche Vereinigung Nordrhein (KVNO). Bedarfsplan zur Sicherstellung der Vertragsärztlichen Versorgung in Nordrhein-nach § 99 SGB V nach Maßgabe der Bedarfsplanungs-Richtlinie (BPL-RL) in der Neufassung vom 20. Dezember 2012 Zuletzt Geändert am 17. Dezember 2020. Erstellt von der Kassenärztlichen Vereinigung Nordrhein im Einvernehmen*mit den Landesverbänden der Krankenkassen und den Ersatzkassen in Nordrhein. 2021. Available online: https://www.kvno.de/fileadmin/shared/pdf/online/amtliche_bekanntmachungen/2022/20220110_bedarfsplan.pdf (accessed on 28 April 2022).
23. Advisory Council on the Assessment of Developments in the Health Care System. Needs-Based Health Care: Opportunities for Rural Regions and Selected Health Care Sectors. Report 2014, Abridged Version. 2014. Available online: https://www.svr-gesundheit.de/fileadmin/Gutachten/Gutachten_2014/Kurzfassung_engl_2014.pdf (accessed on 6 January 2022).
24. Küpper, P.; Steinführer, A. Daseinsvorsorge in Ländlichen Räumen Zwischen Ausdünnung und Erweiterung: Ein Beitrag zur Peripherisierungsdebatte. *Eur. Reg.* **2017**, *23*, 44–60. Available online: https://www.ssoar.info/ssoar/bitstream/handle/document/53589/ssoar-europareg-2017-4-kupper_et_al-Daseinsvorsorge_in_landlichen_Raumen_zwischen.pdf?sequence=1&isAllowed=y&lnkname=ssoar-europareg-2017-4-kupper_et_al-Daseinsvorsorge_in_landlichen_Raumen_zwischen.pdf (accessed on 28 April 2022).
25. Busch, C. *Regulation of Digital Platforms as Infrastructures for Services of General Interest*; WISO Diskurs 09/21 Friedrich-Ebert Stiftung Division of Economic and Social Policy; Division for Economic and Social Policy: Bonn, Germany, 2021. Available online: https://library.fes.de/pdf-files/wiso/17836.pdf (accessed on 28 April 2022).
26. Bezirksregierung Köln. Region+ Wohnen. Verteilung Regionaler Wohnbaulandbedarfe. *Regionalforen* **2018**, *2018*, 57. Available online: https://www.bezreg-koeln.nrw.de/brk_internet/leistungen/abteilung03/32/regionalplanung/ueberarbeitung/regionalplan_ueberarbeitung/region_plus/wohnen/praesentation_koeln.pdf (accessed on 28 April 2022).
27. Droste, B.; van Gemmeren, C. Praktische Erfahrungen mit zASB in der Planungsregion Düsseldorf. In *Neuaufstellung des Zentrale-Orte-Konzepts in Nordrhein-Westfalen*; Greiving, S., Flex, F., Eds.; ARL-Akademie für Raumforschung und Landesplanung: Hannover, Germany, 2016; pp. 156–168. Available online: https://nbn-resolving.de/urn:nbn:de:0156-4035111 (accessed on 28 April 2022).
28. Kaether, J.; Dehne, P.; Neubauer, A. *Regionalstrategie Daseinsvorsorge–Leitfaden für die Praxis*; Bundesministerium für Verkehr und Digitale Infrastruktur (BMVI): Berlin, Germany; Bundesinstitut für Bau-, Stadt- und Raumforschung (BBSR) im Bundesamt für Bauwesen und Raumordnung (BBR): Bonn, Germany, 2016; p. 82. Available online: https://www.bbsr.bund.de/BBSR/DE/veroeffentlichungen/ministerien/bmvi/verschiedene-themen/2016/regionalstrategie-daseinsvorsorge-leitfaden-dl.pdf?__blob=publicationFile&v=4 (accessed on 28 April 2022).
29. Ermes, B.; Follmer, R. Identification of today's and tomorrow's mobility patterns in the Rhenish coal-mining area. *J. Environ. Plan. Manag.* 2022; *submitted*.
30. Küpper, P.; Steinrück, B. Mobilität in Ländlichen Räumen unter Besonderer Berücksichtigung Bedarfsgesteuerter Bedienformen des ÖPNV. Arbeitsberichte aus der vTI-Agrarökonomie. 2010, Volume 2. Available online: https://literatur.thuenen.de/digbib_extern/bitv/dk043302.pdf (accessed on 28 April 2022).
31. EEFA–Energy Environment Forecast Analysis GmbH & Co. KG. Bedeutung der rheinischen Braunkohle–Sektorale und Regionale Beschäftigungs- und Produktionseffekte. Untersuchung im Auftrag der RWE Power AG. 2010. Available online: https://docplayer.org/25874783-Bedeutung-der-rheinischen-braunkohle-sektorale-und-regionale-beschaeftigungs-und-produktionseffekte.html (accessed on 6 January 2022).
32. Gastes, N.; Mahnke, L.; Mickartz, K.; Salden, O.; Walkenbach, N. Betroffenheit des Kreises Euskirchen Durch den Braunkohleinduzierten Strukturwandel im Rheinischen Revier. AGIT Gründen.Ansiedeln.Fördern. 2019. Available online: https://sdnet.kreis-euskirchen.de/sdnetrim/UGhVM0hpd2NXNFdFcExjZY_-a4dMc6sVU_c01xLZBTf2_PHazZvAswtbqRRebwZC/TOP_6_2_Betroffenheitsanalyse_Agit.pdf (accessed on 28 April 2022).
33. Reitzenstein, A.; Popp, R.; Oei, P.Y.; Brauers, H.; Stognief, N.; Kemfert, C.; Kurwan, J.; Wehnert, T. Structural Change in Coal Regions as a Process of Economic and Social-Ecological Transition: Lessons Learnt from Structural Change Processes in Germany. 2022. Available online: https://www.umweltbundesamt.de/publikationen/structural-change-in-coal-regions-as-a-process-of (accessed on 28 April 2022).
34. Weingarten, P.; Steinführer, A. Daseinsvorsorge, gleichwertige Lebensverhältnisse und ländliche Räume im 21. Jahrhundert. *Z. Politikwiss.* **2020**, *30*, 653–665. [CrossRef]

35. Kötter, T.; Berend, L.; Drees, A.; Kropp, S.; Linke, H.-J.; Lorig, A.; Reuter, F.; Thiemann, K.-H.; Voß, W.; Weitkamp, A. Land- und Immobilienmanagement-Begriffe, Handlungsfelder und Strategien. *ZfV* **2015**, *3*, 136.
36. Kersten, J.; Neu, C.; Vogel, B. *Demografie und Demokratie. Zur Politisierung des Wohlfahrtstaates*; Hamburger Edition HIS Verlagsges. mbH: Hamburg, Germany, 2012; p. 78.

land

MDPI

Article

Post-Coal Fantasies: An Actor-Network Theory-Inspired Critique of Post-Coal Development Strategies in the Jiu Valley, Romania

Sebastian Țoc [1] and Filip Mihai Alexandrescu [2,*]

1 Research Institute for Quality of Life, Romanian Academy, National University of Political Science and Public Administration, 012104 Bucharest, Romania; sebastian.toc@politice.ro
2 Research Institute for Quality of Life, Romanian Academy, 050711 Bucharest, Romania
* Correspondence: filip.alexand@iccv.ro

Abstract: Romania is among the countries with a long history of coal mining, which intensified in the context of the industrialization and urbanization policies during the socialist period. During the post-socialist transition, mining units decreased considerably, with those considered economically unprofitable being restructured and eventually closed. This process, which began in the late 1990s, continues today with the increasing discussions about the transition from coal in Romania, the commitment to reduce greenhouse gas emissions, and the signing of the Paris Agreement, which stipulates the abandonment of coal by 2030. The Jiu Valley region is Romania's most important mono-industrial area that developed exclusively around coal extraction. This article proposes an analysis of the social impact of reducing mining activity in this area based on actor-network theory. Recently, the discussions on the "Just Transition" in the Jiu Valley have culminated with the "Jiu Valley Economic, Social, and Environmental Development Strategy" (2021–2030) and the related Action Plan. This paper seeks to address the question: how were people "brought in" for the production of coal during socialism, how were they subsequently "pushed out", and how challenging is it now for the Jiu Valley strategy to reassemble a network to support the post-coal transition? Our analysis helps to understand, in an innovative way, the challenges of the Jiu Valley Strategy as a mismatch between a creative and entrepreneurial city model on the one hand and the ruralized/small urban (former mono-industrial) conditions prevailing in the Jiu Valley on the other. We thus provide a critical analysis of the strategy as one that is conceptually wedded to the large urban center model and applies the just transition framework in a decontextualized way. We emphasize the role of the dismantling of coal production in the narrowing of the space of possibilities of change brought about by the just transition. Using the published literature on the Jiu Valley, we also show what broader lessons can be drawn from this case for similar transition processes in Central and Eastern Europe.

Keywords: actor-network theory; just transition; development strategy; monoindustrial urban area; Jiu Valley

Citation: Țoc, S.; Alexandrescu, F.M. Post-Coal Fantasies: An Actor-Network Theory-Inspired Critique of Post-Coal Development Strategies in the Jiu Valley, Romania. *Land* **2022**, *11*, 1022. https://doi.org/10.3390/land11071022

Academic Editor: Dagmar Haase

Received: 21 May 2022
Accepted: 4 July 2022
Published: 6 July 2022

1. Introduction

The transition away from coal has been the focus of much research and policy discussions, most of them based on the just transition framework [1–8]. Our contribution aims to add to the critical reading of the just transition. First, we follow Brown and Spiegel [9] in recognizing "the complex entanglements of coal within a wider political economy and the symbolic effects it produces in particular places" (p. 151). This means that the starting point of the post-coal transition is not a clean slate but rather a complicated process quite different from the previous era of coal production. On the other hand, decarbonization is itself a challenge, being intrinsically entangled in economics, politics, geography, culture, and knowledge [10]. Our work seeks to bridge the two dominant foci of the current literature—the one that emphasizes the continued but problematic relevance of coal and the other that highlights the challenges of post-coal development—by looking at the *coal-to-post-coal* transition. This allows a better understanding of the "big picture" of the before and after of the coal transition in a particular coal mining area that is the Jiu Valley in Romania. To this end, we use actor-network theory (or ANT, for short). We use ANT in the version inspired by Michel Callon and John Law [11–14]. Our aim is to show what development has meant during the time of intense coal production (up until 1990), how the closure of the mines has dismantled economic life in the Jiu Valley beginning in the late 1990s, and the challenges that the new Jiu Valley post-coal strategy (2021–2030) is likely to face. We identify these three periods as stage I, stage II, and stage III, respectively.

The history of the Jiu Valley, especially its dramatic turn after 1990, illustrates how "the end of the Cold War launched one of the most brutal economic phases of the modern era") [15] (p. 18). This brutal dismantling of manufacturing is massively exemplified by the contraction of the local economy in the Jiu Valley. Saskia Sassen uses the term expulsions to describe the "move from Keynesianism to the global era of privatizations, deregulation, and open borders for some, [which] entailed a switch from dynamics that brought people into dynamics that push people out" [15] (p. 211). The goal of the paper is to explain how people "were brought in" for the intensified production of coal, how they were subsequently "pushed out", and how challenging it is now for the Jiu Valley strategy to reassemble what has been shattered. We trace and explain these processes in terms of the making and unmaking of actor networks based on secondary sources of data and on the available literature.

In short, we argue that the first stage in the development of the Jiu Valley saw a tightly integrated network in which the production of "more coal for the fatherland" [16] (p. 298) played a central role. This network was unraveled when the mines were deemed "unprofitable" or "loss-making" [17]. Unravel means here that the central goal of coal production was progressively reduced (from 66 million tons in 1989 to 1 million in 2021) [18], and the different participants in the network coalesced around new, micro-level networks. This process has taken shape since 1997 and has rendered a regional development network—similar to the one during socialism—increasingly unattainable. From this point of view, the proposed Jiu Valley strategy (henceforth *JVS*) attempts to reassemble a new single network around a post-coal translation of development. Based on actor-network theory, we provide a critique of the *JVS* [19] by showing that it problematically tries to straddle the fragmented networks left in the wake of the collapse of coal production and an untried model of creative urban development fostering a just transition. The *JVS* is located at the intersection of several discursive formations, such as the just transition literature [3,20] but also the creative city [21,22] and the entrepreneurial city [23].

The paper continues with an outline of the theoretical framework based on actor-network theory and tailored to the analysis of the coal/post-coal transition. It then discusses the methodology and describes the case before delving into the results and their interpretation. The latter section is structured into three parts, dealing with the coal-production actor network of the planned economy (until 1990), the contemporary hiatus in development (1997–present), and the future envisioned by the *JVS*. The concluding remarks are provided in the last section of the paper.

2. Theoretical Background: Stages of Actor (Mis)Alignment

This paper relies on a broad conceptual framework inspired by actor-network theory or ANT [12,13,24] to make sense of the development path of the Jiu Valley. This path is divided into three periods, namely socialism, after socialism, and under the current development scenario proposed by the Jiu Valley (post-coal) strategy.

We have chosen the ANT framework given its wide applicability [25] and, in particular, for the attention it pays to the idea of translation. First, translation means placing in specific relationships "heterogeneous entities" to constitute an actor-world by assigning, to each entity, "an identity, interests, a role to play, a course of action to follow, and projects to carry out" [12] (see also [14] for an application to mining). The entities can be social actors, institutions, symbols, but also material objects and infrastructures. The coal-producing region known as the Jiu Valley can thus be seen as a techno-economic network, which is "a coordinated set of heterogeneous actors which interact more or less successfully to develop, produce, distribute and diffuse methods for generating goods and services" [24] (p. 133). The principal good produced by the Jiu Valley techno-economic network has been, over the last 160 years, mostly pit coal (in Romanian, *huilă*). To achieve production, and a given level of production at that, coal deposits, extraction technologies, miners and engineers, roads, and towns had to be brought into specific relationships around the goal of producing coal. This was no easy feat, as the translation was achieved through "creating convergences and homologies by relating things that were previously different" [26] (p. 211). The translation is thus a process of ordering and controlling others, but it is always prone to disruption [14]. To understand what this ordering entails and how it can fail, ANT describes four moments of translation: problematization, interessment, enrolment, and mobilization [11]. Outlining these moments is important in order to explain how development—as a conscious strategy of social actors—is enabled or disabled by various network arrangements.

Problematization occurs when a set of initiating actors, such as politicians and technocrats, define a problem that interests a broader collective. Such a problem may be the need to develop the heavy industry in a largely agrarian economy. In the case under discussion, the problem emerged historically in post-war Romania. Boosting the centers of steel production in this country required the securing of raw materials, especially coal [27]. Over time, a network was established in which the coal mines played an increasingly important, albeit secondary, role in the establishment of the heavy industry as a central goal of economic development. In ANT language, the mines were assigned distinct identities and were attributed specific roles to support the maximization of coal production. As coal became indispensable for industrialization, coal production turned into what ANT-inspired authors call an "obligatory passage point", that is, a "strategic point through which the actor world must pass" [12] (p. 27). Economic and social life became organized around coal production, and its different elements (the mining workforce, the technologies of extraction, the administrative resources, etc.) had to be "persuaded to identify with the network" [28] (p. 361). Despite the mild language, bringing in line different actors of the kind mentioned above always involves a power struggle. "Alliances are normally inequitable [...], with partners relying to different degrees on their relationship" [14] (p. 810).

Getting actors to agree to become jointly involved in intensified coal production is only one side of the task of mine developers. The other side is revealed when testing whether the involvement has actually taken place. This is the second moment of establishing the actor network, that of interessement, which refers to "the group of actions by which an entity [the coal producers] attempts to impose and stabilize the other actors it defines through its problematization" [11] (pp. 207–208). Interessement means interposing effective barriers between the actors enticed to participate in the network and other entities that might lay similar claims on them. Interessement devices can be different intermediaries, such as pieces of law, financial incentives, infrastructures, or money [24], that are meant to dissociate the actors from alternative problematizations. Interessement can take the form of providing incentives for mine workers (such as high salaries and subsidized housing) to decrease the appeal of alternative economies (such as agriculture).

As the third moment of the actor network, enrolment is about defining and coordinating the roles in the network once these have been accepted. "To describe enrolment", Callon states, "is to describe the group of multilateral negotiations, trials of strength, and tricks that accompany the interessement and enable them to succeed" [12] (p. 211). Applied to the production of coal, the different actors need to work together to maximize the production of this raw material. Negotiations are needed because the actors who have been drawn into coal production have given up alternative networks and so need to reach a modus vivendi [12] (p. 213) with the developers and others. Trials of strength are required because actors do not remain in the network by mere inertia but have to be actively kept in. Evidence for this is the history of coal miners' strikes and of their sometimes-brutal suppression, both worldwide and in the case under discussion (as illustrated in Section 4.1).

Finally, mobilization is about setting the actors into motion. Achieving the top production of coal requires the displacement of different actors and their transfer in the space of the mine into mono-industrial towns. It requires a set of complex transformations whereby actors' capacities and needs are adjusted—via specific institutions such as "work battalions"—to the requirements of intensified coal production. Mobilization is important because from the mass of dispersed and relatively inaccessible entities, from this amorphous configuration of actors, the developers need to subordinate everything to the unique goal of the mine.

However, mobilization is by necessity restricted: not all actors' needs and expectations are allowed into the translation. For this reason, it is necessary to establish representatives or spokespersons who use chains of intermediaries going back from the mass of miners and their families to their representatives. If the spokespersons are representative, production can continue unabated. The way the production of coal is organized validates its goal of making the towns truly mono-industrial. This is the last moment of the translation. Callon summarizes all that has happened during this process as follows: "The initial problematization defined a series of negotiable hypotheses on identity, relationships, and goals of the different actors. Now at the end of the four moments described, a constraining network of relationships [an actor network] has been built" [12] (p. 218).

We use ANT to show how, over the course of its industrial history, the Jiu Valley has been constructed in such a way as to translate one of the main goals of socialist Romania, which is heavy industrial development. With the end of the planned economy, this translation has been unraveled in such a way that each of the actors that have previously formed the network has been "lured" [11] (p. 211) to partake in alternative translations away from coal but certainly not beyond coal. In light of these arguments, actor-network theory helps outline a critique of the *JVS* for its failure to provide evidence for how an alternative—post-coal—translation of development can be achieved.

The critique is articulated around the following critical moments of the translation. First, the *JVS* offers a problematization but does not posit a credible obligatory passage point for the post-coal development of the Jiu Valley. This problematization is made in terms of the creative city [21] or the entrepreneurial city [23], which rely on entirely different network configurations than those available in the Jiu Valley. Second, interessement devices are not specified, and moreover, they seem to work in the opposite direction to the way they are supposed to in the *JVS*. The proliferation of smaller actor networks that emerged in the wake and as partial replacements for the collapse of the grand actor network of coal production undermines the idea of a common development strategy for the Jiu Valley. Third, it might be argued, but only hypothetically at this point, that the *JVS* is itself part of the process of selectively developing fragments of the Jiu Valley rather than the region in its entirety. Enrolment and mobilization tend to favor a centrifugal movement rather than the centripetal configuration that would be consistent with a region-wide development.

The processes described are certainly cross-scalar, as they involve the national and the local levels. The actor-network theory was consciously fashioned as a theoretical bridge between the micro- and the macro-social, the local and the national, and the large and the small [29]. The main advantage is that it does not isolate local actions and decisions as explanatory factors but places the weight of the explanation on the articulation of cross-scalar factors. What may appear as problematic or inadequate planning, for example, is the result of a more or less successful translation, which is itself a struggle between unequally powerful actors.

3. Methodology and Case Description

The present analysis focuses on the six towns in the Jiu Valley, which are integrated into an urban network historically developed around mining activity. They are located in Hunedoara county in the development region RO42 West, Macroregion four and include the towns of Petroșani, Petrila, Vulcan, Aninoasa, Lupeni, and Uricani, with approximately 130,000 inhabitants in 2021 (see Table 1 and Figure 1, INS Tempo, indicator POP107D).

Table 1. The distribution of the population from the Jiu Valley (2021).

City	Population	% of Jiu Valey Total
Petroșani	40,319	31%
Vulcan	27,274	21%
Lupeni	25,375	19%
Petrila	23,627	18%
Uricani	9196	7%
Aninoasa	4458	3%

Source: National Institute of Statistics of Romania.

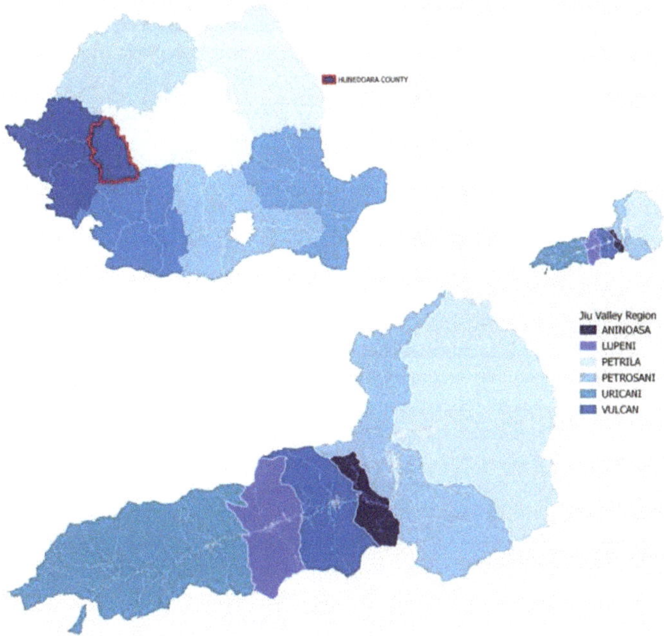

Figure 1. Geographical location of the study area in Romania and the Hunedoara County. Source: https://data.gov.ro/ (accessed on 10 May 2022).

The history of coal mining in the Jiu Valley dates back to the period of the Austro-Hungarian Empire, with the first surface mining starting in 1840 in Vulcan, Petrila, and Petrosani [30]. In the late 1800s and early 1900s, new mines were opened in Uricani and Lupeni. At the beginning of the socialist period, the communist government nationalized all the mines. The mining sector underwent significant development in terms of increased production and the number of employees in the context of rapid industrialization.

After the 1989 Revolution, there was a period of stagnation until 1996–1997, when production collapsed in the context of neoliberal restructuring policies and the start of collective lay-offs. In fact, between 1996 and 1999, the number of employees fell by about 2.2 times, and coal production halved [31]. The decline has continued until the present. In 2010, nearly 2 million tons of coal were daily extracted; in 2015, approximately 1.6 million [31], while in 2022, about one million tons [32].

In 1991, the restructuring of the mining industry began, reducing the number of people working in mining from around 50,000 to 4000 today. Thus, compared to 1990, more than 90% of workers either lost their jobs or retired (many of them early), while opportunities for re-employment were limited and marginal [33]. An initial shock occurred in the late 1990s when the Romanian Government, following the recommendation of the World Bank, launched a restructuring and mine closing program. Between 1997 and 1999, the Government paid compensatory wages for about 20% of the total workforce in the Jiu Valley, thus rendering them, in effect, redundant. Although discursively, compensatory wages were intended to encourage entrepreneurial initiatives, in practice, the aim was to encourage workers to accept the termination of the employment contract [34,35]. After 2000, there was a quasi-total reduction in mining activity, and policies aimed at ensuring the area's redevelopment were far from having the desired effect [35].

Following the decision 787/2010 of the Council of the European Union on state aid to facilitate the closure of uncompetitive coal mines, the *National Society for the Closure of Jiu Valley Mines* (SNIMVJ) was established in 2012. Since then, three out of the last seven mines have been closed. According to the *JVS*, only four out of fifteen mines in 1990 are currently active, and they are set to be closed in the following years. Funding for redevelopment is available in principle, as the Jiu Valley is part of the Just Transition Platform, a network of twenty coal regions that aim to diversify their economic activities and offer residents viable economic alternatives to mining.

In this context, the Romanian Government approved the Jiu Valley Economic, Social, and Environmental Development Strategy for 2022–2030 [19]. The *JVS* is the primary focus of our analysis, which assesses the viability of the proposed development model given the area's historical context. The analysis of the Jiu Valley area has been the subject of numerous case studies aimed at analyzing the impact of post-socialist transition policies on mono-industrial towns. To critically assess the *JVS*, we use secondary literature and data to provide an understanding of the historical context. We describe the development rationale and the network alignment during the planned economy and the subsequent violent misalignments after the Romanian revolution of 1989.

We structured our analysis in three historical stages. In the first stage, we discuss the development project around mining activity in the socialist era using historical literature and data. In the second stage, we analyze the economic and social transformations that took place in the region following the transition from a centrally planned to a market economy. We use secondary data from sociological and anthropological studies conducted in the last 20 years and official data. The aim is to provide a general understanding of the social impact in the region using quantitative and qualitative data from these studies, benefiting from the relatively large number of academic or policy publications on the cities of the Jiu Valley. In the third stage, we conducted a thematic analysis [36] of the main pillars around which the *JVS* was designed. In terms of textual data selection, we followed the main structure of the text (*JVS*). Thereby, we went from the analysis of the main themes of the *JVS*, which are the four pillars on which the strategy is based, to more specific topics and word searches. To further refine our analysis, we also looked at the occurrence of key terms which are part of a recognizable discourse dealing with the creative city [21,22].

4. Results and Discussion

4.1. Network Alignment during the Planned Economy

In the context of a predominantly agrarian country at the beginning of the post-war period, Romania's economic development during the socialist period was premised on the policy of rapid industrialization and urban development. The post-war socialist elites prioritized heavy industry [37]. Studies such as those of Vladimir Pasti [38] illustrate that the development of a socialist society was translated by means of three major economic and social macro units, or *complexe*: rural, socio-industrial, and political-administrative, which aimed to achieve sets of coherent networks within the planned economy. Most of the working population was divided between the first two networks, which will be discussed in the paragraphs below. Both networks functioned as rival translations of development throughout socialism. Rival is meant here as organized as mutually exclusive networks from the point of view of employment and associated infrastructures.

The rural *complex* was organized during the socialist period around industrially operated farms, either in state farms (IAS) or cooperatives of agricultural production (CAP). The modernization of agriculture took place through a long-term cooperativization program from 1949 to 1962, a policy considered almost inevitable in the context of agricultural underproduction, overpopulation, and the lack of mechanization [39]. This political decision was the answer to the problem of fragmentation of agricultural property. According to the 1948 census, 91% of agricultural land was less than 5 hectares and 36% less than 1 hectare, making it impossible to practice modern agriculture [40]. The organization of agricultural production into state farms and cooperatives transitioned to more mechanized agriculture, which reduced the population employed in agriculture, which either moved to urban centers [41] or became commuters to nearby cities to work in the industry. Ivan Szelenyi called this process under-urbanization, arguing that the growth of the urban population in socialist countries was slower than the growth of the industrial proletariat [42].

Socialist industry was the main socio-economic *complex* around which the socialist society was built. Industry provided the industrial worker with social status and the community with essential public services such as education, health, housing, urban administration, culture, and sport [38,43]. In other words, under state socialism, the industry's role was not merely that of producing goods and selling them for profit but as a unit of production with social functions, functioning as the obligatory passage point of the socialist society [43]. At the same time, the state-planned economy and the policy of full employment provided a secure position for the workers. Managers of state-owned enterprises often faced unrealistic targets set by the central authorities in terms of the availability of raw materials and labor or the functioning of the means of production. To achieve the strict requirements of the five-year plan, one of the tactics was to hoard labor and raw materials for when they needed to increase production, indirectly contributing to the policy of guaranteed employment [43].

A particular case of urban communities, called mono-industrial towns or areas, have been established around a dominant industry. While in 1930, there was only one mono-industrial city in Romania, by the end of the socialist era, 43 towns had at least 50% of their population engaged in the same industrial activity [44]. After 1968, localities classified as fulfilling industrial, mining or tourism functions were given city status [45]. As in the case of other former socialist states, "company towns" [46] emerged around factories or mining areas. These towns developed in the context of rapid industrialization in areas where natural resources were available. Their size was under 50,000 inhabitants, and they were often located in areas relatively far from large urban centers [47]. The six localities covered in this study were developed around the mining industry in the Jiu Valley and represent a typical case of company towns. Urbanization in the context of rapid industrialization in the area has mainly meant the construction of new housing blocks and, to a lesser extent, other social facilities specific to urban areas. At the same time, housing conditions are considered worse in the Jiu Valley compared to the national average, with many poorly designed and rapidly built housing, comfort 2 and 3 (for example, in the town of Aninoasa, 50% of the dwellings have shared bathrooms) [44]. Nevertheless, from the mid-1960s until the late 1980s, coal mining experienced rapid growth and was considered essential for sustaining the heavy industry. Working in mining during this period was the primary source of wealth and prestige for workers [48]. Around mining, the state crafted a network of facilities, including housing, access to education, health, culture, sport, and leisure.

In parallel with the development of mining, educational institutions have been established with the aim of providing a skilled workforce in mining [31]. The first higher education institution, the Coal Institute of Petroșani, founded in 1948 (later renamed the Mining Institute of Petroșani), was the leading university-level institution at the national level, training professionals for the mining sector. At the same time, research institutes such as the Institute for Research and Mining Design and the National Institute for Research and Development in Mine Safety and Protection to Explosion were established. Last but not least, vocational schools and high schools have been developed to train a qualified workforce. In fact, by the 1980s, investment in primary and vocational educational infrastructure in Romania had been significant, with the educational system being considered a fundamental element of the planned economy, having the function of preparing the workforce for entry into production [49]. All these institutions tied the educational system in the Jiu Valley to the actor-network of coal production, and their strength was given by the high degree of enrollment into the latter network.

In theoretical terms, the coal industry in the Jiu Valley functioned as an obligatory passage point (OPP) for the local economy, being intimately linked to the national-level OPP represented by the heavy industry [27]. This was tightened even further by the five-year plan. The socialist-era problematization of the Jiu Valley as a coal-producing area was highly successful due to the number and efficiency of interessement devices used by the socialist authorities. These included high salaries for the miners and the provision of public goods, all of which amounted to a strong economic and cultural cohesion around the production of coal:

> *"In Romania's Jiu Valley, the state had long taken a paternal role with the miners, providing them free housing, electricity, heat, water and highly subsidized food in order to assure the promise of upward mobility."* [27] (p. 432)

Apart from receiving special privileges under the socialist translation of coal-based development, privileges that were not enjoyed by other categories of workers [50], the miners maintained a "relatively sympathetic relationship with the state" (Friedman 2007: 432). Enrollment in the actor network was also achieved via "education [which, in addition to] a working class pedigree, meant promise for young people" [27] (p. 437). The coherence of this actor-network was tested during the 1977 strike, which took place in response to planned cuts in compensations for injured miners. After the benefits were restored, alongside several arrests of labor leaders, the socialist interessement devices, managed to realign the network so that it functioned relatively frictionless until the late 1990s [50]. The coal-production regime thus survived the collapse of the centrally planned economy, but not for long.

4.2. The Network Unravels: The Violent Misalignments of the 1990s and Beyond

After 1989, the rural and socio-industrial networks, the *complexe* that supported the planned economy, were dismantled but failed to be replaced by a political project for the organization of the Romanian economy after socialism [38]. As a consequence, if social communities were built around the industrial *complexe*, the disappearance of these *complexe* meant the disappearance of the ties that linked them to local public life in the Jiu Valley. Among the most important social consequences of network disarticulation were the following [38]:

- The shrinking of the working class and migration to rural areas, a phenomenon typical of pre-capitalist societies;
- Insecurity, unemployment, underemployment, and lack of employment opportunities;
- The deterioration of the standard of living and the loss of the working class prestige for (heavy) industrial workers;
- Degradation of the housing stock;
- Degradation of urban infrastructure;
- Reduced access to education and health services.

All these consequences have been felt with greater intensity in the mono-industrial areas. Their tighter integration during socialism into one overarching translation around the OPP of coal production suffered a major blow when this problematization was put in question. The Jiu Valley towns were particularly affected, especially after 1997, when shock therapy policies were implemented in Romania after the failure of the policies proposing a combination of neoliberalism and a developmental state in the early 1990s [51,52]. The tightly coupled actor network of coal production became unraveled, a process that had ramifying consequences for each of the network elements (see Figure 2).

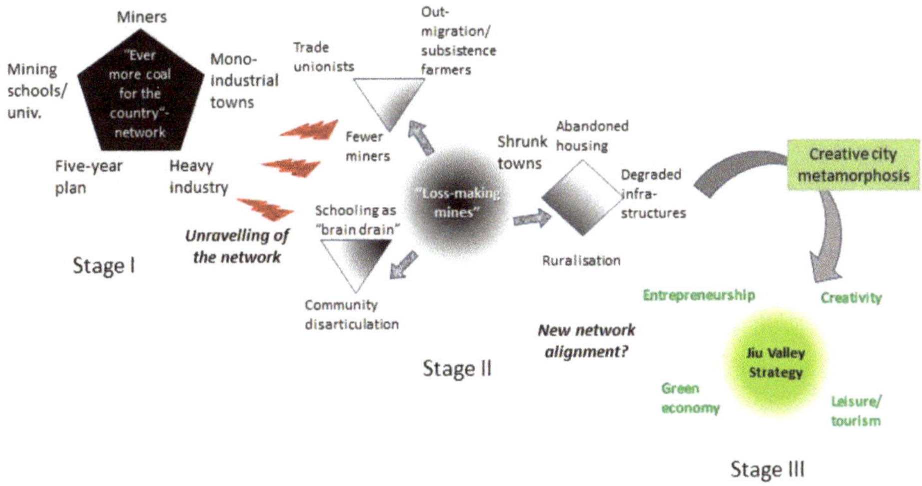

Figure 2. Stages of misalignment of the "Jiu Valley development" translation. Source: the authors based on the work of Callon and Law [13].

First, numerous miners were enticed to leave the translation—being offered between 12 and 20 monthly salaries—to become "voluntarily redundant" [51,53]. Between 1997 and 2007, the population employed in the mining industry decreased by 68% [44]. Many of those who left the mines returned to their places of origin in the Moldova region of Romania [54]. The Jiu Valley area has also become an important source of emigration for work to Western Europe [48]. For a significant part of the inhabitants, seasonal work in agriculture in Western Europe has been the solution that could ensure their subsistence [55] (Muscă and Trifan 2022). Women also became temporary migrant care workers in Italy or Austria [56]. They often live together with the person or persons they care for, which is a demanding type of work with severe consequences for the physical and mental health of the carers [55]. In sum, multiple career trajectories and migration marked the period after the closure of the mines. All these individuals with their families can be said to have been attracted to alternative translations that were not premised on the production of coal—the former OPP of the area—but on individualized survival strategies, either as subsistence farmers (in Moldova) or low-skilled workers (in Western Europe).

In the early 1990s, trade unions, represented by the League of Jiu Valley Miners' Unions, opposed restructuring and wage cuts—in essence, the disarticulation of the coal-based actor network. The level of trust in the unions was high. As Kideckel mentions, "the miners readily transferred their allegiance from state to the union and to its president, Miron Cozma, whom many saw as a savior of their way of life." [54] (p. 49). Other studies discuss the lack of trust in company managers and union leaders, calling them "the people in the offices," who were seen as being complicit in the closure of the mines [35]. Miners see the relationship with unions in terms of "us and them," perceiving that they have no influence on the union [35]. At the same time, since the late 1990s, there has been fragmentation and even conflicts between trade unions as they compete for members and financial gains from mine closures [54] (p. 77). This has shattered the "unity of the mine" and has revealed the "diversity of the miners" [57] and the ways in which the trade unions have provided selective benefits to their miners rather than to the Jiu Valley miners as a class.

The abrupt reduction of the work force set in motion the depopulation process [58], which began to unravel the functioning of the mining towns as urban structures, leading to community disarticulation. Public infrastructure has deteriorated in the context of the impoverishment of local administrations. The smallest town in the Jiu Valley, Aninoasa, went bankrupt in 2010, immediately after the mine closed for good.

The localities where the mining industry has been declining over the last thirty years are described as shrinking towns [44]. Characteristics include demographic decline and aging population, unemployment and reduced employment opportunities, poverty, inadequate living conditions, and declining quality and attractiveness of urban life. Some neighborhoods such as *Dallas*, *Colonia Bosnia*, *Colonia Ianza*, *Venus-Saturn*, and *Sașa* are known as marginalized urban areas with high levels of poverty, deteriorating infrastructure, and utilities, with a population that has little access to education, health, and social services [55]. Being now inhabited mainly by miners laid off after 1997, the transformation of these areas has been radical, going from relatively prosperous areas to neighborhoods with skyrocketing poverty [51]. In theoretical terms, it can be said that the strength of the actor network for the extraction of coal has now turned into its main weakness, in that the collapse of the mine has pushed the workers' living spaces into disarray.

According to an analysis carried out by the World Bank for the Ministry for Development, Public Works and Administration, the six cities of the Jiu Valley are in a challenging situation from a demographic point of view, and forecasts for 2030 are that the demographic evolution will continue the downward trend [59]. In fact, the shrinkage of small towns is also prevalent at the national level. According to Romania's *Territorial Development Strategy* (2016) [60], there is a notable trend of ruralization of towns with less than 20,000 inhabitants, as they are left without urban facilities such as heating or water supply, which have become too costly for the purchasing power of their residents. Bănică and his colleagues use concepts such as "de-urbanization" or "neo-ruralization" to describe how the small towns have become incompatible with the urban status [45]. Additionally, there has been an increase in the population engaged in subsistence farming and a deterioration in living standards in many of these towns. These small towns were the first to feel the shock of the transition from a state-planned economy to a capitalist market economy, with all social structures severely disrupted. A violent misalignment occurred in the mono-industrial towns, which saw their functional ties to the industry abruptly severed. They proved unable to reproduce their economic and social functions in the new economic context where the dominant political discourse was against state interference in the economy and the absence of local resources. With the withdrawal of the state and without private investors, small towns such as those in the Jiu Valley faced multiple crises: economic, housing, infrastructure, demographic, and identity [45,54].

To summarize, stages one and two of the Jiu Valley network allow an understanding of the systemic problems that took the form of radical deindustrialization and de-economization [45] (see Figure 2). The mono-industrial towns of the Jiu Valley were all designed as local instantiations of the translation of coal-based development. In other words, they were almost copycat models of the same mining town template, each of them structured around an individual mine. This can be traced back to the Soviet or socialist model, which imposed "similar functional origins", as has been observed in several Polish towns [61] (p. 7). In Hungary, the city planning principles were decisive in terms of functional (industrial) zones in cities, despite inevitable variations [62]. In the Jiu Valley, they all had similar structures with little complementarity among them (for a counter example, see [63]). They had limited ability to diversify and become cities in the proper sense—as they never pursued an actual path of urbanization [45]. As a result, in the second stage of network misalignment, their de-urbanization pushed them closer to a rural model of development, which had regressive consequences for the whole Jiu Valley area. Under changed deindustrialization policies at the national level, following a neoliberal logic [52], the Jiu Valley actors were forced to assemble an enfeebled local network, which lacked the coherence of the pre-1990 development model.

4.3. *Towards a New Alignment? The Jiu Valley Strategy and the Challenge of a Just Transition*

There were several early attempts to marshal a new development network in the Jiu Valley before the *JVS*, but they mostly failed for the lack of a viable obligatory passage point [11]. Economic investors did not find it profitable to convert old mining sites or invest in new industries, despite the tax facilities offered by the Romanian Government between 1998 and 2008 to all companies that chose to open a business in areas considered disadvantaged, such as the towns of the Jiu Valley. The new investors were exempted from income tax, customs duties, and taxes imposed for changing the use of the land required for the investment [64]. There are studies that argue that policies aimed at improving the economy and living standards in former mono-industrial towns were inconsistent and had a low impact [44]. For example, according to a survey conducted after the end of state aid, most of the business owners who benefited from tax incentives for investments in disadvantaged areas expressed their intention to cease economic activity in those areas [64]. The transformation of the Jiu Valley into a tourist destination has been considered the primary alternative for the area's development in all local development strategies. Still, tourism has been considered an overrated solution with modest results [44]. At the same time, professional reconversion and entrepreneurship support programs, as well as other tax reforms, have had limited success: they were implemented much too late after the closure of the mines and the compensatory payments offered to laid-off workers [44,54]. The limited interest of investors can be read as a form of network opportunism. The state aid offered to the Jiu Valley towns worked as an obligatory passage point, but this was not followed by enrollment and mobilization since much of the local workforce and infrastructures remained outside the network.

A recent qualitative study analyzes how the transition affected women in the Jiu Valley, as social protection measures focused mainly on the miners, most of them men [49]. These women have a rather pessimistic view and low confidence with regard to the area's development. The interviewees mainly invoked the scarcity of private sector jobs, which are also poorly paid and uncertain. Another important reason is the general lack of confidence in authorities. The perception is that central and local authorities have proved incapable of mitigating the brutal economic and social impact of mine closures in the past decades and providing a viable alternative for redevelopment. At the same time, at the local level, there is a perception that migration to larger cities in Romania or Western European countries has become widespread, and young people's aspirations are not linked to the Jiu Valley. Last but not least, for the most vulnerable of those remaining in the Jiu Valley, the interviewees appreciate that the impact of social protection measures has been modest in response to the severe economic problems [55] (pp. 76–82).

The current strategy to revitalize the Jiu Valley is ostensibly built around a just transition framework. According to the International Labor Organization understanding of just transition, which the European Commission also uses, the transition should be based on principles similar to those formulated in the Sustainable Development Goals (SDGs), such as the creation of decent jobs for all, the eradication of poverty, social protection and social inclusion [65]. At the same time, policy measures should be context-sensitive and take into account nine policy areas: macroeconomic and growth policies; industrial and sectoral policies; enterprise policies; skills development; occupational safety and health; social protection; active labor market policies; rights; social dialogue and tripartism [65].

Recognizing the Jiu Valley area as being in economic distress with the mining industry's decline, the Ministry of Investments and European Projects has developed the Jiu Valley Economic, Social, and Environmental Development Strategy (2022–2030). *JVS*'s objective is to provide a framework for the socio-economic development of the area in the context of multiple attempts to complete the transition away from coal and diversify economic activities. At the same time, for the transition to be successful, *JVS* argues that a "radical change of mentality and life as a whole" [19] (p. 5) is needed. The *JVS* vision considers actions such as developing the renewable energy sector, developing year-round tourism, and the professional reconversion of the workforce.

The strategy is organized around four pillars around which an action plan for the next period has been developed. These pillars are the following:

- Improving the quality of life and creating a healthy and sustainable environment for future generations.
- Economic diversification, innovation, and entrepreneurship.
- Sustainable use of local specificity.
- Accessibility, mobility, and connectivity.

The first pillar reiterates that the transition away from coal must be achieved without major social costs. The broad concept of quality of life is operationally defined through a series of measures that facilitate the improvement of human capital (with actions such as modernizing education, health, and social services), boost employment (with activities such as professional reconversion programs), and protect the natural environment (with measures such as reducing pollution and greening the former mining areas). The second pillar focuses on measures to support entrepreneurship (either by supporting local entrepreneurs or attracting businesses to the area) and investment in the green energy sector. At the same time, many activities aim to develop entrepreneurial "skills and culture" in the region. The third pillar focuses on supporting tourism development in the region, with activities such as modernizing tourism infrastructure, services, and activities, supporting creative industries, and organizing cultural, sports, and leisure activities. Finally, the fourth pillar aims to increase accessibility in the six cities in the region and connect them by modernizing road and rail infrastructure to adapt to urban mobility needs.

From a theoretical point of view, the *JVS* can be seen as an actor-network-building exercise that is problematic on several counts. First, all the measures have in common the promotion of a discourse that emphasizes individual responsibility, an element considered essential for the area's redevelopment. In a complementary way, failure is explained in terms of the refusal to take responsibility by the citizens. In this sense, the need for a radical change of "mentality" is often invoked for the transition to take place successfully. In fact, the strategy is considered "an approach to increase the capacity and responsibility of every citizen of the Jiu Valley to propose and participate in the implementation of actions for socio-economic revitalization and transition away from coal" [19] (p. 1). It is noteworthy that a diverse set of *collective problems* are addressed via individualized solutions. Positing a generic "individual" as the obligatory passage point of the network means to diffuse and even atomize the network-building effort. Second, this is based on the naive assumption that the "individual" is simply there and available to assume responsibility for redevelopment. As stage II has shown, however, when faced with the abrupt removal of coal as the OPP of the Jiu Valley development, local families resorted to a variety of micro-network building efforts, leading them away from the Jiu Valley or at least away from the crumbling resources of their local communities.

In terms of problematization as a network-building strategy, the *JVS* states that the transition process is facilitated by the inclusion of the Jiu Valley as a pilot region in the "Coal Regions in Transition" platform and by the possibility of accessing EU funds from the programs such as the "Just Transition Mechanism", "European Green Deal", and "EU Cohesion Policy". At the same time, however, the strategy fails to demonstrate how the Jiu Valley can be made indispensable for each of these EU-wide mechanisms, which have been devised as limited-time opportunities for those regions that can play a role on European markets, be they economic, labor, technology or environmental.

The *JVS* proposes a generic action plan in which institutional and financial responsibility for implementing projects and programs is diffuse. Although funding sources that can be accessed to ensure a just transition are mentioned, it is difficult to address the systemic problems accumulated over the past decades in the Jiu Valley. Moreover, each funding source is issue-specific (e.g., for regional cohesion, for the green economy etc.) and can hardly counter the dismantling of the coal-production network, which was designed to function as a totality.

The "shrinking area" status, which entails a high level of poverty and territorial socio-economic segregation (sometimes overlapping with ethnic segregation), makes it difficult to implement programs aimed at supporting entrepreneurial initiatives, such as developing the creative sectors, green economy, and tourism. Thus, the strategy is rather a general framework for attracting European funds, but these are insufficient in functioning as interessement devices. As a result, attracting funds does not automatically ensure development, as there is rarely a sustainable continuation of the programs after the end of the funding. Rather than creating its actor-network to facilitate local development, the *JVS* enforces the idea that local development needs to pass through the problematizations and obligatory passage points of EU funding requirements. In contrast to the coal-based actor network, the model of development for the Jiu Valley area is not imposed but rather optional.

There is a strong emphasis on creativity in the strategy, which gives a sense of a desired creative-city metamorphosis (see Figure 2). The Romanian words for "creative" or "creativity", *creativ* or *creativitate*, are mentioned 37 times in the *JVS*. The latter (*creativitate*) crops up in connection with "the stimulation of". There is no definition of creativity; however, the only explanatory remark is that the lack of creative stimulation (of school children) is seen as a symptom of poor education. The encouragement of creativity is seen as a source of addressing local ills. The "quality of life" concept figures even more prominently in the *JVS*, with the first "pillar of the strategy" being devoted to "The improvement of the quality of life and the development of a healthy and sustainable environment for future generations." [19] (p. 34). Capturing the gist of the *JVS*, we argue that it seems to be based on a developed urban model focusing on the quality of life and creativity (à la Richard Florida, [21,22] or see the critical work of John Hannigan in his *Fantasy City* [66]). In general terms, this model has been shown to be problematic in Eastern Europe [67,68].

In the particular context of the Jiu Valley, we argue that the creative city is part of an entirely different kind of translation of development, one that includes a critical mass of university-educated urban residents, above-average urban facilities, and above all, the three Ts of technology, talent, and tolerance [22]. These stand in contrast to the working-class culture of the Jiu Valley, and a militant culture at that [50], and the dearth of technological development that goes back to the 1980s. In other words, the local ingredients of an actor network premised on creativity are missing in the Jiu Valley, and thus, there is not much to align through the *JVS*.

Closely related to the quality of life and creativity is "education", which receives more than 100 references in the *JVS*. As part of the actor-network of coal production, education was an important secondary element: the developing structures of coal extraction were supported by a tailored system of technological education. Once the production of coal has been drastically reduced, education is called upon in the *JVS* to play an *independent* role as the driver of development [19] (p. 54). This can hardly be successful since education is efficient when and to the extent to which the whole network works well.

Our analysis is not meant to suggest that the *JVS* for 2021–2030 will be devoid of a positive impact on development. However, we hypothesize that its effects will be rather limited, issue- and place-specific, rather than applicable to the Jiu Valley towns as a whole. We thus expect small and fragmentary so-called "islands of development" rather than a comprehensive development process as the *JVS* purports to offer. An opposite example is the robotics hub organized by the *Planeta Petrila NGO* [55]. Launched as a laudable effort to mobilize educated and tech-savvy youth from the town of Petrila or nearby towns, this initiative tries to connect to transnational trends of development, but for this very reason, it tends to be solipsistic, aloof to local needs [66]. In actor-network terms, it can hardly be said to work as an obligatory passage point for the development of the Jiu Valley.

Similarly, the development of the green economy in the *JVS* is discussed in vague terms. The strategy invokes "attracting" companies from the green economy or the circular economy to "promoting" green energy or smart city systems, but there is no mention of the "primurn movens" [11] (p. 203) of this whole process. Neither the state nor corporate or civil society actors are singled out as the possible first mover of the process of local development. This stands in marked contrast to the coal-based actor network, as it was premised on the overpowering role of the state. The contemporary national state is far from being able to perform this function, but there are also no congeries of actors that can perform an equivalent developmental role. To summarize, the *JVS* proposes a network that lacks the critical focal points (obligatory passage point, enrollment, and mobilization) that would make for a realistic development strategy.

5. Conclusions

In an article published three decades ago in Rural Sociology, William Freudenburg stated that:

> *"The encouragement to develop extractive industries is often coupled with the advice to avoid developing an excessive dependency on a single economic sector.* [*However*] *the very regions and nations having the greatest need to hear such advice may also have the lowest realistic ability to respond to it."* [69] (p. 305)

The history of the Jiu Valley seems to illustrate this claim in exquisite detail, and we have outlined the reasons for this by distinguishing three stages. The first explains the extraordinary degree of network alignment around the ever more intense production of coal during the planned socialist economy in Romania. The second stage captures the profound network misalignment that set in once coal production's obligatory passage point was abruptly removed. This happened following the World Bank and IMF requirements on the Romanian Government to stop subsidizing the "loss-making" mining sector [51,53]. The third stage captures the current attempts to devise an area-wide development strategy for the Jiu Valley.

Our key argument is that the *JVS* and its workability can best be understood and assessed *in close relationship* with both the first and the second stages. The first stage reveals that the production of coal assumed an actor network of such coherence that it essentially suppressed local economic diversity. The miners' extractive activity was the central goal of family and community life [54], so the urban formations from the Jiu Valley were only viable to the extent to which they functioned within a tightly planned and growth-oriented economy. These regions were structurally bereft of diversity for a long time, so it is unrealistic to expect the sort of intrinsic variability of economic life (agriculture, small manufacturing, etc.) that the *JVS* appears to take for granted.

The second stage reveals another facet of the history of the area to set in contrast with the assumptions of the *JVS*. The collapse of the obligatory passage point represented by coal production has thrown the area into disarray, which took the form of a ramifying unraveling of the network. This unraveling reached into every aspect of local life, from housing, food provision, and entertainment to family life, education, and political organization. The result of this were the desperate efforts of the actors to secure access to at least some resources of the local economy, which took the form of micro-level alignments. Almost any sort of collective resource was thus selectively appropriated by individual actors, often to the exclusion of others. The expectation harbored within the *JVS* that actors may be persuaded to rejoin an actor network and share resources that have in the meantime become individualized seems largely unfounded. The second stage thus problematizes the suggestion that the redevelopment of the Jiu Valley can happen as if collective resources are still there where the collapse of the mining industry left them. This, however, is highly problematic, as the second stage has shown.

There are some useful lessons to be learned from this analysis for critical reading of the just transition, at least in Central and Eastern Europe. First, the just transition needs some basic premises in order to be successful, such as a minimally diversified and democratic local economy. This means a reduction in socio-economic inequalities and halting the out-migration flows, which have been on the rise since the contraction of the local economy. Second, the just transition cannot be applied to an area of relative decline—characterized by de-urbanization and de-economization—without first stabilizing the area by providing a minimum of public investment. This does not mean the mere attraction of European funds but a long-term public commitment to ensure predictability and stability over the long term. Third, the just transition concept has been conceived in the developed contexts of Western Europe. We propose a radical learning experience starting from one of the extreme cases, such as the Jiu Valley, rather than from the typical, near-average EU coal regions. The applicability of the just transition framework to cases of extreme post-coal contraction should not take the form of context-insensitive application but rather of careful testing and fine-tuning. Perhaps the main limitation of the paper is the absence of primary empirical data used in the analysis. However, the paper can contribute to the post-coal transition literature due to its wide-range focus on the history of an emblematic coal-mining region of Romania. As a novel contribution to the post-coal transition research, ANT allows a comparison of network alignments across different historical periods. In this way, it provides a heuristic understanding of how actor networks achieved their strength in a certain historical period, how they declined in another period, and how challenging it is for the post-coal strategy to reassemble the network around a far-flung and "footloose" development model of the creative/green economy.

Author Contributions: The authors have contributed equally to each section of the paper. All authors have read and agreed to the published version of the manuscript.

Funding: The authors received no financial support for the research. The APC was funded by the German Federal Ministry of Education and Research (BMBF) project "Querschnittsvorhaben Stadt-Land-Plus", grant number 033L200.

Institutional Review Board Statement: Not applicable.

Informed Consent Statement: Not applicable.

Data Availability Statement: Not applicable.

Conflicts of Interest: The authors declare no conflict of interest.

References

1. Abraham, J. Just transitions for the miners: Labor environmentalism in the Ruhr and Appalachian coalfields. *New Political Sci.* **2017**, *39*, 218–240. [CrossRef]
2. Cocheci, R.M.; Ianos, I.; Sarbu, C.N.; Sorensen, A.; Saghin, I.; Secareanu, G. Assessing environmental fragility in a mining area for specific spatial planning purposes. *Morav. Geogr. Rep.* **2019**, *27*, 169–182.
3. Harrahill, K.; Douglas, O. Framework development for 'just transition' in coal producing jurisdictions. *Energy Policy* **2019**, *134*, 110990. [CrossRef]
4. Jafry, T.; Watson, E.; Mattar, S.D.; Mikulewicz, M. *Just Transitions and Structural Change in Coal Regions: Central and Eastern Europe*; Glasgow Caledonian University: Glasgow, UK, 2022.
5. Kumar, S.; Americo, A.; Billingham, C. *The New Social Contract: A Just Transition*; Foundation for European Progressive Studies: Brussels, Belgium, 2016.
6. Popp, R. A Just Transition of European Coal Regions Assessing Stakeholder Positions Towards the Transition Away from Coal. In *E3G Briefing Paper*; E3G: Brussels, Belgium, 2019.
7. Robins, N.; Brunsting, V.; Wood, D.; Adler, J.; Amin, A.L.; Baines, C. Investing in a Just Transition: Why Investors Need to Integrate a Social Dimension into Their Climate Strategies and How They Could Take Action. 2018. Available online: https://www.lse.ac.uk/granthaminstitute/wp-content/uploads/2018/06/Robins-et-al_Investing-in-a-Just-Transition.pdf (accessed on 20 April 2022).
8. Weller, S.A. Just transition? Strategic framing and the challenges facing coal dependent communities. *Environ. Plan. C Politics Space* **2019**, *37*, 298–316. [CrossRef]
9. Brown, B.; Spiegel, S. Coal, Climate Justice, and the Cultural Politics of Energy Transition. *Glob. Environ. Politics* **2019**, *19*, 149–168. [CrossRef]
10. Sovacool, B.; Martiskainen, M.; Hook, A.; Baker, L. Decarbonization and its discontents: A critical energy justice perspective on four low-carbon transitions. *Clim. Chang.* **2019**, *155*, 581–619. [CrossRef]
11. Callon, M. Some elements of a sociology of translation: Domestication of the scallops and the fishermen of St Brieuc Bay. *Sociol. Rev.* **1984**, *32*, 196–233. [CrossRef]
12. Callon, M. The Sociology of an ActorNetwork: The Case of the Electric Vehicle. In *Mapping the Dynamics of Science and Technology: Sociology of Science in the Real World*; Callon, M., Rip, A., Law, J., Eds.; Macmillan: Basingstoke, UK, 1986.
13. Callon, M.; Law, J. After the individual in society: Lessons on collectivity from science, technology and society. *Can. J. Sociol./Cah. Can. De Sociol.* **1997**, *22*, 165–182. [CrossRef]

14. Horowitz, L.S. Translation alignment: Actor-network theory, resistance, and the power dynamics of alliance in New Caledonia. *Antipode* **2012**, *44*, 806–827. [CrossRef]
15. Sassen, S. *Expulsions. Brutality and Complexity in the Global Economy*; Harvard University Press: Harvard, MA, USA, 2014.
16. Mocanu, D. Marș forțat pe Valea Plângerii [Forced march in the valley of tears]. *Rev. Univers Strateg.* **2020**, *11*, 292–305.
17. Cobârzan, B. Environmental Rehabilitation of Closed Mines. A Chase Study on Romania. *Transylv. Rev. Adm. Sci.* **2008**, *4*, 34–43.
18. ProTv. România, te Iubesc: Înapoi în Mine ["Romania, My Love: Back to the Mines"]. Available online: https://www.youtube.com/watch?v=ZsWHJwBxCag (accessed on 19 April 2022).
19. Jiu Valley Strategy (JVS). Strategia de Dezvoltare Economică, Socială și de Mediu a Văii Jiului (2022–2030) [The Strategy for the Economic, Social and Environmental Development of the Jiu Valley (2022–2030)]. Available online: https://mfe.gov.ro/hg-strategia-valea-jiului/ (accessed on 10 May 2022).
20. Wang, X.; Lo, K. Just transition: A conceptual review. *Energy Res. Soc. Sci.* **2021**, *82*, 102291. [CrossRef]
21. Florida, R. *The Rise of the Creative Class: And How It's Transforming Work, Leisure, Community, and Everyday Life*; Basic Books: North Melbourne, Australia, 2003.
22. Florida, R. *The Rise of the Creative Class–Revisited: Revised and Expanded*; Basic Books (AZ): New York, NY, USA, 2014.
23. Harvey, D. From managerialism to entrepreneurialism: The transformation in urban governance in late capitalism. *Geogr. Ann. Ser. B Hum. Geogr.* **1989**, *71*, 3–17. [CrossRef]
24. Callon, M. Techno-economic networks and irreversibility. *Sociol. Rev.* **1990**, *38*, 132–161. [CrossRef]
25. Liu, H.; Weng, D.; Liu, H. Decoding Rural Space Reconstruction Using an Actor-Network Methodological Approach: A Case Study from the Yangtze River Delta, China. *Land* **2021**, *10*, 1110. [CrossRef]
26. Callon, M. Struggles and negotiations to define what is problematic and what is not. In *The Social Process of Scientific Investigation*; Knorr, K., Krohn, R., Whitley, R., Eds.; Springer: Dordrecht, The Netherlands, 1980; pp. 197–219.
27. Turnock, D. The pattern of industrialization in Romania. *Ann. Assoc. Am. Geogr.* **1970**, *60*, 540–559. [CrossRef]
28. Murdoch, J. The spaces of actor-network theory. *Geoforum* **1998**, *29*, 357–374. [CrossRef]
29. Law, J. After ANT: Complexity, naming and topology. *Sociol. Rev.* **1999**, *47*, 1–14. [CrossRef]
30. Societatea Națională de Închideri Mine Valea Jiului [The Jiu Valley National Mine Closure Society]. Mineritul în Valea Jiului—Repere Istorice [Mining in the Jiu Valley—Key Moments]. Available online: http://www.snimvj.ro/istoric.aspx (accessed on 10 May 2022).
31. Faur, F.; Marchiș, D.; Nistor, C. Evolution of the Coal Mining Sector in Jiu Valley in Terms of Sustainable Development and Current Socio-Economic Implications. *Res. J. Agric. Sci.* **2017**, *49*, 110–117.
32. Digi24. Minele din Valea Jiului au Ajuns la o Producție de Circa 1.000 de Tone de Cărbune pe zi [Jiu Valley Mines Reached a Production of about 1.000 Tons of Coal per Day]. 2022. Available online: https://www.digi24.ro/stiri/economie/energie/minele-din-valea-jiului-si-au-revenit-si-produc-circa-1-000-de-tone-de-carbune-pe-zi-1867199 (accessed on 10 May 2022).
33. Burlacu, R.; Suditu, B.; Gaftea, V. Just Transition in Hunedoara. Economic Diversification in a Fair and Sustainable Way. Available online: https://bankwatch.org/wp-content/uploads/2019/09/just-transition-hunedoara.pdf (accessed on 10 May 2022).
34. Chiribuca, D. The impact of economic restructuring in mono-industrial areas: Strategies and alternatives for the labor reconversion of the formerly redundant in the Jiu Valley, Romania. *Studia Univ. Babes-Bolyai-Sociol.* **1999**, *46*, 23–41.
35. LaBelle, M.C.; Bucată, R.; Stojilovska, A. Radical energy justice: A Green Deal for Romanian coal miners? *J. Environ. Policy Plan.* **2021**. [CrossRef]
36. Braun, V.; Clarke, V. Using thematic analysis in psychology. *Qual. Res. Psychol.* **2006**, *3*, 77–101. [CrossRef]
37. Fisher, J. Planning the City of Socialist Man. *J. Am. Inst. Plan.* **1962**, *28*, 251–265. [CrossRef]
38. Pasti, V. Un deceniu de transformări sociale [A decade of social transformations]. In *Un Deceniu de Tranziție. Situația Copilului și a Familiei în România [A Decade of Transition. The Situation of the Child and Family in Romania]*; Mihăilescu, I., Ed.; UNICEF: Putrajaya, Romania, 2021.
39. Kideckel, D.A. The socialist transformation of agriculture in a Romanian commune, 1945-62. *Am. Ethnol.* **1982**, *9*, 320–340. [CrossRef]
40. Murgescu, B. *România si Europa. Acumularea Decalajelor Economice (1500–2010) [Romania and Europe. The Accumulation of Economic gaps (1500–2010)]*; Polirom: Iași, Romania, 2010.
41. Jigoria-Oprea, L.; Popa, N. Industrial brownfields: An unsolved problem in post-socialist cities. A comparison between two mono industrial cities: Reşiţa (Romania) and Pančevo (Serbia). *Urban Stud.* **2017**, *54*, 2719–2738. [CrossRef]
42. Szelenyi, I. Urban development and regional management in Eastern Europe. *Theory Soc.* **1981**, *10*, 169–205. [CrossRef]
43. Verdery, K. *What was Socialism, and What Comes Next?* Princeton University Press: Princeton, NJ, USA, 1996.
44. Constantinescu, I.P. Shrinking cities in Romania: Former mining cities in Valea Jiului. *Built Environ.* **2012**, *38*, 214–228. [CrossRef]
45. Bănică, A.; Istrate, M.; Tudora, D. (N) ever Becoming Urban? The Crisis of Romania's Small Towns. In *Peripheralization*; Fischer-Tahir, A., Naumann, M., Eds.; Springer VS: Wiesbaden, Germany, 2013; pp. 283–301.
46. Goodrich Lehmann, S.; Ruble, B.A. From 'Soviet' to 'European' Yaroslavl: Changing neighbourhood structure in post-Soviet Russian cities. *Urban Stud.* **1997**, *34*, 1085–1107. [CrossRef]
47. Prokhorova, E. *Reinventing a Russian Mono-Industrial Town: From a Socialist 'Town of MINERS' to a Post-Socialist 'Border Town'*; Itä-Suomen yliopisto: Joensuu, Finland, 2014.

48. Kideckel, D.A. Coal Power: Class, Fetishism, Memory, and Disjuncture in Romania's Jiu Valley and Appalachian West Virginia. *Anuac* **2018**, *7*, 67–88.
49. Tsantis, A.; Pepper, R. *Romania-Industrialization of an Agrarian Economy under Socialist Planning*; The World Bank: Washington DC, USA, 1979.
50. Vasi, I.B. The Fist of the Working Class: The Social Movements of Jiu Valley Miners in Post-Socialist Romania. *East Eur. Politics Soc.* **2004**, *18*, 132–157. [CrossRef]
51. Friedman, J.R. Shock and subjectivity in the age of globalization: Marginalization, exclusion, and the problem of resistance. *Anthropol. Theory* **2007**, *7*, 421–448. [CrossRef]
52. Ban, C. *Ruling Ideas: How Global Neoliberalism Goes Local*; Oxford University Press: Oxford, UK, 2016.
53. Larionescu, M.; Rughiniş, C.; Rădulescu, S.M. *Cu Ochii Minerului: Reforma Mineritului in Romania:(Evaluari Sociologice si Studii de caz) [with Miner's Eyes. Mining Sector Reform in Romania (Sociological Evaluations and Case Studies)]*; Gnosis: București, Romania, 1999.
54. Kideckel, D. Miners and wives in Romania's Jiu Valley: Perspectives on postsocialist class, gender, and social change. *Identities Glob. Stud. Cult. Power* **2004**, *11*, 39–63. [CrossRef]
55. Muscă, M.; Trifan, E. *Participarea Femeilor în Procesul de Tranziție Justă. Analiza Condițiilor Sociale și de Muncă ale Femeilor din Valea Jiului [Women's Participation in the Just Transition Process. The Analysis of Women's Social and Working Conditions in the Jiu Valley]*; Hecate: București, Romania, 2022.
56. Țoc, S.; Guțu, D. Migration and Elderly Care Work in Italy: Three Stories of Romanian and Moldovan Care Workers. *Cent. East. Eur. Migr. Rev.* **2021**, *10*, 71–90.
57. Crăciun, M.; Grecu, M.; Stan, R. *Lumea văii. Unitatea Minei, Diversitatea Minerilor. [The World of the Valley. The Unity of the Mine, the Diversity of the Miners]*; Paidea: Bucharest, Romania, 2002.
58. Constantinescu, I.P.; Dascălu, D.; Sucală, C. An activist perspective on industrial heritage in Petrila, a Romanian mining city. *Public Hist.* **2017**, *39*, 114–141. [CrossRef]
59. World Bank. Creșterea și Contracția Orașelor din România: O Analiză Spațială [Growth and Contraction of Romanian Cities: A Spatial Analysis]. Available online: https://www.mdlpa.ro/userfiles/sipoca711/anexe_livrabil1_2.pdf (accessed on 9 May 2022).
60. Ministry for Development, Public Works and Administration. Strategia de Dezvoltare Teritorială a României [Romania's Territorial Development Strategy]. 2016. Available online: https://www.mdlpa.ro/pages/sdtr (accessed on 9 May 2021).
61. Krzysztofik, R.; Kantor-Pietraga, I.; Kłosowski, F. Between Industrialism and Postindustrialism—the Case of Small Towns in a Large Urban Region: The Katowice Conurbation, Poland. *Urban Sci.* **2019**, *3*, 68. [CrossRef]
62. Germuska, P. Between Theory and Practice: Planning Socialist Cities in Hungary. In *Urban Machinery: Inside Modern European Cities*; Hard, M., Misa, T., Eds.; MIT Press: Cambridge, MA, USA, 2008.
63. Osman, R.; Šerý, O.; Alexandrescu, F.; Malý, J.; Mulíček, O. The establishment of inter-municipal cooperation: The case of a polycentric post-socialist region. *Geogr. Tidsskr.-Dan. J. Geogr.* **2020**, *120*, 2–18. [CrossRef]
64. Competition Council. Un Deceniu de Dezvoltare Regională în România. Studiu Privind Impactul Ajutoarelor de Stat Acordate în Zonele Defavorizate [A Decade of Regional Development in Romania. Study on the Impact of State Aid in Disadvantaged areas], 2010. Available online: http://www.ajutordestat.ro/documente/Brosura%20DRMCAS-Final%206Dec_662ro.pdf (accessed on 9 May 2022).
65. ILO. *Guidelines for a Just Transition towards Environmentally Sustainable Economies and Societies for All*; International Labour Organization: Geneva, Switzerland, 2015.
66. Hannigan, J. *Fantasy City: Pleasure and Profit in the Postmodern Metropolis*; Routledge: England, UK, 2005.
67. Borén, T.; Young, C. Conceptual export and theory mobilities: Exploring the reception and development of the "creative city thesis" in the post-socialist urban realm. *Eurasian Geogr. Econ.* **2016**, *57*, 588–606. [CrossRef]
68. Alexandrescu, F.; Osman, R.; Klusáček, P.; Malý, J. Taming the genius loci? Contesting post-socialist creative industries in the case of Brno's former prison. *Cities* **2020**, *98*, 102578. [CrossRef]
69. Freudenburg, W.R. Addictive economies: Extractive industries and vulnerable localities in a changing world economy. *Rural. Sociol.* **1992**, *57*, 305–332. [CrossRef]

land

Article

The Nexus between Creative Actors and Regional Development

Kamila Borseková [1], Anna Vaňová [2], Janka Šúrová [2], Pavol Kráľ [3], Kamila Turečková [4], Jan Nevima [4] and Stanislav Martinát [5,*]

1 Research and Innovation Centre, Faculty of Economics, Matej Bel University, Cesta na Amfiteáter 1, 974 01 Banská Bystrica, Slovakia; kamila.borsekova@umb.sk
2 Department of Public Economics and Regional Development, Faculty of Economics, Matej Bel University, Tajovského 10, 975 90 Banská Bystrica, Slovakia; anna.vanova@umb.sk (A.V.); janka.surova@umb.sk (J.Š.)
3 Department of Quantitative Methods and Information Systems, Faculty of Economics, Matej Bel University, Tajovského 10, 975 90 Banská Bystrica, Slovakia; pavol.kral@umb.sk
4 Department of Economics and Public Administration, School of Business Administration in Karvina, Silesian University in Opava, Univerzitní náměstí 1934/3, 733 40 Karviná, Czech Republic; tureckova@opf.slu.cz (K.T.); nevima@opf.slu.cz (J.N.)
5 Department of Geography, Faculty of Science, Palacký University Olomouc, 17. Listopadu 12, 771 46 Olomouc, Czech Republic
* Correspondence: stanislav.martinat@upol.cz

Abstract: The paper aims to deepen our understanding of the relationship and the impact of creative actors on regional development and investigates the nexus between them. The novelty of the paper lies in constructing an original criteria matrix of creative actors' indicators which might serve as a basis for further research and policy implications. The original criteria matrix consists of two groups of indicators measuring the creative people and creative capital in the region, their mutual relationships and impact on regional development. We found that creative actors are not distributed equally across the regions in Slovakia. The strong dominance of the region with the capital city (Bratislava) was detected. We also discovered a strong direct positive relationship between representatives of the creative people and a strong direct relationship between the creative people and the emergence of small and medium enterprises (SMEs).

Keywords: creativity; creative actors; criteria matrix; regional development; Slovakia

Citation: Borseková, K.; Vaňová, A.; Šúrová, J.; Kráľ, P.; Turečková, K.; Nevima, J.; Martinát, S. The Nexus between Creative Actors and Regional Development. *Land* **2021**, *10*, 276. https://doi.org/10.3390/land10030276

Academic Editor: Stephan Bartke

Received: 11 February 2021
Accepted: 4 March 2021
Published: 8 March 2021

Publisher's Note: MDPI stays neutral with regard to jurisdictional claims in published maps and institutional affiliations.

1. Introduction

The origin of creative actors originates in the study of Andersson [1]. He defines creativity as the ability to combine knowledge and competence to create something new. This implies that desirable change is at the center of creativity. Moreover, creativity is essentially immaterial and enforceable. There are different types of creativity, e.g., scientific creativity, technological or innovative creativity, economic or entrepreneurial creativity, and artistic or cultural creativity (for more information see Florida [2]). Different types of creativity might be, to a certain extent, mutually dependent in the sense that they may stimulate, enhance, or reinforce each other when located in the same area, particularly an urban region. However, it is well known that artists such as painters, novelists, and poets may develop a high level of creativity also in rural areas and peripheral regions [3].

Creative places and regions have favorable conditions to develop innovative ideas, to design new forms of technology or architecture, to experiment with new business models in order to come up with original design, product, or marketing concepts, and thus to explore new roads to sustainable development [4] on condition that responsible behavior towards the environment is observed and a positive impact on society is achieved [5,6]. However, not all cities, regions, and even countries are able to use the opportunity for sustainable development based on creativity. Slovakia, for example, has currently zero absorption of financial subsidies from the EU funds aimed at the support of creative actors in its regions. In other words, in Slovakia, not enough attention is paid to the importance and impact of creative actors on regional development, both at national and regional level. In addition, the interplay between creative actors remains scarcely investigated, especially from an empirical perspective. This is the research gap that we strive to fill in.

There is a lack of evidence and even empirical data available that might be utilized for identification and quantification of creative actors, their distribution, impact, and the nexus between them. Moreover, the relationship between creative actors in Slovakia has not yet been systematically assessed while just suggesting that the interaction between different types of creative actors favors regional development. Therefore, the objective of this paper is to identify the distribution of creative actors across the regions in Slovakia, to investigate the nexus between creative actors, and to evaluate their impact on regional development in Slovakia.

The paper is structured as follows: Section 2 conceptualizes the theory on creative people and creative capital. Section 3 is dedicated to creative actors and the ways how these can be measured. In this section, the details of our database are shown and the robust methodology is presented. The description of data sources and apparatus of mathematical and statistical methods employed for our research are carefully explained. Section 4 aims at presenting our original empirical research results. Based on the results from the two previous sections, the last section (Section 5) discusses the most important results, concludes policy lessons learnt, and suggests future steps for research.

2. Theoretical Foundations for Creative People and Creative Capital

Creativity in terms of developing creative ideas is the first step of knowledge or innovation production [7–9]. Karlsson [10] defines creativity as an intangible human resource. He adopts the definition from Boden [11] (p. 1) who defines creativity as 'the ability to come up with ideas that are new, surprising, and valuable'. Glaeser [12] labels this ability as creative capital. Holders of creativity are people, creative people, or the so-called creative class. The twin notions of the creative people, or creative class and creative capital, were made popular by Richard Florida in his book The Rise of the Creative Class [2]. Florida claims that the creative class "consists of people who add economic value through their creativity" [2] (p. 68). When considering the characteristic of the creative class, it is stressed that its members engage in professions where the forming of new meaningful forms is typical. The best way of measuring creativity is by considering an individual's profession [13]. Based on this assumption, Mellander and Florida [14] and Florida et al. [15] claim that a particular set of occupations compose the 'creative class', and that this measure of human capital outperforms the conventional use of educational attainment because it accounts for utilized skills rather than mere potential talent [14,16].

The creative class consists of the creative core and creative professionals. The creative core includes scientists and engineers, university professors, poets and novelists, artists, entertainers, actors, designers, and architects, as well as the "thought leadership" of modern society, namely non-fiction writers, editors, cultural figures, think-tank researchers, analysts, together with other opinion-makers. Beyond this core group, the creative class also includes creative professionals who work in a wide range of knowledge-intensive industries such as high-tech sectors, ICT, financial services, the legal and healthcare professions, and business management [2].

There is no doubt that human creativity should be used for the benefit of society as new ideas can generate economic profit [2,17–19], trigger necessary and more environment-centered thinking, and thus contribute to sustainable development. The creative class gives rise to new ideas, mobilizes the creative potential of places in the form of new products, services, information, technological innovations, non-technological processes, and outputs that generate creative capital that is increasingly important for the growth of cities and regions [20,21]. Florida [2] assumes that creativity is a driving force of economic growth and competitiveness of cities and regions. He emphasized that creative people are holders of creativity from which their significant impact on society is derived. Creativity is thus perceived as a "driving force in regional economic growth and prosperity" [2] (p. xxvii).

Howkins [18], in his study, focused on the question how to transform creativity into money, profit, and capital. According to his theory, creativity itself does not have economic value. Therefore, it is necessary to transform creativity into a product that has a market value for buyers and active sellers. At the same time, he underlines that the basic legal rules must be followed. On the contrary, Mundelius [19] characterized creativity as a key factor in creating an economic profit due to an unrestricted comparison with other sources. These ideas resulted in the formation of so-called creative industries, represented mainly by business activities based on individual creativity, skill, and talent. Such an understanding of creative industries primarily include advertising, architecture, art, crafts, design, fashion, film, music, performing arts, publishing, R&D, software, toys and games, TV and radio, and video games [18]. Hence, if cities and regions want to be successful and competitive, they need to create suitable conditions for attracting and retaining creative people producing creative capital as these are considered the principal driver of economic growth and urban spatial development [2,20–23].

Florida's theory of the creative class initiated a huge discussion among academics and was criticized by many of them (inter alia [24–29]). The critique is principally evolving around the impacts of Florida's thoughts on (not) solving but rather increasing inequalities in cities. Peck [24] argues that the creative class theory is lacking causality; other authors talk about Florida´s too excessive attention on large cities [30] and underestimating the concentration of creativity in the lesser-sized cities and their creative milieu. In order to reflect on the criticism of the creative class and its methodology as proposed by Florida [18], the Australian Centre of Excellence for Creative Industries and Innovation developed a concept of creative trident [31]. According to this approach, the creative working population is divided into three groups of workers: workers with a creative job in creative industries, workers with a non-creative job in creative industries, and workers with a creative job outside creative industries (for more information see Throsby [32]). It seems that, as a society, we incline to this approach as creativity is nowadays an inevitable attribute in many professions and industries, and has become a highly-required competence. The creative and innovative potential and its gradual targeted use in cities and regions is becoming a key criterion when deciding on the establishment of new businesses producing higher added value, the formation of clusters and industrial parks, attracting tourists, or locating foreign investment. For example, in terms of locating foreign direct investment, the presence of creative actors in a region can make a positive contribution to the creation of patents, licenses, new technologies, and, in particular, major innovations.

Due to this process, there is also a synergistic effect in the form of so-called sectoral spill-overs. Creative actors themselves become the so-called core of the elements of regional competitiveness and are the bearers of internal competitive advantage. On the contrary, the absence of creative actors slows down these processes, and finally, in the long run, there is a decrease in competitiveness at the regional level [33,34]. All the above-mentioned are closely related to the development of the quaternary sector of the economy [35]. The recognition of the dominant role of creativity is based on radical changes in economic, social, and cultural structures in society [36]. Globalization and the increasing degree of interconnectedness have enabled people and companies to widen their geographical scope and to push product development and service production into new areas (inter alia [2,37–39]. Creative actors are representatives of creativity reflecting the above-mentioned challenges of the current world. For the purpose of this paper, we perceive creative actors as creative people and creative capital, while their identification and applied measures are precisely described in the next two sections.

3. Materials and Methods

3.1. Measuring Creativity

It was only recently that creativity became a relevant research topic. While the scientific debate on this issue is rich from a theoretical point of view, the same cannot be said about the related empirical evidence [40] which is still debatable [41]. A lot depends on definitions and measurement methods. Defining a level of creativity is indeed a very difficult task, essentially as the concept is a bit fuzzy, intangible, and moreover, of a multidimensional nature, and because obviously different types of creativity were defined [42]. The creativity research has mainly followed two pathways, namely the economic growth potential of creative sectors and the urban-geographical pattern of creative actors in cities [43]. There has been a growing debate on how to define creative industries and the creative class and how they develop at local and regional levels [44–46]. The works of several authors (inter alia [45–49]) show that creative industries and the creative class are enhancing the economic development at the local, regional, and consequently national level. However, there is less research trying to identify how different national and regional understandings, economic systems, and geographic and institutional contexts influence the way these industries work. We also do not know much about what type of policies should be implemented to support the development of creative industries and attracting and maintaining the creative class [43].

Florida [2] has also been deeply interested in measuring the creativity of individual geographical areas or cities. As Marlet and van Woerkens [50] (p. 2) state, 'the creative class sets a better standard for measuring human capital'. Florida, together with his colleagues and research fellows, composed a creativity index based on four key components, namely Creative Class concentration, the High Technology Index, the Innovation Index, and the Diversity Index. The creative capital theory says that "regional growth comes from the 3Ts of economic development and to spur innovation and economic growth a region must have all three of them" [50] (p. 11). Numerous critiques of Florida´s approach appeared as the theory does not respect whether it fits the specific regional and urban settings or not [50]. To reflect this concern, the creativity index has undergone several significant changes and variations. Florida and Tinagli [51] extended the concepts and indicators of the Creativity Index. They developed new indicators for the creative class and competitiveness based on the 3Ts of economic development (Technology, Talent, and Tolerance). This index was further improved by Eurostat under the name the Euro-Creativity Index in order to solve shortages in the previous model based on 3Ts. The major shortcomings include a low amount of observed data, and the impossibility to exploit the proposed model for smaller regional units. A general lack of available data is undoubtedly the main flaw. On the other hand, the advantage of this model is its database based on the data from audits conducted by Eurostat which undertakes unified and measurable data for all European countries [52,53].

In order to reflect the lack of data at the regional level and different databases existing in various European countries, we created a modified index of creativity for the purpose of this paper that is, in our opinion, suitable for Slovakia. Our modified index of creativity is based on the 3T index (including the indicators for Talent, Technology, and Tolerance as defined by Florida [2]), and the Euro-Creativity Index. The comparison and modification of indexes including partial indexes is shown in Table 1.

The choice of criteria of the modified index is influenced by the availability of data and the intention to quantify creative actors. It consists of talent partial indexes (Creative Class Index, Human Capital Index); two indexes within technology (Research and Development Index and Innovation Index); and two indexes that belong to tolerance (Diversity Index and Bohemian Index). The modified index of creativity covers a wide spectrum of population related to creativity. Therefore, for the purpose of this paper, we perceive creative people as those whose work is related to creativity. Based on the previous studies, availability of secondary data and primary data, we set a criteria matrix of creative actors' indicators focused on the identification and measurement of creativity in Slovak regions based on the occurrence of creative people and their impact on the regional development via creative capital. We use indicators as follows:

(1) A set of indicators identifying the creative people in a region:
   - The share of the creative class (CC), the share of the population with higher education (PHE), the share of inhabitants representing Bohemians (B).

(2) A set of indicators identifying creative capital in the region, which is divided into three subgroups:
   - Intellectual property (IP) measured via the number of applications for patents (P), utility models (UM), designs (D), and trademarks (T);
   - Entrepreneurship (E) measured via the establishment of small and medium-sized enterprises in creative industries (SMECI) and outside the creative industries, marked as small and medium-sized enterprises in others (SMEOTH);
   - Innovation (I) measured via innovation activities of enterprises, namely innovation enterprises in creative industries (INCI), innovation outside creative industries, marked as innovations in others (INOTH), and innovation in enterprises with more than a 75% share of employees with higher education (INEHE).

Table 1. Modified index of creativity for Slovakia based on creativity indexes.

	3T Index	Euro-Creativity Index	Modified Index of Creativity for Slovakia
Talent	Human Capital Index Creative Class Index	Creative Class Index Human Capital Index Scientific Talent Index	Human Class Index Creative Capital Index
Technology	Innovation Index High-tech Index	Innovation Index High-tech Index Research and Development Index	Research and Development Index Innovation index
Tolerance	Gay Index Bohemian Index Immigration Index	Attitudes Index Value Index Self-expression Index	Diversity Index Bohemian Index

Source: [2,12,51,53–55].

As was pointed out by several authors (inter alia [2,18,19], creative people are important as they give rise to ideas, information, and technology outputs that are important for the growth and urban development of cities and regions. The localization of creative industries is closely related to the location of people who work in creative professions. According to Florida [2], creative people move to creative cities with high-tech industries, fuelling creativity and innovation. The localization factors for creative people are the occurrence of universities, a community of successful entrepreneurs, business angels, and risk investors, and the presence of bohemians, immigrants, and generally a diversity of population. It truly seems that openness, tolerance, and open relationships are a magnet for creative people. Vaňová [56] adds that it is important to support the development of human talents and creativity, and to attract and show tolerance in relation to diversity. This approach forms the basis for the formulation of sustainable competitive and creative urban development strategies of municipalities, cities, or regions.

3.2. Reseach Hypotheses

This paper focuses on the study of the nexus between creative actors (creative people and creative capital), and their impact on regional development. The main objective of this paper is to identify the distribution of creative actors across the regions in Slovakia, to investigate the nexus between creative actors, and to evaluate their impact on regional development in Slovakia. The concept of the paper is based on robust methodology, extensive literature study, data collection and evaluation through a repertoire of mathematical and statistical methods. We set three research hypotheses:

Hypothesis 1 (H1). *The presence of creative actors contributes to regional development in Slovakia.*

Hypothesis 2 (H2). *The distribution of creative people across Slovak regions is not equal.*

Hypothesis 3 (H3). *There is a direct positive relationship between indicators of creative actors.*

3.3. Materials and Data

The formulation of the objective of the research and hypotheses is based on the analysis of human capital [57,58], and identification of its skills and qualities [59] focused on human talent and creativity [2,17–19]. Creative people have the ability to transform talent and creativity into a product or service with a certain value [18,19] and create creative capital in the form of new ideas and entrepreneurial spirit. In addition, creative people are able to respond to new challenges creatively [60]. Therefore, creativity is perceived as a crucial factor in spatial development [2,20,23,61,62]. For the success of a region or a city, it is essential to attract and maintain creative people in the region by using the proper methods and tools (for more information see for example [2,56,63,64].

Determining and setting the criteria for evaluation of the significance and impact of creative people on regional development is based on the works of [2,18,21,65–70] and applied on the condition of the self-government regions in Slovakia (NUTS3 level). Criteria are based on a modified index of creativity, as depicted in Table 1 (please see above). Distribution of creative people in regions is evaluated via the population with a university degree and the creative class divided into three subgroups by ISCO 88 (Creative Core, Creative Professionals, and Bohemians).

By conducting research in Slovakia, we have encountered a similar problem as in many European countries, which is the lack of data at the national and regional level. Therefore, data mining required the usage of several databases, registers, and data sources, namely the state and regional strategies, the Population, home and housing census in the year 2011, several statistical surveys such as a selective statistical survey on the innovations in the years 2010–2012, 2014–2016 (Statistical Office of the Slovak Republic), register of organizations (the Statistical Office of the Slovak Republic), register of the Industrial Property Office of the Slovak Republic, the register and identifier of legal entities, entrepreneurs, and public authorities (the Data Center of the Ministry of Finance of the Slovak Republic). In order to obtain the data in the required form and quality, we also carried out primary research in the form of a survey and we conducted personal consultations at the aforementioned offices between October 2017 and April 2018.

3.4. Methods

As was mentioned in the introductory part, creators of creative capital are creative people with their talents and creativity. As creative people, we understand the creative class, in other words, a share of the population with a higher level of education and the representatives of bohemians. Consequently, we understand creative capital as a result of the activity of creative people. For the purpose of our paper, creative capital consists of indicators related to intellectual property, entrepreneurship, and innovation.

For the measurement of regional disparities or equality in distribution of creative people, we use the coefficient of variation that is a highly suitable method for making comparisons as it is not dependent on the magnitude of the measured values [71] (p. 201). Thus, this coefficient is suitable for measuring various variables and may be utilized. The variation coefficient (Vk) expresses how much of the arithmetic mean value represents the standard deviation. The higher the value, the greater the variability of the given set:

$$Vk = \frac{Sx}{\overline{x}}, \quad (1)$$

where Sx—standard deviation, $\overline{x}$—mean.

In order to investigate the nexus between creative actors, we opt for using correlational analysis, namely the Pearson correlation coefficient (PCC) and the Spearman correlation coefficient (SCC).

Creative capital is a result of the activity of creative people. Identification and quantification of creative capital in Slovakia is based on the criteria matrix as defined above. For the evaluation of the impact of creative people on the regional development, we created a modified index of creativity suitable for the conditions of Slovak regions. For the evaluation of the regions in Slovakia, we employed the methods of the multi-criteria decision analysis consisting of the following steps:

The identification and classification of indicators is based on the criteria matrix of creative actors' indicators as described above. For the purpose of our paper, we use maximization of the evaluation criteria, which indicates that a higher value means a better variant. Before processing the data, it was necessary to adjust the acquired data for the criteria innovation. Since the criteria values in the area of innovation were measured for the NUTS 2 level of the EU´s statistical regions only, we decided to evenly allocate the criteria to each of the NUTS 3 regions.

The importance of the criteria is determined on the basis of the qualitative knowledge-based analysis of the theoretical knowledge and empirical experience in Slovakia and abroad and existing legislation. We determine the importance of the criteria by comparison with other criteria and thus create a ranking based on the sequence ranging from the most important to the least important.

The determination of the criteria weights is based on differentiating the indicators according to their significance. The purpose is to determine the weights of the individual indicators in this step. We opt for using the Saaty method [72]. The Saaty method is a comprehensive approach of estimating the weighting criteria, where the decision-maker compares all possible criteria pairs. We have processed the weights in the online system at https://bpmsg.com/ahp-online-system/ (accessed date: 25 April 2019). The scale ranges from 1 to 9, with a value of 1 indicating that the compared criteria i and j are equally important. A value of 9 indicates that the criterion i absolutely dominates the criterion j in its importance.

In the final stage, we evaluate the results, determine the preferential order of the variants, and select the best option. For this purpose, we chose the TOPSIS (Technique for Order Preference by Similarity to Ideal Solution) and the SAW (Simple Additive Weighting) methods [73]. The TOPSIS method is based on selecting the alternative closest to the ideal value and furthest from the basal value. This method requires that all criteria are maximized. Our data satisfies this assumption, therefore there is no need for their transformation or adjustment for this purpose. The TOPSIS algorithm expresses the following:

(a) a decision matrix normalization—y_{ij} criteria are transformed to r_{ij} values:

$$\frac{y_{ij}}{\sqrt{\sum_{i=1}^{n} y_{ij}^2}}. \tag{2}$$

(b) the specific criteria values in the matrix are multiplied by the weight of the criterion:

$$w_{ij} = v_j r_{ij}. \tag{3}$$

(c) the ideal variant $(h_1, h_2, \ldots, h_n)$ and the basal variant $(d_1, d_2, \ldots, d_n)$ are determined from the elements of matrix **W**:

$$h_j = \max\,(w_{ij}), \tag{4}$$

$$d_j = \min\,(w_{ij}). \tag{5}$$

(d) the distances of the variants from ideal and basal variations are calculated by:

$$d_i^{+} = \sqrt{\sum_{j=1}^{k} (w_{ij} - h_j)^2}, \quad (6)$$

$$d_i^{-} = \sqrt{\sum_{j=1}^{k} (w_{ij} - d_j)^2}. \quad (7)$$

(e) the regions are then ranked from the perspective of human capital creativity based on descending values of an indicator c_i defined as follows:

$$c_i = \frac{d_i^{-}}{d_i^{-} + d_i^{+}}. \quad (8)$$

As the comprehensive data for all criteria are only available for the year 2011, it is not possible to monitor the regional ranking in a longer period.

The TOPSIS method, which is focused on a relative comparison, is supplemented by the SAW method, which helps to compare the regions absolutely. The SAW method uses the degree of goal fulfilment in terms of criterion. We use the same weights as determined by using the Saaty and the scoring methods. In the case of using the SAW method, we replace the value of the criteria by the points from 0 to 10 which express our expert opinion, as the value of the criterion meets the ideal status of what the region should achieve in the given criterion. In other words, how the region meets the sub-target in terms of the criteria value. One point is interpreted as fulfilling 10% of the objective from the criteria point of view. Ten points thus represent a 100% fulfilment in terms of the criteria. A ten-point scale, in our opinion, corresponds to the sensitivity with which an expert is able to quantify the fulfilment of the sub-objective.

To process the collected data and provide calculations as indicated below, we used statistical systems and the following statistical and analytical software: R 3.5.1 for Statistical Computing (R Core Team 2018), MS Excel, AHP online, Openxlsx [74], Topsis [75], and Factoextra [76]. The following Figure 1 sums up the comprehensive methodology. The methodology used in our research is as robust as possible exploiting several scientific and statistical methods.

3.5. Summary of the Methodology Applied

The main goal of the paper was—as was mentioned before—to identify the distribution of creative actors across the regions in Slovakia, to investigate the nexus between creative actors, and to evaluate their impact on the regional development in Slovakia. The exploratory scientific research on creativity in Slovak regions is based on a consistent literature review, statistical analysis, comparisons, and evaluation of comprehensive data sets. Firstly, we set a criteria matrix of creative actors' indicators, which consists of two groups of indicators identifying the creative people and creative capital in a region. The creative people part of the matrix consists of creative class (CC), population with higher education (PHE), and Bohemians (B). Creative capital consists of indicators related to Intellectual Property (IP), Entrepreneurship (E), and Innovation (I) and their sub-indicators as defined above. Verification of the Hypothesis 2 (H2) is based on a comparison of indicators which belong to the creative population via the Variation coefficient (Vc). For the verification of the Hypothesis 1 (H1) and the Hypothesis 3 (H3) we employed the Pearson and Spearman correlation coefficients (PCC, SCC), the TOPSIS and SAW method. The robustness of methodology used in this paper and the exploitation of the repertoire of mathematical and statistical methods is the main valuable contribution of this paper.

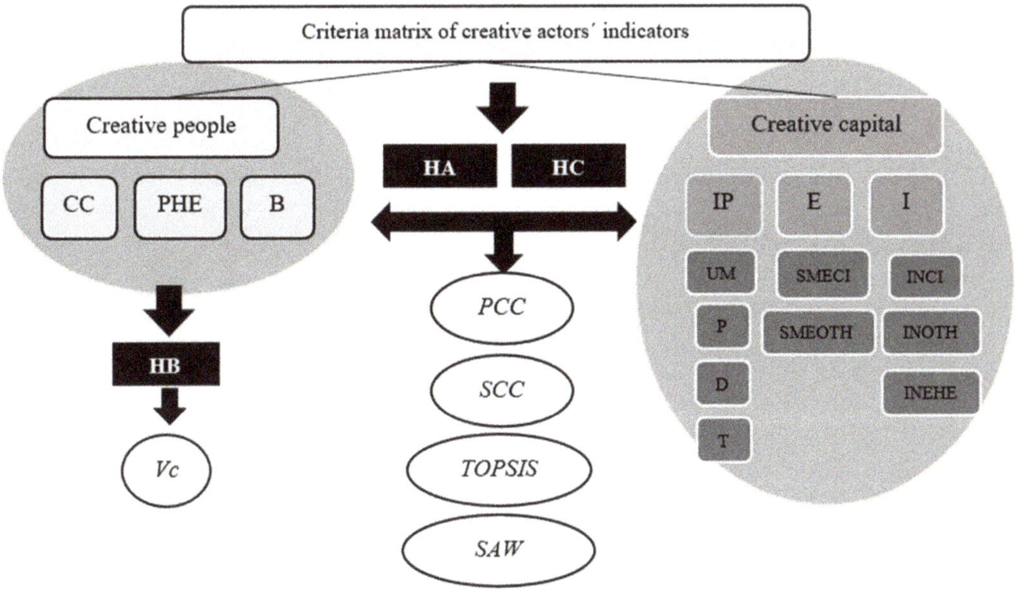

Figure 1. Methodological approach applied.

4. Results

The following section is divided into two sub-sections. The first, Section 4.1 aims to accomplish the first part of the scientific objective of this paper which is to identify the distribution of creative people across the regions in Slovakia (and to verify the Hypothesis 2 (H2)). The second, Section 4.2 is focused on the execution of the second part of the objective which is to evaluate the impact of creative actors on the regional development in Slovakia, and to investigate the nexus between creative actors (and to verify the Hypothesis 1 (H1) and the Hypothesis 3 (H3)). In this section, the research results on the distribution of creative people in the regions of Slovakia are presented together with our finding on the nexus between creative actors and their impact on the regional development measured via creative capital and the ranking of the Slovak regions.

4.1. Distribution of Creative People in the Regions of Slovakia

The analysis on the distribution of creative people is based on the evaluation of indicators that characterize and define creative people: the share of the creative class (CC) in the selected sectors of the economically active population, the share of the population with higher education (PHE) and the share of inhabitants representing Bohemians (B). Table 2 shows the distribution of creative people according to the above-mentioned indicators in the regions of Slovakia (all numbers included are in percentages). In addition, this table depicts the equality, or rather the inequality in the distribution of creative people in Slovak regions based on the above-mentioned indicators.

Data on the representation of creative people in the self-government regions (NUTS 3 regions), as displayed in Table 2, shows the dominant position of the Bratislava Region where 9.20% of economically active people work in creative sectors. This region involves the metropolitan area of the capital city and the share of people working in creative industries is almost three times higher in comparison to all other regions of Slovakia. The average share of university-educated people in Slovakia was 14.03% in 2011, dominating the Bratislava Region with a share of 26.17%. The rest of the seven Slovak regions reached similar values with the lowest share in the Trnava Region. The share of the population with higher education is, as in the previous case, highest in the Bratislava Region. This is caused by a higher concentration of universities in the capital city but also by the metropolitan character of the city that is highly attractive for creative people and higher-educated people. This confirms the results of the study by Hansen and Niedomysl [67] that correlation between people with higher education and creative people is high. The share of Bohemians in Slovak regions truly brings an interesting result. In the case of five Slovak regions (Bratislava, Nitra, Presov, Kosice, and Banska Bystrica), the share of Bohemians is higher than the national average. This indicator, along with Florida´s statement (Bohemians play an important role in acquiring other creative categories and are a sign of urban cultural tolerance), suggests a potential for the development of appropriate regional development strategies to attract and maintain creative people in lagging areas [66]. Presov, Kosice, and Banska Bystrica regions belong to the regions with the lowest regional GDP, higher unemployment, and the highest labor outflow. When comparing the four selected indicators, we can see inequalities in the distribution of creative class (0.46) and the population with higher education (0.33) across Slovak regions (see Table 2).

Table 2. Inequalities in distribution of creative people in Slovak regions (in %).

Region	Creative Class	People with Higher Educational	Bohemians
BA	9.20	26.17	6.99
TT	3.63	11.50	5.23
TN	3.04	12.51	4.65
NR	3.44	11.99	6.10
ZA	3.69	12.96	4.74
BB	3.71	12.33	5.87
PO	3.01	11.70	6.32
KE	3.86	13.06	6.06
Slovakia	4.19	14.03	5.74
Standard deviation	1.9135	4.6196	0.7601
Average	4.1947	14.0279	5.7445
Vk	0.45617	0.3293	0.1323

Legend: BA: Bratislava Region, BB Banska Bystrica Region, TT Trnava Region, TN Trencin Region, PO Presov Region, NR Nitra Region, KE Kosice Region, ZA Zilina Region. **Source:** own processing based on the data from the Statistical Office of Slovak Republic from the year 2011.

Based on the previous results on the distribution of creative people, we opt to examine the dependencies between the creative class and a set of selected indicators as in the previous cases based on the correlation analysis by using the Pearson correlation coefficient. The results of the correlation analysis are described in Table 3 below. The Pearson correlation coefficient is usually regarded as an appropriate method for measuring the linear association between two sets of variables of interest as it is based on the method of a covariance. It provides the information about the magnitude of the association, or the correlation, as well as the direction of the mutual relations.

Table 3. Dependencies between the occurrence of creative people and a set of selected indicators in Slovakia.

Indicators	Creative Class	People with Higher Education	Bohemians
Pearson coefficient	1	0.958 **	0.993 **
Research areas	8	8	8

Source: own processing by using R 3.5.1 for Statistical Computing (R Core Team 2018). ** correlation is significant at the 0.01 level (2-tailed).

We can observe a strong direct positive relationship between the population with a higher education and the occurrence of creative class (with a correlation coefficient of 0.958). There is a strong direct positive relation between the size of the creative class and the number of Bohemians. Our research results also showed a strong direct positive nexus between the representatives of creative people.

The secondary data on the occurrence of creative people allowed us to quantify and measure their representation in Slovakia. The research results and their evaluation verify the Hypothesis 2 (H2) that states that the distribution of creative people across the Slovak regions is not equal. The research results also show that there is a strong dominance of the Bratislava Region, while the remainder of the regions are more or less at the same level. In the next section, we focus on presenting the nexus between creative actors and their impact on regional development.

4.2. Creative Capital and Regional Development in Slovak Regions

In the previous section, we identified and quantified the occurrence of creative people and their spatial distribution across the regions of Slovakia. In order to meet the objective of this paper, we further need to identify the relationship between the occurrence of creative people and creative capital. For this purpose, we opt for the deployment of the Spearman correlation coefficient and we use all the indicators as defined in the criteria matrix in Section 3. The results are displayed in Table 4.

Table 4. Nexus between the occurrence of creative people and creative capital in Slovakia in the year 2011.

	SMEOTH	SMECI	P	UM	D	T	INOTH	INCI	INEHE
CC	0.857	0.786	−0.119	0.311	0.049	0.095	0.000	0.000	0.346
B	0.833	0.738	−0.310	0.072	0.049	−0.095	−0.037	0.210	0.037
PHE	0.881	0.548	−0.119	0.503	0.272	0.000	−0.210	0.037	0.037

Source: own processing in MS Excel. **Legend:** creative population consists of the following indicators: the share of the creative class (CC), the share of the population with higher education (PHE), the share of inhabitants representing Bohemians (B). Creative capital consists of the following indicators: a number of applications for patents (P), utility models (UM), designs (D) and trademarks (T), small and medium-sized enterprises in creative industries (SMECI) and outside the creative industries, marked as small and medium-sized enterprises in others (SMEOTH), innovation enterprises in creative industries (INCI), innovation outside creative industries, marked as innovations in others (INOTH) and innovation in enterprises with more than 75% share of employees with higher education (INEHE).

The investigation of the nexus between the creative people and the selected indicators of creative capital in Table 4 has shown a strong direct relationship between the occurrence of creative people and the emergence of SMEs in both creative industries and other industries. Moreover, each of the indicators of the occurrence of creative people shows a strong direct relationship to the emergence of SMEs. The research results also showed a moderate positive correlation between the occurrence of people with a higher education and utility models, and a weak direct correlation of utility models and the creative class. Our research results regarding the weak negative correlation between the creative class and the occurrence of the patents are consistent with the results from the study of Boschma and Fritsch [66] (p. 417). This might probably be caused by the fact that the creative class works primarily in cultural and creative industries that are not technology-intensive. A more in-depth study on the distribution of the copyrights seems to be needed here as it might serve a more appropriate and relevant indicator.

The following part of this section is dedicated to the ranking of Slovak regions as far as their creative capital is concerned. The measurement of creative capital in Slovak regions is based on a criteria matrix of creative actors´ indicators as defined in Section 3 and consists of indicators covering entrepreneurship, intellectual property, and innovation.

The creation of the ranking is based on the TOPSIS method exploitation. Due to the need for comprehensive data including those from the population census, we must use the most recent data from the year 2011 when the last census was organized. Before applying the TOPSIS method, we need to determine the weights of the criteria. For this purpose, we use the Saaty method as indicated in Section 3 (The procedure of assigning weights through the Saaty method can be seen at https://bpmsg.com/ahp-online-system/ (accessed date: 25 April 2019) under the name of the Creativity Index project using the email address and password that we can provide in the case of any interest.).

In order to apply the more comprehensive view on the identification and measurement of creativity in Slovak regions, we additionally used the SAW method which reflects the order of the regions from the actual values of the examined indicators. In this method, we replaced the absolute values with points. When assigning points to the individual (NUTS 3) regions and to the absolute values of variables, we used the comparison with available values (the average values) measured in the European Union. The weights for the variables are the same as described by the TOPSIS method (the Saaty method). Assigning the points to absolute values according to the available EU data is based on a set of indicators identifying the occurrence of creative people and creative capital.

A set of indicators identifying the occurrence of creative people in the regions of Slovakia is based on multiple data sources as follows:

- The share of the creative class (CC), for which we used the map of creative class localization in Europe created by the European Spatial Planning Observation Network (ESPON) (For more information, see https://www.espon.eu/search/node/creative%20workforce (accessed date: 25 April 2019));
- The share of the population with a higher education (PHE) that is based on the data from the applied research report of ESPON (For more information, see https://www.espon.eu/topics-policy/publications/synthesis-reports/third-espon-2013-synthesis-report (accessed date: 25 April 2019));
- The share of inhabitants representing Bohemians (B) is based on work by Boschma and Fritsch [66] (p. 404).

A set of indicators identifying creative capital in the region is divided into three subgroups:

- The intellectual property measured via the number of patent applications (P), utility models (UM), designs (D), and trademarks (T). All variables occurring in this group were compared with the data available from the research and ranking of the European commission which evaluates Slovakia as a moderate innovator (For more information, see https://ec.europa.eu/growth/industry/innovation/facts-figures/scoreboards_en (accessed date: 25 April 2019)).
- Entrepreneurship measured via the establishment of small and medium-sized enterprises in creative industries (SMECI) and outside the creative industries, marked as small and medium-sized enterprises in others (SMEOTH). We have analyzed both variables based on the data from EC (For more information, see https://ec.europa.eu/digital-single-market/en/policies/startup-europe (accessed date: 25 April 2019)) and the world business monitor [70]. The available documents contain the data on an existing start-up ecosystem but do not include country ranking. For this reason, based on the comparison and deduction of available data, we have assigned points for the SME variables.
- Innovation measured via innovation activities of enterprises, namely innovation enterprises in creative industries (INCI), innovation outside creative industries, marked as innovations in others (INOTH), and innovation in enterprises with more than a 75% share of employees with higher education (INEHE). We used the data from the European Innovation Ranking in the group of indicators on intellectual property.

We further evaluated the Slovak regions by using the TOPSIS method and the SAW method. In addition, we added the rank of the regions according to indicators of the creative population to bring a comprehensive view on the distribution of creativity in Slovak regions. Table 5 shows the ranking of the Slovak regions based on the criteria matrix of indicators focused on the identification and measurement of creativity (as defined in Section 3).

Table 5. Ranking of creativity in Slovak regions based on the criteria matrix of creative actors′ indicators.

Region	TOPSIS Method		SAW Method		Creative People Rankings		
	Score	Rank	Score	Rank	Rank CC	Rank PHE	Rank B
BA	1	1	0.688	1	1	1	1
TT	0.0839	8	0.469	7	5	8	6
TN	0.1253	6	0.438	8	8	4	8
NR	0.1451	3	0.499	4	7	6	4
ZA	0.1403	5	0.482	6	4	3	7
BB	0.1156	7	0.496	5	3	5	5
PO	0.1435	4	0.509	3	8	7	3
KE	0.1812	2	0.559	2	2	2	2

Source: own processing in MS Excel and the R programme. **Legend:** BA Bratislava Region, BB Banska Bystrica Region, TT Trnava Region, TN Trencin Region, PO Presov Region, NR Nitra Region, KE Kosice Region, ZA Zilina Region. **Comment:** Creative population consists of the following indicators: the share of creative people (CP), the share of the population with higher education (PHE), the share of inhabitants representing Bohemians (B).

The research results summarized in Table 5 show that the Bratislava Region is closest to the ideal value and at the same time furthest from the basal value according to the TOPSIS method. It is followed by the Kosice Region, the Nitra Region, the Presov Region, the Zilina Region, and the Trencin Region.

The research results also indicate the dominant position of the Bratislava Region leading in all inspected indicators. The second-best position we found for the Kosice Region that reached the second-best results in all investigated indicators. It is noteworthy that the Kosice Region includes the second biggest city of Slovakia (Kosice) which was awarded the title of the European Capital of Culture in 2013. Kosice invested more than 71 million euros in the development of the city and its urban infrastructure. This caused a significant increase in the number of facilities offering cultural products, services, and events, including the Kulturpark, Kunsthalle, and the SPOT community art centers. It is evident that hundreds of new jobs were created as a synergy effect of the title of the European Capital of Culture [77]. Before receiving the title, the Kosice Region belonged to the least developed regions in Slovakia, with the highest unemployment rate and the lowest regional GDP. Synergies related to the preparations, reaching and benefiting from being the European Capital of Culture started a creative destruction run by young artists and activists leading to a new European image of the city and the expansion of creative industries [78]. It is also noteworthy that the Kosice Region belongs to the peripheral region of the EU on the borders with Ukraine. As confirmed by our research results, the localization of creative people indeed assists and helps to produce creative capital. It is thus obvious that with sufficient financial and organizational support, it is possible to enhance the regional development via creativity. The following section aims to discuss and conclude the most interesting research results and to propose policy implications accordingly.

5. Discussion and Conclusions

The main objective of this paper was to identify the distribution of creative actors across the regions in Slovakia, to investigate the nexus between creative actors, and to evaluate their impact on the regional development in Slovakia.

To achieve this goal, we set three hypotheses. The Hypothesis 2 (H2) related to the inequality in the distribution of creative people across Slovak regions was verified in Section 4.1 (see Table 2). The research results show the dominant position of the Bratislava Region, followed by the Kosice Region. Our research results are in compliance with recent studies of [79–82] that highlight the prominent positions of the regions including nation's capital cities. If we compare our results with the average in the European Union, all Slovak regions are above the average values. The best evaluation was detected in the case of the Bratislava Region with a score of 69% of the EU average. The Kosice Region and the Presov Region reached a score above 50% of the EU level. The Banska Bystrica Region and the Nitra Region are just slightly below 50%, and the rest of the regions are below 50% of the European average (see Table 5 in the Section 4.2, row no. 3).

The Hypothesis 3 (H3) was tested through the investigation on the nexus between creative actors and the nexus between creative people and selected indicators of creative capital. Our research results indicate a strong direct positive nexus between the representatives of creative people and a strong direct dependence between the creative people and the emergence of SMEs in both creative industries and other industries. We can observe a moderate positive correlation between the occurrence of the people with a higher education and utility models, a weak direct correlation of utility models and creative class, and a weak negative correlation between creative class and the occurrence of the patent applications. Therefore, the Hypothesis 3 (H3) was verified only partially.

The Hypothesis 1 (H1) stating that the presence of creative actors contributes to the regional development in Slovakia was also verified only partially by the example of the Kosice Region. The Kosice Region includes the second biggest city of Slovakia (Kosice), which was awarded the title of the European Capital of Culture in 2013. The Kosice Region, especially the City of Kosice, heavily invested in the creative infrastructure. As a synergy effect of reaching the title of the European Capital of Culture, there is a shift to the concept of culture-led urban regeneration [83,84], which is stimulated by the young organizers and entrepreneurs. These young creative people brought knowledge from other European cultural centers and implemented them in Kosice [81]. The position of the Presov Region may be the consequence of positive externalities and desired synergic effects, given the proximity to the neighboring Kosice Region. This result points to a spatial dependence and the existence of a spatial autocorrelation among the regions, as it seems that the region's creative performance may be linked to the level of creativity of its neighbors (for more details please see [82,85]). This is a piece of evidence for the political representatives of the regional authorities that the investments into culture and creativity might indeed bring a significant impact on regional development. It seems to be clear that the support of creativity leads to the regional development, as was confirmed by both our research results and evidenced by the theory and practice from all over the world.

It should be noted here that the dynamics of regional development in a given region consist of many mutually inter-related elements, creating their own original regional mix of human, intangible, and physical resources. Some of these elements, such as the natural environment and the location, are in some respects 'given' or 'fixed', while other resources, such as the population structure and the structure of the economy, are highly variable and a have great potential to be influenced and directed by the target support or a suitable regional development strategy [86–90]. In this sense, enormous space for the sustainability transitions could be found in the forming of the regional economic structures. Moreover, the changes in the social and economic organization of cities and regions may lead to shifts in whole sectors of the economy where the people are able to produce creative capital.

The peripheral regions of Europe situated on its eastern border (two of them, the Kosice Region and the Presov Region, are located in Slovakia) are importantly challenged by company closures, job losses, and outward migration, which creates financial and social challenges for many peripheral municipalities. Therefore, it is of the utmost importance to better understand how to attract, maintain, and develop creative people and generally the creativity. The study of Hansen and Aner [91] on the location dynamics of highly educated people migrating to peripheral regions of Denmark shows that it is important to promote jobs, highlight the quality of jobs, and the higher levels of responsibility that creative people may experience in a job in peripheral or lagging regions. The effort to retain newly graduated people in the regions appears to be important as they can be seen as an important and undervalued resource for local employers and the future of these regions. It has been proved that these people are more interested in developing their activities, are more loyal and regionally patriotic if they are considered to be a valuable part of the local community. Their dynamic and flexible approach may spill over into the host enterprise or organization and enhance the creation of new SMEs in or outside the creative industries. In the broader local or regional environment, this might help to increase regional creativity and competitiveness. Therefore, the main role of places, no matter if they are metropolitan regions or peripheral and rural regions, is to create conditions for attracting and retaining creative people who are producing creative capital.

In most of the EU member states, increased attention is being paid to promoting creativity, creative industries, and attraction of creative people. The benefits of this policy are visible and mapped. There is strong competition among the regions to attract creative people which causes positive effects for more creative regions, and on the contrary, causes outmigration of creative people from the lagging regions which is not so positive. In Slovakia, minimal attention has been paid so far to the support of creative industries and attraction or support oriented towards creative people. This statement seems to be also confirmed by our research results. The inability of Slovakia to absorb the financial support from the European Union aimed at enhancing creativity in the regions is manifested by all Slovak regions lagging behind the EU average in creativity and attracting creative people. On the other hand, investment in culture, support of creative industries, attracting and retaining creative people in the region brought a significant impact on the regional development, as investigated in the case of the peripheral Kosice Region. This is in compliance with the recent study of Yum [92] that creative infrastructure and culture are important elements of creativity. As pointed out by Florida in his last book The New Urban Crisis [93], we can still believe in the value of the creative class as an engine for economic prosperity in cities and regions, but not as an isolated strategy. Its successful implementation would require a broader change in the way cities and regions are planned, designed, and developed.

We have ascertained that creative people are not distributed equally across the regions of Slovakia. The objective of our research is fulfilled as our empirical research results showed that indicators of the occurrence of creative people and creative capital are critical parameters for the development of the regions, ceteris paribus. We investigated the nexus between creative actors, and we discovered the strong direct positive relationship between the representatives of creative people and the strong direct dependence between the creative people and the emergence of SMEs in both creative industries and other industries. The comprehensive methodological approach for measurement of the nexus between creative actors and regional development, as proposed in this paper, helps to depict the impact of creative actors on regional development and might serve as an inspiration for further research or policy implications.

We propose the regions to implement the support for creative actors in integrated strategies of regional development as the potential of creative actors and their contribution to development of regions is still vastly underrated. We also suggest the regions follow the empirical evidence gathered, for example by Rodrigues and Franco [94], that regional strategies should concentrate increasingly on developing the networks and partnerships. Together with the authors of this study we believe that open collaborations between all the regional actors involved might lead to the absorption of the mutual synergies that collaborations provide [95]. In this sense, the performance of collaborations might be maximized and lead to promising results.

Author Contributions: Conceptualization, K.B., P.K., J.N.; methodology, K.B., P.K., J.Š., K.T. and A.V.; formal analysis, K.B., P.K., J.Š., S.M. and A.V.; investigation, K.B., P.K., J.Š., J.N. and A.V.; resources,

K.B., K.T. and J.N.; data curation, K.B., P.K., J.Š., S.M. and A.V.; writing—original draft preparation, K.B., P.K., J.Š. and A.V.; writing—review and editing, K.B., K.T., J.N. and S.M.; visualization, K.B., J.Š., P.K. and K.T. All authors have read and agreed to the published version of the manuscript.

Funding: The paper is a partial output of the project VEGA 1/0380/20 Innovative approaches to the development of small and medium-sized cities. This research paper was also written as a part of a research project supporting a long-term inter-sectoral collaboration "Smart technologies for the improvement of quality of life in cities and regions", identification number CZ.02.1.01/0.0/0.0/17_049/0008452. This project is funded by the European Union Social Fund, the Operational Programme "Research, Development and Education" led by the Ministry of Education, Youth and Sports of the Czech Republic.

Conflicts of Interest: The authors declare no conflict of interest. The funders had no role in the design of the study; in the collection, analyses, or interpretation of data; in the writing of the manuscript, or in the decision to publish the results.

References

1. Andersson, Å.E. Creativity and Regional Development. *Pap. Reg. Sci. Assoc.* **1985**, *56*, 5–20. [CrossRef]
2. Florida, R. *The Rise of the Creative Class*; Basic Books: New York, NY, USA, 2002.
3. Karlsson, C.; Johansson, B.; Stough, R.R. The Regional Economics of Knowledge and Talent. In *The Regional Economics of Knowledge and Talent*; Edward Elgar Publishing: Northampton, MA, USA, 2012; pp. 27–62.
4. Neal, Z.P. *The Connected City*; Routledge: London, UK, 2012.
5. Turečková, K.; Nevima, J. The Cost Benefit Analysis for the Concept of a Smart City: How to Measure the Efficiency of Smart Solutions? *Sustainability* **2020**, *12*, 2663. [CrossRef]
6. Tafel-Viia, K.; Terk, E.; Lassur, S.; Viia, A. Creative industries in the capital cities of the Baltic States: Are there innova-tions in urban policy? *Morav. Geogr. Rep.* **2015**, *23*, 47–58. [CrossRef]
7. Griliches, Z. Issues in assessing the contribution of R&D to productivity growth. *Bell J. Econ.* **1979**, *10*, 92–116.
8. Romer, P.M. Endogenous technical change. *J. Political Econ.* **1990**, *98*, S71–S102. [CrossRef]
9. Bode, E.; Villar, L.P. Creativity, education or what? On the measurement of regional human capital. *Pap. Reg. Sci.* **2015**, *96*, S51–S67. [CrossRef]
10. Karlsson, C. Clusters, networks and creativity. In *Handbook of Creative Cities*; Andersson, D.E., Andersson, Å.E., Mellander, C., Eds.; Edward Elgar: Cheltenham, UK, 2011; pp. 95–114.
11. Boden, M.A. Précis ofThe creative mind: Myths and mechanisms. *Behav. Brain Sci.* **1994**, *17*, 519–531. [CrossRef]
12. Glaeser, E.L. Review of Richard Florida's The Rise of the Creative Class. *Reg. Sci. Urban Econ.* **2005**, *35*, 593–596. [CrossRef]
13. Faggian, A.; Comunian, R.; Jewell, S.; Kelly, U. Bohemian Graduates in the UK: Disciplines and Location Determinants of Creative Careers. *Reg. Stud.* **2013**, *47*, 183–200. [CrossRef]
14. Mellander, C.; Florida, R. *The Creative Class or Human Capital? Explaining Regional Development in Sweden*; Electronic Working Paper Series in Economics and Institutions of Innovation Paper, No. 79; Royal Institute of Technology, CESIS-Centre of Excellence for Science and Innovation Studies: Stockholm, Sweden, 2007.
15. Florida, R.; Mellander, C.; Stolarick, K. Inside the black box of regional development—Human capital, the creative class and tolerance. *J. Econ. Geogr.* **2008**, *8*, 615–649. [CrossRef]
16. Faggian, A.; Comunian, R.; Li, Q.C. Interregional migration of human creative capital: The case of "Bohemian gradu-ates". *Geoforum* **2014**, *55*, 33–42. [CrossRef]
17. Hartley, J.; Potts, J.; Cunningham, S.; Flew, T.; Keane, M.; Banks, J. *Creative Industries*; SAGE Publications Pvt Ltd.: Newbury Park, CA, USA, 2013; pp. 59–62.
18. Howkins, J. *The Creative Economy: How People Make Money from Ideas*; Allen Lane: London, UK, 2001.
19. Mundelius, M. *The Reliance of Berlin's Creative Industries on Milieus*; VDM Verlag dr. Mueller: Saabrücken, Germany, 2008.
20. Batabyal, A.A.; Nijkamp, P. Creative capital in production, inefficiency, and inequality: A theoretical analysis. *Int. Rev. Econ. Financ.* **2016**, *45*, 553–558. [CrossRef]
21. Florida, R. Cities and the Creative Class. *City Community* **2003**, *2*, 3–19. [CrossRef]
22. Florida, R. *Cities and the Creative Class*; Routledge: New York, NY, USA, 2005.
23. Landry, C.H. *The Creative City: A Toolkit for Urban Innovators*; Earthscan: London, UK, 2008.
24. Peck, J. Struggling with the Creative Class. *Int. J. Urban Reg. Res.* **2005**, *29*, 740–770. [CrossRef]
25. Pratt, A.C. Creative cities: The cultural industries and the creative class. *Geogr. Ann. Ser. B Hum. Geogr.* **2008**, *90*, 107–117. [CrossRef]
26. Moss, I.D. Deconstructing Richard Florida. Available online: http://createquity.com/2009/04/deconstructing-richard-florida.html (accessed on 11 December 2020).
27. Hoyman, M.; Faricy, C.H. It Takes a Village: A Test of the Creative Class, Social Capital, and Human Capital Theories. *Urban Aff. Rev.* **2009**, *44*, 311–333. [CrossRef]

28. Markusen, A. Urban Development and the Politics of a Creative Class: Evidence from a Study of Artists. *Environ. Plan. A Econ. Space* **2006**, *38*, 1921–1940. [CrossRef]
29. Krätke, S. *The Creative Capital of Cities: Interactive Knowledge Creation and the Urbanization Economies of Innovation*; Wiley-Blackwell: Chichester, UK, 2011.
30. Montgomery, J. Beware 'The creative class': Creativity and wealth creation revisited. *Local Econ.* **2005**, *20*, 337–343. [CrossRef]
31. Higgs, P.; Cunningham, S. *Australia's Creative Economy: Mapping Methodologies*; ARC Centre of Excellence for Creative Industries & Innovation (CCI): Brisbane, Australia, 2007.
32. Throsby, D. The concentric circles model of the cultural industries. *Cult. Trends* **2008**, *17*, 147–164. [CrossRef]
33. Nevima, J. *Konkurenceschopnost Regionů Visegrádské Čtyřky (Teoretické a Empirické Přístupy k Vymezení, Měření a Hodnocení)*; Professional Publishing: Praha, Czech Republic, 2014.
34. Novotná, J.; Novotný, L. Industrial clusters in a post-socialist country: The case of the wine industry in Slovakia. *Morav. Geogr. Rep.* **2019**, *27*, 62–78. [CrossRef]
35. Turečková, K. Quaternary sector as a source of growth and competitiveness in the EU. In Proceedings of the 2nd International Conference on European Integration Ostrava, Ostrava, Czech Republic, 15–16 May 2014; pp. 723–730.
36. Gwosdz, K.; Domański, B.; Bilska-Wodecka, E. Localised capabilities as an intermediating factor in the transition from an old to a new development path: The case of post-socialist industrial towns. *Morav. Geogr. Rep.* **2020**, *28*, 123–135. [CrossRef]
37. Boix, R.; Hervás-Oliver, J.L.; De Miguel-Molina, B. Micro-geographies of creative industries clusters in Europe: From hot spots to assemblages. *Pap. Reg. Sci.* **2015**, *94*, 753–772. [CrossRef]
38. Boix-Domenech, R.; Soler-Marco, V. Creative service industries and regional productivity. *Pap. Reg. Sci.* **2015**, *96*, 261–279. [CrossRef]
39. Tinagli, I.; Florida, R.; Ström, P.; Wahlqvist, E. Sweden in the Creative Age. 2007. Available online: https://creativeclass.com/articles/Sweden%20in%20the%20Creative%20Age.pdf (accessed on 17 November 2020).
40. Lazzeretti, L.; Capone, F.; Boix, R. Reasons for clustering of creative industries in Italy and Spain. *Eur. Plan. Stud.* **2012**, *20*, 1243–1262. [CrossRef]
41. Marrocu, E.; Paci, R. Regional development and creativity. *Int. Reg. Sci. Rev.* **2013**, *36*, 354–391. [CrossRef]
42. Cerisola, S. Creativity and local economic development: The role of synergy among different talents. *Pap. Reg. Sci.* **2018**, *97*, 199–215. [CrossRef]
43. Kourtit, K.; Nijkamp, P. Creative actors and historical–cultural assets in urban regions. *Reg. Stud.* **2018**, *53*, 977–990. [CrossRef]
44. Chapain, C.; Clifton, N.; Comunian, R. Understanding Creative Regions: Bridging the Gap between Global Discourses and Regional and National Contexts. *Reg. Stud.* **2013**, *47*, 131–134. [CrossRef]
45. Chapain, C.; Comunian, R. Enabling and Inhibiting the Creative Economy: The Role of the Local and Regional Dimensions in England. *Reg. Stud.* **2009**, *44*, 717–734. [CrossRef]
46. de Figueiredo, J.L.; de Jesus, D.S.V.; Tavares Robaina, D.; Lohmann Couri, C. The development potential index of creative economy for Brazilian federal state capitals. *Creat. Ind. J.* **2019**, *12*, 185–203. [CrossRef]
47. Evans, G. Creative Cities, Creative Spaces and Urban Policy. *Urban Stud.* **2009**, *46*, 1003–1040. [CrossRef]
48. Romein, A.; Trip, J.J. Theory and practice of the creative city thesis: The cases of Amsterdam and Rotterdam. Presented at the ACSP-Aesop 4th Joint Congress 'Bridging the Divide: Celebrating the City, Chicago, IL, USA, 6–11 July 2008.
49. Storper, M.; Scott, A.J. Rethinking human capital, creativity and urban growth. *J. Econ. Geogr.* **2008**, *9*, 147–167. [CrossRef]
50. Marlet, G.; van Woerkens, C. The Dutch creative class and how it fosters urban employment growth. *Urban Stud.* **2007**, *44*, 2605–2626. [CrossRef]
51. Florida, R.; Tinagli, I. *Europe in the Creative Age*; Demos: London, UK, 2004.
52. Asheim, B.; Hansen, H.G. Knowledge bases, talents, contexts: On the usefulness of the creative class approach in Sweden. *Econ. Geogr.* **2009**, *85*, 425–442. [CrossRef]
53. Kloudová, J. Kreativní ekonomika a její měření. *Ekon. Časopis* **2009**, *57*, 247–262.
54. Kloudová, J.; Chwaszcz, O. New way of analysis of creative centres within Europe. *Econ. Manag.* **2011**, *16*, 197–206.
55. Vaňová, A.; Kozáková, V.; Miškovičová, A. Modifikácia indexu kreativity pre Slovensko. *Banská Bystrica Reg. Direct* **2013**, *1*, 58–79.
56. Vaňová, A. Kreatívna ekonomika a rozvoj územia (z pohľadu marketingu územia). In *Kloudová el al. Kreativní Ekonomika*; EUROKÓ DEX: Bratislava, Slovakia, 2010; pp. 59–84.
57. Schultz, T.W. Investment in Human Capital. *Am. Econ. Rev.* **1961**, *51*, 1–17.
58. Becker, G.S. *Human Capital*; University of Chicago Press: Chicago, IL, USA, 1993.
59. Fitzenz, J. *ROI of Human Capital: Measuring Economic Value*; AMACOM: New York, NY, USA, 2000.
60. Nijkamp, P. E Pluribus Unum. *Reg. Direct* **2009**, *2*, 567–575.
61. Kourtit, K.; Nijkamp, P.; Lowik, S.; Van Vught, F.; Vulto, P. From islands of innovation to creative hotspots. *Reg. Sci. Policy Pract.* **2011**, *3*, 145–161. [CrossRef]
62. Stimson, R.J.; Stough, R.; Nijkamp, P. (Eds.) *Endogenous Regional Development: Perspectives, Measurement and Empirical Investigation*; Edward Elgar Publishing: Northampton, MA, USA, 2011.
63. Montgomery, J. Creative industry business incubators and managed workspaces: A review of best practice. *Plan. Pr. Res.* **2007**, *22*, 601–617. [CrossRef]

64. Sørensen, F.; Fuglsang, L.; Sundbo, J. Experience economy, creative class and business development in small Danish towns. *Urban Res. Pract.* **2010**, *3*, 177–202. [CrossRef]
65. Florida, R.; Mellander, C.; Stolarick, K.; Silk, K.; Matheson, Z.; Hopgood, M. Creativity and Prosperity: The Global Creativity Index. Available online: http://martinprosperity.org/wp-content/uploads/2011/10/Creativity-and-Prosperity_The-Global-Creativity-Index.pdf (accessed on 12 November 2020).
66. Boschma, R.A.; Fritsch, M. Creative Class and Regional Growth: Empirical Evidence from Seven European Countries. *Econ. Geogr.* **2009**, *85*, 391–423. [CrossRef]
67. Hansen, H.K.; Niedomysl, T. Migration of the creative class: Evidence from Sweden. *J. Econ. Geogr.* **2008**, *9*, 191–206. [CrossRef]
68. Šprocha, B. Vnútorná migrácia osôb s vysokoškolským vzdelaním do 35 rokov. *Progn. Práce Sav* **2011**, *3*, 299.
69. Rehák, Š. Kreatívna trieda a kreatívne mestá. Teoretický koncept a jeho limity. *Sociológia* **2014**, *46*, 599–619.
70. GEM. Global Entrepreneurship Monitor. Available online: http://www.gemconsortiumorg/report (accessed on 17 November 2020).
71. Ira, V.; Garitan, M. Ekonomická aktivita v tvorivých odvetviach: Postavenie Bratislavy v rámci Slovenska a vnútromestské rozdiely. *Geogr. Slovaca* **2014**, *28*, 197–209.
72. Saaty, T.L. How to make a decision: The analytic hierarchy process. *Eur. J. Oper. Res.* **1990**, *48*, 9–26. [CrossRef]
73. Velasquez, M.; Hester, P.T. An analysis of multi-criteria decision making methods. *Int. J. Oper. Res.* **2013**, *10*, 56–66.
74. Walker, A. Openxlsx: Read, Write and Edit XLSX Files. R Package Version 4.1.0. Available online: https://CRAN.R-project.org/package=openxlsx (accessed on 1 December 2020).
75. Yazdi, M.M. Topsis: TOPSIS Method for Multiple-Criteria Decision MAKING (MCDM). R Package Version 1.0. Available online: https://CRAN.R-project.org/package=topsis (accessed on 1 December 2020).
76. Kassambara, A.; Mundt, F. Factoextra: Extract and Visualize the Results of Multivariate Data Analyses. R Package Version 1.0.7. Available online: https://CRAN.R-project.org/package=factoextra (accessed on 3 December 2020).
77. Borseková, K.; Vaňová, A.; Vitálišová, K. Building Košice European Capital of Culture: Towards a Creative City? In *Tourism in the City*; Springer: Cham, Switzerland, 2017; pp. 193–205.
78. Hudec, O.; Remoaldo, P.C.; Urbančíková, N.; Cadima Ribeiro, J.A. Stepping Out of the Shadows: Legacy of the European Capitals of Culture, Guimarães 2012 and Košice. *Sustainability* **2019**, *11*, 1469. [CrossRef]
79. Montalto, V.; Moura, C.J.T.; Langedijk, S.; Saisana, M. Culture counts: An empirical approach to measure the cultural and creative vitality of European cities. *Cities* **2019**, *89*, 167–185. [CrossRef]
80. Miguel, B.S.; Herrero-Prieto, L.C. Reliability of creative composite indicators with territorial specification in the EU. *Sustainability* **2020**, *12*, 3070. [CrossRef]
81. Correia, C.M. Measuring creativity in the EU member states. Investigaciones Regionales. *J. Reg. Res.* **2014**, *30*, 7–26.
82. Petríková, K.; Vaňová, A.; Borsekova, K. The role of creative economy in Slovak Republic. *AI Soc.* **2013**, *30*, 271–281. [CrossRef]
83. Bosák, V.; Nováček, A.; Slach, O. Industrial culture as an asset, barrier and creative challenge for restructuring of old industrial cities: Case study of Ostrava (Czechia). *GeoScape* **2018**, *12*, 52–64. [CrossRef]
84. Bosák, V.; Slach, O.; Nováček, A.; Krtička, L. Temporary use and brownfield regeneration in post-socialist context: From bottom-up governance to artists exploitation. *Eur. Plan. Stud.* **2019**, *28*, 604–626. [CrossRef]
85. Hudec, O.; Džupka, P. Culture-led regeneration through the young generation: Košice as the European Capital of Culture. *Eur. Urban Reg. Stud.* **2016**, *23*, 531–538. [CrossRef]
86. Boix, R.; Lazzeretti, L.; Cappone, F.; Sánchez-Serra, D. The geography of creative industries in Europe: Comparing France, Great Britain, Italy and Spain. In *Creative Industries and Innovation in Europe. Concepts, Measures and Comparative Case Studies*, 1st ed.; Lazzeretti, L., Ed.; Routledge: Abingdon, UK, 2012.
87. Bertacchini, E.E.; Borrione, P. The Geography of the Italian Creative Economy: The Special Role of the Design and Craft-based Industries. *Reg. Stud.* **2013**, *47*, 135–147. [CrossRef]
88. Yum, S. The relationship between creative industries and the urban economy in the USA. *Creative Ind. J.* **2020**, *13*, 95–116. [CrossRef]
89. Boal, I.; Herrero, L.C. Where are the artists? Analysing economies of agglomeration in Castile and León, Spain. *Pap. Reg. Sci.* **2017**, *97*, 995–1016. [CrossRef]
90. Heikkilä, E.; Pikkarainen, M. Differential population development in the regions of Finland. *Popul. Space Place* **2009**, *16*, 323–334. [CrossRef]
91. Hansen, H.K.; Aner, L.G. On the location dynamics of highly educated people migrating to peripheral regions of Denmark. *Popul. Space Place* **2017**, *23*, e2076. [CrossRef]
92. Yum, S. How Can We Measure the Magnitude of Creative Cities? A New Creativity Index: 3Ci. *Creat. Res. J.* **2020**, *32*, 174–183. [CrossRef]
93. Florida, R. *The New Urban Crisis: How Our Cities Are Increasing Inequality, Deepening Segregation, and Failing the Middle Class—And What We Can Do about It*; Basic Books: New York, NY, USA, 2017.
94. Rodrigues, M.; Franco, M. Composite index to measure cities' creative performance: An empirical study in the Portu-guese context. *Sustainability* **2019**, *11*, 774. [CrossRef]
95. Hansen, T.; Coenen, L. The geography of sustainability transitions: Review, synthesis and reflections on an emergent re-search field. *Environ. Innov. Soc. Transit.* **2015**, *17*, 92–109. [CrossRef]

land

MDPI

Article

Urban-Rural Dependencies and Opportunities to Design Nature-Based Solutions for Resilience in Europe and China

Ellen Banzhaf [1,*], Sally Anderson [2], Gwendoline Grandin [3], Richard Hardiman [4], Anne Jensen [5], Laurence Jones [6], Julius Knopp [1], Gregor Levin [5], Duncan Russel [7], Wanben Wu [1,8], Jun Yang [9] and Marianne Zandersen [10]

1 Department of Urban and Environmental Sociology, UFZ-Helmholtz Centre for Environmental Research, 04318 Leipzig, Germany; julius.knopp@ufz.de (J.K.); wanben.wu@ufz.de or 19110700121@fudan.edu.cn (W.W.)
2 Department of Educational Anthropology, Danish School of Education, Aarhus University, 2400 Copenhagen, Denmark; saan@edu.au.dk
3 Paris Region Biodiversity Agency, Paris Region Institut, CEDEX 15, 75740 Paris, France; gwendoline.grandin@institutparisregion.fr
4 Department of Asian Studies, Faculty of Humanities, Hebrew University of Jerusalem, Jerusalem 9190401, Israel; richard.hardiman@mail.huji.ac.il
5 Department of Environmental Sciences, Aarhus University, 4000 Roskilde, Denmark; aj@envs.au.dk (A.J.); gl@envs.au.dk (G.L.)
6 UK Centre for Ecology & Hydrology, Environment Centre Wales, Deiniol Road, Bangor LL57 2UW, UK; lj@ceh.ac.uk
7 Department of Politics, University of Exeter, Exeter EX4 4QG, UK; d.j.russel@exeter.ac.uk
8 Ministry of Education Key Laboratory for Biodiversity Science and Ecological Engineering, Coastal Ecosystems Research Station of the Yangtze River Estuary, Shanghai Institute of EcoChongming (SIEC), Fudan University, Shanghai 200433, China
9 Ministry of Education Key Laboratory for Earth System Modeling, Department of Earth System Science, Institute for Global Change Studies, Tsinghua University, Beijing 100084, China; larix@tsinghua.edu.cn
10 Department of Environmental Science & Aarhus University Interdisciplinary Centre for Climate Change (iCLIMATE), Aarhus University, 4000 Roskilde, Denmark; mz@envs.au.dk
* Correspondence: ellen.banzhaf@ufz.de; Tel.: +49-(0)341-2351738

Citation: Banzhaf, E.; Anderson, S.; Grandin, G.; Hardiman, R.; Jensen, A.; Jones, L.; Knopp, J.; Levin, G.; Russel, D.; Wu, W.; et al. Urban-Rural Dependencies and Opportunities to Design Nature-Based Solutions for Resilience in Europe and China. *Land* **2022**, *11*, 480. https://doi.org/10.3390/land11040480

Academic Editors: Stephan Bartke and Sigrun Kabisch

Received: 22 February 2022
Accepted: 23 March 2022
Published: 26 March 2022

Abstract: Interrelationships between urban and rural areas are fundamental for the development and safeguarding of viable future living conditions and quality of life. These areas are not well-delineated or self-sufficient, and existing interrelations may privilege one over the other. Major urban challenges facing China and Europe are related to changes in climate, environment, and to decision-making that makes urban and rural landscapes more susceptible to environmental pressures. Focusing on the six European and Chinese cities and surrounding rural areas, under study in the joint EC and MOST-funded REGREEN project, we examine how nature-based solutions (NBS) may assist in counteracting these pressures. We explore urban-rural dependencies and partnerships regarding NBS that can enhance resilience in Europe and China. We analyse differences between European and Chinese systems of governance, reflecting on the significance of the scale of research needed to understand how NBS provide benefits. We highlight interactions between differently delineated sheds (watershed, airshed, natureshed, and peopleshed), which influence the interrelationships between urban and rural areas. There may be one-way or two-way interdependence, and the impact may be uni or multi-directional. The European and Chinese solutions, exemplified in this article, tackle the nexus of environmental and peoplesheds. We discuss complex human interactions (and how to model them) that may, or may not, lead to viable and equitable partnerships for implementing NBS in cities within Europe and in China.

Keywords: nature-based solutions (NBS); green infrastructure; urban planning; airshed; watershed; natureshed; peopleshed; rural–urban fringe; social and biophysical scales

1. Introduction

Urbanisation is one of the most challenging processes we face in the 21st century. This calls for new thinking to create robust collaboration across urban-rural areas that fosters sustainable development in both [1]. While many interrelations between urban and adjacent rural areas exist, these may be skewed in ways that favour one over the other [2,3]. To ensure that such interrelations contribute to a balanced regional partnering from which both areas benefit, interdependencies must be made explicit, with potential asymmetries sounded out [4,5]. Forging sustainable and resilient urban-rural partnerships requires formulating central questions and acknowledging challenges [6]. Especially when designing nature-based solutions (NBS), we must understand that water, air, and nature provide ecosystem services for people, both in and across the areas in which they live and work [7]. Hence, the partnering of urban-rural interests, in relation to watersheds, airsheds, and naturesheds in urban-rural spaces, should be beneficial for an entire peopleshed, which is understood, here, to encompass all who live and work in and across a particular rural and urban area [8,9]. By renewing our understanding of these interlinkages, we may re-adjust our strategies for forging well-functioning and equitable partnerships.

Urban and rural areas are not well-delineated, bounded, or self-sufficient spaces [10]. Rather, comprising areas of different land cover and land use, as well as the different mobility patterns this entails, they are closely intertwined and intensively dependent on each other [11,12]. In a physical sense, they are simultaneously sources and sinks in the resource cycle, especially when it comes to nutrition, water, energy, waste, and housing [1,13,14]. Their interrelationship is, thus, fundamental for the development and safeguarding of viable living conditions and the quality of life in a particular region. Viability, here, relates to jobs and livelihoods, housing markets, recreation, education, and basic supply for residents in a well-balanced way across rural and urban areas [15]. The United Nations' global Sustainability Development Goals (SDGs) propose, in SDG target 11a, that urban-rural links are a critical nexus for planning well-functioning regions [16]. Such a nexus must be based on a more sustainable design of economic, social, political, and environmental urban-rural interaction.

Drawing on the ongoing REGREEN project (https://www.regreen-project.eu/, accessed on 22 March 2022), this article presents examples from China and Europe, which help to understand the importance of different local and cultural contexts for restoring, enhancing, and creating urban and rural, as well as green and blue, infrastructure. We consider cultural histories, societal challenges, topographies, climate, governance systems, and economic and social structures as underlying factors that influence resilience and social cohesion in urban and rural areas. We use and combine these factors to illustrate good examples on both continents. The aim is to generate transferable knowledge of sustainable urban development under different conditions, which can be directly applicable to other cities. Interactions between ecosystems and human societies are highly complex [17–19], with multiple forms of interaction and feedback loops that constantly change both natural and societal systems. For this reason, we follow a holistic social-ecological framework that outlines interrelationships between ecological and social issues to restore and rehabilitate urban ecosystems, as well as enhance sustainable urbanisation in China and Europe.

Major urban challenges, pertinent to both China and Europe, are related to changes in climate and environment. They are also related to historical and current decision-making and management approaches that make urban and adjacent rural landscapes more susceptible to pressures such as flooding or noise, resulting in loss of urban biodiversity, green areas, and open streams. Urbanisation and mass transport have a persistent impact on air quality and health. Different population densities steer accessibility to greenspace and induce asymmetries between urban and rural areas. We will examine how NBS may assist in counteracting these challenges by focusing on six focal European and Chinese cities, and their surrounding rural areas, that are our study sites in the above-mentioned EC and MOST funded project. Acknowledging differences between European and Chinese systems of governance, we learn from exploring the cultural contexts in and between China and Europe. The study seeks to understand urban-rural dependencies and partnerships, regarding NBS, which contribute to amending resilience in Europe and China. The novelty of this research is in the interlinkage of various "sheds" to foster NBS in urban-rural areas. To do so, we put emphasis on conceptualising "people" in urban-rural interlinkages to assess urban-rural NBS, the broad benefits to people, and the challenge of how best to manage the use of, and people's interaction with, the natural environment in urban and rural areas. Our paper seeks to address the following questions:

- How can we explore viable NBS at different scales, and how may they differ in China and Europe?
- How does governance contribute to balancing the specific needs of urban and rural areas in China and Europe?
- How can we tackle the complex human interactions in urban-rural areas with regard to implementing NBS?

The novelty of this research contribution lies in the comparison of governance structures and NBS implementation strategies, between European and Chinese cities with a regional focus ... With a view to the conceptual contribution, the novelty is highlighting the concept of peopleshed, as a non-metric size, and linking it to the NBS scales of watershed, airshed, and natureshed to bring in a broader and more differentiated view on the benefits of NBS.

2. Conceptual Understanding of Urban-Rural Interaction in Selected European and Chinese Regions

2.1. Study Area Selection

The selected cities are all participants in the EC- and MOST funded REGREEN project. The advantage of this selection is manifold. First, to actively participate in this project, these cities have committed themselves to implement NBS for a more equitable, greener, and healthier city. With departments of these cities as active project members, we have a common understanding of all the research efforts we undertake to foster NBS. Second, to comprehend and chart urban areas at various extents, it is important to exemplify representative types. Hence, we have chosen to include megacities such as Shanghai, Beijing, and the Paris Region, as well as typical medium-sized cities, such as Ningbo in China and Aarhus in Europe, and even a small Croatian town of Velika Gorica, in the study. Finally, this selection affords valuable data and information, such as scale dependencies, functional areas, governance structures, knowledge about different 'scales', and initial experiences, that we are using to improve urban-rural partnerships through NBS. Dependent on the illustration of urban-rural partnerships in the different cases (see Section 3), we may relate to the city's boundary or to the functional urban area in which interlinkages with and within peoplesheds occur. Finally, we also have the relevant expertise to illustrate and elaborate the different aspects of the significant European and Chinese situations.

2.2. Developing NBS at Various Urban-Rural Boundaries

Following the call for new integrated territorial planning approaches, as well as people and place-based development at urban-rural interfaces, aligned with localised SDGs, we present and discuss recent attempts to develop such approaches. Integrated planning accounts for the manifold interrelations and linkages between urban and rural spaces.

During periods of rapid urbanisation, intricate, historically evolved urban-rural metabolisms are often at risk, exacerbating environmental, social, and economic sustainability challenges. In this context, NBS offer a useful conceptual framework in which to readdress contemporary multidimensional and multi-scalar challenges—climate change resilience, biodiversity loss, and social inequity—facing both urban and rural areas. To enable the implementation of NBS, governance structures must be understood and addressed. These structures vary between China and Europe, as well as within Europe, at both national and regional levels. Although research on urban governance has increased in recent years, it still lacks analyses of spatial concern, spatial effectiveness, and further-reaching approaches that address space-based solutions for integrated food and substance circuits, as well as innovative settlement models in—particular, cultural landscapes. Steering urban-rural space calls for new forms of institutional collaboration, and here, governance research is still at the beginning (Section 2.3).

When targeting sustainable regional development and, more specifically, how best to achieve an integrated governance of urban-rural relationships, our descriptions and analyses must address conceptual, methodological, and empirical issues. Balancing interests between urban and rural areas in ways that respect their particularities is important for achieving equitable qualities of life. While such questions of urban-rural dependencies and parities are not yet resolved, it is imperative to address them to encourage and promote ways of "partnering" that provide attractive and viable living conditions in an entire region. A myriad of bottlenecks has become obvious and discussed [20–23]. Overcoming such bottlenecks may lead to novel ways of partnering between urban and rural communities to create viable and lasting solutions. Efficient land-use management plays a significant role when it comes to developing regionally integrated patterns of settlement. We thus need to exemplify mechanisms and best practices that may lead to implementing processes of resilient urban development, which are also transferable to other locations. For example, interventions for air quality purification can be planned either in suburban or rural areas in order to decrease air pollution in cities [24,25]. The optimum locations for planting new woodlands can be evaluated and planned using modelling approaches, which evaluate where there will be lowered concentrations of pollutants (the benefit-plume) as a result of planting new woodland. These show the calculated change in $PM_{2.5}$ concentration that result from pollution removal by woodland in that location, taking into account real-world emission sources and meteorology. Such model applications can be used to reduce the $PM_{2.5}$ concentrations in urban areas as well as providing aesthetic values in the rural area. Additionally, city-regional strategies can align political priorities across urban-rural divides and provide green and blue corridors, as well as local water retention areas, to alleviate risks of flooding in cities as well as in rural villages and farmlands (see Section 3.2) [26].

Tackling the issue of whether urban-rural interactions are continuous or dis-continuous raises questions around spatial attributes [27]. We may define urban-rural space as a physical continuity delimited by a catchment area (see Section 2.4) or by other (topographical) features that restrict and channel movement. We may also define urban-rural space as bounded administrative units, well-delineated to manage spatial planning. More striking, perhaps, is the less-used spatial defining tool of "peopleshed", referring, in general, to people who reside and work in and across an urban-rural area, who make social demands on, and undertake recreational and cultural activities in both, and thus set socio-cultural boundaries at another dimension. Juxtaposing these defining criteria reveals the blurred and dynamic quality of urban-rural demarcation. In contrast to the urban footprint, a peopleshed has no hard boundary. What constitutes the urban and rural is, thus, much more than just the spatial extent or administrative assignment. We must explore static and dynamic spatial patterns over time to gain a deeper understanding of their dynamics, especially in terms of land-use management and conflicts resulting from this. Interactions between urban and rural areas are not stable over time. For this reason, we might need to forge a more complex definition of what comprises urban and rural by delineating their mutual functionalities, the movements and flows between them, variations in nightlight intensities, and their physical, administrative, conceptual, and social boundaries. While current thinking focuses on how biophysical structures and processes (natural capital) support services that benefit people, it is important to consider how they are co-produced across social strata.

2.3. Developing NBS at Multiple Scales

Spatial scales are key to understanding how NBS deliver benefits. Several principles underlie this claim. First, the ecological processes, which deliver benefits, operate at different scales. The cooling of heat island effects by urban greenspace is primarily a local benefit, felt within a greenspace area such as a park, and typically up to 200–300 m away [28]. At the other end of the scale, carbon sequestration benefits are transferred globally, due to atmospheric mixing and the long timescales involved in climate change [29]. Even for a single benefit, such as cooling by a park, there are scale effects—the direct shading from the sun only occurs under the tree canopy, while evapotranspirative cooling of the local air may extend a few hundred metres [30,31]. Second, those processes provide benefits to people only when there are people to receive them; this is also called "realised ecosystem services" [32] or demand [33]. Ecological processes, such as water flows and air mass movement, together with the daily or seasonal patterns of human mobility, produce complex spatial interactions in the supply and delivery of ecosystem service benefits. Given that urban and rural areas do not exist in isolation of each other, inter-relations between the two can be used to help enhance the supply of NBS benefits in both directions. Hutchins et al. [34] propose multi-domain relationships, which explain both size-dependency effects of NBS and distance-decay relationships for the benefit received from a given NBS (see Figure 1). These relationships are often non-linear and vary by type of benefit provided. Knowing how these processes work, and the scales (and directions) at which they operate, it is possible to define "sheds", which are differently delineated areas with various kinds of interactions between urban and rural areas (Figure 1). They interdepend in a one-way or two-way flow, they may be uni or multi-directional, and they will differ for the respective domains such as water, air, nature, and people. These domains can be referred to as watersheds, airsheds, or naturesheds, and beyond these, most importantly, as peoplesheds (see Section 2.4).

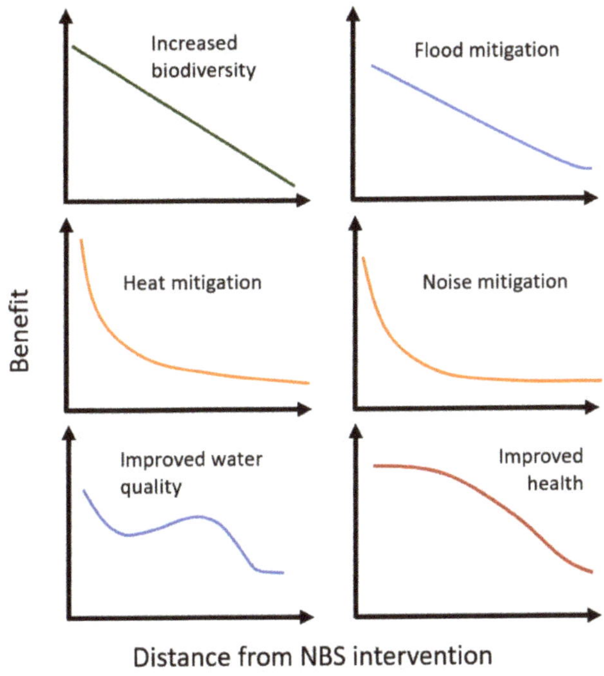

Figure 1. Distance-decay functions for the influence of benefit with distance from the NBS. Modified from [7].

Watersheds, airsheds, and peoplesheds seldom completely overlap. They are differently delineated and characterized by different drivers, structures, and dynamic patterns of movement and flow. Watershed, airshed, and natureshed are defined by their regional extent, their temporal limitation, and their environmental importance for urban-rural interaction. In the urban-rural interface, peoplesheds comprise biophysical, socio-demographic, socio-economic, and socio-cultural scales of human influence, as well as mobility aspects, with varying boundaries, trajectories, and flows of people (Section 2.4).

2.4. Governance of Urban-Rural Interaction across China and Europe

Various regulatory and market-based instruments exist to govern urban-rural relationships that enhance resilience. Urban strategies, land use planning, and zoning are efficient instruments to coordinate, prepare, and design the (multi-)functionality of areas and sectoral activities. Such urban strategies and planning laws, and especially how they are conceived and implemented, differ across European countries and between Europe and China, accentuated by different governance and planning systems, governing capacities, and resources. Furthermore, the extent, speed, and intensity of urban sprawl varies greatly between European and Chinese urban areas. While case-study cities in Europe have grown at a modest pace (11.6%) over the last 20 years, significant growth rates (84%) in Chinese case-study cities have, during the same period, led to a massive expansion into adjacent rural regions (Figure 2).

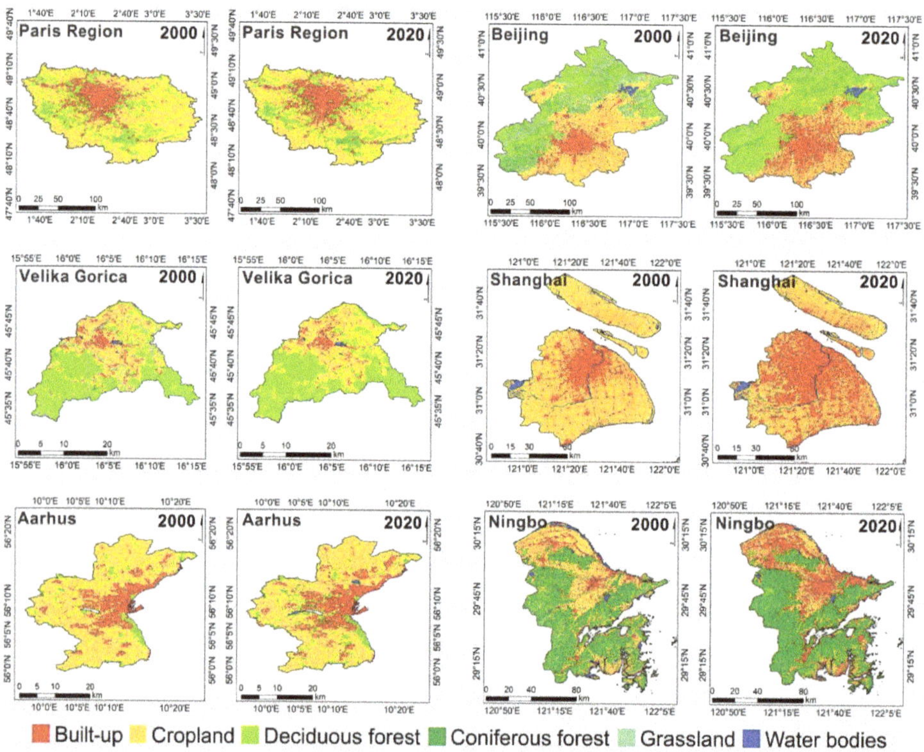

Figure 2. Monitoring land-cover changes in three European and three Chinese cities over the last 20 years. Data modified after Banzhaf et al., 2021 [35].

Governance is influential in limiting the process of urban sprawl, stimulating change in land use functions, and fostering a more sustainable development of urban and rural areas, even deepening their interactions. With the growing challenges of climate hazards, notably flooding and biodiversity degradation, additional land management instruments in urban-rural zones will often be needed. Policy instruments to manage urban-rural interactions involve command and control (e.g., deciding planning and zoning laws), market-based compulsory financial mechanisms (e.g., charges on landowners if their land does not meet requirements to retain water locally), and market-based voluntary incentives (e.g., land acquisition and pre-acquisition, easements, subsidies/compensation, or conditional performance contracting). Changes in policymaking and planning procedures, as well as altered political priorities, may also affect the urban-rural interactions, leading, e.g., to coordinated climate adaptation interventions [36] and influencing NBS-based business development [37] across the urban-rural divide. Depending on a society's reference point for determining whether a landowner's practices represent a negative or a positive externality [38], the urban-rural relationship may be dominated by regulations and taxes (e.g., on landowners, both urban and rural, who fail to meet environmental standards, in land management, that protect urban and rural areas) or dominated by incentives (e.g., to landowners who change land use practices towards natural flood management beyond regulations).

The Chinese government has proposed many urban-rural interaction policies since the 1990s [39–41] (Figure 3). In 1997, the central government issued the policy of "Requisition-compensation Balance of Farmland (ECBF)", which is a strict policy aimed at controlling farmland conversion and replenishing farmland loss caused by urban expansion. In 2012, the concept of "ecological civilization" was formally adopted as a national development strategy, and "Beautiful China" was set up as the main goal. Written into the Chinese constitution in 2018, the concept of ecological civilisation has become the guiding policy for China to achieve a balance between environmental protection and development. A central theme of ecological civilisation is to realize an integrated development between urban and rural areas. In 2013, "Ecological Red Lines" was implemented as a national policy aimed at reforming eco-protection management and promoting ecological civilisation. In 2014, China issued the "New-type Urbanization Plan (NUP) (2014–2020)", which is directed to a more sustainable urbanisation, expanding domestic demand, and rural-urban coordination. Since 2018, the government proposed the "Rural Re-vitalization Strategic Plan (2018–2022)" in which terms such as "green development", "strengthening resource conservation and utilization", "clean agricultural production", and "rural ecological protection and restoration" are explained in detail. At the city level, the three urban agglomerations, Beijing, Shanghai, and Ningbo, have proposed various policies related to urban-rural interactions (Figure 3). For example, the Beijing government issued the "Beijing Urban Master Plan (2016–2035)", which aimed to build a new type of city with green wisdom, distinctive features, liveability, and workability. Shanghai's Ecological Space Construction and Urban Environment Optimization "14th Five-Year" Plan (2020–2025) targeted the building of an "ecological city", with the aim of improving the ecological quality and functions of the urban and rural park system. Ningbo's Ecological Environmental Protection "14th Five-Year" Plan also defined air quality as "fresh", water quality as "clear", and introduced various soil quality concepts to improve urban-rural interactions.

The governance situation in the European Union (EU) is very different from that seen in China. In the European Union, decision-making competencies and governance, surrounding urban-rural interactions and NBS, reside primarily with member states. How urban-rural relationships work in the European Union is, therefore, very much dependent on the constitutional composition of the member state, urban strategies, and localised planning policies. This means that some urban municipalities (e.g., Paris) in some member states are fairly strong political actors in and of themselves, risking rather fragmented governance between urban areas and their rural counterparts. Other urban municipalities—for example, in more federal systems—can be less powerful and subordinate to a higher tier of government that is the principal regional decision-maker able to enact policy approaches that cut across urban-rural divides. This is not to say that there are no policies at the EU level that seek to address urban-rural relations and NBS, but unlike the situation in China, these tend to be soft, non-binding policies, and they are less explicitly focused on urban-rural relationships in relation to NBS. This is mainly due to the EU subsidiarity principle that prompted the formulation and adoption of the European Spatial Development Perspective, with the aim of providing a common basis for member states' planning. The

Perspective has pushed spatial policies to be placed under EU cohesion policy [42,43], which implies that planning is a member-state competence, while many adjacent policy areas, such as climate mitigation, water policy, and air quality, are subject to EU policies and regulations. In 1999, the European Spatial Development Perspective had a specific objective relating to the development of urban-rural partnerships [44]. While non-binding, this objective provides guidelines to all tiers of planning authorities. More recently, an informal meeting of member state planning ministers agreed on the Territorial Agenda 2030 (https://territorialagenda.eu/ accessed on 22 March 2022) to help promote territorial cohesion in European regions, including cohesion between urban and rural areas. Moreover, Cohesion Funds, aimed at programmes seeking to reduce disparities between regions in the EU, may also play a role in facilitating stronger urban-rural relationships.

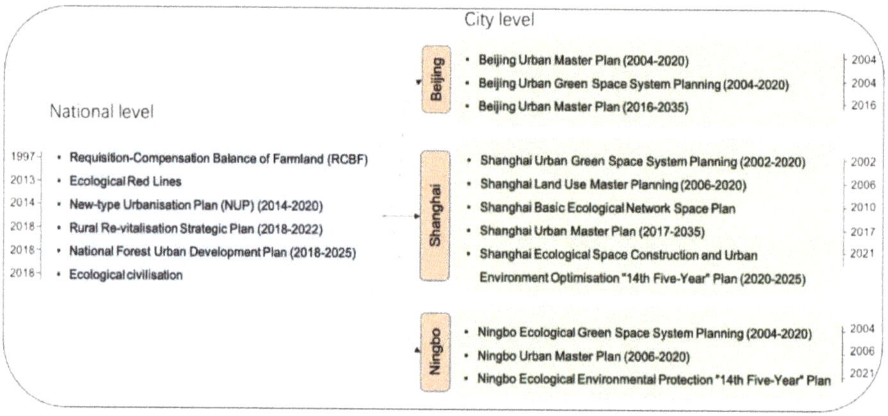

Figure 3. Policy evolution in China, on urban-rural interactions, from 1997 to 2021. To the left, the evolution is explained at the national level; to the right, it is exemplified in more detail, for the three urban areas of the REGREEN project, at city level.

In relation to NBS, the European Union's Urban Agenda [45] highlights the importance of NBS and related biodiversity for adapting to climate change, but it does not explicitly deal with urban-rural relationships in this context. The EU initiated a Research and Innovation policy agenda on NBS to include the re-naturing of cities and territorial resilience [46]. Preceding this EU agenda on NBS, the EU committed to a GI strategy for the protection of natural capital [47], and the Resource Efficiency Roadmap [48] identified investment in Green Infrastructure as an important step towards protecting natural capital. Over the past decade, the EU has placed the use of nature and ecosystem-based approaches at the centre of several policies with urban-rural implications, including climate change adaptation [47], flood management [49], water retention [50], biodiversity [48], and disaster risk management [51,52]. More recently, the ambitions of working with nature have been strengthened in the EU Biodiversity Strategy for 2030 [53], with legally binding nature restoration targets, the Strategy on Adaptation to Climate Change [54], and the Forest Strategy [54]. These policies represent key pillars of the European Green Deal [55], relying on NBS to, i.a., increase resilience to climate change impacts by preserving and restoring ecosystems and managing forests sustainably. Given that these policies are generic, they can be applied at a scale that encompasses urban-rural linkages, and, indeed, they would require enhanced collaboration between urban and rural communities.

The patchwork of EU policies related to NBS and urban-rural relationships, coupled with the fact that decision-making on these aspects rests largely with member states, means that governance structures in this area are rather fragmented. It is, therefore, not surprising that the approaches taken in the three European cities studied in REGREEN are very different. For instance, Velika Gorica (Croatia) is still in the early stages of NBS planning;

with an interest primarily in green roofs, planning is more focused on the urban built-up area. In contrast, the Paris region (France) and Aarhus (Denmark) have more encompassing polices in place, which, among other things, involve interventions on the urban-rural boundaries (see Section 3).

2.5. Supply and Demand of ES for Urban-Rural Interdependencies for Various "Sheds"

Direct regional interdependencies between urban and rural areas become very clear when modelling ecosystem services such as flood regulation. Localised urban-rural regions are also part of a larger space in which several urban and rural areas are interconnected. In the following, we discuss how watersheds, airsheds, naturesheds, and peoplesheds exemplify this larger spatial context in different ways. Due to the spatial context for the urban-rural interconnectedness, the origin of impacts, as well their effects, differ in space and time.

Watersheds are a central concept in hydrology and simply illustrate the principle that activities in a catchment, upstream of a rural area or a city, will affect the volume and timing of water flows, as well as the sediment load and agrochemical content [56] of the water flowing through the settlement. A watershed may be understood as an area from which all precipitation flows to a single stream or set of streams [57]. Its drainage divides natural, as well as built, areas in their respective boundaries: surface water, ground water, as well as larger and smaller subunits. In contrast to rural areas, the large impermeable space of cities is much more exposed to extreme events, such as flash floods. Together, the surface and sub-surface drainage infrastructure, designed to rapidly shift water away from the city, will also affect water flows into rural areas downriver and, possibly, other cities further downstream [58]. Similarly, industrial processes, traffic, air pollution, and waste management infrastructures are likely to substantially affect water quality downstream [59]. Watersheds neatly illustrate how inter-connected rural and urban areas are at multiple scales.

The same principles apply to *airsheds*. They may be understood as a highly dynamic "catchment area" above land, the extent of which varies on any given day, according to the current weather situation. Comprising a much larger area than watersheds, airsheds are differently bounded with different patterns of flow, transportation, and deposition. Air pollutant emissions from urban areas upwind, agricultural emissions of ammonia compounds, and bare ground, leading to the entrainment of dust particles, are all potential air pollutants that then travel over cities and affect the background atmospheric pollution load [60], which, in turn, affects the health of city residents as much as the locally generated air pollutants in the city. Air masses moving over a city will pick up additional pollutants from the city and transport them over downwind rural areas. Ground-level (tropospheric) ozone concentrations are typically higher in rural areas [61] because the precursor chemicals generated in urban areas are transported downwind, while some of the more reactive chemicals, such as nitric oxide, which reduce ozone concentrations in cities, only travel much shorter distances before they are broken down.

Naturesheds, or biodiversity sheds, can be seen as the circulating interactions of plants and animals across rural and urban areas. These interactions may, for example, include the use of water bodies in cities as stopping points for migratory wildfowl, particularly in arid areas. There are also daily foraging or roosting patterns for species, which might roost or lay up in rural areas while sleeping but move into cities during the day (or night) to feed, taking advantage of gardens or waste as food sources. Some species show the opposite behaviour—for example, bats that roost under bridges but move out into the surrounding rural areas to feed at night.

As with air, water, flora, and fauna, people are both situated and mobile. *A peopleshed* may, thus, be delineated by patterns of human dwellings and mobility across broader, interdependent urban-rural areas in ways that variously impact them. Such patterns encompass daily commutes, weekend and holiday travel, recreational activities, and configurations of public and private space. They may be driven by basic needs (livelihood, shelter, education) or social organisation, forms of relatedness, and cultural understandings (identity, well-being, social networks, place attachment, and aesthetic preference). More importantly, people are not, in essence, urban or rural. They may reside and work in, or across, urban and rural areas. This mobility may be patterned by age (children and the elderly are often more sedentary than working adults) and by life cycles, as people move to urban areas for education/work and "back to the country" to raise families or retire. Finally, a peopleshed may be delineated as a "community type", a linked social and biophysical spatial entity invested by ownership, stewardship, social relations, and, most importantly, by values placed on socio-physical landscape characteristics [62–64] (see Sections 3.3 and 3.4). Peoplesheds, thus delineated, place social demands on ecosystems, water, woodland, air, and biodiversity, which may be both ecologically and socially beneficial [64] (see Section 3.4).

In conceptualizing urban-rural interaction, and in modelling the "reach" of various NBS benefits, we must consider interdependencies forged by flows of air, water, animals, plants, goods, and people. Taken together, watersheds, airsheds, naturesheds, and peoplesheds allow for a comprehensive assessment of urban-rural NBS, the totality of its benefits to people, and how best to manage people's use of, and engagement with, natural environments in urban and rural areas (Figure 4).

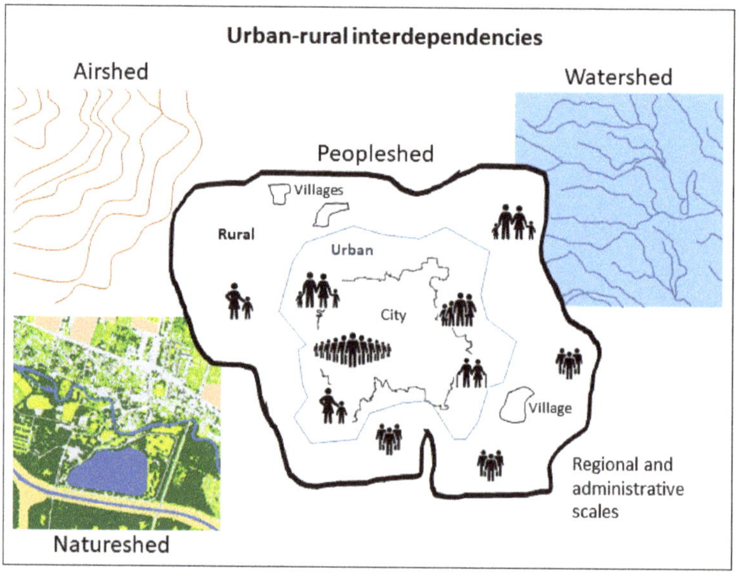

Figure 4. Interdependencies of various "sheds". Own sources.

2.6. Defining and Mapping Urban and Rural Areas

To better understand specific urban-rural interactions, consistent mapping approaches are needed [65]. This mapping should include all the mentioned interrelated factors if they can be spatially explicit. Aside from gridded population data and maps of human settlements, which both imply a peopleshed, available global or continental land cover products do not distinguish precisely between urban and rural areas. They are somewhat lacking, as their goal is either to map an urban landscape in the functional urban area, including peri-urban and suburban areas, as well as commuting zones [66], but leaving out surrounding rural areas, or to create maps of a more global nature.

Thus, different thematic and spatial resolutions do not necessarily represent both the urban and rural land cover [67]. Most importantly, the focus and resolution of the thematic contents vary widely; they may, for example, either favour urban land-cover classes (e.g., a built-up area) or generalise them for a more homogeneous output linked to the spatial resolution of the respective "shed" [68,69].

These biases can be partially overcome by studying the differences in existing information in urban and rural peoplesheds. Different spatial extents picture administrative delineations and model other representations, such as the functional urban area that encompasses a peopleshed, delineated by human mobility. This functional urban area stretches further beyond the city and the municipal limits. The different extents enhance the understanding of the benefits that NBS of different scales provide (Figure 5). By comparing different markers, i.e., (i) administrative boundaries, (ii) urban extent, and the above-mentioned (iii) watershed, airshed, natureshed, and peopleshed, we attempt to find balance between the size of the area of interest and the spatial extent needed to study urban-rural interactions. For this reason, we added a 20 km buffer to undertake analyses in the different sheds.

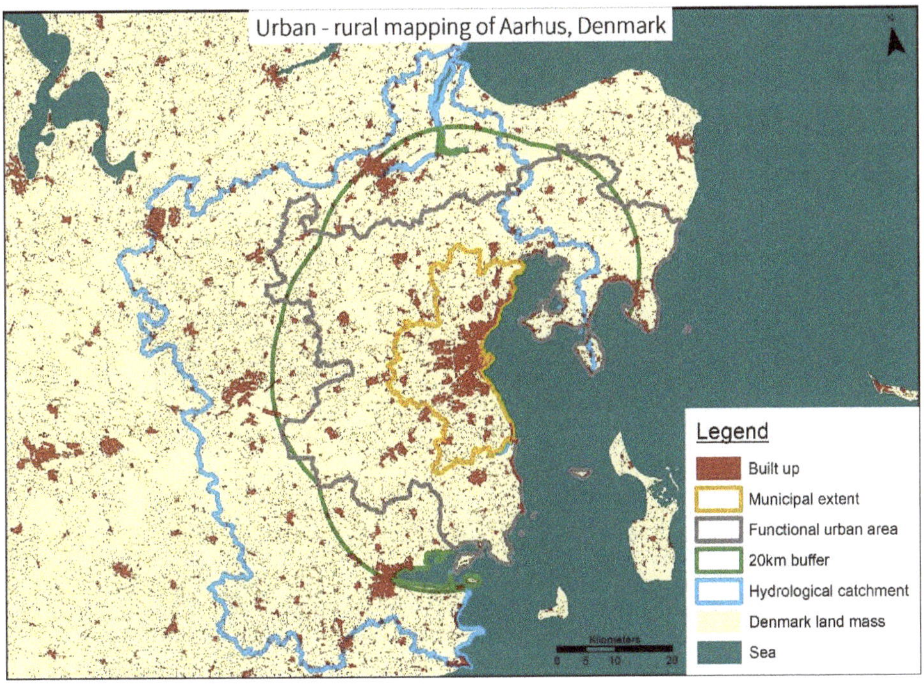

Figure 5. Different extents of the urban area, including the functional urban area, according to the EC [66] and boundaries not set by humans, e.g., watershed or airshed, and a 20 km buffer zone around the municipal border.

To map different structures and patterns of urban and rural areas, we need to define the appropriate scale, as the influence of scale on mapping land cover has long been documented [70]. The scale is usually adapted to the smallest element, which can vary extremely between urban and rural settings. Therefore, finding an appropriate scale that best pictures an urban and rural area can be challenging. To avoid losing any details, we suggest a smallest common denominator approach: choosing the best-fitting scale based on the smallest element of interest.

In the REGREEN project, we made these considerations when defining goals for mapping urban-rural interactions. Figure 6 illustrates the differences in represented land cover classes, based on the chosen scale, and the resolution of the input data influencing the thematic resolution. The low-resolution mapping focuses on the extent of municipal boundaries. A buffer was extended for the mapping of Sentinel 1 and 2 data to cover the airshed of a given city. As for the high-resolution mapping, a smaller buffer was created around the municipal boundaries to include rural areas. Such a buffer was mostly restricted by computational time, but it could be extended further in the future.

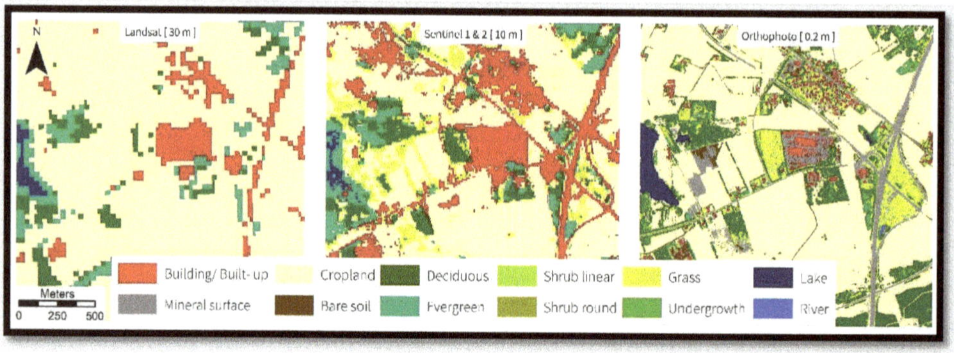

Figure 6. Scale-dependent aggregation of spatial knowledge consisting of land-cover and land-use data obtained by remote sensing techniques. The change in distribution of the different land cover and land use can be seen along the gradient of the spatial resolution. While Landsat and Sentinel 2 are optical remote sensing satellites, Sentinel 1 uses radar for imaging. The orthophotos were taken using airplanes.

The novelty of this study is to connect the different "sheds", such as watersheds, airsheds, and naturesheds, with peoplesheds. All of them tackle different scales in the urban-rural extents. Consequently, we neither relate to just one urban scale, nor do we determine one single spatial extent for all NBS implementations (Section 3). We profit from the myriad of data we have at hand, at the various scales, to conceptualise our approach and illustrate it with initial solutions.

3. Solutions for Improving Urban-Rural Partnerships through NBS to Increase Resilience

In this section, we highlight cases from Europe and China that show the employment of NBS, through different forms of urban-rural collaboration, with the aim of mitigating environmental challenges to urban areas. In many of these cases, rural partners (stakeholders representing a particular livelihood) have received economic compensation for their part in creating and maintaining NBS (forest, wetlands) that benefit urban dwellers, regarding challenges posed by air pollution and flooding. The presentation is followed by a brief discussion of the potential contribution of relevant peopleshed models to NBS design and collaboration.

3.1. Collaborative NBS Solutions for Addressing Airshed Problems by Tackling Resource Dependency—The Million-Mu Afforestation Project around Beijing, China

Urban expansion often takes a toll on rural areas. Farmlands, wetlands, and other lands of important ecological value are converted into built-up areas. The impact is the strongest at the urban-rural interface. In addition to land loss, these interfaces are often sites for placing waste, generated in landfills, in the urban fringe. Polluted environment conditions cause a decrease in crop yields and lead to health problems, such as a rising incidence of chronic diseases among rural residents [71]. While farmers may have a strong desire to tackle the problem, they often lack the skill and resources to act alone. Solutions originating from urban areas, on the other hand, often do not take the social and economic welfare of farmers into consideration. For example, the idea in China to turn illegal dumping sites on farmlands into suburban parks may solve the pollution problem, but it simultaneously creates livelihood issues for rural residents.

The "Million-mu Afforestation Project" (one million mu ≈ 66,667 ha), initiated in Beijing, China, provides an example of how authorities may work together, across the rural-urban interface, to address environmental problems in both zones while simultaneously maintaining livelihoods in rural areas. The project, initiated in 2012, aimed to plant one million mu tree plantations in Beijing over a five-year period. Most plantations are located on fallowed farmlands and waste lands located at the current urban-rural interface [72]. The planting programme was a response to deteriorating urban environment conditions, as Beijing has just experienced the worst smog problem in its history. In addition, each spring, the city suffers from heavy dust storms blowing in particulate matter from remote bare land or from the desert. During this same period, the city experienced rapid expansion. Between 2000 and 2020, built-up areas increased from 1640 km^2 to 2860 km^2—a 74% increase [35]. It was urgent to control the rapid urban sprawl to protect residents in both urban and rural areas. The planting programme thus served the dual purposes of improving the environment and containing rapid urban expansion. Unlike other planting programmes in China, the "Million-mu Afforestation Project" set up a mechanism to financially benefit rural residents. Instead of expropriating farmland, the city government rents the land, thereby providing farmers with a stable source of income. In addition, the Beijing government hires farmers to care for the plantations, providing them with yet another income source.

With the support of farmers, the five-year project was completed in less than four years. In 2018, the government initiated another "Million-mu Afforestation Project" for the period of 2018–2022. So far, the project has been credited for a series of ecosystem services, which include removing air pollutants, alleviating the urban heat island, improving urban biodiversity, and providing recreational opportunities [73,74]. There is also evidence that farmers benefit, both ecologically and economically, from the project [75]. The example shows that understanding urban and rural residents, as an integrated peopleshed delineated by their common dependency on clear air, can lead to ways of addressing environmental problems without sacrificing the social and economic welfare of one group of residents and landowners over the other. Urban authorities compensate rural landowners and residents through land rent and job opportunities associated with the project, and both areas receive ecosystem services in return. Urban and rural areas share, more equitably, the responsibility of creating a better environment at the urban-rural interface. Moreover, the million-mu afforestation project has laid an important foundation for green, intelligent, and distinctive urban planning in Beijing that maintains, and even enhances, the quality of life in the city and the surrounding rural areas.

3.2. NBS Solutions for Addressing Watershed Risks around the Yangtze River, China and the Need to Consider Peoplesheds

Changes in rural land cover and land management, leading to habitat destruction, can cause significant flood risks at the watershed level for both urban and rural communities. During July–August of 1998, the mid and lower reaches of the Yangtze River basin experienced the worst flooding in 40 years. People living in around 100,000 square kilometres were evacuated, and 13.3 million houses were damaged or destroyed [76], resulting in 3656 lives lost, 15 million homeless, and $44 billion in economic losses [77]. Had these severe measures not been taken by opening the dykes and flooding rural farmland and villages, urban destruction would have been far more serious, resulting in an even greater loss of life and infrastructure damage.

Subsequent investigation into the cause of the floods came to two main conclusions: (i) Extensive deforestation in the upper reaches of the Yangtze River basin resulted in reduced infiltration of rainwater into soils and increased transportation of sediment and deposition in the mid and lower reaches of the river, thereby reducing discharge flow and channel capacity, which resulted in flooding [78]; (ii) reduction in the area of buffer lakes, especially the Poyang and Dongting Lakes, in the lower reaches of the Yangtze. These lakes and wetlands act as a safety valve for overflow water from the Yangtze River during the monsoonal rains and thereby mitigate the extent and impact of flooding. However, during the past 50 years, these lake areas have been reduced significantly due to farmland encroachment. Through the construction of small dams, dykes, and polders, farmers replaced the lake wetlands with agriculture and settlements [79].

To mitigate and prevent the future flooding of both rural and urban areas, the Chinese government undertook three initiatives: (i) a total logging ban and reafforestation programme, initiated in 1998 in 18 of China's 31 provinces and extended to regions outside the Yangtze river basin, including northeast China; (ii) a "Grain for Green" programme, running from 1999 until 2004, converting agricultural land on slopes above 25 degrees to forests, for which farmers received, in compensation, 150 jin (75 kg) grain and 150 RMB per mu/year (€450/ha/year) for 8 years [80]; (iii) restoration of Poyang and Dongting Lake areas by breaking down the dams, dykes, and polders as well as converting farmland back to lake wetlands. The ongoing restoration process has been hampered by the completion of the Three Gorges Dam, which also provides flood mitigation measures. Restoration efforts have also been met with some resistance from farmers occupying the lake areas, demanding higher compensation. According to Liang et al., 2012 [80], compensation offered to farmers in the Poyang Lake area to restore farmland to wetland was insufficient to induce change (See peopleshed as a community type in Sections 2.4 and 3.1).

The "Grain for Green" programme and wetland restoration measures have greatly contributed to the construction of China's ecological civilisation and promoted a design that creates more sustainable land use in Chinese cities and the surrounding rural areas. The example shows that, while large-scale strategic implementation to reverse non-sustainable land-use practices, by introducing NBS upstream and across at the watershed level, does reduce flood risk, it is not entirely without conflict between rural stakeholders and the government. A major drawback is that the application of NBS in rural areas, in the floodplain of the Yangtse River, had a direct impact upon the livelihoods of villagers by affecting local enterprises, which was a disadvantage that was not fully taken into account during the policymaking and planning period. In the Grain for Green programme, it has become clear that farmers who gave up their agricultural land for tree planting have, in the long-term, lost their means of generating household income [80]. Similarly, logging bans in forest areas have caused significant loss of livelihood that could not be entirely compensated by undertaking tree planting activities. Restoration of the wetland areas of Poyang and Dongting lakes also met resistance from local farmers, resulting in a compromise situation in which only a small part of the lakes was restored, reducing their effectiveness in acting as buffer lakes against flooding [81].

This example demonstrates the import of considering the nexus between watersheds and peoplesheds when planning and designing NBS, as well as forging viable partnerships across urban and rural interests. In this specific case, delineating peoplesheds, either by means of resource dependency or by specific community type, may have helped mitigate the loss of lives, livelihood, and concomitant local resistance (See Section 2.4).

3.3. Farmers as Water Managers in Europe—Creating a New Type of Contract Connecting Peopleshed and Watershed

Urban areas located around rivers and streams are prone to fluvial flooding during extreme rain events or sudden snow melts, disrupting traffic and damaging buildings and other assets. This risk to urban areas is exacerbated due to modifications and constrictions of watercourses, in both rural and urban areas, not leaving enough space in the landscape for water to be retained and slowed down. Urban areas can, to some degree, adapt to increasing peak flows but are highly dependent on solutions in peri-urban and rural areas to create more space for water in the landscape and, thus, reduce the flood risks in urban areas. An example of how urban and rural communities have partnered together to alleviate urban fluvial flood risks is the municipality of Holstebro, Denmark, which has negotiated a common contract with 57 landowners, a peopleshed upstream of the local river, to allow their land to be flooded for a short period of time, on average, every five years. The flooding would be controlled over 148 ha of agricultural land (mostly permanent grass out of rotation) through the construction of a dam and a sluice that can withhold up to a 1:100-year flood event. Farmers are paid a one-off compensation of 1300 EUR/ha for entering the contract. Specific compensation may be paid for any losses during flooding, and farmers will have the opportunity to file compensation claims if flooding causes other documented damages [82]. This example shows how urban residents, property owners, and businesses, represented by the municipality, are, to a large extent, dependent on rural landowners to help solve an urban flood risk problem, which would be far more expensive if the hazard were to be handled uniquely within the urban area. This example of municipal partnering with rural landowners, who were equitably compensated for any damages to productive rural land within the watershed, illustrates a novel approach that recognises the resource interdependency of an urban-rural peopleshed.

3.4. Multifactorial Land Distribution in Europe—Seeking Win-Win Solutions between Farmers and Municipal Planners to Solve Watershed and Livelihood Problems

In the past, substantial amounts of low-lying peat lands across Europe have been drained to provide more farmland. Such farmlands, located in peri-urban landscapes, could provide significant multiple benefits, and at multiple scales, if ecologically restored, but livelihood concerns represent a significant barrier. Aarhus Municipality, Denmark, works with upland farmers to redistribute land, such that low-lying peat land owned by the farmers is exchanged with high quality farmland owned by the municipality or other landowners interested in selling their land. Through this voluntary land redistribution programme, farmers obtain better land, often located nearer to their farms, while the municipality and residents stand to gain several benefits: (1) protection of groundwater against pesticides and nutrients, (2) improved water quality, (3) more areas available in the landscape for water storage, (4) reduction in the risk of urban flooding, and (5) the creation of higher quality natural sites, which, in combination with investment in recreational infrastructure, provide recreational and well-being benefits to all residents. Finally, the re-wetting of low-lying peatlands will stop the release of carbon emissions, which a global benefit. This example shows the mutually beneficial outcome of multifunctional land distribution, forged in common by rural landowners and municipal planners, and providing benefits that operate at multiple scales, from global (carbon sequestration) to watershed (adaptation, water quality) and peopleshed (livelihood, recreation, landscape aesthetics, well-being).

3.5. NBS Solutions Combining Watersheds, Naturesheds and Peoplesheds to Provide Multiple Benefits in Paris, Europe

The residential area named "Le Vignois" of Gonesse city, in the Paris Region, France, was frequently flooded, due to the channelling and concreting of the Croult riverbed and to large amounts of rainwater. During a storm, the level of the river could rise from 20 centimetres to 2 metres in 10 to 15 min. In 2019, to mitigate flooding in the Vignois district, the local water authority SIAH (Syndicat Intercommunal d'Aménagement Hydraulique of Croult and Petit Rosne) led an ecological river restoration project on former agricultural land to create a natural floodplain with a capacity of 55,000 m^3.

The ecological restoration project covered 12.8 ha of privately owned agricultural land, where 28 farmers cultivated horticulture in the upstream area and cereal crops in the downstream area on 19 plots. SIAH, with the support of the city of Gonesse, negotiated with the farmers to acquire the necessary land. After negotiations, SIAH bought the land from 23 farmers but had to expropriate 2.5 ha from another five farmers. Governmental agencies such as SIAH can resort to expropriation, requiring a "déclaration d'utilité publique", i.e., a declaration of public interest, to justify the public benefit. In this case, the public interest was about protecting citizens from flooding. The total cost of land acquisition was 1 million euros.

The course of the Croult was restored over 800 linear metres, such that it regained its former riverbed, with meanders and planted banks. The mono-bioculture, in which the river used to flow, was transformed into a 12-hectare wetland, high in biodiversity, while absorbing floodwater and runoff. During heavy rainfall, the river can again overflow into this wetland, preventing flooding in the residential area. In addition to the hydrological process, SIAH focused on the ecological aspects. The floodplain (banks, stream, and water bodies) has been designed to offer new habitats for species to rest and find shelter or food.

To ensure the ecological monitoring of flora and fauna, the Paris Region Biodiversity Agency works with SIAH, as well as with the Office for Insects and their Environment and the natural research office EcoloGIE. One year after the last work, the first results of on-site inventories show a good recovery with 14 wildlife habitats, 110 species of flora, 27 species of dragonflies, 24 species of rhopalocera, 70 species of heterocera, 13 species of orthopterans, 3 species of amphibians, 66 species of birds, and 8 species of chiropterans (Figure 7).

Figure 7. Newly created wetlands: river expands to a larger floodplain, thus preventing flooding in neighbouring settlement while providing habitats for many species © SIAH: Syndicat Intercommunal d'Aménagement Hydraulique of Croult and Petit Rosne.

This example shows how the implementation of a peri-urban NBS, to regulate watercourses and prevent urban flooding (watershed), can and should incorporate biodiversity (natureshed). In this multifunctional approach, new high-value habitats were created while providing opportunities for recreation (peopleshed).

3.6. *Configuring Peoplesheds in Urban-Rural Partnering*

Most of the above examples illustrate partnerships forged between municipalities in Europe and China, represented by local and national authorities, as well as individual rural landowners/farmers in specific watersheds and airsheds. These asymmetric partnerships aim to implement nature-based solutions that increase resilience and provide benefits—e.g., protection against pollution or flooding—for urban as well as rural residents. The examples demonstrate the importance of considering how urban-rural interlinkages between air, water, nature, and people are understood, prioritised, and spatially represented. They also demonstrate the need to carefully consider how "people", differently categorised as "residents", "landowners", "populations", "authorities", or "citizens", are represented in analyses of urban-rural interlinkages. Finally, they demonstrate the need to examine the different forms of national and local governance, and the power relations at play, in forging rural–urban partnerships to implement nature-based solutions.

Precisely deployed, the concept of peopleshed helps conceptualise and spatially delineate "people" in specified rural-urban areas regarding: (1) who contributes to (un)sustainable land use and management, (2) who benefits from NBS, and (3) who resists or advances viable partnerships that promote NBS. Peoplesheds, spatially delineated by human resource dependency (as depicted in Figure 4), ideally take rural and urban populations into account equitably. This model is useful for gaining broad spatial understanding of resource use and demand, as well as the potential reach of NBS benefits, when rearranging the rural land cover land use and management typology (see Sections 3.2 and 3.3).

Differences in resource dependency across space and livelihood point out potential conflicts of interest. To better understand resistance to NBS, it is useful to delineate peoplesheds by 'community type' [65]. A rural peopleshed, for example, models resource dependency (productive farmland or forest access), as well as the socio-spatial extent of valued landscapes, multigenerational relations, local knowledge, and environmental strategies experienced by rural landowners and residents. Peoplesheds delineated by community type contribute more in-depth spatial and sociocultural understanding regarding conflicts of interest, value clashes, and incommensurate losses. Including this model could conceivably contribute to the forging of more robust partnerships and mutually beneficial NBS.

4. Discussion

This paper explores urban-rural dependencies and partnerships regarding NBS that can enhance resilience in Europe and China. It addresses the research questions on the contribution of governance, on scale dependent solutions, and on tackling the complex human interactions in urban-rural areas The following sub-sections address these questions [83].

4.1. Governance Contributing to Balancing the Specific Needs of Urban and Rural Areas in China and Europe

Policies and corresponding measures are important to ensure that NBS can be implemented in a systemic, sustained, and viable manner. China and Europe display significant differences in governance, policy making, and policy implementation systems and cultures. Chinese policy implementation in regional, urban, and local policies is consistent and responsive from the national level to the city scale, which provide the basis for direct coordination of the implementation of NBS policies. An example of this is the national Sponge City Programme that has translated into numerous municipal sponge city plans [84]. In centralised governance systems such as the Chinese, this direct link can lead to large-scale strategic NBS implementation, as in the cases of flood management in the Yangtze River Basin and the Million-mu Afforestation Project around Beijing mentioned above. Partnering between this centralised governance system and rural landowners may, however, not always be optimal in terms of securing sustainable livelihoods during and after the implementation of NBS.

Governance of and with NBS in EU member-states differs greatly from China's strong centralisation. This is due to both the national systems and cultures of governing and to the character of EU policy making and planning under the auspice of EU's territorial cohesion policy and the European Spatial Development Perspective [85]. The European countries exemplified how each have governance systems for implementing NBS to promote environmentally, socially, and economically viable urban-rural interactions and cohesion, as well as the spatial and land-use plans to sustain these. Governance systems that vary across national and local borders may impede progress in developing and harvesting the range of multiple benefits flowing from NBS—especially when seeking NBS solutions that cross urban-rural boundaries [86]. To address this limitation requires, not only deeper understanding of specific NBS but also of the political policy contexts in which urban nature is introduced as a solution to major challenges. The above-listed references witness hardly any recent studies with a focus on both Europe and China, which is central to this study.

Despite the differences in urban development stages and governance structures in China and Europe, both illustrate ways towards synergistic urban and rural development. Building on previous work in this field, this research has shifted the emphasis to a direct comparison of Europe and China, in a differentiated way, rather than looking at the global level as a whole [87]. All the included governance systems have, to a varying extent, an inclination towards evidence-based policy making and planning, while also building on theoretical frameworks of NBS to solve the mounting problems of flooding, air pollution, climate hazards, health, biodiversity, and the well-being of citizens living in both urban and rural areas.

In this process, governmental intervention through coordinated policies, capacity building, and citizen and stakeholder participation are essential prerequisites for successful NBS implementation. Particularly when constructing frameworks for NBS that may foster urban-rural partnerships, it is valuable to target the comprehensive identified benefits of NBS programmes. These include the common interests of urban and rural residents, social inclusion, and progress in urban nature restoration, air and water quality, climate adaptation, as well as the combined benefits of environmental protection and sustainable economic development.

4.2. Exploring Solutions for NBS at Different Scales in Europe and China

When investigating urban-rural interactions to design NBS for resilience, the spatial scale, at which the study can be mapped, is a crucial factor. While it is possible to choose a reasonable scale at which both urban and rural features are represented, the question of scale remains central, as different methods and models need different spatial resolutions. When considering different sheds, as we do in this paper, it becomes clear that they can either be a subset (partial quantity) of each other or an intersection. As such, the proposed workflow is still far from all-encompassing. However, the exploration of NBS has been demonstrated at different scales to offer some first solutions for fostering urban-rural interactions.

To tackle challenges such as air pollution, flood risks, or water quality, it is necessary to operate at a large extent that requires urban-rural spatial integration. Since larger extents are usually mapped at a coarser resolution, details are inevitably neglected. When implementing NBS on the ground, detailed information and an analysis are brought in and nested in a multi-scale plan. If one decides to work with the drawbacks that are inherent to aggregation, such as losing details at a coarse resolution, individual scales and spatial extents can be set for tackling targeted environmental constraints, such as air pollution, water quality, and access to green spaces.

However, considering different dependencies and opportunities of watersheds, with upstream and downstream impacts, as well as directional and non-directional effects, selecting an appropriate multifunctional NBS is a challenge. With this in mind, a subdivision is likely to be more in order, as each aspect can be studied at the scale and spatial extent which is the most fitting. The scale at which the study takes place thus remains a central, yet only partially solved, question.

This paper shows the difficulties, but also the opportunities, inherent in bringing together natural boundaries, administrative boundaries, conceptual boundaries, and boundaries constructed by patterns of settlement, land use, mobility, and community. Spatial representations of cities, based on jurisdictional boundaries, are too narrowly defined and lack the human aspect of peoplesheds, as well as ecological aspects, such as ecosystem service provision and natureshed. Considering that, this article expands the discussion on urban boundaries towards a more nature and human-centred perspective [88,89]. As complex NBS can span multiple administrative units, exchanges between neighbouring administrative units and levels, in terms of relevant peoplesheds, play a key role in finding and taking appropriate measures. Such exchanges are necessary, for instance, in the case of downstream and upstream effects when dealing with watersheds. As the NBS can be part of the urban-rural intersection, and cover both rural and urban areas, this exchange becomes even more important.

4.3. Fashioning Equitable Urban-Rural Interactions and Partnerships

Unsustainable land management, in both Europe and China, provokes or exacerbates existing environmental risks. This includes uncontrolled urban sprawl, logging steep-sloped woodlands, and draining and reclaiming wetlands. Polluting landfills, located at the urban fringe, with waste from urban centres also cause local health and livelihood problems. Reversing such unsustainable land management practices requires integrated governance of urban-rural interlinkages and partnerships that takes respective singularities and interests into account. Such equitable urban-rural partnerships can build on experiences from urban areas [90].

While mutual interaction and learning across urban and rural spaces supports viable urban-rural partnering, to productively conceptualise these multi-dimensional partnerships, we need to consider not only the flow—e.g., of water or money—but also of people, ideas, attachments, habits, and policies regarding areas designated as urban and rural. Moreover, we must not lose sight of the dyadic conceptualisations, ideologies, and economic realities of governance that may favour one "partner" over the other [27,91] (Section 3.3).

Urban-rural interaction is often grounded in valorised conceptual binaries and asymmetric power relations that can lead to political prioritisation of one over the other [91,92] (Section 3.3). With increasing urbanisation, both planned and unplanned, it is important to understand whether, and how, asymmetric relations, and the valorised imaginaries they draw on and sustain, impinge on policies and planning for resilience and sustainability. Awareness-raising and dialogue that considers the viewpoints and perceptions of individuals and local communities is crucial to developing on-the-ground support for viable and efficient NBS. Without appropriate recognition of the conditions and needs of rural, as well as urban, landowners and livelihoods, implementation of NBS in rural areas, for the benefit of mainly urban residents, may not reach the desired scale or quality to ensure sufficient resilience. Interestingly, present discourses of resilience and sustainability may be pushing an understanding of urban-rural partnerships between equivalent, interdependent areas and stakeholders—a move which, in time, may drive a more symmetrical rural-urban valorisation [83].

4.4. Conceptualizing "People" in Urban-Rural Interlinkages and Partnerships

Including socio-cultural contexts for implementing and living with NBS is essential for grasping their complex interplay in and across rural and urban areas. Although not yet well developed, the concept of peopleshed is useful for modelling, designing, and negotiating NBS across rural-urban interfaces. Unlike watersheds, peoplesheds do not just exist 'out there'. They are spatially modelled on different dimensions, such as patterns of human mobility and dwelling, mutual resource dependency, or community type. Selecting which dimensions are most relevant for modelling peoplesheds in specific cases is important for processes of rural-urban partnering, largely because partnerships are forged by people [93] with different experiences, knowledge, senses of place, environmental concerns, degrees of investedness, and political power [94] (Section 3.2). Mapping the relationship between a particular NBS and a specifically delineated peopleshed points out the mutual embeddedness of rural-urban environments and the different ways NBS may play into everyday lives and existing practices.

Considering peoplesheds is crucial for indexing human factors that impact local environments and resource management. A peopleshed delineated by mutual resource dependency may, for example, help ensure more equitable partnering across rural and urban areas, interests, and population segments by taking different kinds of dependencies into account. A peopleshed delineated by human mobility that indexes flows of people across landscapes highlights specific channels, routes, and hubs that affect the environment and resource management. A peopleshed depicting the extent of human mobility in and near the coastal Danish city of Aarhus, a REGREEN study site, would, for example, include commuters from the city's rural "hinterlands" and from an array of other inland and coastal urban areas, as people in this region move easily across multiple watersheds and urban-rural areas. Finally, a peopleshed delineated by community type [64] couples the biophysical investigation of perennial practices with social landscape analysis to explore stakeholders' sense of place and valuation of "their" landscape. This knowledge is important for landscape changes in areas that are privately owned.

Whereas scientific study and municipal planning are commonly delineated by administrative and ecological boundaries, various urban and rural stakeholders may envision "their" rural countryside, or urban or peri-urban neighbourhood, as a linked social and biophysical spatial entity, delimited by landscape characteristics, lifestyles and aesthetics, local communities, and social networks [95]. It is thus useful to consider the dimension of community, together, with resource dependency when delineating peoplesheds of relevance to viable and robust rural-urban partnering.

The concept of peopleshed allows us to explore links between biophysical and social scales as well as, of equal import, to explore how diverse stakeholders themselves link these scales based on concerns with global environments, livelihoods, the resources necessary to sustain them, recreational consumption, land ownership and stewardship, and aesthetics associated with a sense of place [94,95]. Pursuing urban-rural partnerships for implementing NBS may be the first step in conceptualizing urban peoplesheds and naturesheds in their own right—ones that do not just draw ecosystem benefits from rural partners (Sections 3.2 and 3.3) but that reciprocate with ecosystem services of their own.

Finally, more in-depth analysis of (social) consequences of NBS implementation, in and across rural and urban areas, is required. This is associated with follow-up monitoring and evaluation to ensure that both the desired physical and social results are achieved in the long-term. A key feature is transdisciplinary involvement in the policy and planning process, as well as the subsequent monitoring.

5. Conclusions

The different perspectives on NBS, in the urban-rural interfaces addressed in this paper, give rise to novel and necessary questions concerning understanding, designing, and adopting multifunctional NBS. Attempting to balance specific needs across rural and urban land areas, interests, and communities reveals manifold, complex, and cross-disciplinary linkages of interventions to implement beneficial NBS. To understand opportunities, obstacles, and resistance, we need to consider issues of governance and relevant peopleshed dimensions to forge partnerships less constrained by technical fixes and administrative lock-ins. This will lead to new avenues for appraising efforts undertaken, and benefits achieved, at the level of multifunctional and multipurpose interventions. It will also lead to understandings of peoplesheds with much less a priori rural-urban divide.

Planning for NBS in rural areas by regulating ecosystem services in airsheds, watersheds, and naturesheds, to reduce risks of pollution and flooding, enhance biodiversity, and improve opportunities for recreation, health, and well-being, can substantially improve resilience, although primarily for urban residents. An increasing number of strong examples where this has taken place are emerging in both China and Europe. Differences in the application of NBS between China and Europe become especially evident in the scale of intervention of NBS, e.g., afforestation and flood plain restoration. With the strong centralist focus in China, compared to the rather fragmented governance structures in Europe, in relation to NBS and urban-rural relationships, Chinese NBS initiatives can move both faster and further than is tradition, and feasible, in Europe. With the regional focus on NBS implementation in many Chinese cases, the potential for trade-offs and conflicts between environmental or disaster risk-reduction benefits and the livelihoods of local people becomes particularly apparent. The differences in NBS application are closely linked to the territorial division and how urban-rural relationships are governed. Drawing on a number of novel practices and good examples of implementing NBS at a landscape scale in Europe and China, this article also shows some of the pitfalls when policies are not equitable enough for rural residents.

For designing sustainable and effective NBS in the urban-rural interface, we find the following dependencies and opportunities. First, the ecological processes of NBS that deliver benefits to people operate at different scales, ranging from very local benefits to global benefits. Moreover, these NBS benefits depend on the size and connectedness of NBS. Both the size-dependency effects of NBS and the distance-decay relationships for benefits received from a given NBS are central to planning implementation. Second, the use of nature, air, and water "sheds" is key to clarifying interdependencies, directions of ecological processes, and the level of needed urban-rural interaction when designing NBS at landscape scale. Third, peoplesheds are instrumental spatial tools for tackling complex human interaction when implementing large-scale NBS in or across urban-rural areas. Mapping the different dimensions of peoplesheds—human resource dependency, community type, and patterns of human dwelling and mobility—provides a key opportunity for forging mutually beneficial partnerships in the urban-rural interface that extends beyond stereotypical administrative or biophysical boundaries. Finally, regardless of how centralised or decentralised a governance system is, policies that incorporate fair and equitable processes, instruments, and compensatory measures are far more likely to lead to higher acceptance of NBS and a more optimal scale of implementation.

This study is an initial investigation of the urban-rural connection and its importance for implementing sustainable and efficient NBS. Most studies and projects take a purely urban perspective or a larger regional perspective. Very few studies investigate the intricate relationship, dependencies, and conflicts between urban and rural areas with regard to NBS. By introducing "peoplesheds" into the analysis, we propose an important linkage to the typical "airshed", "natureshed" or "watershed" analyses. Peoplesheds and their connotations are often overlooked, and this aspect brings in the necessary added value.

It is generally understood that there is a myriad of imbalances and biased dependencies between rural and urban areas. Thus, this article highlights promising practices for successful partnering, around the implementation of NBS, which make both urban and rural areas more resilient.

Author Contributions: Conceptualization, E.B., L.J., S.A. and M.Z.; methodology, E.B., L.J., S.A. and M.Z.; investigation, E.B., W.W., J.K., M.Z., S.A., R.H., D.R., J.Y., G.L., G.G., A.J. and L.J.; resources, E.B., J.K., G.G., J.Y., R.H. and M.Z.; data curation, J.K. and G.L.; writing—original draft preparation, E.B., S.A. and M.Z.; writing—review and editing, E.B., W.W., J.K., M.Z., S.A., R.H., D.R., J.Y., G.L., G.G., A.J. and L.J.; visualization, J.K., G.L., E.B. and W.W.; project administration, E.B. and M.Z.; funding acquisition, E.B., M.Z., S.A., R.H., D.R., J.Y., G.L., G.G., A.J. and L.J.; formal analysis, n.a.; software n.a.; supervision, n.a.; validation, n.a. All authors have read and agreed to the published version of the manuscript.

Funding: This research was funded by the European Commission, Horizon 2020 programme, grant number 821016, and the National Key R&D Program Intergovernmental Cooperation in International Science and Technology Innovation from Ministry of Science and Technology of China (Grant no. 2021YFE93100).

Institutional Review Board Statement: Not applicable.

Informed Consent Statement: Not applicable.

Data Availability Statement: No new data were created or analyzed in this study. Data sharing is not applicable to this article.

Acknowledgments: We greatly appreciate research and discussions with practitioners through the EC REGREEN Nature-based Solutions project (regreen-project.eu/) which enabled our inter- and transdisciplinary collaboration. This project has received funding from the European Union's Horizon 2020 research and innovation programme under grant agreement No 821016, and the National Key R&D Program Intergovernmental Cooperation in International Science and Technology Innovation from Ministry of Science and Technology of China (Grant no. 2021YFE93100).

Conflicts of Interest: The authors declare no conflict of interest.

References

1. UN-Habitat. *The Value of Sustainable Urbanization*; United Nations Human Settlements Programme: Nairobi, Kenya, 2020.
2. Zhu, Y.-G.; Reid, B.J.; Meharg, A.A.; Banwart, S.A.; Fu, B.-J. Optimizing Peri-URban Ecosystems (PURE) to re-couple urban-rural symbiosis. *Sci. Total Environ.* **2017**, *586*, 1085–1090. [CrossRef] [PubMed]

3. Jamshed, A.; Birkmann, J.; Feldmeyer, D.; Rana, I.A. A Conceptual Framework to Understand the Dynamics of Rural–Urban Linkages for Rural Flood Vulnerability. *Sustainability* **2020**, *12*, 2894. [CrossRef]
4. Preiss, P.; Charão-Marques, F.; Wiskerke, J.S. Fostering sustainable urban-rural linkages through local food supply: A transnational analysis of collaborative food alliances. *Sustainability* **2017**, *9*, 1155. [CrossRef]
5. UN-Habitat. HABITAT III the New Urban Agenda 2016. Available online: https://uploads.habitat3.org/hb3/NUA-English.pdf (accessed on 7 December 2021).
6. UN-Habitat. *Implementing the New Urban Agenda by Strengthening Urban-Rural Linkages—Leave No One and No Space Behind*; United Nations Human Settlements Programme: Nairobi, Kenya, 2017.
7. Gebre, T.; Gebremedhin, B. The mutual benefits of promoting rural-urban interdependence through linked ecosystem services. *Glob. Ecol. Conserv.* **2019**, *20*, e00707. [CrossRef]
8. Bencardino, M.; Nesticò, A. Demographic changes and real estate values. A quantitative model for analyzing the urban-rural linkages. *Sustainability* **2017**, *9*, 536. [CrossRef]
9. Khan, S.; Hwang, G.J.; Azeem Abbas, M.; Rehman, A. Mitigating the urban–rural educational gap in developing countries through mobile technology-supported learning. *Br. J. Educ. Technol.* **2019**, *50*, 735–749. [CrossRef]
10. Scheuer, S.; Haase, D.; Volk, M. On the Nexus of the Spatial Dynamics of Global Urbanization and the Age of the City. *PLoS ONE* **2016**, *11*, e0160471. [CrossRef]
11. Sharma, A. Urban proximity and spatial pattern of land use and development in rural India. *J. Dev. Stud.* **2016**, *52*, 1593–1611. [CrossRef]
12. Duvivier, C.; Li, S.; Renard, M.-F. Are workers close to cities paid higher nonagricultural wages in rural China? *Appl. Econ.* **2013**, *45*, 4308–4322. [CrossRef]
13. Cook, P. Infrastructure, rural electrification and development. *Energy Sustain. Dev.* **2011**, *15*, 304–313. [CrossRef]
14. Lerner, A.M.; Eakin, H. An obsolete dichotomy? Rethinking the rural–urban interface in terms of food security and production in the global south. *Geogr. J.* **2011**, *177*, 311–320. [CrossRef]
15. Rutherford, J.; Coutard, O. *Urban Energy Transitions: Places, Processes and Politics of Socio-Technical Change*; Sage Publications: London, UK, 2014.
16. De Jong, M.; Joss, S.; Schraven, D.; Zhan, C.; Weijnen, M. Sustainable–smart–resilient–low carbon–eco–knowledge cities; making sense of a multitude of concepts promoting sustainable urbanization. *J. Clean. Prod.* **2015**, *109*, 25–38. [CrossRef]
17. Gatzweiler, F.W.; Boufford, J.I.; Pomykala, A. Harness urban complexity for health and well-being. In *Urban Planet Knowledge towards Sustainable Cities*; Cambridge University Press: Cambridge, UK, 2018; pp. 113–129. [CrossRef]
18. Gulsrud, N.M.; Hertzog, K.; Shears, I. Innovative urban forestry governance in Melbourne?: Investigating "green placemaking" as a nature-based solution. *Environ. Res.* **2018**, *161*, 158–167. [CrossRef] [PubMed]
19. Li, L.; Cheshmehzangi, A.; Chan, F.K.S.; Ives, C.D. Mapping the research landscape of nature-based solutions in urbanism. *Sustainability* **2021**, *13*, 3876. [CrossRef]
20. Iojă, C.I.; Niţă, M.R.; Vânău, G.O.; Onose, D.A.; Gavrilidis, A.A. Using multi-criteria analysis for the identification of spatial land-use conflicts in the Bucharest Metropolitan Area. *Ecol. Indic.* **2014**, *42*, 112–121. [CrossRef]
21. Bontje, M.; Musterd, S.; Pelzer, P. *Inventive City-Regions: Path Dependence and Creative Knowledge Strategies*; Routledge: London, UK, 2011.
22. Borzacchiello, M.T.; Nijkamp, P.; Koomen, E. Accessibility and urban development: A grid-based comparative statistical analysis of Dutch cities. *Environ. Plan. B Plan. Des.* **2010**, *37*, 148–169. [CrossRef]
23. Mahendra, A.; Seto, K.C. *Upward and Outward Growth: Managing Urban Expansion for More Equitable Cities in the Global South*; World Resources Institute: Washington, DC, USA, 2019.
24. Vieno, M.; Heal, M.R.; Twigg, M.M.; MacKenzie, I.A.; Braban, C.F.; Lingard, J.; Ritchie, S.; Beck, R.; Móring, A.; Ots, R. The UK particulate matter air pollution episode of March–April 2014: More than Saharan dust. *Environ. Res. Lett.* **2016**, *11*, 044004. [CrossRef]
25. Jones, L.; Vieno, M.; Fitch, A.; Carnell, E.; Steadman, C.; Cryle, P.; Holland, M.; Nemitz, E.; Morton, D.; Hall, J. Urban natural capital accounts: Developing a novel approach to quantify air pollution removal by vegetation. *J. Environ. Econ. Policy* **2019**, *8*, 413–428. [CrossRef]
26. Kythreotis, A.P.; Jonas, A.E.; Howarth, C. Locating climate adaptation in urban and regional studies. *Reg. Stud.* **2020**, *54*, 576–588. [CrossRef]
27. Heathcott, J.; Rogan, K. Landscape Entaglements: Toward a descriptive project for Planning. *Berkeley Plan. J.* **2020**, *31*, 68–99. [CrossRef]
28. Hamada, S.; Ohta, T. Seasonal variations in the cooling effect of urban green areas on surrounding urban areas. *Urban For. Urban Green.* **2010**, *9*, 15–24. [CrossRef]
29. Fisher, B.; Turner, K.; Zylstra, M.; Brouwer, R.; De Groot, R.; Farber, S.; Ferraro, P.; Green, R.; Hadley, D.; Harlow, J. Ecosystem services and economic theory: Integration for policy-relevant research. *Ecol. Appl.* **2008**, *18*, 2050–2067. [CrossRef] [PubMed]
30. Lin, W.; Yu, T.; Chang, X.; Wu, W.; Zhang, Y. Calculating cooling extents of green parks using remote sensing: Method and test. *Landsc. Urban Plan.* **2015**, *134*, 66–75. [CrossRef]
31. Yu, Z.; Fryd, O.; Sun, R.; Jørgensen, G.; Yang, G.; Özdil, N.C.; Vejre, H. Where and how to cool? An idealized urban thermal security pattern model. *Landsc. Ecol.* **2021**, *36*, 2165–2174. [CrossRef]

32. Jones, L.; Norton, L.; Austin, Z.; Browne, A.; Donovan, D.; Emmett, B.; Grabowski, Z.; Howard, D.; Jones, J.P.; Kenter, J. Stocks and flows of natural and human-derived capital in ecosystem services. *Land Use Policy* **2016**, *52*, 151–162. [CrossRef]
33. Tallis, H.; Mooney, H.; Andelman, S.; Balvanera, P.; Cramer, W.; Karp, D.; Polasky, S.; Reyers, B.; Ricketts, T.; Running, S. A global system for monitoring ecosystem service change. *Bioscience* **2012**, *62*, 977–986. [CrossRef]
34. Hutchins, M.G.; Fletcher, D.; Hagen-Zanker, A.; Jia, H.; Jones, L.; Li, H.; Loiselle, S.; Miller, J.; Reis, S.; Seifert-Dähnn, I. Why scale is vital to plan optimal nature-based solutions for resilient cities. *Environ. Res. Lett.* **2021**, *16*, 044008. [CrossRef]
35. Banzhaf, E.; Wu, W.; Luo, X.; Knopp, J. Integrated Mapping of Spatial Urban Dynamics—A European-Chinese Exploration. Part 1—Methodology for Automatic Land Cover Classification Tailored towards Spatial Allocation of Ecosystem Services Features. *Remote Sens.* **2021**, *13*, 1744. [CrossRef]
36. Russel, D.; Castellari, S.; Capriolo, A.; Dessai, S.; Hildén, M.; Jensen, A.; Karali, E.; Mäkinen, K.; Ørsted Nielsen, H.; Weiland, S. Policy Coordination for National Climate Change Adaptation in Europe: All Process, but Little Power. *Sustainability* **2020**, *12*, 5393. [CrossRef]
37. Mayor, B.; Toxopeus, H.; McQuaid, S.; Croci, E.; Lucchitta, B.; Reddy, S.E.; Egusquiza, A.; Altamirano, M.A.; Trumbic, T.; Tuerk, A. State of the art and latest advances in exploring business models for nature-based solutions. *Sustainability* **2021**, *13*, 7413. [CrossRef]
38. Bromley, D.W.; Hodge, I. Private property rights and presumptive policy entitlements: Reconsidering the premises of rural policy. *Eur. Rev. Agric. Econ.* **1990**, *17*, 197–214. [CrossRef]
39. Ye, X.; Christiansen, F. China's urban-rural integration policies. *J. Curr. Chin. Aff.* **2009**, *38*, 117–143. [CrossRef]
40. Chen, K.; Long, H.; Liao, L.; Tu, S.; Li, T. Land use transitions and urban-rural integrated development: Theoretical framework and China's evidence. *Land Use Policy* **2020**, *92*, 104465. [CrossRef]
41. Wu, W.-B.; Ma, J.; Meadows, M.E.; Banzhaf, E.; Huang, T.-Y.; Liu, Y.-F.; Zhao, B. Spatio-temporal changes in urban green space in 107 Chinese cities (1990–2019): The role of economic drivers and policy. *Int. J. Appl. Earth Obs. Geoinf.* **2021**, *103*, 102525. [CrossRef]
42. Faludi, A. A historical institutionalist account of European spatial planning. *Plan. Perspect.* **2018**, *33*, 507–522. [CrossRef]
43. Richardson, T.; Jensen, O.B. Discourses of mobility and polycentric development: A contested view of European spatial planning. *Eur. Plan. Stud.* **2000**, *8*, 503–520. [CrossRef]
44. Zonneveld, W.; Stead, D. European territorial cooperation and the concept of urban–rural relationships. *Plan. Pract. Res.* **2007**, *22*, 439–453. [CrossRef]
45. European Commission. *The Urban Agenda for the EU*; European Commission: Publications office of the European Union: Luxembourg, 2016.
46. Cecchi, C. *Towards an EU Research and Innovation Policy Agenda for Nature-Based Solutions & Re-Naturing Cities. Final Report of the Horizon 2020 Expert Group on Nature-Based Solutions and Re-Naturing Cities*; European Commission: Publications office of the European Union: Luxembourg, 2015.
47. European Commission. *Communication from the Commission to the European Parliament, the Council, the European Economic and Social Committee and the Committee of the Regions. An EU Strategy on Adaptation to Climate*; COM(2013) 216; European Commission: Publications office of the European Union: Luxembourg, 2013.
48. European Commission. *Our Life Insurance, Our Natural Capital: An EU Biodiversity Strategy to 2020. Communication from the Commission to the European Parliament, the Council, the Economic and Social Committee and the Committee of the Regions Our Life Insurance, Our Natural Capital: An EU Biodiversity Strategy to 2020*; COM(2011)244; European Commission: Publications office of the European Union: Luxembourg, 2011.
49. European Union. *Directive 2007/60/EC of the European Parliament and of the Council of 23 October 2007 on the Assessment and Management of Flood Risks*; European Union: Publications office of the European Union: Luxembourg, 2007.
50. European Commission. *A Blueprint to Safeguard Europe's Water Resources. Communication from the Commission to the European Parliament, the Council, the European Economic and Social Committee and the Committee of the Regions.*; COM(2012)673; European Commission: Publications office of the European Union: Luxembourg, 2012.
51. European Commission. *Towards Better Environmental Options for Flood Risk Management*; European Commission: Publications office of the European Union: Luxembourg, 2011.
52. European Environmental Agency. *Nature-Based Solutions in Europe: Policy, Knowledge and Practice for Climate Change Adaptation and Disaster Risk Reduction*; EEA Report No 01/2021; Publications office of the European Union: Luxembourg, 2021.
53. European Commission. *Communication from the Commission to the European Parliament, the Council, the European Economic and Social Committee and the Committee of the Regions*; COM(2020) 380 Final; European Commission: Publications office of the European Union: Luxembourg, 2020.
54. European Commission. *Communication from the Commission to the European Parliament, the Council, the European Economic and Social Committee and the Committee of the Regions—Forging a Climate-Resilient Europe—The New EU Strategy on Adaptation to Climate Change*; European Commission: Brussels, Belgium, 2021.
55. European Commission. *Communication from the Commission to the European Parliament, the Council, the European Economic and Social Committee and the Committee of the Regions—The European Green Deal*; European Commission: Brussels, Belgium, 2019.
56. Foley, J.A.; DeFries, R.; Asner, G.P.; Barford, C.; Bonan, G.; Carpenter, S.R.; Chapin, F.S.; Coe, M.T.; Daily, G.C.; Gibbs, H.K. Global consequences of land use. *Science* **2005**, *309*, 570–574. [CrossRef]

57. Brooks, P.D.; Chorover, J.; Fan, Y.; Godsey, S.E.; Maxwell, R.M.; McNamara, J.P.; Tague, C. Hydrological partitioning in the critical zone: Recent advances and opportunities for developing transferable understanding of water cycle dynamics. *Water Resour. Res.* **2015**, *51*, 6973–6987. [CrossRef]
58. Miller, J.D.; Hutchins, M. The impacts of urbanisation and climate change on urban flooding and urban water quality: A review of the evidence concerning the United Kingdom. *J. Hydrol. Reg. Stud.* **2017**, *12*, 345–362. [CrossRef]
59. McGrane, S.J. Impacts of urbanisation on hydrological and water quality dynamics, and urban water management: A review. *Hydrol. Sci. J.* **2016**, *61*, 2295–2311. [CrossRef]
60. Lelieveld, J.; Evans, J.S.; Fnais, M.; Giannadaki, D.; Pozzer, A. The contribution of outdoor air pollution sources to premature mortality on a global scale. *Nature* **2015**, *525*, 367–371. [CrossRef] [PubMed]
61. Chan, C.Y.; Chan, L.Y. Effect of meteorology and air pollutant transport on ozone episodes at a subtropical coastal Asian city, Hong Kong. *J. Geophys. Res. Atmos.* **2000**, *105*, 20707–20724. [CrossRef]
62. Lidén, H. Common neighbourhoods–diversified lives: Growing up in urban Norway. In *Children's Places*; Routledge: London, UK, 2013; pp. 123–141.
63. Gullestad, M. Symbolic "fences" in urban Norwegian neighbourhoods. *Ethnos* **1986**, *51*, 52–70. [CrossRef]
64. Flotemersch, J.E.; Shattuck, S.M.; Aho, K.B.; Cox, C.E.; Cairns, M.R. Factors influencing social demands of aquatic ecosystems. *Ecol. Soc. J. Integr. Sci. Resil. Sustain.* **2019**, *24*, 1–9. [CrossRef]
65. Cattaneo, A.; Nelson, A.; McMenomy, T. Global mapping of urban–rural catchment areas reveals unequal access to services. *Proc. Natl. Acad. Sci. USA* **2021**, *118*, e2011990118. [CrossRef]
66. European Commission. *Mapping Guide v6.2 for a European Urban Atlas*; Publications office of the European Union: Luxembourg, 2020; Volume 6.2.
67. Buchhorn, M.; Lesiv, M.; Tsendbazar, N.-E.; Herold, M.; Bertels, L.; Smets, B. Copernicus global land cover layers—Collection 2. *Remote Sens.* **2020**, *12*, 1044. [CrossRef]
68. Treitz, P.M.; Howarth, P.J.; Gong, P. Application of satellite and GIS technologies for land-cover and land-use mapping at the rural-urban fringe: A case study. *Photogramm. Eng. Remote Sens.* **1992**, *58*, 439–448.
69. Liu, L.; Zhang, X.; Gao, Y.; Chen, X.; Shuai, X.; Mi, J. Finer-Resolution Mapping of Global Land Cover: Recent Developments, Consistency Analysis, and Prospects. *J. Remote Sens.* **2021**, *2021*, 5289697. [CrossRef]
70. Feng, Y.; Liu, Y. Fractal dimension as an indicator for quantifying the effects of changing spatial scales on landscape metrics. *Ecol. Indic.* **2015**, *53*, 18–27. [CrossRef]
71. Li, C.; Sun, G.; Wu, Z.; Zhong, H.; Wang, R.; Liu, X.; Guo, Z.; Cheng, J. Soil physiochemical properties and landscape patterns control trace metal contamination at the urban-rural interface in southern China. *Environ. Pollut.* **2019**, *250*, 537–545. [CrossRef] [PubMed]
72. Yao, N.; van den Bosch, C.C.K.; Yang, J.; Devisscher, T.; Wirtz, Z.; Jia, L.; Duan, J.; Ma, L. Beijing's 50 million new urban trees: Strategic governance for large-scale urban afforestation. *Urban For. Urban Green.* **2019**, *44*, 126392. [CrossRef]
73. Pei, N.; Wang, C.; Jin, J.; Jia, B.; Chen, B.; Qie, G.; Qiu, E.; Gu, L.; Sun, R.; Li, J. Long-term afforestation efforts increase bird species diversity in Beijing, China. *Urban For. Urban Green.* **2018**, *29*, 88–95. [CrossRef]
74. Tang, X.; Pan, Y.; Gao, B.; Gao, Y. Evaluation of ecosystem service value of plain afforestation in Beijing. *Acta Sci. Nat. Alium Univ. Pekin.* **2016**, *52*, 274–278.
75. Xue, W.; Fan, Y. The afforestation project in plain area of Beijing to boost rural revitalization in the capital region. *For. Econ.* **2018**, *6*, 54–59. [CrossRef]
76. Spignesi, S.J. *Catastrophe!: The 100 Greatest Disasters of All Time*; Citadel Press: New York, NY, USA, 2004.
77. Biswas, A.K.; Tortajada, C. *Counting the Costs of Floods in China*; Asia & The Pacific Policy Society. Available online: http://www.policyforum.net/counting-costs-floods-china/ (accessed on 24 March 2016).
78. Lang, G. Forests, floods, and the environmental state in China. *Organ. Environ.* **2002**, *15*, 109–130. [CrossRef]
79. Shankman, D.; Davis, L.; De Leeuw, J. River management, landuse change, and future flood risk in China's Poyang Lake region. *Int. J. River Basin Manag.* **2009**, *7*, 423–431. [CrossRef]
80. Liang, Y.; Li, S.; Feldman, M.W.; Daily, G.C. Does household composition matter? The impact of the Grain for Green Program on rural livelihoods in China. *Ecol. Econ.* **2012**, *75*, 152–160. [CrossRef]
81. Lin, H. *Policy Impact on the Relationships among Environment, Public Health, and Regional Economic Development in Poyang Lake, China*; University of Michigan: Ann Arbor, MI, USA, 2005.
82. Zandersen, M.; Oddershede, J.S.; Pedersen, A.B.; Nielsen, H.Ø.; Termansen, M. Nature Based Solutions for Climate Adaptation-Paying Farmers for Flood Control. *Ecol. Econ.* **2021**, *179*, 106705. [CrossRef]
83. Derkzen, P.; Franklin, A.; Bock, B. Examining power struggles as a signifier of successful partnership working: A case study of partnership dynamics. *J. Rural Stud.* **2008**, *24*, 458–466. [CrossRef]
84. Meng, M.; Dabrowski, M.; Stead, D. Shifts in spatial plans for flood resilience and climate adaptation: Examining planning procedure and planning mandates. *Sustainability* **2020**, *12*, 105. [CrossRef]
85. Gemmer, M.; Wilkes, A.; Vaucel, L.M. Governing Climate Change Adaptation in the EU and China: An Analysis of Formal Institutions. *Adv. Clim. Change Res.* **2011**, *2*, 1–11. [CrossRef]
86. Biesbroek, G.R.; Swart, R.J.; Carter, T.R.; Cowan, C.; Henrichs, T.; Mela, H.; Morecroft, M.D.; Rey, D. Europe adapts to climate change: Comparing National Adaptation Strategies. *Glob. Environ. Chang.* **2010**, *20*, 440–450. [CrossRef]

87. Biesbroek, R.; Wright, S.J.; Korswagen Eguren, S.; Bonotto, A.; Athanasiadis, I.N. Policy attention to climate change impacts, adaptation and vulnerability: A global assessment of National Communications (1994–2019). *Clim. Policy* **2022**, *22*, 97–111. [CrossRef]
88. Taubenböck, H.; Weigand, M.; Esch, T.; Staab, J.; Wurm, M.; Mast, J.; Dech, S. A new ranking of the world's largest cities—Do administrative units obscure morphological realities? *Remote Sens. Environ.* **2019**, *232*, 111353. [CrossRef]
89. Acedo, A.; Johnson, P.A. Home range and habitat: Using platial characteristics to define urban areas from the bottom up. *Trans. GIS* **2020**, *24*, 819–841. [CrossRef]
90. van Ham, C.; Klimmek, H. Partnerships for Nature-Based Solutions in Urban Areas—Showcasing Successful Examples. In *Theory and Practice of Urban Sustainability Transitions Nature Based Solutions to Climate Change Adaptation in Urban Areas*; Kabisch, N., Korn, H., Stadler, J., Bonn, A., Eds.; Springer: Cham, Switzerland, 2017. [CrossRef]
91. Kelly-Reif, K.; Wing, S. Urban-rural exploitation: An underappreciated dimension of environmental injustice. *J. Rural Stud.* **2016**, *47*, 350–358. [CrossRef]
92. Hommes, L.; Boelens, R.; Harris, L.M.; Veldwisch, G.J. Rural–urban water struggles: Urbanizing hydrosocial territories and evolving connections, discourses and identities. *Water Int.* **2019**, *44*, 81–94. [CrossRef]
93. Basquill, E.J. Saving the Planet through People. *Leadersh. Manag. Eng.* **2007**, *7*, 158–162. [CrossRef]
94. Flotemersch, J.; Aho, K. Factors influencing perceptions of aquatic ecosystems. *Ambio* **2021**, *50*, 425–435. [CrossRef]
95. Atwell, R.C.; Schulte, L.A.; Westphal, L.M. Landscape, community, countryside: Linking biophysical and social scales in US Corn Belt agricultural landscapes. *Landsc. Ecol.* **2009**, *24*, 791–806. [CrossRef]

land

MDPI

Article

Resilience and Circularity: Revisiting the Role of Urban Village in Rural-Urban Migration in Beijing, China

Ran Liu * and Yuhang Jia

College of Resource Environment and Tourism, Capital Normal University, No. 105 West 3rd Ring Road North, Beijing 100048, China; 2200902020@cnu.edu.cn
* Correspondence: liu.ran.space@hotmail.com

Abstract: Recent policies in China have encouraged rural-urban circular migration and an "amphibious" and flexible status of settlement, reacting against the recent risks of economic fluctuation in cities. Rural land, as a form of insurance and welfare, can handle random hazards, and the new Land Management Law guarantees that rural migrants who settle in the city can maintain their rights to farmland, homesteads, and a collective income distribution. Existing studies have pointed out that homeland tenure can reduce migrants' urban settlement intentions (which is a self-reported subjective perception of city life). However, little is known about how the rural-urban circularity and rural tenure system (especially for those still holding hometown lands in the countryside) affect rural migrants' temporary urban settlements (especially for those preferring to stay in informal communities in the host city). The existing studies on the urban villages in China have focused only on the side of the receiving cities, but have rarely mentioned the other side of this process, focusing on migrants' rural land tenure issues in their hometowns. This study discusses the rationale of informality (the urban village) and attests to whether, and to what extent, rural migrants' retention of their hometown lands can affect their tenure security choices (urban village or not) in Chinese metropolises such as Beijing. Binary logistic regression was conducted and the data analysis proved that rural migrants who kept their hometown lands, compared to their land-loss counterparts, were more likely to live in a Beijing urban village. This displays the resilience and circularity of rural-urban migration in China, wherein the rural migrant households demonstrate the "micro-family economy", maintaining tenure security in their hometown and avoiding the dissipation of their family income in their destination. The Discussion and Conclusions sections of this paper refer to some policy implications related to maintaining the rural-urban dual system, protecting rural migrant land rights, and beefing up the "opportunity structure" (including maintaining the low-rent areas in metropolises such as Beijing) in the 14th Five Year Plan period.

Keywords: rural land tenure; resilience; circular migration; rural-urban migrants; Beijing; urban village

Citation: Liu, R.; Jia, Y. Resilience and Circularity: Revisiting the Role of Urban Village in Rural-Urban Migration in Beijing, China. *Land* **2021**, *10*, 1284. https://doi.org/10.3390/land10121284

Academic Editor: Hualou Long

Received: 29 September 2021
Accepted: 20 November 2021
Published: 23 November 2021

1. Introduction

The "urban village" has become a hot topic in relation to informal settlements in rapidly expanding metropolises in developing countries, including those in the Chinese context [1,2]. The "urban village" (namely, *chengzhongcun*) is, in the context of land use and urban planning provisos, incompatible with official norms, and its participants suffer from substandard housing, inferior quality in the built environment, and a lack of facilities and infrastructure. It is acknowledged that the "urban village" has reduced the cost of urbanization and facilitated China's rural-to-urban migration in the past four decades, by acting as a "transitory shelter", providing cheap rental housing for low-wage migrant workers and creating rental income for local landless peasants. Urban villages are rural patches of land scattered in the city area, surrounded by or adjacent to urbanized and developed landscapes [1,2]. The rapidly industrializing and urbanizing areas of developing countries, including those in China, are faced with similar challenges concerning the inadequate supply of public housing for migrants, such as a shortfalls in government budgets to fund public services and inequitable access to urban welfare. The use of informal habitats to house migrant workers and other low-wage earners is popular in these areas. The urban village has been referred to as a new way to plan/build mixed-use communities in a recent literature review [3].

However, on the national circular migration scale and from the perspective of migrant tenants, the existing interpretation of the urban village is fragmented into geographical and rural-urban relations. In China, land is state-owned and commodified in the cities. However, the rural land is collectively owned by the village committee or township government and cannot be traded in most circumstances (unless designated by the government for commercial uses). This dual rural-urban land system dates back to Mao's central planning system, when China's urbanization level lagged far behind the developing world and free labor migration was a forbidden area until the 1980s. In Mao's central planning system, the rural-urban schism was apparent, and the rural *hukou* was managed through collective ownership of rural land, run by the People's Communes and aiming at self-sufficiency. The *hukou* system is a residency permit system in China that has separated the rural population from the urban population, and the local population from the non-local population. The *hukou* system confers a wide range of welfare benefits on the urban population. Migrants maintain their non-local residency status and practically enjoy relatively few welfare benefits in the host city [4]. This rural and urban dualism has been challenged since market-oriented reforms and the gradual *hukou* reforms facilitated labor mobility. However, restrictions are still imposed on the usage and transactions regarding rural land or housing. In integrated rural-urban planning and governance processes that are top-down, the administrative redefinition of rural land to urban land through farmland expropriation and resident status conversion is the only legal way of "rural land urbanization" from a rural land status (that is, collectively owned by the village and uncommodifiable) to an urban one (that is stated-owned and able to be traded in the land market) [1]. To date, the leasing and sales of self-built housing on rural lands remain extralegal, according to the Land Management Law in China. The boundaries between rural and urban land use are the administrative, regulatory, and tenure divisions in China. China's urban-rural household registration system (*hukou*) and the inherited rural-urban schism since Mao's central planning system have artificially divided rural and urban areas throughout the country, forming the dual rural vs. urban social-economic structure and land-use management system [4–6].

Interestingly, rural migrants' choice of informal settlement (e.g., urban villages) goes beyond the boundaries between the dualistic rural-urban economic and land use structure. More specifically, rural migrants attach more importance to their hometown's tenurial security but they do not care much about whether their urban shelter is formal or informal in their host city. Urban villages (as extralegal housing clusters) are rampant in their host cities. Migrant tenants' "perceived tenurial security" is strong enough to support exchanges of informal property and has shaped a huge informal housing market in host cities. Migrant households demonstrate a "micro-family economy", keeping tenure security in their hometown and avoiding the dissipation of family income in destinations, and therefore choosing to stay in urban villages. According to Fan (2021), Asian urban-bound migration is part of the "split households" strategy to diversify the sources of family income [7,8]. We have taken the phrase "micro-level family economy" from previous studies on developing countries to highlight the role of the "family utility maximization strategy" to explain migrants' tenure choices [9,10]. Breadwinners enter cities to accumulate savings for their family, and city migrants still retain their village residency status, allowing them to continue to enjoy rights and shares to the farmland and homestead back home. The migrants are not left out in benefit shares from rural collective income distribution and compensations when their farmland is acquired and converted to other uses [11,12]. The Strategic Plan for Rural Revitalization (SPRR, 2018–2022) also encourages the amphibious status of rural-urban migration (i.e., migration between urban destinations and rural origins, rather than settling in cities) in response to the recent risks of economic fluctuation. Rural migrants are free to go to a city or back to their hometown and thus become more adaptive to the uncertainties involved in urbanization [13]. One focus of the new Land Management Law is to improve the three-layer system of rural land rights by maintaining collectivity in ownership rights, ensuring contract rights, and making management/use rights transferrable. Rural migrants who settle in the city are also allowed to maintain their land rights to their farmland, homestead, and the collective income distribution that is attached to their rural residency status.

However, urban villages are regarded as informal habitats to house rural migrant workers and other low-wage earners. The destruction or maintenance of such habitats by the governments of migrant-receiving cities has become an issue of good or bad governance related to the city's imaging and high-tech, aesthetic, and greening pursuits [1]. This study examines the rationale behind the informality of the urban village structure and attests to whether, and to what extent, the keeping of hometown lands by rural migrants can affect their tenure security choices (urban village or not) in Chinese metropolises such as Beijing. We tentatively propose the assumption that rural migrants who keep hometown lands are more likely to choose informality in their urban destinations. The following sections examine (a) the role of urban villages in rural-urban circular migration in China; (b) empirical studies and statistical analysis on the tenure choices of rural migrants in Beijing; and (c) the policy implications from the discussion of the rural-urban continuum in a circular migration context.

2. Informality or Resilience? The Linkage between Urban Villages and Migration

2.1. Urban Village as Informality from a Dualistic Rural vs. Urban Land Use Perspective

Unlike the Western community development plan for urban villages to preserve local neighborhood atmosphere [14], the urban village in China is not built to revive the nostalgic dreams and positive aspects of traditional village life in the city. The urban village is normally interpreted as a transitory shelter for low-wage migrant workers in an urbanizing China. Given migrants' low wages and lack of access to subsidized public housing, the low-cost informal rental housing sector of the urban village has surfaced to support the large demand among these migrants [15–18]. Similarly to India and Southeast Asian countries, the urban village in China refers to informal settlements in expanding metropolises [19]. The urban village is seen as a pathology of the rapid urban expansion in an urbanizing Asia and has negative connotations, including low-quality informal tenements, poor infrastructure and facilities, overcrowding, and a lack of planning regulations and management. The city governments in Beijing, Guangzhou, and Shenzhen have low tolerance for dilapidated illegal constructions on rural land and have taken action to clear up the urban village housing that has sheltered migrant tenants.

Is the urban village a disordered and unregulated informal settlement that is foreign to the planned city? The recent comparative studies on this bottom-up urban environment by Oostrum (2021) have proved that urban villages do not develop haphazardly; instead, their transformation of density, mixing, and access is adaptive and interdependent [19]. As discussed by Liu et al. (2020) in their study on the peri-urban mosaic formation, it is fragmentary and piecemeal land acquisition that produces mosaic-like urban villages—an land process that is informal in nature through the lens of the rural-cum-urban land development process (instead of a dualistic rural-versus-urban land-use conversion process) [20]. According to Oostrum (2021), the urban village in Asia demonstrates recurring patterns in a spatial configuration that embody urban design principles and are attuned to residents' needs. The urban village is not an example of pure informality that transgresses land use regulations and building codes [19].

Existing studies have also explained the reason why urban villages are deemed informal in the land management system. Ambiguous land ownership in rural villages is the main reason for the persistence of informality in a rapidly urbanizing China [21]. Differing from other developing countries, the urban village in China grows from a distinct rural-urban dual land system, wherein urban land is state-owned and land-use rights are transferred by city governments, whereas rural land is owned collectively by farmers while being unalienable [20]. Rural land is used exclusively within agricultural communities, and cannot be converted into urban uses without going through state-initiated rural land expropriation procedures. Tenements built on rural land for migrants' transitory lodgings are therefore illegal. It has been proven that urban villages are the outcome of grassroots land speculation and rentiers, under the impact of market forces and state regulation. Informal land uses can create tensions between governments, real estate developers, and local communities over land interests, and between intra-community neighbors and family members over issues of land ownership and distribution [22]. In urban villages under the rural-urban dual land system in China, the perceived tenurial security is, however, strong enough to support exchanges of informal property and shape a huge informal housing market for migrant tenants [23,24]. Perceived tenure security has been proven to be important, affecting the supply and pricing of urban village housing [24,25].

2.2. *Urban Village as Resilience from a Rural-Urban Circular Migration Perspective*

The online Oxford dictionary defines resilience as "the ability of a substance or object to spring back into shape" or "the capacity to recover quickly from difficulties". In this study, resilience is interpreted as the multiple-choice capacity to sustain rural-urban migration in the face of stress and adversity in everyday city life (such as economic risks and uncertainties, poverty, institutional barriers, housing unaffordability, and public housing inaccessibility). Low-rent urban village housing can provide more housing choices for low-wage migrants and beef up the resilience of migrants to city life (such as the migrants' ability to adapt successfully in the face of stress in the city). Informal means are devised to settle down and serve low-income rural migrants, at a low cost.

The access of migrant workers to the urban village, however, is not only restricted in terms of affordability, but also on grounds of legality, as the urban village in the great metropolises has been "cleared-up" and rebuilt into commercial areas, high-tech industrial parks, and talent apartments to curb the "urban sprawl" and intensify land-use efficiency [1]. However, can formalization address the urban village phenomenon, if the key to the issue of informal housing on rural lands is undefined property rights over rural land and houses? We have underestimated the demand side, which is the migrants' tenurial choices between village homes and city places. Existing studies on the spatiality and informality of the urban village, however, have failed to prove the feasibility of "amphibious warfare" for rural-urban migrants, straddling their family "utility maximization" rationalities of (a) hometown tenure security and (b) thrifty consumption of urban housing, taking the form of informal settlement to avoid the dissipation of the family income. The amphibious status of rural migrants and their preference for hometown tenure security cannot be underestimated.

C. Cindy Fan (2021) conducted a series of surveys on the "split households" arrangement in receiving places and sending places for rural-urban migration and proved the prevalence of circular migration in China, similar to that in Africa, Asia, and Latin America [7]. Circular migration can be a step towards permanent urban settlement, and also probably towards returning to one's hometown. Fan's surveys have given some proof of familial utility maximization strategies among rural migrants in China: (a) the couple migration may simply be a strategy to maximize household labor power devoted to their urban earnings, rather than a transition toward permanent settlement in the city; (b) migrants are not willing to spend their hard-earned income in cities (instead, settling in the crowded and cheap urban village housing), not because of unaffordability, but because they can save earnings as a remittance to rural homes; (c) young migrant workers continue to build large houses in their home villages to attract their potential brides and facilitate patrilocal residence; and (d) many migrants are unwilling to give up the rural *hukou* due to access and benefits that are tied to their rural *hukou*, including farming and housing land

and compensation for a land requisition that are considered increasingly more valuable than urban *hukou* [7,11–13,18,26].

Rather than attracting the rural population to buy urban housing, the recent policy in China has begun to guarantee that rural migrants are free to go forward to the city or back to their hometown, and is becoming more adaptive to the uncertainties in urbanization [13]. The 14th Five Year Plan has reaffirmed the importance of protecting the migrants' land rights to farmland, homestead, and a collective income distribution, which are attached to their rural residency status, even though some rural migrants have a permanent urban settlement. Return migration is indeed a component of the circular movement through which rural returnees are expected to overcome the economic risk and inefficient market in cities. Return migration is prevalent in the cities of developing countries with precarious job opportunities and insufficient social security systems [8].

We assume the urban village to be an urban issue, but it is also a part of the migration trajectory, bridging a whole rural tenure system between the village home and the city. We can hardly predict the outcome of the demolition of the urban village if we remain blind to this circular migration and "amphibious warfare" institution for rural-urban migrants. As elaborated by Wu et al. (2013), since the demolition of the urban village fails to tackle the root demand for unregulated living and working space, demolition only leads to "the replication of informality in more remote rural villages" and other kinds of neighborhoods [21]. Informality in China is not only created by the dual rural-urban land market and management system and by an under-provision of migrant housing, it has also become part of micro-migration strategies to maximize familial utility on the move. In this sense, a "transitory" stay in the urban village and circular migration is typical of many breadwinners who decide to avoid family income dissipation.

3. Rural Migrants' Transitory Stays in Beijing Urban Villages

3.1. Research Area

We tentatively proposed the assumption that rural migrants who keep their hometown lands are more likely to choose the informality of urban destinations. Data analysis on Beijing was conducted to attest to this hypothesis. Beijing was chosen for this empirical study because it is the largest destination for migrant workers in North China, and it is the city that enacts the strictest *hukou* control and bans on sales of rural land and housing in China. At the same time, Beijing has a strong desire to compete for a place in the global city roster, thus initiating city re-imaging movements and having a low tolerance for dilapidated illegal constructions, as shown in Figure 1 [27–30]. The scholars, however, regard informal housing as a model for "inclusive and pro-poor urbanization" in China [31], since it has reduced the cost of urbanization and facilitated China's rural-urban migration in the last four decades. According to Peter Ho's (2014, 2017) "credibility thesis", titling or formal registration of these informal housing systems may not matter from the perspective of migrant tenants [32–34]. Unlike De Soto's advocates of formal title, Ho and Sun (2021) pointed out that actors may feel perfectly secure without formal rights; and, on the contrary, formal rights may not improve people's access to credit [31]. Ho also pointed out some factors that can affect "perceived tenure security", such as the duration of occupation, the settlement's size, the level of service provision, the perception of past and current policies, the cohesion of community organization, and the employment opportunities. The urban village constitutes a credible and affordable alternative for rural-urban migrants, for whom urban formal rental housing is out of reach, or for whom social housing belongs to a closed-off arena due to the *hukou* system [31]. Migrant tenants' fear of eviction has been proven to be important to reduce the "perceived security of tenure" [35]. However, even when faced with demolition and eviction, the urban village is still "partially credible", a better choice than "empty" institutions such as unaffordable formal rental housing or inaccessible social housing.

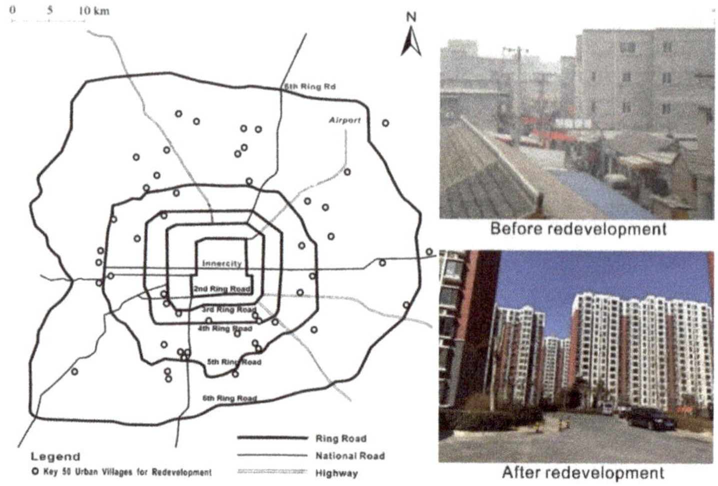

(**a**) Beijing map (**b**) Tangjialing

Figure 1. The 50 key urban villages identified for redevelopment in Beijing since 2010 (Source: the name list of the 50 key urban villages is from News. Focus [27]; field trip by authors).

To date, formalization has been the main top-down government approach to redeveloping urban villages and legalizing their property rights. Rural lands are either converted into state-owned lands for urban uses or become a "rural-urban integrative demonstration project" to house talented migrants on the preserved rural lands. As reflected by the demolition of the famous "Ant Tribe" Tangjialing Village (Figure 1b) for low-income IT workers' apartments in Zhongguancun Science Park, the informality and the everyday life of low-wage migrants were disrupted by the state-initiated formalization schemes, and some dislocated tenants moved to more remote villages in suburban Changping [21,35]. Beijing's urban village is being rebuilt into an urban community, but this is leading to the gentrification process, replacing low-wage migrant tenants [1,36]. As pointed out by Wu et al. (2013), the "beautification and modernization narrative" of urban village redevelopment is rarely balanced with the opportunity cost to the local economy of "removing the main source of low-cost homes" [21]. This particular contextual setting and policy application in Beijing has the potential to (a) inspire the new rural-urban nexus framework (not merely on the side of urban regeneration); (b) provide more temporary affordable housing in a transitional period of urbanization; (c) offer continuous feedback of housing needs in the loop of rural-urban circular migration; and (d) create rural-cum-urban mixed functions that are not dualistic anymore. The next section introduces the data source and analysis methods used to attest whether, and to what extent, rural migrants' keeping of their hometown lands can affect their tenure security choices (urban village or not) in Beijing.

3.2. Data and Methods

The urban village is not merely an informal urban destination, but also represents a strategy of migrant family economy based on principles of resilience and circularity. This study refers to residence status in the urban village, but does not consider the rent levels, which are full of uncertainties and vary with different tenure security on rural plots [24]. This section uses the data from the 2017 Migrant Dynamics Monitoring Survey (MDMS) to look into whether, and how, rural migrants' tenure status in their places of origin would influence their tenure security choices in their destinations. The data on migrants' stays in the urban village were derived from the 2017 Migrant Dynamics Monitoring Survey (MDMS), conducted by the National Health Commission of China. The eligible migrants in this national survey were those who moved across a county (*xiàn*) or city (*shì*) boundary from their registered household and who had stayed in the current destination for more than one month. The data collected on the residential committee units were based on the probability proportional to size (PPS) sampling schemes from each sub-district and town in the city unit. In the PPS sampling, the probability of the selection for a sampling unit is directly proportional to a size measure.

In the total Beijing migrant samples, 2992 cases of the migrant labor force with rural origins were valid for statistical studies, and the remaining retired or unemployed data were discarded. The data demonstrate a male-dominated migration, because 55.8% were male; 37.1% of rural migrants were in the below-30 age group, 33.0% were in the 31–40 age group, and 29.9% were in the above-40 age group. Our focus was on the rural migrants' preference for the informal tenure type in the urban village, which could represent an integrative rural tenurial continuum, bridging their origins and destinations.

Logistic regression was used to analyze relationships between rural migrants' choices of the urban village in their destination or not (dichotomous dependent variable) and the explanatory variables. The explanatory variables included (a) migrant household profiles; (b) migratory status; (c) employment and housing expense stress in destinations; (d) landholding in rural origins (categorical and continuous independent variables). The logistic regression approach was used to combine these independent variables to estimate the probability that the migrants' transitory stay in the urban village would occur.

With a focus on the role of hometown rural tenurial status, we first consider the tenurial distribution of hometown farmland and homesteads: (a) holding both farmland and a homestead; (b) farmland only; (c) homestead only; and (d) land loss (neither farmland nor homestead). The migrants' tenure choices in their destinations were composed of two types: (a) urban village, and (b) others. It was hypothesized that, for migrants, the keeping of rural lands in their hometown could predict a higher odds of choosing an insecure housing tenure in the urban village of their destination. Binary logistic regression was conducted in SPSS, and the regression analysis results are listed below.

3.3. Results

According to the MDMS data, 49.5% of rural migrants in Beijing had chosen to live in the urban village, especially the inner- and outer-suburban villages. Of these, 50.5% chose to settle in the urban formal community in Beijing. As shown in Table 1, the majority of those in the urban village were tenants (99.3%), and migrant homeowners in the urban village held the long-term rent that was issued by the village committee or the township government [24]. In the urban formal community, 86.0% were tenants, 13.3% were private homeowners in the housing market, and 0.6% were homeowners of the government-subsidized housing. When living in the urban village, rural migrants can lower their housing expenses in Beijing but suffer from substandard housing and inferior quality in their built and living environment. When choosing the urban formal community, the land use and construction are compatible with the urban planning provisos and norms, but tenants are faced with high rent costs. According to He, even young professionals are burdened with higher housing costs and precarity in megacities such as Beijing and Shanghai in China, due to the financialization of rental housing [37].

Table 1. Rural migrants' housing choices in Beijing, 2017.

	Home Owning	Subsidized Home Owning	Small Property-Owning	Renting	Total
Urban formal community	13.3%	0.6%	0.1%	86.0%	100.0%
Urban village	0.4%	0.1%	0.3%	99.3%	100.0%
Total	6.9%	0.3%	0.2%	92.6%	100.0%

Source: 2017 MDMS.

The binary logistic regression analysis demonstrated the significant impact of the keeping of hometown land on the likelihood that rural migrants would choose the urban village or not in their urban destinations. The full model contains four groups of predictors—household profiles, migratory status, employment and housing expense stress in destinations, and landholding in rural origins.

Table 2 shows the binary regression results. Omnibus tests of model coefficients indicate that the binary regression model is highly significant, χ^2 (43, N = 2,992) = 1019.732, $p < 0.001$, indicating that the model can distinguish between rural migrants who stay and do not stay in the Beijing urban village. The Hosmer–Lemeshow goodness of fit test (higher than 0.05) indicates that the model fits the data well, with no significant differences between the observational data and the forecast data. The Cox and Snell R-squared value (0.289) and the Nagelkerke R-squared value (0.385) indicate the amount of variation in the dependent variable explained by the model. However, these are described as pseudo-R-squared statistics, rather than the true R-squared values in the multiple regression output. The predicted correct percentage (74%) indicates how well the model can predict the correct category (choosing the urban village or not) for each case.

Table 2. Binary logistic regression of the migrants' choices of the Beijing urban village, 2017.

	Migrants' Choices Regarding the Informal Tenure of the Urban Village (Ref = Urban Formal Market)		
Predictors	*B*	*SE*	*Exp(β)*
Household profile			
Age	−0.054 ***	0.006	0.947
Gender (ref = female)	0.108	0.096	1.114
Marriage (ref = unmarried)	0.448 ***	0.165	1.565
Education level (ref = college and above)			
Primary and below	0.460 **	0.217	1.583
Junior secondary	0.492 ***	0.157	1.636
Senior/technical secondary	0.192	0.153	1.211
Logged annual family income	−3.954 ***	0.241	0.019
Migration status			
Household size in urban destinations	0.021	0.050	1.021
Place of origin (ref = East China)			
North-eastern region	0.531 ***	0.193	1.701
North-western region	0.005	0.256	1.005
North China	0.188	0.118	1.207
Central China	0.182	0.125	1.200
South China	−1.048	0.815	0.351
South-western region	0.389 *	0.203	1.475
Second generation of rural migrant workers (ref = others)	−0.192 *	0.114	0.825
Employment in urban destinations			
Occupation (ref = blue collar)			
Cadre, manager or head	0.040	0.642	1.041
Technician/professional	−0.099	0.215	0.906
Staff/clerk	−0.665	0.421	0.514
Service worker	−0.291	0.181	0.748
Agricultural worker	0.562	0.822	1.755
Other	−0.829 ***	0.281	0.437
Industry (ref = labour-intensive manufacturing)			
Primary industry	−0.364	0.742	0.695
Non-manufacturing sectors in secondary industry	0.118	0.223	1.125
Capital-intensive manufacturing	0.574 *	0.329	1.775
Skill-intensive manufacturing	0.134	0.253	1.144
Other types of manufacturing	0.541	0.371	1.718
Producer services	−0.047	0.199	0.954
Public services	−0.599 **	0.245	0.550
Consumer services	−0.086	0.163	0.918
Employer type (ref = privately-owned/joint stock)			
State-owned	−0.448 **	0.187	0.639
Collective-owned	−0.253	0.381	0.776
Foreign-invested/joint venture	−0.055	0.260	0.947
Family- or individually-owned	−0.071	0.146	0.932
Non-profit organization	−1.879	1.269	0.153
Other	0.596 ***	0.196	1.814

Table 2. *Cont.*

	Migrants' Choices Regarding the Informal Tenure of the Urban Village (Ref = Urban Formal Market)		
Predictors	*B*	*SE*	*Exp(β)*
Employment status (ref = stable employees)			
Temporary employees (no contract)	−0.059	0.205	0.942
Employer	−0.656 ***	0.238	0.519
Self-running	0.163	0.152	1.177
Other	−0.296	0.389	0.744
Housing pressure in urban destinations			
Housing expense-to-income ratio	−5.795 ***	0.408	0.003
Landed status in rural origins			
If still holding rural lands in hometown (ref = land loss)			
Holding both farmland and homestead	0.738 ***	0.123	2.092
Homestead only	0.861 ***	0.131	2.365
Farmland only	0.661 ***	0.197	1.936
Constant	16.957 ***	1.001	
N	2992		
df	43		
λ^2	1019.732 ***		
−2 Log Likelihood	3127.760		
Cox and Snell R^2	0.289		
Nagelkerke R^2	0.385		
Percent correctly classified	74%		

Note: Significant at * 0.1; ** 0.05; *** 0.01 level.

Exp(β) values are the odds ratios (OR) for each of the explanatory variables, showing "the change in odds of being in one of the categories of the outcome when the value of a predictor increases by one unit" [38] (p. 461). The rural migrants with hometown lands, compared to their land-loss counterparts (ref), were more likely to live in the urban village (OR = 2.092 for farmland and homestead holders, 2.365 for mere homestead holders, and 1.936 for mere farmland holders), controlling for all other factors in the model. A rural homestead in a hometown is a more prominent pulling factor than farmland in regard to returning home eventually, since migrants holding homesteads (OR = 2.365) were more likely to choose a transitory stay in the urban village of their destinations than those merely holding farmland (OR = 1.936). It was evident that alleviating housing expense pressure is a strong motivation for staying in the urban village (OR = 0.003), as its regression coefficient was the strongest of all these explanatory variables (Table 2). Income level was listed in the top two of the coefficient values.

As shown in Table 2, in the explanatory factors concerning household profiles, only age, marriage, education, and income variables made a statistically significant contribution to the model. Rural migrants who were younger, married, less educated, and low earners were more likely to stay in the urban village, controlling for all other factors in the model. Migrants from the north-eastern and south-western region were 1.701 and 1.475 times more likely to stay in the Beijing urban village, compared to those from the more advanced East China (ref), when controlling for all other factors in the model. Those from south China were the least likely to choose the urban village. The place of origin matters, as south China is more advanced and affluent than most of the listed regions, and those from south China can afford to stay elsewhere. Additionally, second-generation rural migrant workers, compared to those from other backgrounds (ref), were (OR = 0.825) less likely to choose the urban village.

When it comes to employment in urban destinations, rural migrants who have other occupation types (such as freelance workers in different kinds of metropolitan industries), get involved in public services and state-owned sectors and assume the role of employers, and are less likely to stay in the urban village (compared to blue collar workers, especially those engaged in labor-intensive manufacturing and the private economy as stable employees). Migrants in capital-intensive manufacturing are more inclined to choose the urban village (OR = 1.775), compared to those in labor-intensive manufacturing (ref), due to the suburbanization of manufacturing clusters (such as electronic and equipment manufacturing) and the migrant workers' preference for the nearby villages in the suburb.

4. Discussion: Revisiting a Synthesis between the Rural-Urban Dual System

We performed exploratory regression analysis to obtain convincing statistical results about the relationship between the holding of hometown tenure and rural migrants' choices as to their housing tenure (informally residing in the urban village or not) in their destinations. Rural land status in rural origins was proven to exert a statistically significant contribution to the model. Those holding homesteads in their places of origin were most likely to choose the urban village in their destinations, displaying a hypothesized "micro-family economy", keeping tenure security in their hometown and avoiding the dissipation of their family income in their destinations. The assumed "permanent urban settlement" in the previous studies is questioned in this study, and doubts are therefore raised as to its underestimation of the "amphibious" and flexible status of rural-urban migration in transitional economies such as China. The "amphibious warfare" for rural-urban migrants includes (a) the hometown landholding in their rural places of origin, and (b) the multiple-choice of urban housing in their destinations (settlement or a transitory stay and the thrifty consumption of informal housing).

4.1. Different Rural Land Institutions, Different Urbanization Paths in Developing Countries

First, urban villages are regarded as informal settlements built on rural lands in host cities. However, they are also a manifestation of China's unique urbanization path and rural land institutions. In the 1950s, Brazil began its agricultural modernization at the expense of small-scale farming, explaining why landless rural workers had to migrate to the rapidly industrializing cities that provided the only prospects for them [39]. Such squatting is, however, discouraged or even forbidden in urban China. This migration policy difference between different countries in the developing countries is derived from their disparate urbanization paths in different contexts.

Studies on China's rural-urban divide have highlighted a series of *hukou* policies against migrants' permanent stays in cities [40]. However, these have neglected the long-lasting divide in the rural vs. urban welfare distribution system, which is a kind of "complementary balance" in the use of a rural vs. urban dualist land system to overcome the economic risks, if any, in the transitional period. As though situated in a rural-cum-urban "amphibious" status, migrants can access the labor and capital market for efficiency in urban destinations, and also ensure the continuity of their family and lineage, maintaining security in their rural places of origin. The importance of small-scale farming and the strong agricultural collective entity (in both origins and destinations) cannot be understated in a mobile and urbanizing Asia. In recent years in China, some harsher residency controls were enforced on lower-skilled migrant workers and illegal village constructions, and the urban villages were ended with demolitions. The non-*hukou* migrant tenants are neglected and become victims to some extent as a result of informal tenement demolition and soaring rent in the city. The urban village as the main source of low-rent housing cannot be understated.

The statistical analysis presented in Table 2 revealed some interesting findings. Second-generation rural migrant workers were found to be less likely to choose the urban village, compared to other backgrounds. This implies that second-generation rural migrant workers have a great desire to become urban settlers, and they can gain access to opportunities and resources (compared with others with rural origins). This assimilation process is important for rural-urban migration, as second-generation migrants tend to settle down in formal communities and gradually integrate into the receiving society. In addition, the blue collar workers among rural migrants (especially those in capital-intensive manufacturing jobs) are more inclined to choose the urban village. The suburbanization of manufacturing in the cities in China has created the residential suburbanization of migrant workers. Some of them choose to live in the nearby villages in the suburb as a result. For instance, in the

China–Singapore Suzhou Industrial Park (SIP), low-skilled migrants live in the low-end rental housing market, including informal ones in the nearby villages [41].

4.2. Maintaining the Rural-Urban Dual System and Improving the Opportunity Structure for "Amphibious" Migrant Workers

There have been long-lasting academic debates on the "structural" power of China's rural-urban dual system on its migrant workers. Chan and Wei elaborated on the impacts of the rural-urban dual system in China on internal migration. They analyzed the mechanisms of the rural-urban dual system, and how it has helped to generate low-cost migrant labor for urban markets [4]. As argued earlier, rural migrants in China are embedded firmly in a mega-land institution: hometown lands as welfare and security, urban lands as a speculative market (if they become home-occupiers), and a consumptive use (as tenants), and the urban village as their affordable though transitory shelter. Chan and Wei's (2019) thesis on the rural-urban dual system recognized the asymmetry of the power structure, highlighting that the dual system provides the "opportunity for discrimination" against rural migrant workers [4]. On the contrary, Chen and Fan's (2016) thesis on the split household strategy for economic calculus ignored this structural asymmetry but highlighted the "opportunities for risk reduction" for microeconomic rationalities [12]. How can we interpret the divergence in the opinions of Chan and Fan?

As argued by Chan and Wei (2019), China's rural-urban dual system is a "mega dual system", different from the more limited and smaller dualistic structure in the Lewis model [4]. To fix the rural-urban gap and the plight of "split households" (such as migrant children and left-behind children), central governments in China enacted the National New Urbanization Plan (NUP, 2014–2020) and the Strategic Plan for Rural Revitalization (SPRR, 2018–2022) to grant more welfare benefits for migrants in urban destinations and more economic opportunities for returnees to their rural places of origin. However, the rural land system has not been removed from a transitional China, as reaffirmed in the new version of the 2021 Land Management Law in China.

However, policy-makers have neglected the fact that transitory stays and the need for cheap housing are still the primary concerns of migrants. The demolition of urban villages is validated according to security and public interest concerns, but this has changed the housing opportunity structure for low-wage migrants. It is an interesting question whether the peri-urban migrant enclaves will become regulatory areas, or if they will provide low-rent housing opportunities for risk reduction [35]. The more committed institutions of land use are still engaged in debate regarding how to regulate and serve these urban floaters within China's mega rural-urban dual system and development model.

5. Conclusions

In this study, we have revisited the concept of the urban village in metropolitan China, which does not represent pure informality in a destination, but it is also a part of a rural-urban circular migration trajectory, since rural migrants who keep their hometown lands are more likely to choose a transitory stay in an urban village. Migrants' straddling of their village homes and city living-places has been underestimated in previous studies and policy-making. To date, few resettlement sites have been provided for these dislocated migrant tenants.

We have also discussed the policy debates on the Chinese rural-urban dual system—does the rural-urban dual system provide "opportunities for discrimination" (for investors and urban governments), or "opportunities for risk reduction" (for the migrant family economy)? Chinese governments are implementing macro policies (NUP and SPRR) that beef up the opportunity structure for "amphibious" migrant workers. However, there is still a long way ahead, regarding the questions of how to perverse low-rent housing areas for temporary migrant workers, and how to provide more public services for the new trends of the increasing "familization of migration".

Another implication in land use policy is the importance of a welfare-oriented land system. Optimal land use is deemed good as it can generate higher economic returns, but the value of resilience for social welfare and integration is underestimated. The demolition of urban villages is the result of long-term land-use conflicts between city governments, property developers, low-wage migrant workers, and other disadvantaged residents. In this study, resilience is conceived as a capacity that can develop through village-based stakeholder interactions to mediate the housing stress of migrants (e.g., the urban village). More research is needed to look into forms of community property and collective land ownership that are still robust in the 21st Century [42].

Author Contributions: Conceptualization and original draft preparation, R.L.; data curation, Y.J. All authors have read and agreed to the published version of the manuscript.

Funding: This research was funded by the grant from the National Natural Science Foundation of China (NSFC project number: 42071201, 41701188), and a grant from the Beijing Philosophy and Social Science Planning Program and Social Science Research Key Program of Beijing Municipal Commission of Education (project number: SZ202010028016).

Data Availability Statement: Data use of the 2017 Migrant Dynamics Monitoring Survey (MDMS) is supported and approved by the Migrant Population Service Center, the National Health Commission of China. https://www.chinaldrk.org.cn/wjw/#/home (accessed on 11 November 2021).

Conflicts of Interest: The authors declare no conflict of interest.

References

1. Liu, R.; Wong, T. Urban village redevelopment in Beijing: The state-dominated formalization of informal housing. *Cities* **2018**, *72*, 160–172. [CrossRef]
2. Wu, F. Housing in Chinese urban villages: The dwellers, conditions and tenancy informality. *Hous. Stud.* **2016**, *31*, 852–870. [CrossRef]
3. Wang, C. Urban village on a global scale: Diverse interpretations of one label. *Urban Geogr.* **2020**, 1–22. [CrossRef]
4. Chan, K.W.; Wei, Y. Two systems in one country: The origin, functions, and mechanisms of the rural-urban dual system in China. *Eurasian Geogr. Econ.* **2019**, *60*, 422–454. [CrossRef]
5. Zhou, Y.; Li, X.; Liu, Y. Land use change and driving factors in rural China during the period 1995–2015. *Land Use Policy* **2020**, *99*, 105048. [CrossRef]
6. Standing Committee. *Land Administrative Law of the People's Republic of China. The 9th National People's Congress of the People's Republic of China (1st Revision)*; Standing Committee: Beijing, China, 1998.
7. Fan, C.C. Householding and split households: Examples and stories of Asian migrants to cities. *Cities* **2021**, *113*, 103147. [CrossRef]
8. Stark, O.; Bloom, D.E. The new economics of labor migration. *Am. Econ. Rev.* **1985**, *75*, 173–178.
9. Hugo, G.J. Effects of International Migration on the Family in Indonesia. *Asian Pac. Migr. J.* **2002**, *11*, 13–46. [CrossRef]
10. Rahman, M.M. Emigration and the Family Economy: Bangladeshi Labor Migration to Saudi Arabia. *Asian Pac. Migr. J.* **2011**, *20*, 389–411. [CrossRef]
11. Chen, C. Why Migrant Workers in China Continue to Build Large Houses in Home Villages: A Case Study of a Migrant-Sending Village in Anhui. *Mod. China* **2019**, *46*, 521–554. [CrossRef]
12. Chen, C.; Fan, C.C. China's *hukou* puzzle: Why don't rural migrants want urban *hukou*. *China Rev.* **2016**, *16*, 9–39.
13. Han, J. *Opinions on Promoting the Connection between Small Farmers and the Development of Modern Agriculture*; Ministry of Agriculture and Rural Affairs: Beijing, China, 2019. Available online: http://www.moa.gov.cn/hd/zbft_news/xnhxdnyfz/wzzb/ (accessed on 29 September 2021).
14. Aldous, T. *Urban Villages: A Concept for Creating Mixed-Use Urban Developments on a Sustainable Scale*; Urban Villages Group: London, UK, 1992.
15. Lefebvre, H. *The Urban. Revolution*; University of Minnesota Press: Minneapolis, MN, USA, 2003.
16. Liu, Y.; Li, Z.; Liu, Y.; Chen, H. Growth of rural migrant enclaves in Guangzhou, China: Agency, everyday practice and social mobility. *Urban Stud.* **2015**, *52*, 3086–3105. [CrossRef]
17. Zhang, L. *Strangers in the City: Reconfigurations of Space, Power, and Social Networks Within China's Floating Population*; Stanford University Press: Stanford, CA, USA, 2002.
18. Zheng, S.; Long, F.; Fan, C.C.; Gu, Y. Urban Villages in China: A 2008 Survey of Migrant Settlements in Beijing. *Eurasian Geogr. Econ.* **2009**, *50*, 425–446. [CrossRef]
19. Van Oostrum, M. Access, density and mix of informal settlement: Comparing urban villages in China and India. *Cities* **2021**, *117*, 103334. [CrossRef]
20. Liu, R.; Wong, T.; Liu, S. The peri-urban mosaic of Changping in metropolizing Beijing: Peasants' response and negotiation processes. *Cities* **2020**, *107*, 102932. [CrossRef]
21. Wu, F.; Zhang, F.; Webster, C. Informality and the development and demolition of urban villages in the Chinese peri-urban area. *Urban Stud.* **2013**, *50*, 1919–1934. [CrossRef]

22. Kan, K.; Chen, X. Land speculation by villagers: Territorialities of accumulation and exclusion in peri-urban China. *Cities* **2021**, *119*, 103394. [CrossRef]
23. Kiddle, G.L. Key theory and evolving debates in international housing policy: From legalisation to perceived security of tenure approaches. *Geogr. Compass* **2010**, *4*, 881–892. [CrossRef]
24. Zhang, M.; Zhao, P. The determinants of informal housing price in Beijing: Village power, informal institutions, and property security. *Cities* **2018**, *77*, 117–129. [CrossRef]
25. Li, B.; Tong, D.; Wu, Y.; Li, G. Government-backed "laundering of the grey" in upgrading urban village properties: Ningmeng Apartment Project in Shuiwei Village, Shenzhen, China. *Prog. Plann.* **2021**, *146*, 100436. [CrossRef]
26. Fan, C.C. Settlement intention and split households: Findings from a survey of migrants in Beijing's urban villages. *China Rev.* **2011**, *11*, 11–42.
27. News.Focus. Beijing Jinnian Dongqian Cunzhuang Chao 300 Ge, Yi Tengtui Tudi Yue 600 Gongqing [Beijing Dismantling 300 Villages for Redevelopment]. 2010. Available online: http://news.focus.cn/bj/2010-11-19/1105994.html (accessed on 29 September 2021).
28. Bao, J.; Liu, Y.; Li, X. Urban spectacles as a pretext: The hidden political economy in the 2010 Asian Games in Guangzhou, China. *Urban Geogr.* **2019**, *40*, 409–427. [CrossRef]
29. Shin, H.B. Contesting speculative urbanisation and strategising discontent. *City* **2014**, *18*, 509–516. [CrossRef]
30. Shin, H.B.; Li, B. Whose games? The costs of being "Olympic citizens" in Beijing. *Environ. Urban* **2013**, *25*, 559–576. [CrossRef]
31. Ho, P.; Sun, L. A model for inclusive, pro-poor urbanization? The credibility of informal, affordable "single-family" homes in China. *Cities* **2021**, *97*, 102465.
32. Ho, P. The "credibility thesis" and its application to property rights: (in)secure land tenure, conflict and social welfare in China. *Land Use Policy* **2014**, *40*, 13–27. [CrossRef]
33. Ho, P. *Unmaking China's Development: Function and Credibility of Institutions*; Cambridge University Press: Cambridge, UK, 2017.
34. Grabel, I. The political economy of "policy credibility": The new-classical macroeconomics and the remaking of emerging economies. *Camb. J. Econ.* **2000**, *24*, 1–19. [CrossRef]
35. Liu, R.; Wong, T.; Liu, S. Low-wage migrants in northwestern Beijing, China: The hikers in the urbanisation and growth process. *Asia Pac. Viewp.* **2013**, *54*, 352–371. [CrossRef]
36. Wu, W.; Wang, J. Gentrification effects of China's urban village renewals. *Urban Stud.* **2017**, *54*, 214–229. [CrossRef]
37. He, S. Loft-living under state-led financialisation: Adaptive reuse of industrial buildings for long-term rental apartments in Beijing, China. FinGeo Seminar, 2021/06/29. Available online: https://www.fingeo.net/ (accessed on 11 November 2021).
38. Tabachnick, B.G.; Fidell, L.S. *Using Multivariate Statistics*; Pearson: New Jersey, NJ, USA, 2007.
39. Haroldo, T.; Humberto, A.; De Oliveira, M.A. São Paulo peri-urban dynamics: Some social causes and environmental consequences. *Environ. Urban* **2007**, *19*, 207–223.
40. Chan, K.W. *Urbanization with Chinese Characteristics: The Hukou System and Migration*; Routledge: London, UK, 2018.
41. He, S.; Chang, Y. A zone of exception? Interrogating the hybrid housing regime and nested enclaves in China-Singapore Suzhou-Industrial-Park. *Hous. Stud.* **2020**, 1–25. [CrossRef]
42. Wily, L.A. Collective land ownership in the 21st Century: Overview of global trends. *Land* **2018**, *7*, 68. [CrossRef]

land

MDPI

Article

Quantification of Resilience Considering Different Migration Biographies: A Case Study of Pune, India

Ann-Christine Link [1], Yuanzao Zhu [2] and Raphael Karutz [3,*]

1 Department of Economic Geography and Location Research, Philipps-Universität Marburg, 30537 Marburg, Germany; ann-christine.link@uni-marburg.de
2 UFZ—Helmholtz Centre for Environmental Research, Department of Economics, 04318 Leipzig, Germany; yuanzao.zhu@ufz.de
3 UFZ—Helmholtz Centre for Environmental Research, Department of Urban and Environmental Sociology, 04318 Leipzig, Germany
* Correspondence: raphael.karutz@ufz.de; Tel.: +49-341-2354771

Abstract: Urbanization proceeds globally and is often driven by migration. Simultaneously, cities face severe exposure to environmental hazards such as floods and heatwaves posing threats to millions of urban households. Consequently, fostering urban households' resilience is imperative, yet often impeded by the lack of its accurate assessment. We developed a structural equation model to quantify households' resilience, considering their assets, housing, and health properties. Based on a household survey (n = 1872), we calculate the resilience of households in Pune, India with and without migration biography and compare different sub-groups. We further analyze how households are exposed to and affected by floods and heatwaves. Our results show that not migration as such but the type of migration, particularly, the residence zone at the migration destination (formal urban or slum) and migration origin (urban or rural) provide insights into households' resilience and affectedness by extreme weather events. While on average, migrants in our study have higher resilience than non-migrants, the sub-group of rural migrants living in slums score significantly lower than the respective non-migrant cohort. Further characteristics of the migration biography such as migration distance, time since arrival at the destination, and the reasons for migration contribute to households' resilience. Consequently, the opposing generalized notions in literature of migrants either as the least resilient group or as high performers, need to be overcome as our study shows that within one city, migrants are found both at the top and the bottom of the resilience range. Thus, we recommend that policymakers include migrants' biographies when assessing their resilience and when designing resilience improvement interventions to help the least resilient migrant groups more effectively.

Keywords: livelihood resilience; migration; urbanization; extreme events; India

Citation: Link, A.-C.; Zhu, Y.; Karutz, R. Quantification of Resilience Considering Different Migration Biographies: A Case Study of Pune, India. *Land* **2021**, *10*, 1134. https://doi.org/10.3390/land10111134

Academic Editors: Stephan Bartke and Sigrun Kabisch

Received: 30 September 2021
Accepted: 22 October 2021
Published: 25 October 2021

1. Introduction

The megatrend of urbanization has largely become a phenomenon of the Global South [1]. The growth of urban centers is not only governed by natural population growth but rural to urban and urban to urban migration (For our analysis, we adopt IOM's general definition of migrants: "A person who moves away from his or her place of usual residence, whether within a country or across an international border, temporarily or permanently and for a variety of reasons" [2]). India is a prominent example of the wide practices of migration and associated urbanization. It features the world's largest emigration and domestic migration population [3], favored by persisting vertical and spatial inequality [4]. Simultaneously, India is among the countries most prone to natural disasters, placing an additional burden on these already stressed urban infrastructures [5]. Although there is little consensus on how environmental factors impact economic and socio-political crises and drive migration, changes in variabilities and anomalies of rainfall and rapid-onset disasters have been shown in the past to significantly impact migration [6].

Ultimately, diverse and complex systems of push-pull factors, including those associated with urbanization processes, are influencing domestic migration throughout India [7].

After decades of problematizing and attempting to reduce domestic migration [8], a more recent understanding of (domestic) mobility has been established where it is not seen as an outcome of failing governmental or developmental strategies but as a normal aspect of life [9]. Hence, migration can be conceptualized as a livelihood, adaptation, and resilience strategy to diverse external and internal stressors (political, social, economic, and environmental) to improve the situation of migrants and their families [10].

To ensure sustainable economic, social, and demographic growth of fast-growing cities in the face of stressors, households need to be resilient. By actively building the resilience of urban households, the negative impacts of extreme weather events can be alleviated. Effective strategies to increase resilience, however, require a detailed understanding of the status quo. As a highly context-specific concept, resilience needs to be well-conceptualized and, in a second step, quantified.

Resilience is a complex and contested concept with no universally agreed-upon definition [11]. First appearing in the ecological discourse, resilience was referred to as the magnitude of the disturbance that can be absorbed by a system before it redefines its structure [12]. Many other disciplines took up and expanded the concept since then. Particularly in the last decade, calls for a more integrated understanding of resilience have gained prominence, in particular from socio-political and economic perspectives [13–16].

For our analysis, we build on the concept of livelihood resilience, defined as the capacity of individuals to maintain and enhance their livelihood opportunities and well-being despite stressors [17]. Livelihood resilience is strongly linked to social resilience concepts, typically viewed as the "ability of human communities to withstand external shocks to their social infrastructures such as environmental variability or social, economic, and political upheaval." [18]. Livelihoods can be described as systems that comprise all capabilities, assets (material and social), and activities needed for living-making [19]. Consequently, livelihood resilience does not only describe the resources that people own but also the strategies that they develop to adapt, making livelihood resilience a dynamic approach to resilience.

People are viewed as the central actors in resilience, adaptation practice, and policy since it is the individuals themselves who are making a living, trying to meet their economic and social needs while coping and responding to diverse uncertainties and opportunities [20]. Furthermore, livelihood resilience acknowledges the importance of considering scale and uncertainty since livelihoods are actor-, place-, and context-specific [21]. Speranza et al. [17] developed an analytical framework where livelihood resilience is determined by an actor's buffer capacity, ability to self-organize, and capacity to learn, serving as the conceptual basis for our resilience model (RM).

The quantification of resilience is challenging due to the concept's nature since resilience is considered a latent variable that cannot be directly observed [16]. Most approaches to resilience quantification and modeling follow the understanding that households' or communities' resilience is shaped by socio-economic and organizational factors, which enable them to respond and adapt to stressors [11].

The Food and Agriculture Organization of the United Nations (FAO) pioneered the quantification of resilience, focusing on food shocks [22]. They conceptualize households' options for living making as defined by so-called pillars: access to social safety nets (food assistance, social security), basic services (water, healthcare, electricity), and assets (land, livestock, house). A household's resilience is described as the status and supply of all pillars over time and is linked to the pillar's stability [23]. Since 2008, the FAO developed a statistical model known as the resilience index and measurement analysis (RIMA) [24]. Based on factor analyses (FA) and structural equation models (SEM), RIMA produces a resilience capacity index (RCI) and a resilience structure matrix (RSM). The RCI identifies households most at risk and specifies the areas of resilience weaknesses whereas the RSM serves as a descriptive tool to explain how each resilience pillar relates to the resilience capacity and how each observed variable correlates to its pillar [25,26]. RIMA has since evolved with RIMA-I building on mixed-effects models [27] and RIMA-II on multiple cause multiple indicator models (MIMIC) allowing for resilience predictions over time and the inclusion of indirect resilience such as speed of recovery or loss extent [28]. The different RIMA generations have been applied widely in various developing countries to either quantify current or predict future household resilience status to food shocks [27,29,30].

All RIMA approaches focus on resilience to food shocks, primarily targeting households in rural settings, and are hence not directly applicable to the purposes of this study regarding quantifying resilience in an urban setting. Even though the body of knowledge on urban resilience has been expanding rapidly [31], the majority of studies focus on urban areas in countries of the Global North and assess resilience at an aggregated (e.g., municipal), rather than household-scale [32]. Moreover, many urban RMs and indices require detailed time-series of socio-economic and biophysical data, which is often lacking in the Global South, especially at the household scale [33]. Examples of urban quantification indices include the socio-ecological index for flood events [34], the climate disaster resilience index [35], the urban resilience index [36], and the integrated resilience index [37].

To our knowledge, studies that quantify the resilience of migrant groups compared to non-migrant households from an urban destination perspective are lacking [38]. Previous studies set in an urban-migration context in India focus predominantly on the resilience to environmental stressors such as mountain hazards [39], climate change [40–43], and the combination of specific extreme events such as droughts, floods, heatwaves, and storm surges [44,45]. The methodological approaches of these studies range from primarily mixed-method designs based on interviews and household surveys [40,43,44] to literature reviews [39–42,44,46]. Rarely modelling approaches are pursued [41].

To that end, we address the aforementioned knowledge gaps by quantifying and comparing urban resilience of non-migrants and migrants at one of the main Indian migration destinations [47], namely the western-Indian city Pune while accounting for migration biographies. So far migration and resilience scholars have broadly highlighted that the resilience of migrants is related to the "circumstances of the journey" and "the reception in the receiving community" [48]. In this study, we investigate more deeply how migration biographies impact the resilience of migrants. We consider the following four specific migration characteristics to implement migrants' diverse socio-economic and cultural backgrounds as well as trajectories: time since arrival in Pune, reasons for migration, migration distance, and environmental exposure (flood, heat). The analyses are structured along six sub-groups based on current residence zone (formal urban, slum) and migration origin (non-migrants, urban origin, rural origin). This analysis is critical to unravelling the opposing generalized notions of migrants defined as possessing either extremely low [49] or extremely high resilience at the migration destination [50], and improve the targeting and effectiveness of resilience interventions.

2. Materials and Methods

2.1. Study Site

Pune Metropolitan Region (in the following: Pune) comprises Pune Municipal Corporation (PMC) and Pimpri Chinchwad Municipal Corporation (PCMC), as well as peri-urban areas. Pune has the eighth-largest metropolitan economy and the sixth-largest per capita income in India [51]. Known as the "new industrial and educational hub" with burgeoning manufacturing and IT industry as well as many renowned higher education facilities [52], the city has witnessed substantial population growth. Since the last decade, 50% of the population growth has been due to migration [53]. Pune district ranks second in Maharashtra's migration destinations with most of the arrivals coming from within Maharashtra (>80%) and many (>40%), even, from within the same district (Figure 1).

Located in the east of India and close to the Arabian Sea it is specifically prone to environmental stressors such as monsoonal floods and extreme heat [54,55]. Pune's households experienced three large flood events in the last three consecutive years (2018,

2019, and 2020), increasing in size and damage, involving dozens of casualties [56–58]. In these last three years, several heat warnings had to be issued during the summer months of April and May [59]. In 2019, Pune recorded 42.9 °C which is the highest temperature in 36 years [60].

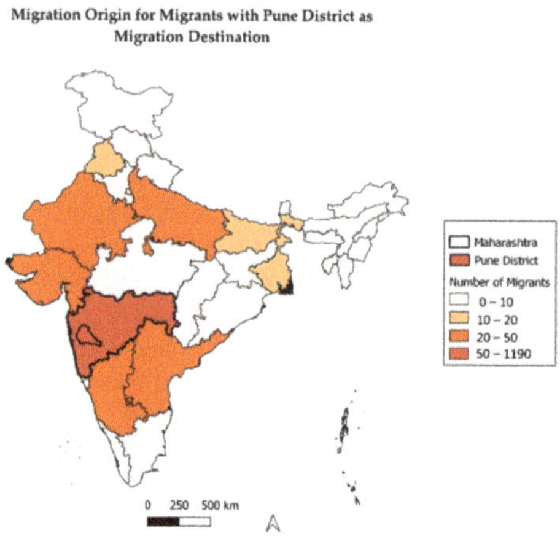

Figure 1. Total number of Pune District migrants per state of origin in 64th NSSO round 2008 (n = 1438) [61].

2.2. Data

To build a reliable resilience model (RM), large and detailed data sets of relevant socio-economic and demographic variables are required. Therefore, our RM is based on a general sample of Pune's households (n = 1872), aiming at representativeness by randomly selecting households within pre-defined spatial and socio-economic clusters in central and peripheral Pune. The household survey (HHS) includes information on water and energy use, source and expenditure, housing conditions, income and food expenditure, and migration. The households were interviewed in the year 2020 based on an interview template (cf. SM data collection). About 30% of the households (n = 553) have a migration background. If available, migration information was gathered for the responder and partner. Response variables in the HHS are categorical, nominal, ordinal, interval, or text.

2.3. Methods

In this study, an RM based on socio-economic factors and the livelihood resilience concept was developed. It serves the purpose of quantifying and comparing the resilience of Pune migrants and non-migrants. Since resilience is a latent variable that cannot directly be observed, our RM is based on so-called pillars which are influenced by many other observable variables which contribute proportionally to their pillar. This structure makes our RM a hierarchical model (cf. SM, Figure S5). Due to the multidimensionality of resilience, this RM is specifically designed for the study region Pune to properly account for all relevant contexts.

The spreadsheet program Microsoft Excel 2019 was used to pre-process the HHS data. This step involved excluding all variables with more than 30% missing data. Variables directly linked to the household size such as income, space, or rooms were turned into per capita variables. Due to the lack of predictor variables in the HHS, RIMA based on a structural equation model (SEM) was pursued. However, contrary to Alinovi et al. [24], an iterative-inductive approach of determining principal components was chosen to avoid subjectivity and arbitrariness.

The selection of pillars and variables as well as the model building was implemented with the statistical program RStudio (cf. SM, R code). The variable and pillar selection was built on insights of the RIMA models [22,24,25,62], urban resilience models [34,36,37,63], statistical relationships (principal component analysis, PCA), and the criteria of only including variables with factor loading scores higher than 0.65 (cf. SM, model). A categorical PCA (PRINCALS) was applied not only to condense and scale the data because most of the HHS variables are nominal or ordinal while measured at different scales but also to produce weights for each variable, depending on the variation and covariation of the data matrix. Since PRINCALS was chosen to determine the principal components of the data set, continuous variables were categorized based on quartiles or quantiles where low values represent low performances and high variables represent high performances of the households.

This approach led to the determination of three foundation pillars: *assets*, *housing*, and *health*. The first pillar *assets* comprises of the variables; yearly income (per person, pp), years of education, and food–income ratio. The second pillar *housing* is composed of the variables; zone type, dwelling type, space (pp), and rooms (pp). The third pillar *health* encompasses the availability of flush toilets, electricity backup, number of cooling devices, and total water storage (Figure 2; cf. SI, Table S1).

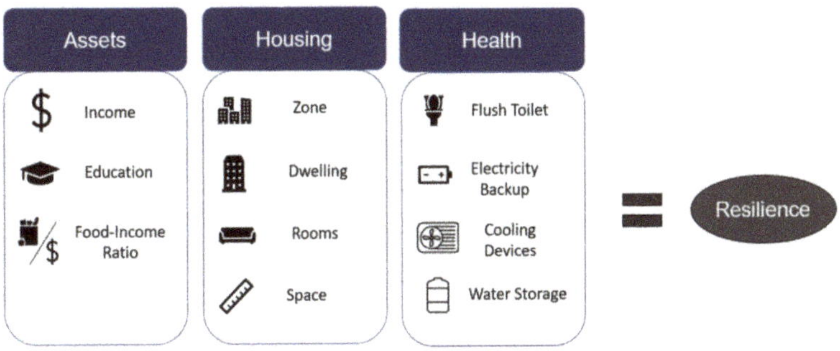

Figure 2. Resilience model framework describing how resilience is calculated based on the three pillars: assets, housing, and health as well as their associated variables.

The weightings of each observable variable within the three pillars (RSM) are displayed using radar charts (cf. SM, Figure S7). The radar charts are valuable to understand the importance of each variable within a pillar. For the *assets* pillar, income has the highest influence, followed by the food–income ratio, and education. For the *housing* pillar, zone and dwelling contribute equally strongly, followed by space and rooms per capita. Then for the *health* pillar, flush toilets and electricity backup have the highest impact, followed by total cooling devices and total water storage (cf. SM, pillars).

In the next step, the object scores of the first component of each pillar were implemented into the SEM to explain the latent variable *resilience*. The path diagram of the SEM reveals that the three pillars, *assets* (0.78), *housing* (0.77), and *health* (0.69), contribute equally strongly to the latent variable resilience (cf. SM, Figure S6). All tests for the goodness of fit return satisfying results (CFI > 0.9, RMSEA < 0.08, and SMR < 0.08). The object scores of the SEM represent the resilience (RCI) of each household.

For more differentiated resilience comparisons, the results were organized using four migration characteristics: time since arrival, reasons for migration, migration distance, and environmental exposure (flood and heat exposure). For each migration characteristic, the results are presented in a matrix, employing six sub-groups based on the current residence zone (formal urban, slum) and migration origin (rural, urban; Figure 3). These are called in the following: non-migrant-urban, non-migrant-slum, urban-urban, urban-slum, rural-urban, and rural-slum.

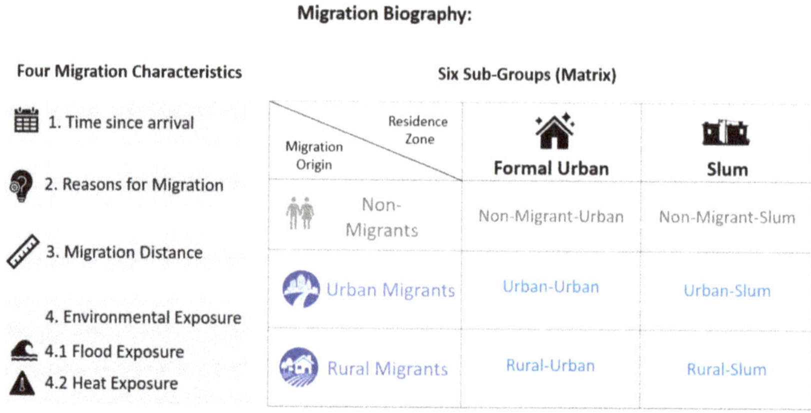

Figure 3. Explanation of migration biography including four migration characteristics and six sub-groups from matrix.

In the following, the categorization and methodology of the migration characteristics are described in more depth. The *time since arrival* migration characteristic was organized into three classifications, namely $\leq$ 10, 10–20, and >20 years. The *reasons for migration* characteristic included in the HHS survey and this paper are work, living conditions, education, and family (cf. SM, Figure S1). The *migration origin* characteristic is structured using three classifications: same district (Pune), same state but other districts than Pune, and other states than Maharashtra. For the *flood exposure* characteristic, the sub-groups' resilience with and without flood experience based on the HHS was calculated. For the *temperature exposure* characteristic, we used moderate resolution imaging spectroradiometer (MODIS) satellite imagery on day surface temperature and normalized vegetation difference index (NDVI). The latter was included due to the moderating effects of vegetation on heat stress and other extreme events [64–66], expecting a positive correlation with households resilience. The remote sensing data was processed in the geoinformation software QGIS 3.10 and analyzed accordingly. The 90th percentile of day temperatures of each year was calculated. Then, the exceedance of this 90th percentile for each household was assessed based on its location and split into two categories, namely low (<7) and high heat exposure ($\geq$7; cf. SI, migration characteristics). Additionally, the number of extreme heat exposures was interpolated to display spatial patterns in Pune (cf. SM, Figure S12). For the NDVI, an inclusion in the RM did not return any statistically significant results and was thus excluded (cf. SM, experiments with environmental data).

To test whether the sub-groups display significant resilience differences, suitable t-tests were applied (cf. S; data processing). The RCI results are displayed using kernel density estimations (kde) and a grid-cell map. In the kde, for comparison reasons, the average RCI of each sub-group is highlighted in form of a vertical line. In the grid-cell map (Figure 4), Pune's household resilience is displayed spatially using QGIS, applying a fishnet approach where the RCI values were averaged on 1000 m grids.

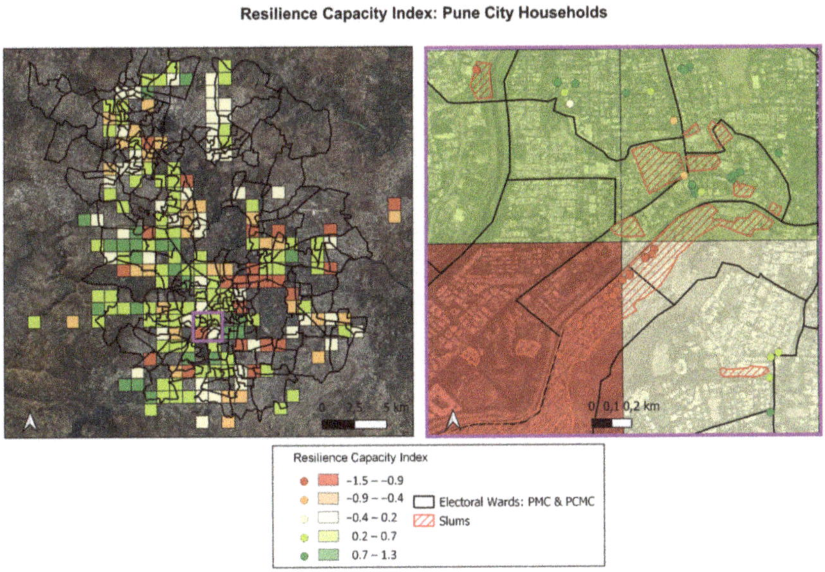

Figure 4. Resilience capacity index of Pune households using 1000 m grids. In the zoomed-in section (purple), slums are shown hatched in red. Dots correspond to individual households. For data security reasons, locations correspond to the next street crossing, not the exact building.

3. Results

The results of the RM are presented in increasing differentiation. We first illustrate the mean resilience score of the total sample spatially distributed across Pune. Second, we contrast the resilience of the six sub-groups with a specific focus on migrants and non-migrants. Third, we compare the resilience of the six sub-groups for each of the four migration characteristics to determine the role of migration biography on quantifying resilience.

3.1. Overall Resilience Map

The generated resilience map (Figure 4) shows that the north of Pune (Pimpri Chinchwad) and the west are associated with middle to high resilience, whereas the center, south, and east have mixed to low resilience. In general, resilience shows to be strongly associated with the residence zone since households living in formal urban areas tendentially have higher resilience than households living in slums. Additionally, small-scale spatial patterns become clear, with high-resilience areas often located next to low-resilience areas.

3.2. Migrants vs. Non-Migrants

The resilience of migrants and non-migrants is displayed using a kde (Figure 5). Comparing the resilience of migrants and non-migrants (Table 1), we see that on average, Pune's migrants are significantly more resilient than non-migrants ($p < 0.05$). The shape of the kde-curves is similar for both groups. However, the kde-curve of migrants is shifted to the right, indicating that they have fewer households with low resilience and more households with high resilience compared to non-migrants.

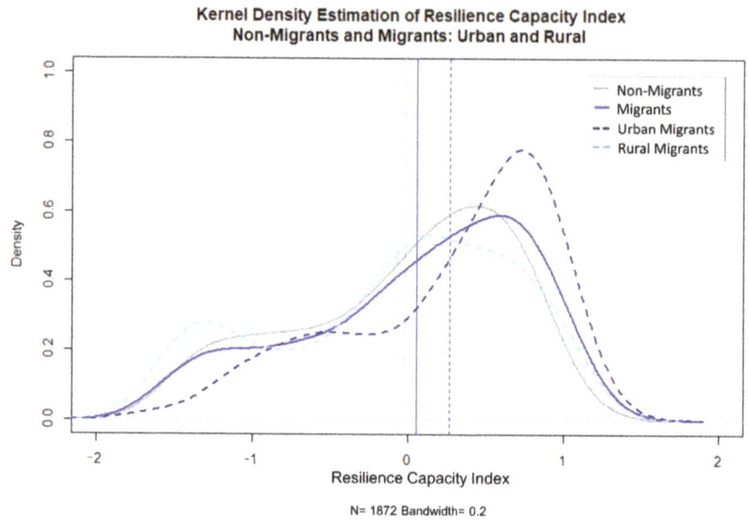

Figure 5. Kernel density estimation of RCI: migrants (urban, rural origin) and non-migrants.

Table 1. RCI and average object scores of 1st component of each RM pillar for migrants (rural, urban origin) and non-migrants.

Group	RCI	Assets	Housing	Health	*n*
Migrants	0.0562	0.0401	0.0431	0.1458	553
Urban	0.2652	0.2849	0.3291	0.3625	193
Rural	−0.0769	−0.1261	−0.1988	0.1097	239
Non-Migrants	−0.0236	−0.0168	−0.0181	−0.0611	1319

When differentiating between migrants who moved to Pune from an urban origin and those that migrated from a rural origin, urban migrants are significantly more resilient ($p < 0.001$) than rural migrants. Moreover, rural migrants have an average RCI that is even lower than that of non-migrants (Figure 5, dotted lines).

The average object scores of the first component of each pillar, as well as the descriptive analysis of the observed variables underlying each pillar for migrants and non-migrants, support the results of our RM (Table 1).

Regarding the first pillar *assets*, fewer migrants are found in the lowest income category and more migrants are found in the highest income category. On average, migrants earn 39,000 Rs a year more than non-migrants, which represents a 19% higher income of migrants compared to non-migrants. The difference between urban migrants and non-migrants is even larger (115,000 Rs, 40% more). While the average educational attainment of the three focus groups is relatively similar (migrants: 12.40 years, urban migrants: 13.64 years, rural migrants: 11.54 years, and non-migrants: 12.53 years), urban migrants still have the highest educational attainment while rural migrants have the lowest. Rural migrants do not only have the highest share of illiterates but also the lowest share in the highest education category (cf. SM, Figure S8, Table S2).

For the second pillar *housing*, the proportion of migrants compared to non-migrants living in formal urban residence zones, associated with a middle to high standard of living, is greater (migrants: 84%, urban migrants: 92%, rural migrants: 77%, and non-migrants: 81%). Moreover, the numbers of rooms and space per capita are higher for migrants; however, again only due to the higher values of urban migrants (rooms: migrants (0.93), urban migrants (0.94), rural migrant (0.70), non-migrant (0.9); space: migrants (177 ft^2), urban migrants (215 ft^2), rural migrants (191 ft^2), and non-migrants (163 ft^2); cf. SM, Figure S9, Table S3).

The third pillar *health* provides a similar picture to the previous two pillars, except for the availability of electricity backup: It is the case again that migrants are scoring higher in all four categories *water storage* (migrants: 17,777 liters, urban migrants: 32,767 liters, rural migrants: 11,479 liters, and non-migrants: 11,580 liters), the number of total *cooling devices* (migrants: 1.3, urban migrants: 1.3, rural migrants: 1.4, and non-migrants: 1.2), percentage of households with *flush toilets* (migrants: 51%, urban migrants: 54%, rural migrants: 45%, and non-migrants: 39%), and percentage of households with *electricity backup* (migrants: 30%, urban migrants: 17%, rural migrants: 28%, and non-migrants: 24%). However, in the *electricity backup* category, the higher results of migrants are not due to the higher availability of electricity backups of urban migrants but rural migrants instead (cf. SM, Figure S10, Table S4).

When including the additional layer of complexity by considering residence zone and migration origin, significant differences between the six previously defined sub-groups become visible (Figure 6).

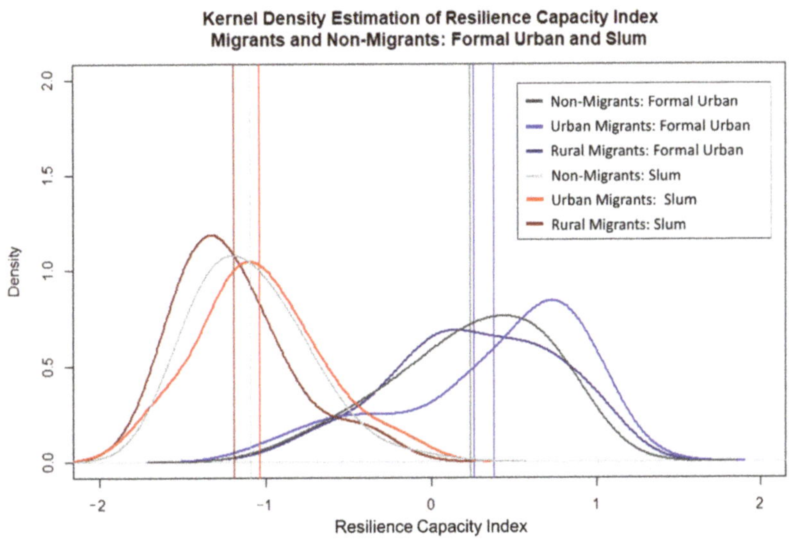

Figure 6. Kernel density estimation of RCI for the six analyzed sub-groups.

Slum residents have a significantly lower resilience compared to their formal urban counterparts (p-value < 0.001). Additionally, urban migrants have a higher resilience in both residence zones compared to rural migrants and non-migrants. Migrants are on average slightly more resilient than non-migrants in the formal urban category, which does not apply to the slum category where it is the other way around (Table 2).

Table 2. RCI of migrants (urban, rural origin) and non-migrants.

Group	Formal Urban	*n*	Slums	*n*
Migrants	0.2849	463	−1.1205	90
Urban	0.3749	178	−1.0363	15
Rural	0.2568	184	−1.1933	55
Non-Migrants	0.2324	1064	−1.0915	255

3.3. *Intra-Migrant Comparions*

Literature and HHS data provide ample evidence for the high socio-economic diversity among different migrant groups within the city [67]. Therefore, in the next section, we contrast the resilience scores against four key characteristics of migration: (1) time since arrival, (2) reasons for migration, and (3) distance travelled from the migration origin. We further analyzed (4) how the exposure to extreme weather (flood, heat) at the migration destination correlates with the households' resilience. To allow for better comparison, we include non-migrants into the plots of the four migrant sub-groups.

3.3.1. Time since Arrival

The resilience of migrants based on the following three categories is assessed: <10, 10–20, and >20 years since arrival in Pune. When classifying migrants based on their time since arrival in Pune, distinct differences between the sub-groups appear (Figure 7, Table 3). Compared to the non-migrant sub-groups, recently arrived migrants are in all cases more resilient, whereas those longest in Pune have lower resilience throughout the four sub-groups.

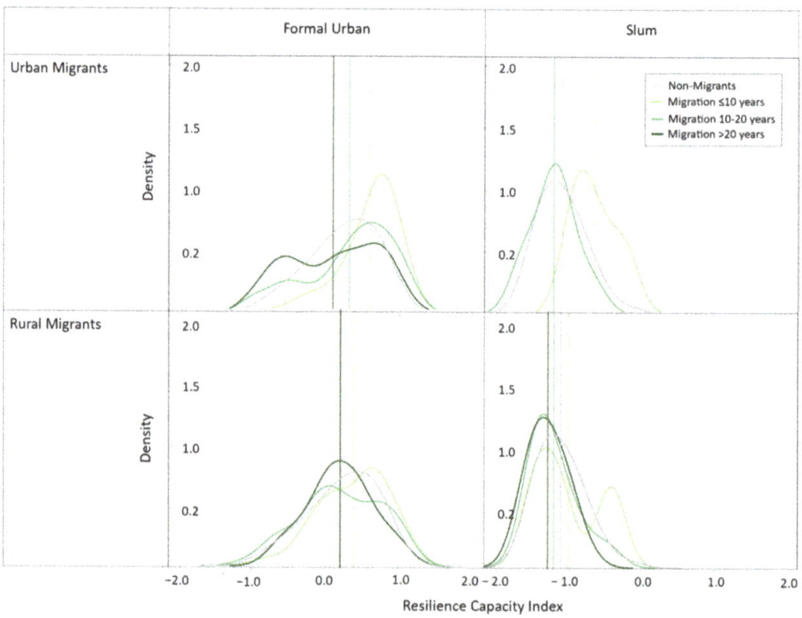

Figure 7. Resilience matrix displaying kernel density estimation of RCI of sub-groups based on the migration characteristic time since arrival.

Table 3. RCI of sub-groups based on the migration characteristic time since arrival (* low sample size).

Time SINCE Arrival	Formal Urban	*n*	Slums	*n*
Non-Migrant	0.2324	1064	−1.0915	255
Urban Migrant	**0.3717**	**178**	**−1.0363**	**15**
≤10	0.5974	75	−0.6444	4 *
10–20	0.3138	52	−1.2051	9 *
>20	0.0933	50	NA	2 *
Rural Migrant	**0.2568**	**184**	**−1.1933**	**55**
≤10	0.3841	59	−0.9845	9 *
10–20	0.1975	69	−1.1874	20
>20	0.1957	56	−1.2702	26

3.3.2. Reasons for Migration

The following reasons for migration are considered in this analysis: work, living conditions, education, and family (Figure 8, Table 4). For the two sub-groups urban-urban and rural-urban, those who moved to Pune for education reasons have the highest resilience. In contrast, among both migrant groups living in slums (urban-slum and rural-slum), those who moved because of family reasons have the highest resilience. While most migrants moved to Pune for work reasons, their resilience tends to be one of the lowest, especially for migrants living in slums.

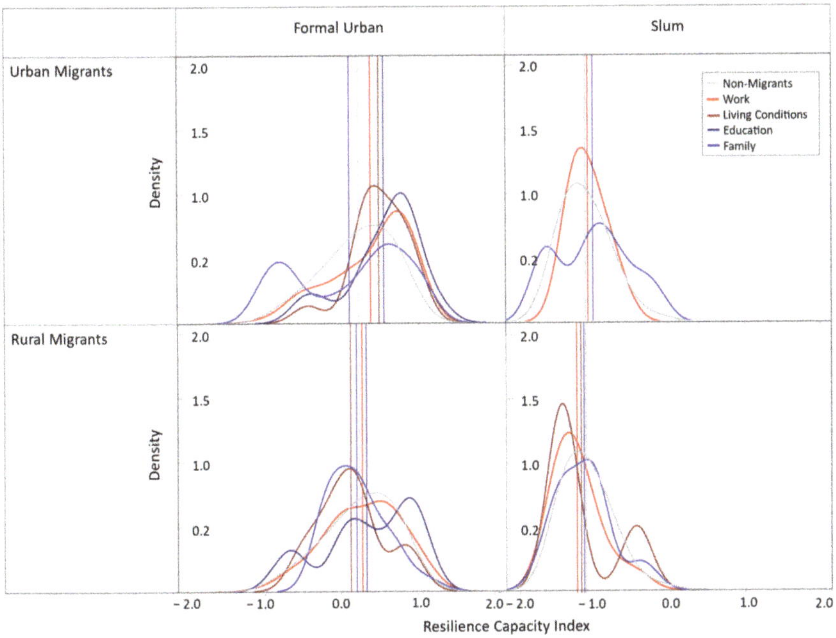

Figure 8. Resilience matrix displaying kernel density estimation of RCI of sub-groups based on the migration characteristic reasons for migration.

Table 4. RCI of sub-groups based on the migration characteristic reasons for migration (* low sample size).

Reasons for Migration	Formal Urban	*n*	Slums	*n*
Non-Migrant	0.2324	1064	−1.0915	255
Urban Migrant	**0.3749**	**178**	**−1.0363**	**15**
Work	0.3947	98	−1.0601	7 *
Living Conditions	0.5041	15	NA	0 *
Education	0.5664	28	NA	1 *
Family	0.1116	36	−0.9935	7 *
Rural Migrant	**0.2568**	**184**	**−1.1933**	**55**
Work	0.2641	123	−1.2148	40
Living Conditions	0.1151	11	−1.1586	4 *
Education	0.3228	27	NA	1 *
Family	0.1927	22	−1.1185	9 *

3.3.3. Migration Distance

The resilience of migrants based on their migration distance is compared using the categories same district, same state but other districts, and other states (Figure 9, Table 5). For the urban-urban sub-group, migrants from other states have the highest resilience whereas rural-urban migrating from other states have the lowest resilience. For both sub-groups of rural migrants, those who migrated shorter distances (same district, same state) have a higher resilience than migrants who migrated longer distances (other states).

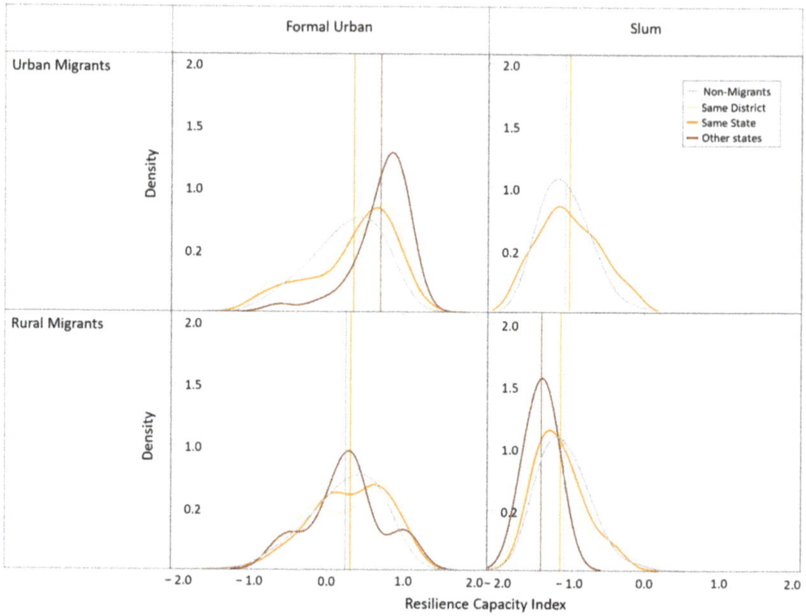

Figure 9. Resilience matrix displaying kernel density estimation of RCI of sub-groups based on the migration characteristic migration distance.

Table 5. RCI sub-groups based on the migration characteristic migration distance (* low sample size).

Migration Origin	Formal Urban	*n*	Slums	*n*
Non-Migrant	0.2324	1064	−1.0915	255
Urban Migrant	0.4155	178	−1.0434	15
Same district	0.3413	16	NA	1 *
Same state	0.3403	111	−1.0324	11
Other states	0.7047	33	NA	2 *
Rural Migrant	**0.2957**	**184**	**−1.2047**	**55**
Same district	0.2954	18	−1.1715	10
Same state	0.3049	113	−1.1433	31
Other states	0.2353	17	−1.4077	11

3.3.4. Environmental Exposure: Flood and Heat

The resilience of migrants to environmental exposure is assessed based on (1) whether they experienced a flood event or not and (2) on their exposure to extreme heat events (high or low exposure) in the last three years in Pune.

Regarding floods, according to the HHS data (cf. SM, Figure S11), a higher share of migrants (5.6%) compared to non-migrants (2.1%) has been exposed to floods in the past. Again, it is especially the rural migrants who are prone to floods (rural migrants: 7.5%, urban migrants: 3.1%). Furthermore, the RM shows that flood exposure tends to reduce the resilience of migrants and non-migrants considerably, except for the urban-urban sub-group, where the opposite effect is found (Figure 10, Table 6).

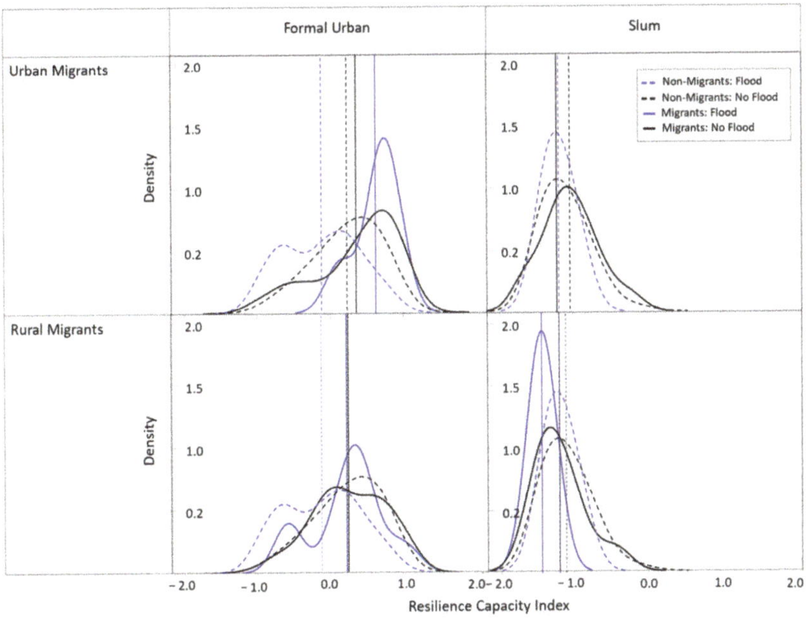

Figure 10. Resilience matrix displaying kernel density estimation of RCI of sub-groups based on the migration characteristic environmental exposure, displaying flood exposure.

Table 6. RCI of sub-groups based on the migration characteristic environmental exposure, displaying flood exposure (* low sample size).

Flood Exposure	Formal Urban	*n*	Slums	*n*
Non-Migrant	**0.2324**	**1064**	**−1.0915**	**255**
Flood	−0.1109	21	−1.1772	7 *
No-Flood	0.2393	1043	−1.0891	248
Urban Migrant	**0.3749**	**178**	**−1.0363**	**15**
Flood	0.6369	5 *	NA	1 *
No-Flood	0.3673	173	−1.0242	14
Rural Migrant	**0.2568**	**184**	**−1.1933**	**55**
Flood	0.2333	15	−1.4215	3 *
No-Flood	0.2589	169	−1.1801	52

Regarding heat stress, the interpolated extreme day surface temperature map of Pune shows distinct spatial and temporal differences (cf. SM, Figure S12) with heat islands as well as areas of low extreme temperature exposure. Overall, the west of Pune tends to experience more extreme day temperatures compared to the east.

A higher share of migrants (23%) compared to non-migrants (15%) experiences high heat exposure (cf. SM, Figure S13). Among migrants, it is rural migrants who are more prone to high heat exposure than urban migrants (rural migrants: 24%, urban migrants: 14%). In general, a higher heat exposure leads to a lower resilience except for the non-migrant-urban sub-group where the opposite effect can be observed. Especially for rural-slum migrants, higher heat exposure is associated with a significantly lower resilience ($p < 0.01$) compared to low heat exposure (Figure 11, Table 7).

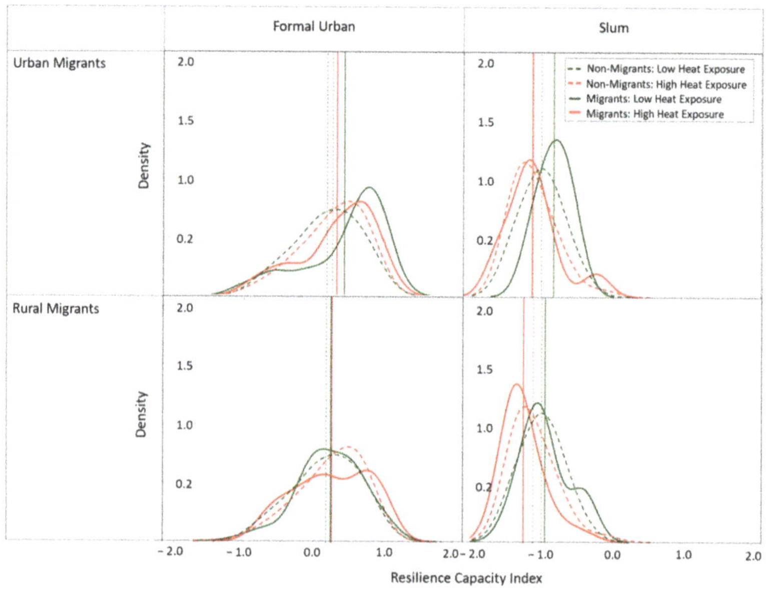

Figure 11. Resilience matrix displaying kernel density estimation of RCI of sub-groups based on the migration characteristic environmental exposure, displaying heat exposure.

Table 7. RCI of sub-groups based on the migration characteristic environmental exposure, displaying heat exposure (night; * low sample size).

Heat Exposure (Night)	Formal Urban	*n*	Slums	*n*
Non-Migrant	**0.2324**	**1064**	**−1.0915**	**255**
Low	0.1850	451	−1.0089	78
High	0.2672	613	−1.1279	177
Urban Migrant	**0.3749**	**178**	**−1.0363**	**15**
Low	0.4357	80	−0.8328	5 *
High	0.3253	98	−1.1380	10
Rural Migrant	**0.2568**	**184**	**−1.1933**	**55**
Low	0.2498	105	−0.9581	14
High	0.2660	79	−1.2736	41

4. Discussion

The results of our RM show how the livelihood resilience of urban households is shaped by migration biographies. Several findings are of particular interest. *First*, resilience can vary greatly on small spatial scales, from neighborhood to neighborhood, which is linked with high differences in the availability of livelihood opportunities [68] and the nested, often informal built environment in cities such as Pune [69]. *Second*, comparing the total migration and non-migration cohorts, we noted a significantly higher resilience of those that have moved to Pune than those that were born in the city. On the one side, this finding is consistent with the comparisons of migrants and non-migrants by Revathy et al. [50] and Haan [70] but contrary to Foresight's [9] assessment of migrants to be a particularly vulnerable urban group. The prior found that migrants tend to perform well in urban labor markets mainly due to higher levels of education and advantageous age cohorts [71]. The latter argues that they often lack the human, social, or financial capital to protect themselves from the environmental risks in high-density settlements where they predominantly live. This apparent contradiction is reflected in our *third* main finding, showing that the high resilience of migrants is accounted for only by urban migrants while rural migrants have a slightly lower resilience than the non-migrant sub-group. Acknowledging this difference is critical, albeit often neglected in research and policy—as it points to the need to differentiate between migrant biographies.

The higher resilience of urban migrants compared to rural migrants is linked with higher levels in terms of income, education, residence zone, and housing quality which relates to research by Oberai [72] who presents evidence that the poor have a comparatively greater propensity to out-migrate from rural areas than the poor from urban areas. Additionally, according to Srivastava [73] nearly half of rural migrants are found at the bottom six consumption deciles and work predominantly as casual wage employed or as self-employed in the informal sector. As a result, they tend to have no civic identity and citizenship at the destination, limited access to housing and basic amenities, poor entitlements, deficient workplace conditions, and experience labor market discrimination [74]. Moreover, we found that rural migrants perform better compared to urban migrants in one of the input variables of the RM, namely in the availability of electricity backup. On the one side, this can be interpreted as a sign of increased resilience but also a sign for the higher necessity of having electricity backup since electricity is unreliable [75].

For the investigation of *the four migration characteristics*, we disaggregated the sample into six sub-groups displayed in the result matrices:

First, we found an inverse relationship between *time since arrival* at the destination and resilience for all migrant sub-groups. This contradicts the widespread assumption that migrants need time at their new migration destination to settle and establish themselves [76]. A possible explanation for this pattern may be past peaks of rural-urban migration to Pune (the last around the year 2000) due to severe droughts. These caused a major, unplanned migrant influx to Pune, especially to the city's slums (cf. SM, Figure S2). The HHS data suggests that more recently, the decision to leave the rural origin and move to the city may have been better planned and less displacement-driven (cf. SM, Figure S3). Nevertheless, Revi [77] highlights that climate change is expected to increase the occurrences and intensities of droughts in western India which may force more landless, small, and marginal farmers to migrate to the city's informal neighborhoods. These rural-urban migrants are specifically vulnerable as previously described since they tend to have low human and financial capital while predominantly living in slums which are associated with higher environmental risks.

Second, marked differences between the sub-groups are examined regarding the correlation between *reasons for migration* and resilience: formal urban migrants score highest when motivated by *education* and slum-based migrants when migrating due to *family* reasons. We see that the prior sub-group is dominated by young migrants with high financial capital that allows them to pursue further educational attainment (human capital).

The importance of *family* for the latter sub-group's resilience supports the argument that networks through relationships of friendship and mutual origin are critical [78].

Third, comparing the effects of *migration distance* across the sub-groups, we find that while urban-urban migrants from other states feature the highest resilience, the exact opposite is the case for the rural-slum sub-group, where migrants from the same district have the highest resilience. The HHS (cf. SM, Figure S4) and evidence from literature suggest an inversely proportional relationship between migration flows and the distance between regions [79]. Possible earnings, distances, transportation costs, information on job opportunities, and psychic costs are weighed against each other. The higher financial and human capital of urban-urban migrants allows them to travel farther distances whereas rural migrants, especially those that live in the slums of Pune, have low financial and human capital and thus fewer means to travel long distances. This may serve as an explanation for these oppositional findings.

Fourth, considering Pune's increasing exposure to extreme weather events it is alarming to see how all migrant sub-groups have greater *environmental exposure* than the non-migrant sub-groups, especially because exposure to floods or heat is generally correlated with lower resilience scores. It should be noted that the most resilient migrant group (urban-urban) is not significantly affected by extreme events. On the contrary, we see the paradoxical effect of higher resilience when exposed to floods. An explanation for this could be that they received compensation after the floods, irrespective of their losses, which may have been comparatively low. This, however, would need further investigation. Moreover, the rural-slum sub-group, on the other side, was severely affected both by floods and extreme heat, underlining its particularly high vulnerability. These results are supported by literature since rural-urban migrants tend to predominantly live-in slums which are linked with lower housing quality. Therefore, this sub-group's housing does not only provide less protection during extreme weather events such as strong winds or floods but is also destroyed more quickly and easily [49,80].

In summary, the subdivision of migrants in matrices by residence zone and migration origin yields two contrasting archetypical migrant groups in Pune: First, those that live in formal urban areas with urban migration origin, and second, those that live in slums with rural migration origins. On the one side, we find that in the first, most resilient group, the highest-scoring ones have moved recently to Pune and came from far (outside Maharashtra). This group was attracted by Pune's excellent education and the potentially good living conditions—the city has recently been ranked most livable in India by the Ministry of Housing and Urban Affairs [81]. On the other side, we find that those households have the lowest resilience who have moved from villages outside Maharashtra to Pune's slums long ago and came to find work in the city. Other, less clear-cut archetypes may be slum households that recently came from another city within Maharashtra for family or work reasons (relatively high resilience), or those settling in slum areas after leaving the village due to living conditions (relatively low resilience).

Furthermore, it needs to be highlighted that the livelihood resilience concept serves as an important input for the RM since it provides relevant insights into quantifying the resilience of different sub-groups with and without migration backgrounds. Its detailed conceptual and analytical framework makes it particularly suitable for translation into a quantitative model on urban migrants' resilience. Related concepts such as social resilience [18] or translocal social resilience [82], so far, either lack relevant aspects regarding the migration context or are too complex to be implemented in an RM [17].

Overall, the data shows how diverse and complex migration patterns and the associated household resilience are. While the clustering into stylized archetypes may help for a better understanding of qualitatively different groups, a categorization can never be absolute and should be treated with caution. Additional limitations of our RM include a limited sample size of the HHS, incomplete data on water and electricity access, quality, and quantity, as well as a lack of predictor variables. All the above resulted in both a limited number of pillars and prevented a quantification of indirect resilience. Moreover, two specific methodological shortcomings of quantitative resilience approaches are recognized: first, a possible arbitrariness in the selection of input and predictor variables, and second, the challenges of accounting for the socio-cultural contexts in which resilience occurs. Furthermore, a statistical model can never capture the multi-facetted resilience complexity of urban households completely and we need to be cautious with causal inferences. Despite these limitations, we argue that our presented RM can provide valuable insights for policymakers to set and implement targeted resilience interventions.

5. Conclusions

This study shows that on average households with migration backgrounds are more resilient than households without. More importantly, we show that not migration, as such, but the migration biography matters. In particular, migrants' residence zone at the migration destination, migration origin, migration distance, time since arrival at the destination, and the reasons for migration strongly influence households' resilience and affectedness by extreme weather events. Only by including characteristics from the migration biographies, we detect the significant differences between rural and urban migrants and the impact of the residence zone, time since arrival or distance migrated on resilience. Our results thus paint a multi-faceted picture of Pune's migrant population which contradicts simplistic notions of migrants as socio-economic "high-performers" or "problem children."

The resilience of households in rapidly growing cities to environmental hazards is rising in importance: The exposed population is growing as well as the frequency and intensity of extreme weather events. Especially for emerging megacities across the world, efforts towards actively increasing urban household resilience are needed to allow for sustainable economic and demographic growth. While our results describe the situation in Pune, the approach can be applied to any other city facing large migration influx, urban informality, and natural hazards. This study provides valuable information for policymakers on the resilience structure of urban households with a specific focus on migrants, enhancing the targeting and effectiveness of resilience interventions. On the one side, our RM fosters the understanding of policymakers on the resilience of particular population groups and the reasons for it. On the other side, it can be used to estimate the resilience effects of specific proactive and reactive interventions.

Therefore, in the future, resilience research in Pune and other cities of the Global South should be continued using similar surveys to establish a time-series data set while addressing the previously established limitations. This can be achieved, for example, by increasing the sample size to decrease standard deviation and uncoverage bias. Additionally, expanding the comprehensives of the RM by including important insights from, for example, the social resilience and translocal social resilience concepts as well as environmental factors (vegetation) from remote sensing data could provide valuable insights into resilience dynamics. Moreover, future research should aim to implement a mixed-method design since the combination of quantitative and qualitative data allows for a more layered analysis at varying depths of meaning. The qualitative data could allow to test the accuracy of the RM results as well as reveal which additional parameters might have to be included in the RM.

Finally, we advocate that both policymakers and scientists account for migration biographies in assessments of resilience and adjacent concepts such as vulnerability and livability. The socio-economic and cultural properties are too diverse to justify treating migrants as one homogeneous group. Neglecting migration biographies can lead to the implementation of ineffective resilience interventions and an associated misallocation of scarce funds. Endeavors such as the RM presented in this work can lay the ground for sustaining the resilience of urban households in rapidly-growing cities despite the dual pressure of elevating urbanization and climate-change-induced hazards.

Supplementary Materials: The following are available online at https://www.mdpi.com/article/10.3390/land10111134/s1, Supplementary Material: 1. HHS, 1.1 Data Collection, 1.2 Reasons for Migration, 1.3 Descriptive Statistics, 2. Model, 2.1 Data Processing, 2.2 Structure, 2.3 Testing Resilience Interventions with Model, 2.4 Pillars, 2.4.1 Assets, 2.4.2 Housing, 2.4.3 Health, 3. Migration Characteristics, 3.1 Flood Exposure, 3.2 Heat Exposure, and 3.3 Experiments with environmental data. Figure S1: Reasons for migration categories used in HHS, Figure S2: (a) Yearly number of migrants in Pune and (b) yearly number of migrants in Pune's formal urban areas and slums (1924–2019) based on HHS, Figure S3: Share of Urban and Rural Migrants for years since arrival categories, Figure S4: Origin of migration of Pune migrants based on HHS, Figure S5: Path diagram of hierarchical Resilience Model based on assets (ass), housing (hos), health (hlt) where circles represent latent variables, rectangles observable variables, Figure S6: Path diagram of Structural Equation Model displaying how each pillar relates to resilience, Figure S7: Resilience Structure Matrix of three pillars assets, housing, and health, Figure S8: Figures of descriptive statistics of assets pillar (income, education, and food–income ratio) for migrant, urban migrant, rural migrant, and non-migrant, Figure S9: Figures of descriptive statistics of housing pillar (zone, dwelling, rooms, and space) for migrant, urban migrant, rural migrant, and non-migrant, Figure S10: Figures of descriptive statistics of *health* pillar (flush toilet, total water storage, total cooling devices, and electricity backup) for migrant, urban migrant, rural migrant, and non-migrant, Figure S11: Percentage of flood-affected households (migrant, urban migrant, rural migrant, and non-migrant), Figure S12: Interpolation of extreme temperature experiences per household (90th percentile), Figure S13: Heat exposure (low, middle, high) experienced by different focus groups (migrant, urban migrant, rural migrant, and non-migrant), Table S1: Categories and Indicators of the Resilience Model, Table S2: Tables of descriptive statistics of assets pillar (income, education, and food–income ratio) for migrant, urban migrant, rural migrant, and non-migrant (* low sample size), Table S3: Tables of descriptive statistics of housing pillar (zone, dwelling, space, and rooms) for migrant, urban migrant, rural migrant, and non-migrant (* low sample size), Table S4: Figures of descriptive statistics of health pillar (flush toilet, total water storage, total cooling devices, and electricity backup) for migrant, urban migrant, rural migrant, and non-migrant (* low sample size).

Author Contributions: Conceptualization, A.-C.L.; Data curation, A.-C.L., Y.Z. and R.K.; Formal analysis, A.-C.L.; Investigation, A.-C.L.; Methodology, A.-C.L.; Software, A.-C.L.; Supervision, R.K.; Visualization, A.-C.L.; Writing—original draft, A.-C.L. and R.K.; Writing—review & editing, Y.Z. and R.K. All authors have read and agreed to the published version of the manuscript.

Funding: This work was conducted as part of the Belmont Forum Sustainable Urbanisation Global Initiative (SUGI)/Food-Water-Energy Nexus theme for which coordination was supported by the US National Science Foundation under grant ICER/EAR-1829999 to Stanford University. UFZ received funding from the Federal Ministry of Education and Research (BMBF) under grant 033WU002. Any opinions, findings, and conclusions, or recommendations expressed in this material do not necessarily reflect the views of the funding organizations.

Institutional Review Board Statement: Ethical review and approval were waived for this study. This study does not involve animals. The involvement of humans occurred during the survey data collection process. The data collection was conducted strictly in accordance with the European General Data Protection Regulation. Only non-sensitive personal data was collected and analyzed. All data has been treated confidentially.

Informed Consent Statement: Informed consent was obtained from all subjects involved in the study verbally. No written informed consent is available since the survey interviews were collected anonymously. All the interviews were conducted under the condition that verbal informed consent of the respondent was obtained.

Data Availability Statement: The data presented in this study are available on request from the corresponding author. The data are not publicly available due to privacy regulations with regards to the households survey results.

Acknowledgments: The authors are grateful to Christian Kuhlicke, Eberhard Rothfuß, and Sigrun Kabisch for their supervision and feedback. The authors would also like to extend their gratitude to Vishal Gaikwad and his team from the Gokhale Institute of Politics and Economics, Pune for data collection and cleaning, and two anonymous reviewers for helpful feedback.

Conflicts of Interest: The authors declare no conflict of interest. The funders had no role in the design of the study; in the collection, analyses, or interpretation of data; in the writing of the manuscript, or in the decision to publish the results.

References

1. United Nations Department of Economic and Social Affairs. *World Urbanization Prospects: The 2014 Revision: Highlights*; United Nations: New York, NY, USA, 2014. Available online: https://population.un.org/wup/Publications/Files/WUP2014-Highlights.pdf (accessed on 31 May 2021).
2. United Nations Migration Agency. Who Is a Migrant? Available online: https://www.iom.int/who-is-a-migrant (accessed on 31 May 2021).
3. International Migration Organization. *World Migration Report 2020*; United Nations: Geneva, Switzerland, 2019; ISBN 9789290687894.
4. Perch-Nielsen, S.L.; Bättig, M.B.; Imboden, D. Exploring the link between climate change and migration. *Clim. Chang.* **2008**, *91*, 375–393. [CrossRef]
5. Eckstein, D.; Künzel, V.; Schäfer, L. Global Climate Risk Index 2021: Who Suffers Most from Extreme Weather Events? Weather-Related Loss Events in 2019 and 200-2019. 2021. Available online: https://germanwatch.org/sites/default/files/Global%20Climate%20Risk%20Index%202021_2.pdf (accessed on 22 June 2021).
6. Hoffmann, R.; Dimitrova, A.; Muttarak, R.; Crespo Cuaresma, J.; Peisker, J. A meta-analysis of country-level studies on environmental change and migration. *Nat. Clim. Chang.* **2020**, *10*, 904–912. [CrossRef]
7. Kainth, G.S. *Push and Pull Factors of Migration: A Case of Brick Kiln Industry of Punjab State*; Asia Pacific Journal of Social Sciences: Bengaluru, India, 2009.
8. WBGU. *World in Transition: Climate Change as a Security Risk; Summary for Policymakers*; Wissenschaftlicher Beirat d. Bundesregierung Globale Umweltveränderungen: Berlin, Germany, 2007; ISBN 978-3-936191-20-2.
9. Foresight. *Migration and Global Environmental Change (Final Project Report)*; The Government Office for Science: London, UK, 2011.
10. Mallick, B. The Nexus between Socio-Ecological System, Livelihood Resilience, and Migration Decisions: Empirical Evidence from Bangladesh. *Sustainability* **2019**, *11*, 3332. [CrossRef]
11. Cutter, S.L.; Barnes, L.; Berry, M.; Burton, C.; Evans, E.; Tate, E.; Webb, J. A place-based model for understanding community resilience to natural disasters. *Glob. Environ. Chang.* **2008**, *18*, 598–606. [CrossRef]
12. Holling, C. Resilience and Stability of Ecological Systems. *Annu. Rev. Ecol. Syst.* **1973**, *4*, 1–23. [CrossRef]
13. Hayward, B.M. Rethinking Resilience: Reflections on the Earthquakes in Christchurch, New Zealand, 2010 and 2011. *Ecol. Soc.* **2013**, *18*, 6. [CrossRef]
14. Brown, K. Global environmental change I: A social turn for resilience? *Prog. Hum. Geogr.* **2014**, *38*, 107–117. [CrossRef]
15. O'Brien, K. Responding to environmental change: A new age for human geography? *Prog. Hum. Geogr.* **2011**, *35*, 542–549. [CrossRef]
16. Weichelsgartner, J.; Kelman, I. Geographies of resilience: Challenges and opportunities of a descriptive concept. *Prog. Hum. Geogr.* **2015**, *39*, 249–267. [CrossRef]
17. Speranza, C.I.; Wiesmann, U.; Rist, S. An indicator framework for assessing livelihood resilience in the context of social–ecological dynamics. *Glob. Environ. Chang.* **2014**, *28*, 109–119. [CrossRef]
18. Adger, N.W. Social and ecological resilience: Are they related? *Prog. Hum. Geogr.* **2000**, *43*, 347–364. [CrossRef]
19. Chambers, R.; Conway, G. *Sustainable Rural Livelihoods: Practical Concepts for the 21st Century*; Institute of Development Studies: Sussex, UK, 1992.
20. de Haan, L.J.; Zoomers, A. Exploring the Frontier of Livelihoods Research. *Dev. Chang.* **2005**, *36*, 27–47. [CrossRef]
21. Marschke, M.J.; Berkes, F. Exploring Strategies that Build Livelihood Resilience: A Case from Cambodia. *Ecol. Soc.* **2006**, *11*, 16. [CrossRef]
22. Pingali, P.; Alinovi, L.; Sutton, J. Food security in complex emergencies: Enhancing food system resilience. *Disasters* **2005**, *29* (Suppl. 1), S5–S24. [CrossRef]
23. FAO. Measuring Resilience: A Concept Note on the Resilience Tool. *Food Secur. Inf. Decis. Mak. Concept Note* **2012**, 1–4. Available online: https://www.fao.org/resilience/resources/resources-detail/en/c/317275/ (accessed on 22 October 2021).
24. Alinovi, L.; Mane, E.; Romano, D. Measuring Household Resilience to Food Insecurity: Application to Palestinian Households. In *Agricultural Survey Methods*; Benedetti, R., Ed.; John Wiley & Sons: Chichester, UK, 2008; pp. 1–35. ISBN 9780470665480.
25. D'Errico, M.; Di Giuseppe, S.; Pietrelli, R.; Grazioli, F. *Resilience Analysis in Mali: FAO Resilience Analysis No. 4*; FAO: Rome, Italy, 2015.
26. D'Errico, M.; Di Giuseppe, S.; Pietrelli, R.; Grazioli, F. *Resilience Analysis in Niger: FAO Resilience Analysis No. 3*; FAO: Rome, Italy, 2015.
27. Ciani, F.; Romano, D. Testing for Household Resilience to Food Insecurity: Evidence from Nicaragua 1053-2016-85948. 2014. Available online: https://ageconsearch.umn.edu/record/172958/ (accessed on 10 September 2021).
28. FAO. Resilience Index Measurement and Analysis II-RIMA-II. *Food Agric. Organ. United Nations 2016* **2016**, 1–80. Available online: https://www.preventionweb.net/publication/rima-ii-resilience-index-measurement-and-analysis-ii (accessed on 22 October 2021).
29. Kozlowska, K.; D'Errico, M.; Pietrelli, R.; Di Giuseppe, S. *Resilience Analysis in Burkina Faso 1998–2003*; FAO: Rome, Italy, 2015.
30. D'Errico, M.; Romano, D.; Pietrelli, R. Household resilience to food insecurity: Evidence from Tanzania and Uganda. *Food Sec.* **2018**, *10*, 1033–1054. [CrossRef]

31. Dhar, T.K.; Khirfan, L. A multi-scale and multi-dimensional framework for enhancing the resilience of urban form to climate change. *Urban Clim.* **2017**, *19*, 72–91. [CrossRef]
32. Gonzales, P.; Ajami, N.K. An integrative regional resilience framework for the changing urban water paradigm. *Sustain. Cities Soc.* **2017**, *30*, 128–138. [CrossRef]
33. Chun, H.; Chi, S.; Hwang, B. A Spatial Disaster Assessment Model of Social Resilience Based on Geographically Weighted Regression. *Sustainability* **2017**, *9*, 2222. [CrossRef]
34. Kotzee, I.; Reyers, B. Piloting a social-ecological index for measuring flood resilience: A composite index approach. *Ecol. Indic.* **2016**, *60*, 45–53. [CrossRef]
35. Joerin, J.; Shaw, R.; Takeuchi, Y.; Krishnamurthy, R. The adoption of a climate disaster resilience index in Chennai, India. *Disasters* **2014**, *38*, 540–561. [CrossRef] [PubMed]
36. Suárez, M.; Gómez-Baggethun, E.; Benayas, J.; Tilbury, D. Towards an Urban Resilience Index: A Case Study in 50 Spanish Cities. *Sustainability* **2016**, *8*, 774. [CrossRef]
37. Abdrabo, M.A.; Hassaan, M.A. An integrated framework for urban resilience to climate change—Case study: Sea level rise impacts on the Nile Delta coastal urban areas. *Urban Clim.* **2015**, *14*, 554–565. [CrossRef]
38. Gaillard, J.C.; Walters, V.; Rickerby, M.; Shi, Y. Persistent Precarity and the Disaster of Everyday Life: Homeless People's Experiences of Natural and Other Hazards. *Int. J. Disaster Risk Sci.* **2019**, *10*, 332–342. [CrossRef]
39. Gardner, J.S.; Dekens, J. Mountain hazards and the resilience of social–ecological systems: Lessons learned in India and Canada. *Nat. Hazards* **2006**, *41*, 317–336. [CrossRef]
40. Gautam, Y. Seasonal Migration and Livelihood Resilience in the Face of Climate Change in Nepal. *Mt. Res. Dev.* **2017**, *37*, 436–445. [CrossRef]
41. Brenkert, A.L.; Malone, E.L. Modeling Vulnerability and Resilience to Climate Change: A Case Study of India and Indian States. *Clim. Chang.* **2005**, *72*, 57–102. [CrossRef]
42. Deshingkar, P. Environmental risk, resilience and migration: Implications for natural resource management and agriculture. *Environ. Res. Lett.* **2012**, *7*, 15603. [CrossRef]
43. Das, S.; DSouza, N.M. Identifying the local factors of resilience during cyclone Hudhud and Phailin on the east coast of India. *Ambio* **2020**, *49*, 950–961. [CrossRef]
44. Duncan, J.M.; Tompkins, E.L.; Dash, J.; Tripathy, B. Resilience to hazards: Rice farmers in the Mahanadi Delta, India. *Ecol. Soc.* **2017**, *22*, 3. [CrossRef]
45. Birthal, P.S.; Hazrana, J. Crop diversification and resilience of agriculture to climatic shocks: Evidence from India. *Agric. Syst.* **2019**, *173*, 345–354. [CrossRef]
46. Nayak, P.K.; Oliveira, L.E.; Berkes, F. Resource degradation, marginalization, and poverty in small-scale fisheries: Threats to social-ecological resilience in India and Brazil. *Ecol. Soc.* **2014**, *19*, 13. [CrossRef]
47. Census of India. D3—Migrants by Place of Last Residence and Reasons for Migration-2011. 2011. Available online: https://censusindia.gov.in/DigitalLibrary/TablesSeries2001.aspx (accessed on 1 April 2021).
48. Mayer, B. *The Arbitrary Project of Protecting Environmental Migrants. Environmental Migration and Social Inequality*; Springer: Cham, Switzerland, 2016; pp. 189–200.
49. Hari, V.; Dharmasthala, S.; Koppa, A.; Karmakar, S.; Kumar, R. Climate hazards are threatening vulnerable migrants in Indian megacities. *Nat. Clim. Chang.* **2021**, *11*, 636–638. [CrossRef]
50. Revathy, N.; Thilagavathi, M.; Surendran, A. A Comparative Analysis of Rural-Urban Migrants and Non-Migrants in the Selected Region of Tamil Nadu, India. *Econ. Aff.* **2020**, *65*, 23–30.
51. Census of India. Pune City Population Census 2011–2021: Maharashtra. Available online: https://www.census2011.co.in/census/city/375-pune.html (accessed on 26 May 2021).
52. Krishnamurthy, R.; Mishra, R.; Desouza, K.C. City profile: Pune, India. *Cities* **2016**, *53*, 98–109. [CrossRef]
53. Pune Municipal Corporation. Fire Hazards Response and Mitigation Plan. 2011. Available online: http://www.pmc.gov.in/informpdf/Fire_Hazards/3annexurefinal.pdf (accessed on 22 June 2021).
54. Cronin, V.; Guthrie, P. Community-led resettlement: From a flood-affected slum to a new society in Pune, India. *Environ. Hazards* **2011**, *10*, 310–326. [CrossRef]
55. Shinde, K.A. Disruption, resilience, and vernacular heritage in an Indian city: Pune after the 1961 floods. *Urban Stud.* **2017**, *54*, 382–398. [CrossRef]
56. PTI. Intense Rain Pounds Pune; 17 Killed, Nearly 16,000 Rescued. *Economic Times [Online]*, 26 September 2019. Available online: https://economictimes.indiatimes.com/news/politics-and-nation/flooding-in-pune-after-heavy-rain-7-killed-500-rescued/articleshow/71306297.cms(accessed on 11 March 2021).
57. ET Online. Mutha Canal Wall Collapse Floods Dandekar Bridge Area in Pune. *Economic Times [Online]*, 27 September 2018. Available online: https://economictimes.indiatimes.com/news/politics-and-nation/mula-canal-wall-collapse-floods-dandekar-bridge-area-in-pune/articleshow/65978344.cms(accessed on 11 March 2021).
58. Sharma, S.; Mujumdar, P. Increasing frequency and spatial extent of concurrent meteorological droughts and heatwaves in India. *Sci. Rep.* **2017**, *7*, 15582. [CrossRef]

59. Correspondent, H. IMD Issues Heatwave Warning for Pune. *Hindustan Times [Online]*, 20 May 2019. Available online: https://www.hindustantimes.com/pune-news/imd-issues-heatwave-warning-for-pune/story-uk4ZYpQDwjA8hGyID1eitK.html(accessed on 19 March 2021).
60. Gautam, A. At 42.9 Degrees Pune in Maharashtra Breaks Record of Last 36 Years. *Skymet Weather Services [Online]*, 28 April 2019. Available online: https://www.skymetweather.com/content/weather-news-and-analysis/at-42-9-degrees-pune-in-maharashtra-breaks-record-of-last-36-years/(accessed on 19 March 2021).
61. National Sample Survey Office. India—Employment, Unemployment and Migration Survey: NSS 64th Round. 2008. Available online: http://www.icssrdataservice.in/datarepository/index.php/catalog/33/data-dictionary (accessed on 8 October 2021).
62. Alinovi, L.; D'Errico, M.; Mane, E.; Romano, D. Livelihood strategies and household resilience to food insecurity: An empirical analysis to Kenya. In Proceedings of the Conference on "Promoting Resilience through Social Protection in Sub-Saharan Africa, Dakar, Senegal, 28–30 June 2010.
63. Oriangi, G.; Bamutaze, Y.; Mukwaya, P.I.; Musali, P.; Di Baldassarre, G.; Pilesjö, P. Testing the Proposed Municipality Resilience Index to Climate Change Shocks and Stresses in Mbale Municipality in Eastern Uganda. *Am. J. Clim. Chang.* **2019**, *8*, 520–543. [CrossRef]
64. Das, P.; Mudi, S.; Behera, M.D.; Barik, S.K.; Mishra, D.R.; Roy, P.S. Automated Mapping for Long-Term Analysis of Shifting Cultivation in Northeast India. *Remote Sens.* **2021**, *13*, 1066. [CrossRef]
65. Ghaderpour, E.; Ben Abbes, A.; Rhif, M.; Pagiatakis, S.D.; Farah, I.R. Non-stationary and unequally spaced NDVI time series analyses by the LSWAVE software. *Int. J. Remote Sens.* **2020**, *41*, 2374–2390. [CrossRef]
66. Hazaymeh, K.; Hassan, Q.K. Remote sensing of agricultural drought monitoring: A state of art review. *AIMS Environ. Sci.* **2016**, *3*, 604–630. [CrossRef]
67. Purkayastha, B. Migration, migrants, and human security. *Curr. Sociol.* **2018**, *66*, 167–191. [CrossRef]
68. Jimmy, E.N.; Martinez, J.; Verplanke, J. Spatial Patterns of Residential Fragmentation and Quality of Life in Nairobi City, Kenya. *Appl. Res. Qual. Life* **2020**, *15*, 1493–1517. [CrossRef]
69. Hernbäck, J. *Influence of Urban Form on Co-Presence in Public Space: A Space Syntax Analysis of Informal Settlements in Pune, India*; DiVA: Uppsala, Sweden, 2012.
70. de Haan, A. Rural-Urban Migration and Poverty: The Case of India. *IDS Bull.* **1997**, *28*, 35–47. [CrossRef]
71. Rajendran, D.; Ng, E.S.; Sears, G.; Ayub, N. Determinants of Migrant Career Success: A Study of Recent Skilled Migrants in Australia. *Int. Migr.* **2020**, *58*, 30–51. [CrossRef]
72. Oberai, A.S. *Problems of Urbanisation and Growth of Large Cities in Developing Countries: A Conceptual Framework for Policy Analysis*; International Labour Organization: Geneva, Switzerland, 1989. Available online: https://econpapers.repec.org/paper/iloilowps/992668663402676.htm (accessed on 10 September 2021).
73. Srivastava, R. Labour migration in India: Recent trends, patterns and policy issues. *Indian J. Labour Econ.* **2011**, *54*, 411–440.
74. Sabates-Wheeler, R.; Waite, M. *Migration and Social Protection: A Concept Paper*; Institute of Development Studies: Sussex, UK, 2003.
75. Aklin, M.; Bayer, P.; Harish, S.P.; Urpelainen, J. Quantifying slum electrification in India and explaining local variation. *Energy* **2015**, *80*, 203–212. [CrossRef]
76. Pennix, R. *Integration of Migrants: Economic, Social, Cultural and Political Dimensions*; UNECE: Geneva, Switzerland, 2005.
77. Revi, A. Climate change risk: An adaptation and mitigation agenda for Indian cities. *Environ. Urban.* **2008**, *20*, 207–229. [CrossRef]
78. Brettell, C.B.; Hollifield, J.F. *Migration Theory: Talking across Disciplines*, 3rd ed.; Taylor and Francis: Hoboken, NJ, USA, 2014; ISBN 9781317805984.
79. Boyle, P.; Halfacree, K.H.; Robinson, V. *Exploring Contemporary Migration*; Taylor and Francis: Hoboken, NJ, USA, 2014; ISBN 9781317890874.
80. Desai, R.; Sanghvi, S. *Circular Migration and Urban Housing. Handbook of Internal Migration in India*; SAGE Publications Pvt Ltd.: New Delhi, India, 2020; pp. 462–475.
81. Ministry of Housing and Urban Affairs. Ease of Living. 2018. Available online: http://amplifi.mohua.gov.in/eol-landing (accessed on 26 September 2021).
82. Sakdapolrak, P.; Naruchaikusol, S.; Ober, K.; Peth, S.; Porst, L.; Rockenbauch, T.; Tolo, V. Migration in a changing climate. Towards a translocal social resilience approach. *DIE ERDE–J. Geogr. Soc. Berl.* **2016**, *147*, 81–94. [CrossRef]

land

MDPI

Article

Spatial Diffusion of E-Commerce in China's Counties: Based on the Perspective of Regional Inequality

Fan Wang [1,2], Mingfeng Wang [1,2,*] and Shichen Yuan [1,2]

[1] The Center for Modern Chinese City Studies, East China Normal University, Shanghai 200062, China; 52183902011@stu.ecnu.edu.cn (F.W.); 51203902023@stu.ecnu.edu.cn (S.Y.)
[2] School of Urban and Regional Science, East China Normal University, Shanghai 200241, China
* Correspondence: mfwang@re.ecnu.edu.cn

Abstract: In recent decades, China has been on a new journey toward a digital economy of which e-commerce accounts for a substantial proportion. Despite some controversy, the innovation diffusion hypothesis and efficiency hypothesis of online shopping have been tested in research on the urban–rural dual structure. However, research on the spatial diffusion model of online business is sparse. Based on the online business and online shopping index released by the Ali Research Institute, this article compares the spatial diffusion model of online shopping and online business in the core–periphery structure based on the inequality between the eastern and western regions of China. Our study suggests that online business trends are in line only with the innovation diffusion hypothesis, with marginal counties having lower levels of online business. Online shopping, on the other hand, is in line with the innovation diffusion hypothesis and partially with the efficiency hypothesis, with a higher index of online shopping in the core regions and some peripheral counties. The discrepancy in the spatial diffusion mode is due to the differences in aims and supporting elements between online business and online shopping. Apart from infrastructure, the diffusion of online business is largely constrained by the regional industrial base, while online shopping is influenced by income and savings levels, which is the main reason for the differences in the spatial diffusion of online business and online shopping. We argue that the diffusion of online business has not led to the ability to balance regional inequalities at the national scale, while online shopping has the potential to bridge core and peripheral disparities better than online businesses, not in terms of the ability to bridge economic disparities, but in terms of the potential to reduce spatial consumption inequalities and welfare gains.

Keywords: online business; online shopping; spatial diffusion model; innovation diffusion hypothesis; efficiency hypothesis; regional inequality

Citation: Wang, F.; Wang, M.; Yuan, S. Spatial Diffusion of E-Commerce in China's Counties: Based on the Perspective of Regional Inequality. *Land* **2021**, *10*, 1141. https://doi.org/10.3390/land10111141

Academic Editors: Stephan Bartke and Sigrun Kabisch

Received: 29 September 2021
Accepted: 22 October 2021
Published: 27 October 2021

1. Introduction

Since the advent of information and communication technologies (ICTs), there have been ongoing debates about the spatio-temporal effects of the spatial expansion of the internet. The impact of the spread of the internet on the geographic location of economic activity has also been the subject of much scholarly attention [1,2]. In response to the technical characteristics of the internet, the theory of "time-space compression" was proposed [3]. Moreover, Cairncross [4] declared the "death of distance", and Friedman [5] stated that "the world is flat" in the internet era. It is indisputable that the application of the internet overcomes the friction caused by distance and reduces the communication, transportation and search costs for economic activities [6], and the development of networks creates decentralization in the core–periphery structure.

However, different views on the impact of the internet remain. In the urban–rural dichotomy, the central city is considered to be an important distribution area for internet technologies. Compared with urban areas, rural areas are obviously high-cost markets.

Thus, investment in new technologies will continue to be made in urban regions, and this urban orientation has led to a gap in digital diffusion between urban and rural regions [7]. Furthermore, differences in social and economic characteristics, such as knowledge, skills, and income, have also led to social exclusion in the adoption of internet technologies [8]. Thus, the polarizing trend of the core area within the core–periphery structure of urban and rural regions has also been repeatedly emphasized in the adoption of the internet.

Research on the spatial expansion of the internet between urban and rural regions has focused on the commercial application of the internet—e-commerce—as a breakthrough. Anderson et al. [1] proposed two mutually targeted hypotheses regarding whether the adoption of online business might affect the spatial structure, namely, the innovation diffusion hypothesis, based on the innovation hypothesis, and the efficiency hypothesis, based on the demand hypothesis. Compared with residents in other regions, people living in urban regions tend to use e-commerce because of their higher education and internet skills [9]. However, the adoption of online shopping in remote areas and sparsely populated areas lacking stores is particularly valuable in overcoming spatial friction [2]. There has been an interesting debate as to whether e-commerce usage in rural areas will exceed that in urban areas [6].

Previous testing of the spatial diffusion hypothesis of e-commerce has focused on the municipal level. Two of Anderson's hypotheses on the spatial diffusion of online shopping are tested by comparing the differences in online shopping adoption between urban areas and suburban or rural regions. It is also common in research to equate e-commerce directly with online business or online shopping in order to study the differences between urban and rural areas, without clearly distinguishing between online business and online shopping. Our study contributes to the growing body of research on the diffusion of e-commerce by providing a different perspective for understanding the expansion of e-commerce at the national level. For example, we distinguish between online shopping and online business and extend the core–periphery structure from the urban scale to the regional scale.

This article takes mainland China as the research scale and counties, except for municipal districts, as the research unit. Compared with the city unit, the number of county units is greater, allowing the precise identification of more detailed features of diffusion. We adopt the core–periphery structure based on the regional inequality between the eastern and western regions of China. The eastern part of China is economically developed and technologically advanced, while the remote areas in the west have low population and economic density. There is a gap in technological adaptability between the western and eastern regions. Therefore, we adopt the core–periphery structure with the east as the core area and the west as the marginal area based on the socioeconomic gradients of different regions in China. Previous tests of the innovation diffusion hypothesis and efficiency hypothesis have mainly been based on the diffusion of online shopping, using regression methods to analyze the econometric correlation between online shopping and urban or rural areas as a way of testing the diffusion hypothesis. Starting from the two opposites of "retail" and "consumption" in e-commerce, this paper tests and compares the applicability of online retail and online shopping in the innovation diffusion hypothesis and efficiency hypothesis. In this process, we discard the regression method and return to the essence of diffusion by using the diffusion gradient layout characteristics to determine which hypothesis best explains the large-scale diffusion of e-commerce. The mechanisms underlying the differences in the spatial diffusion patterns between online business and online shopping are also analyzed.

The following section introduces the literature on the diffusion of the internet and the spatial diffusion of e-commerce. Our review focuses on the digital divide caused by internet diffusion, the mechanism of internet diffusion, and the spatial diffusion of e-commerce. Section 3 elaborates on the data sources. Section 4 compares the spatial diffusion model of online business and online shopping. Section 5 uses the ordinary least squares (OLS) method to analyze the diffusion mechanism of online shopping and online business in different samples. Conclusions are drawn and discussed in Section 6.

2. Literature Review

2.1. Internet Diffusion and the Spatial Digital Divide

The digital divide between regions plays a considerable role in the proliferation of e-commerce, and in particular, the layout of digital infrastructure can significantly influence the adopting of e-commerce in different regions. The term "digital divide" is used to describe the difference between the "haves" and "have-nots" of digital infrastructure, referring to the wide social and regional differences in people's capacity to access and adopt digital equipment and services as well as their capacity to access the internet in terms of physical connectivity and use of facilities [10,11]. OECD (2001) described the "digital divide" as the gap between groups of different socioeconomic levels in terms of access to ICTs and the use of the internet for various activities. The digital divide has also been described as disparities in internet access, usage, search strategies, social support, information evaluation, and user diversity [12].

To conceptualize the digital divide, the dichotomy of the digital divide between different scales was a valuable concept at the beginning of the development of the internet, and most of the policy discussions and research on telecommunications issues have focused on the issue of access to technical facilities [13–15]. However, as the digital infrastructure divide has narrowed, research in recent years has diversified this binary concept, and new intricacies have been recognized in internet-access disparities [16,17]. Access to ICT is not the same as adoption or, more importantly, effective use. As a result, statements such as redefining the digital divide and transcending access rights have emerged [18,19]. Numerous empirical studies have emphasized the importance of distinguishing between access and use, as having access to an internet infrastructure does not mean that people will use it [20]. The redefinition of the concept of the digital divide focuses on the inequality of internet use, which is referred to as the deepening digital divide [21,22]. Following discussions on adoption, the issue of digital usability has become the third level of the digital divide, where dissimilar skill levels lead to new disparities [23]. This third level of the digital divide is related to the knowledge gap, and therefore, a change in the concept of the digital divide is necessary. The digital divide is not only a technological issue, but also a social issue that reflects broader social, economic, cultural, and learning inequalities [24].

Apart from the conceptual issues of the digital divide, research on the inter-regional digital divide is more oriented toward spatial scale studies and different objects. The digital divide coexists in different dimensions and forms, including digital divisions at different spatial scales and with different socio-demographic characteristics, and they are also linked to the ownership of ICT. Unlike the divide that stems from technology, people in marginalized areas are also affected by their own social exclusion factors, such as geographic location and social characteristics. The existing social divide needs to be reproduced in studies of digital inequality, which is a more pressing issue than the digital infrastructure gap in marginalized areas.

Digital inequalities involving spatial scales, including between countries, between urban and rural areas, and within cities and villages, have been explored by scholars. Early studies of inequality on the internet dealt with the inequalities in connectivity between developed and developing countries, and how these differences might affect society [10]. Compared to individuals in suburban and urban areas, studies have found that rural residents are less likely to use internet technology for economic and other everyday activities, a relationship that is a product of the slow diffusion of advanced technology to rural areas [25]. ICT adopters can be categorized according to their stage of adoption, and there is evidence that marginal areas will continue to lag behind as technologies continue to iterate [26,27]. Not only do marginal communities have less broadband internet usage than metropolitan communities, but also the degree of broadband availability varies within marginal communities [28].

Factors contributing to differences in ICT systems, classified according to socio-demographic characteristics, are socio-economic status in relation to the state of the technological infrastructure, including age, income, highest level of education attained, ethnicity, etc. Looking back at rural issues a decade ago, low levels of education were a major barrier to ICT development in marginal areas [29]. Nowadays, education remains a huge factor in the internet use divide in marginalized communities [30]. For people living in marginal areas, rural communities may not be able to afford superfast broadband connections on their incomes, and even regular broadband can be no small challenge [31,32]. There is also a digital divide between the sexes, with the act of sharing online being very different in terms of gender, with men more likely to be involved [33,34]. Social factors affecting the digital divide also include age, with older people generally having lower access to ICT [35,36], and persistent gaps between different racial and ethnic groups [37].

2.2. Mechanism of Internet Diffusion

The diffusion of innovation is a long-term and continuous process, during which the share of companies adopting and applying new technologies grows [38]. Hägerstrand considers distance to be the main influencing factor for the diffusion of innovation, especially distance from innovation centers [39]. A number of empirical studies have proven that distance is negatively related to the degree of innovation diffusion [40,41]. Spatial proximity, cognitive proximity, social proximity, organizational proximity and institutional proximity have also received extensive attention from geographers in the study of innovation diffusion [42–44].

In terms of the diffusion mechanism, Hargittai summarized that the economic level, human capital, institutional environment and existing technology accumulation are all important elements in the diffusion of the internet in a country or city [45]. The spatial diffusion of the internet in the United States and the European Union is based on the theory of neoliberal economics [46]. Notably, the diffusion of internet access is a process driven by the market and profit, and cities always take the lead in internet diffusion [47]. The spatial diffusion of digital networks needs to create value for all major stakeholders. In particular, only when all participants can obtain appropriate value in the form of economy, efficiency, legitimacy, knowledge or social benefits can the network be successfully spread [48]. Guillen and Suarez conducted a survey using national-level data and demonstrated that internet adoption is a complex phenomenon and that the spatial diffusion of the internet is more influenced by per capita income, infrastructure, English language proficiency and entrepreneurial conditions than by the impact of public policy on telecommunications [49]. Additionally, Beilock and Dimitrova identified that per capita GDP is the most important factor affecting the rate of internet adoption [50]. According to the research of Lin et al., the proliferation of the internet is positively correlated with economic growth, with spillover effects being more pronounced in developed regions [51].

2.3. Spatial Diffusion of E-Commerce

From the perspective of "retail" and "consumption", the combination of the internet and commerce has resulted in two types of technology applications: business-oriented online business and consumer-oriented online shopping, which are collectively referred to as e-commerce [1]. Research on the spatial diffusion of e-commerce has focused on comparing the diffusion order and diffusion mode between core and marginal areas.

Most of the research on the spatial diffusion of online shopping has aimed to verify the two mutually targeted hypotheses proposed by Anderson. Farag et al., for example, indicated the spatial distribution of internet users and online buyers in the Netherlands from 1996 to 2001 and the impact of spatial variables on online shopping [2]. They found that the innovation diffusion hypothesis and the efficiency hypothesis coexist in the diffusion of e-commerce. On the one hand, residents living in highly urbanized regions are more likely to search or purchase online. On the other hand, residents living in less urbanized or non-urbanized areas with lower accessibility to stores can purchase products online more often. As De Blasio noted, the internet cannot reduce the role of distance, with urban consumers using the internet and e-commerce much more frequently than consumers in weakly urbanized areas [52]. According to the survey of Motte-Baumvol et al., the efficiency hypothesis is partially validated by findings that suburban and urban households have very different patterns and that online purchases can remove spatial constraints on access to tangible goods [53]. Cao et al. examined the relationship between spatial attributes and online shopping and found supporting evidence for the innovation diffusion hypothesis, arguing that internet users living in cities or areas with high shopping convenience are inclined to shop online more frequently than those in other regions, as the former are more educated and use the internet more than the latter [54]. Beckers et al. highlighted the fact that there is no significant effect of the level of urbanization of a region on the probability of online shopping by analyzing data from a survey conducted by the Belgian Retail Federation with over 1500 respondents [55].

Research on the spatial diffusion hypothesis of e-commerce has long been centered on online shopping. However, does the spread of online business follow the innovation diffusion hypothesis and efficiency hypothesis? Do residents in marginal areas adopt the internet in order to overcome the distance barrier to sell their agricultural products? Previous studies of the spatial diffusion of online business included two main issues: the extent to which the diffusion of online business affects the number of visits to offline stores, and the comparison of the order of diffusion in urban and rural areas [6]. We focus primarily on the latter question. Retailers in urban areas are more likely to take on an online sales strategy, as urban areas typically have more advanced telecommunications infrastructures, stronger internet innovation capabilities and knowledge spillover, and more internet-related production services [56]. Clarke et al., for example, investigated a large-scale commercial consumer survey in the U.K. to explore the expansion of e-commerce in British retail and to examine the spatial differences in e-commerce adoption [6]. According to their research, e-commerce initially spread from London and major cities but was not just an urban trend. Due to the improvement of broadband service quality, usage in rural areas is increasing.

2.4. Digital Inequality and Spatial Diffusion of E-Commerce in China

The distribution of internet users in China is characterized by significant spatial differences, with the eastern coastal regions accounting for more than 50% of the country's internet users. China is the world's largest ICT market in terms of the number of mobile devices used, the number of internet users and the number of broadband users, but there are significant numerical inequalities within and between provinces, municipalities and counties [57]. The growth of e-commerce in China exhibits national inequalities that are constrained by local economic, political and infrastructural conditions. The growth of e-commerce exhibits a spatial mix that relies on the structure of cyberspace and physical space [58]. As a rural e-commerce cluster with Chinese characteristics, there are many studies on the spatial distribution of Taobao villages that involve the spatial diffusion of e-commerce. Similar to the regional development of the ICT industry, Taobao Villages first emerged in more developed regions and then gradually spread to less developed regions, showing a tendency of agglomerative diffusion [59]. As Liu et al. noted, more than 90% of Taobao villages are located in East China, while Central China has the second highest number of Taobao villages and the least in the West [60]. The study by Geng et al. also shows that, overall, the concentration of e-commerce in the northern and western regions of China is decreasing, with a stable trend of geographical concentration of e-commerce activity in the eastern and southern regions [61]. The use of e-shopping technology is also very unevenly distributed across regions, with more developed cities in the east being higher than those in the less developed west, and cities at higher administrative levels being higher than those at lower administrative levels [62]. The level of online shopping in China is trending downwards from the eastern coastal areas to the western rural mountainous areas [57].

3. Data and Method

The data in this article are from the 2015 China County E-commerce Index, which is the online business index and online shopping index of all the counties in China published on the website of the Ali Research Institute. These indexes are based on the e-commerce data of Taobao.com (accessed on 15 September 2019), China's largest e-commerce website, and are very comprehensive. The Online Retail Index (OBI) and Online Shopping Index (OSI) are constructed by Ali Research to characterize the level of e-tailing and online shopping in each county. The formulae are as follows:

$$OBI = 0.6 \times \alpha + 0.4 \times \beta \tag{1}$$

$$OSI = 0.6 \times \gamma + 0.4 \times \rho \tag{2}$$

where OBI stands for Online Sales Index and OSI represents Online Shopping Index; α stands for e-tailing density, that is, the ratio of the number of e-retailers to the population; β indicates e-tailing trade level, which is the ratio of the number of e-retailers with an annual turnover of over RMB 240,000 to the total number of e-retailers; γ stands for online shopping density, that is, the ratio of the number of online consumers to the population; ρ means online shopping consumption level, which is the ratio of online shopping consumption to the d represents the level of online shopping consumption, that is, the ratio of online shopping consumption to the number of online shoppers. The values range from 0 to 100, with larger values indicating a higher level of online shopping development. The socioeconomic data used in the regression model are partly derived from the China County Statistical Yearbook and partly crawled from Tianyancha.com (accessed on 23 April 2020). The offline retailer, ICT service providers and logistics service provider data obtained from Tianyancha.com (accessed on 23 April 2020). The length of high-grade roads, per capita GDP, employment in service industry, industrial output value above designated size, number of manufacturing enterprises, population density, and per capita savings data come from the China County Statistical Yearbook. After sorting out the data, they are all unified to the county level for analysis of the mechanism.

This paper conducts an exploratory geographic data analysis based on ArcGIS software to explore the distribution patterns of spatial data and visualize the exponential grading of online business and online shopping. We calculate simultaneously the local autocorrelation index of online business and online shopping to obtain their cold and hot spots distribution and identify their local spatial autocorrelation characteristics. The average index per hundred kilometers is calculated to verify the spatial diffusion characteristics of e-commerce. Finally, a least squares regression model is used to explore the factors influencing the diffusion of online business and online shopping.

4. Differences in Spatial Diffusion

4.1. Comparison of Spatial Diffusion Characteristics

We use the core–periphery structure based on China's "east–west" structure. Due to the large number of county units, in verifying the spatial diffusion pattern of online business and online shopping, we can make judgements based on the transition characteristics from core to peripheral counties. If the spatial diffusion from east to west conforms to the pattern of gradient diffusion from core to edge, then it is in line with the innovation diffusion hypothesis at the county scale. Conversely, there is the efficiency hypothesis. Since the cross-sectional data of each year can be seen as the result of the previous diffusion, we use the 2015 online business and online shopping data as the diffusion results of the previous years to study the spatial diffusion of e-commerce.

The distribution of county-level online business in Figure 1 shows that the high value of the index is distributed in eastern coastal China, where high-value contiguous areas are formed. The central region, as a transitional region between the eastern and western regions, has a lower index than the eastern region. Online business in the western counties is dominated by a low-value distribution. The county-level online business index has obvious gradient distribution characteristics. High values are distributed in the eastern coastal areas from Bohai Bay to the Pearl River Delta. The mid-level is North China, Central China and South China, and the vast western area is the outermost level. The eastern region has a higher level of economic development and is also a center of technological innovation. The general forms of technology diffusion are hierarchical and contagious diffusion. The eastern region, being at the heart of online business technology innovation, is the first to be exposed to online business technology and therefore, has an advantage in online business adoption.

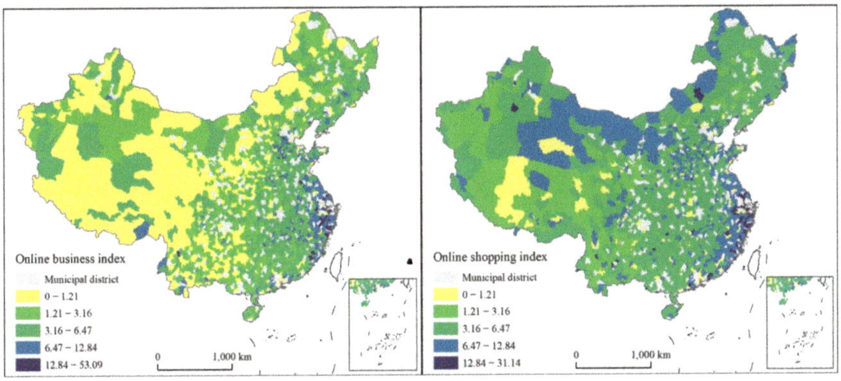

Figure 1. County-level online business and online shopping index in 2015.

The state of online shopping in counties shows that high-value areas are more widely distributed than are online businesses. High values are distributed not only in the eastern coastal areas, but also in the North China Plain and northern regions. Low values are located in the western region and are distributed over a smaller area. Comparing the spatial characteristics of online business and online shopping in counties, it can be seen that online shopping in general conforms to the same gradient distribution characteristics. However, there are many high values in the remote areas of Northeast and Northwest China, as well as on the North China Plain. The presence of these areas seems to diverge from the gradient distribution characteristics of the core–periphery structure from east to west. Compared to the eastern region, the peripheral counties have larger administrative districts but are more sparsely populated. Retail services are mainly located in administrative premises and most residents have difficulty accessing these services. The widespread distribution of internet infrastructure and high penetration of mobile internet in China, as well as the low technical threshold for online shopping, has led residents to switch from offline consumption to online shopping, with higher indices in some marginal areas.

4.2. Spatial Agglomeration Characteristics

This analysis of the distribution of spatial diffusion provides a preliminary perception of the diffusion characteristics of online business and online shopping. In the core and peripheral areas, we find some high values that presumably reflect agglomeration, but whether these areas are truly high-value clusters still needs to be explored using more scientific spatial analysis. In Figure 2, hot-spot analysis is used to explore the high-value agglomeration of online business and online shopping. The analysis of the online business index shows that the hot spots are concentrated in the eastern part from Bohai Bay to south-western Fujian, with a significance level of more than 90%. Most of China's top 100 counties in terms of GDP are located in the eastern region, which have a good manufacturing base, a complete retail supply chain and abundant human capital. As a result, the high value areas for online business are clustered in the eastern regions. Because of the mixed distribution of high and low values in North China, Central China, and South China, the analysis results are not significant. Most of the areas in the west are cold-spot clusters with a significance level of more than 90%, supplemented by the presence of some small patches of mixed high and low values in the west, which show non-significant results in the analysis of cold and hot spots. The hot-spot analysis shows the characteristics of the gradient distribution of the online business index more clearly, but there is a non-significant area in the northwest region, and further demonstration is needed.

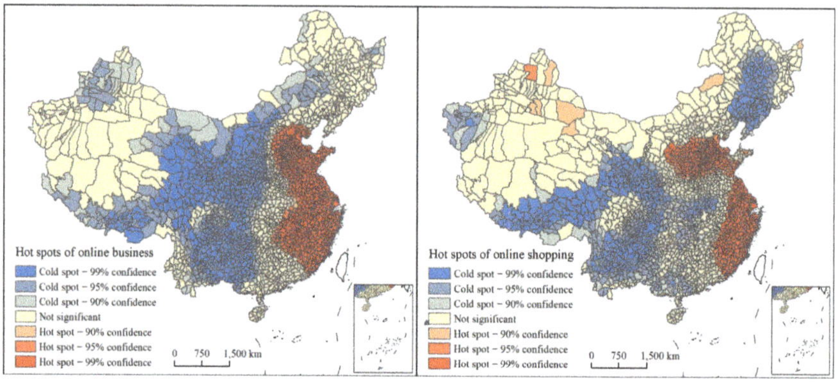

Figure 2. County-level hot spots of online business and online shopping indexes in 2015.

The online shopping index has formed a very large hot spot in the eastern region and its hinterland, North China. We observe only that there are high values in North China in the previous analysis, but we do not know whether this area represents a "marginal reversal". The analysis here suggests that the high-value area in North China is not a "marginal reversal" but is part of the greater whole of the eastern core area. However, this result also needs further verification. Additionally, there are six small hot spots in the northern region. Whether there is a "revolution" in the peripheral regions also needs further investigation. The northern fringe regions are mostly mountainous, grassland or desert terrain, with poor accessibility to retail services and a strong incentive for residents to shop online. Moreover, the industries in the northern regions are mostly resource extraction and animal husbandry, where the inhabitants have a high per capita income and a strong spending power, thus creating high-value agglomerations.

4.3. *Differences in the Same County*

Having analyzed the spatial diffusion patterns of online shopping and online business indexes in different counties, we will explore which pattern is more dominant within the same county. Especially in remote regions, if one of these indexes has a greater advantage and is widely distributed, this can indicate the existence of marginal reversal. Therefore, the difference between the online business and online shopping indexes is used to represent the divide in Figure 3. Online shopping has an absolute advantage in the northern region, including the fringe areas from the northeast to the northwest, which form an agglomeration. There are few regions in eastern China, where online shopping has an absolute advantage except for the region on the west coast of the Taiwan Straits. This finding shows that peripheral areas have a demand for technology but are better adapted to online shopping and less well adapted to online business. Combined with the previous analysis, it can be seen that the eastern coastal region takes into account the application of business-oriented and consumption-oriented technology, which is a relatively comprehensive technical adaptation. Marginal areas are subject to regional endowments, industrial foundations and knowledge thresholds, resulting in poorer application of business-oriented technology, while the application of consumption-oriented technology has an advantage.

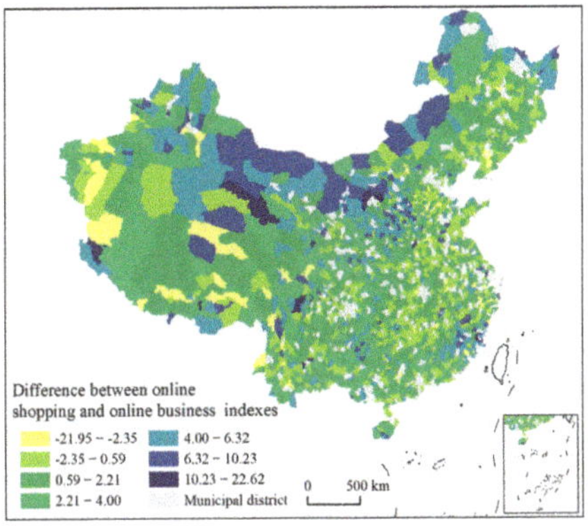

Figure 3. County-level differences in online business and online shopping indexes in 2015.

4.4. Diffusion Features in the Core-Periphery Structure

From the perspective of spatial distribution, online business and online shopping have different diffusion features, and it can be concluded that there are differences in their spatial diffusion patterns. To examine this result in more depth, the eastern coast (from Bohai Bay to the Pearl River Delta) is selected as the baseline, and the average values of online business and online shopping indexes within a radius of 100 km from the baseline are calculated. At the same time, the three largest cities in China—Beijing, Shanghai, and Guangzhou—are selected as benchmark points for cross-comparison with the baseline. The eastern region, as the center of innovation for ICT and e-commerce in China, is the starting point for the proliferation of e-commerce. Therefore, we have chosen these two sets of baselines and benchmark points.

According to the average value per 100 km in Figure 4, the spatial spread of online business basically shows a decreasing trend from the core area to the edge. There are only two peaks around A (2500 km) and B (3200 km), but neither of them is sufficiently sustained and robust. On the whole, online shopping presents characteristics of decreasing from east to west, but there are some regions where this decreasing pattern is broken. Two crests appear in regions C (1200–1800 km) and D (2300–3200 km), both of which remain in a longer range and exhibit good robustness.

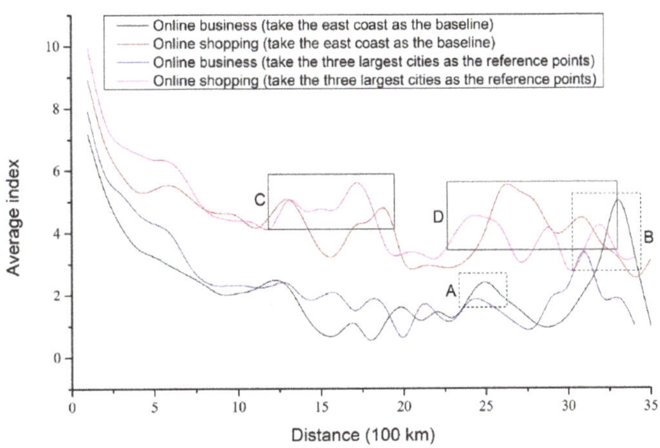

Figure 4. Average index of online business and online shopping decaying with distance. A, B, C and D are crests of online business and online shopping.

Summarizing all the diffusion characteristics analyzed above, we can affirm that online business basically decreases in a gradient from the core area to the peripheral area, which is in line only with the innovation diffusion hypothesis. In addition to the diffusion of innovation from high-tech areas to low-tech areas, the phenomenon of low-tech areas preferentially using technology in response to their own needs also exists in online shopping, generally conforming to the innovation diffusion hypothesis, and some areas also conform to the efficiency hypothesis. Thus, the analysis in this section essentially verifies the previous analysis.

5. Why Are There Differences?

As the previous analysis shows that the spatial diffusion patterns of online shopping and online business differ, regression methods are used to explore the reasons. We take the county-scale online business index and online shopping index as dependent variables to explore the mechanism of the spatial diffusion difference. As Coe and Yeung noted, many social and cultural aspects of e-commerce development are embedded in specific places and environments [63]. Innovative applications of e-commerce cannot appear in a vacuum but need to combine sufficient internet infrastructure, human capital, entrepreneurship, and supportive finances, logistics and government agencies. Boschma and Weltevreden also found that the adoption of the internet is affected by the unique characteristics of enterprises, network relationships and geographic locations [64]. When selecting elements that affect the adoption of online business, the possibility of offline retail transforming into online business is considered, and infrastructure elements, such as ICT services, logistics services, and the length of high-grade roads, are also included [65]. Areas with good infrastructure are in a better environment to adopt online business and are more likely to adopt new technology [66]. In terms of the offline retail, the socioeconomic environment, per capita GDP, employment in the service industry, the output value of industrial enterprises above a designated size, and manufacturing enterprises are selected as indicators. The greater the number of offline retailers, the better the entrepreneurial climate in the region. Previous research has found that a significant amount of online business has been transformed from offline retail [67]. Economic growth accompanied by a good institutional environment and highly skilled human capital have a catalytic effect on the use of online business [68]. A quality labor market can reduce labor costs for online business companies, and the matching of labor markets with companies is smoother [69]. The higher the output value of industrial enterprises above the scale and manufacturing enterprises, the richer the variety of products in the region, supporting online business development in terms of production in the supply chain [70].

As Farag et al. stated, shopping accessibility, that is, the number of stores within a fixed region, is an important variable affecting residents' online shopping [2]. Cao et al. incorporated internet infrastructure, education level, and income into the regression model to investigate the spread and use of online shopping, and the results were also significant [54]. Clarke et al. suggested that age and income are crucial demographic discriminators of e-commerce usage [6]. Since the number of offline retail options reflects local shopping demand to a certain extent, the number of offline retail options is used as a variable. Infrastructure elements, such as ICT services, logistics services, and the density of high-grade roads, are also considered independent variables, similar to online business. The total per capita savings and per capita GDP that affect residents' consumption decisions are also selected as an independent variable. These two factors are related to spending power. The higher the GDP per capita and total savings per capita, the higher the likelihood of online consumption [71]. Counties with low population density indicate remote areas, and the population density variable is added to verify whether remote areas have higher consumer demand. Descriptive statistics are provided in Table 1.

Table 1. Descriptive statistics.

Variable	Min.	Max.	Mean	Std. Dev.
Offline retailer	27	80,913	6190.84	6007.36
ICT service providers	2	48,051	378.67	1293.33
Logistics service providers	1	2018	61.85	64.49
Length of high-grade roads	0	1510	200.17	205.48
Per capita GDP	12.95	17,023.24	1173.90	1116.26
Employment in service industry	68	1,407,234	69,589.93	74,135.89
Industrial output value above designated size	0	7852.39	254.07	463.22
Number of manufacturing enterprises	7	59,024	2692.67	5032.81
Population density	0.12	2062.5	295.41	282.15
Per capita savings	316.8	156,913	19,832.61	14,248.6

Stata was used to perform OLS regression analysis. Table 2 shows that the fitting coefficients of the model are all above 0.45, and the model is in line with expectations. In the regression model of online business, the amount of offline retail has a significant positive effect on the development of online business. As a relatively new retail form in the Internet era, online business is also rooted in and transformed from offline retail. Strong offline retailing is a necessary foundation for the development of online business [72]. In terms of infrastructure, the number of information technology service providers and the density of high-grade roads have a significant positive impact on the development of online business, while the regression coefficient of logistics service providers is not significant. The development of online business is inseparable from the growth of the internet, logistics, and transportation. Infrastructure is an essential condition for the growth of online business [73]. In the macrosocial environment dimension, except for the industrial output value above a designated amount, other variables, including per capita GDP, service industry population, and number of manufacturing companies, have significant positive effects. These factors are indispensable for the development of online business; in particular, the local manufacturing industry can provide the necessary commodity supply chain for the development of online retail, and the human capital of the service industry can provide relevant technical labor and promote the spatial diffusion of online business [63].

Table 2. Ordinary least squares (OLS) estimation.

	Online Business	Online Shopping
Constant	1.512 ***	1.711 ***
Offline retailer	7.53×10^{-5} ***	4.24×10^{-5} ***
Telecom service providers	0.000 **	0.000 ***
Logistics service providers	0.001	0.003 ***
Length of high-grade roads	0.001 **	0.001 ***
Per capita GDP	0.000 ***	0.000 ***
Employment in service industry	1.49×10^{-6}	
Industrial output value above designated size	-7.19×10^{-8} ***	
Number of manufacturing enterprises	0.000 ***	
Population density		0.002 ***
Per capita savings		0.000 ***
R2	0.467	0.59
Ad R2	0.464	0.589
Observations	1919	1919

Significance at the 10%, 5%, and 1% levels is indicated by *, **, and ***, respectively.

In the online shopping regression model, the number of offline retailers in the region has a positive effect on online shopping. It is generally believed that the demand for online shopping in remote areas is due to the lack of offline retail services, which seems to contradict our results. In fact, the demand for online shopping does not increase without a foundation. Most online shopping is transformed from offline demand, and the growth of online shopping also develops from the offline economy. However, online shopping is also restricted and affected by its own economic capabilities, whether in core or remote areas. Regions with higher per capita GDP and per capita total savings have higher demand for online shopping [54]. The results show that the regression coefficients of per capita GDP and per capita savings are both significant positive effects, which supports the previous research conclusions. Among the infrastructure elements, the regression results of information technology services, logistics services and the length of high-grade roads are significant positive effects, which are important supporting conditions for promoting the development of online shopping. The regression results show that population density has a positive effect on online shopping. This result is different from the existing assumptions that the lower the population density is, the greater the demand for online shopping in remote areas will be. This discrepancy may be because the demand for online shopping in areas with low population density is also bounded by a composite of affordability and infrastructure.

6. Discussion

With the advancement of ICT, there is ongoing debate about its impact on core and peripheral areas. Especially with the increasing popularity of e-commerce, research on the spatial diffusion mode of e-commerce has been the focus of geographical research. The main focus has been on the spatial diffusion model of online shopping in different regions. Over the past few decades, marginal areas, including rural areas, have undergone a process of de-agriculturalization and post-productivism, gradually transitioning from an agricultural economy to more diversified economic activities, including handicrafts, retail and tourism [74,75]. However, scholars have still conducted little research on online shopping in peripheral areas and have paid more attention to the spatial diffusion mode of online business and its impact on peripheral areas. Online business in rural China has developed rapidly with the support of e-commerce platforms and the government. In particular, the development of Taobao Village has resulted in the accumulation of much experience and seems to be a "marginal revolution" realized through internet technology in remote areas [76,77].

From a regional perspective, we have clarified the differences in the diffusion characteristics of core and peripheral regions through a comparative study of online business and online shopping diffusion in China. At the same time, we have a clearer understanding of the different factors that influence spatial diffusion. The diffusion of online business in Chinese counties is in line with the hypothesis of innovation diffusion. The dominant area for the development of online business is the eastern region, which not only has a relatively complete network infrastructure but also benefits from a strong industrial foundation that provides the necessary supply chain for the development of online business. The efficiency hypothesis of online business is not yet supported in peripheral areas, and the effect of using internet technology to overcome spatial friction is not obvious in these areas. County-level online shopping conforms to the efficiency hypothesis, which has been confirmed in previous "urban–rural" comparative studies, but overall, online shopping is in line with the innovation diffusion hypothesis. Rural and peripheral areas have the motivation to adopt online shopping, but its transformation into real shopping behavior is restricted by the level of income and savings. Affordability is, therefore, the main reason why some marginal areas have achieved a "marginal revolution" in terms of online shopping.

Online business was once considered to be a momentous way to narrow the gap between the peripheral and core regions, especially at small scales, and such findings can often be obtained [60]. From the findings of this paper's country perspective, it is clear that there is a large gap between online business development in peripheral regions and core regions, and that online business development is still higher in core regions with first-mover advantages. The development of online business has widened the regional gap to a greater extent [61]. Due to differences in regional endowments and industrial bases, industries in peripheral regions have difficulty supporting retail development. There may be a few exceptional cases in marginal areas of high online business indices, but at the national level, this assumption of trying to rely on online business to narrow regional economic differences is difficult to achieve. Online shopping is more capable of bridging core and peripheral differences than online business, a capability that refers not to the ability to bridge economic differences, but to the potential to reduce spatial consumption inequality and the welfare gains [71]. As a new trade and distribution technology, online shopping can increase urban–rural and inter-regional trade and alleviate spatial consumption inequalities. The marginal areas may not have the industrial base to support the development of online business, but the reverse gradient diffusion of online shopping shows the possibility of increased accessibility of shopping for the residents of the marginal areas, and online shopping shows the potential to significantly improve the convenience and quality of life of the residents of the marginal areas.

We acknowledge that the development of online business will lead to the growth of some marginal and rural areas, and there are many success stories in China, Europe, Africa and South Asia [52,78–80]. However, on a national scale, it is clearly unrealistic to try to rely on online business as a means of reducing regional economic disparities. What needs to be known is that not every village is suitable for online business development and that villages in core regions will always have a comparative advantage in developing online business. As a local development policy, therefore, online business is a better way to achieve regional economic revitalization. However, online business is not suitable as a policy to balance regional differences at the national level for either developing or developed countries. In European countries in particular, the economic function of the countryside is not particularly emphasized, and the countryside is used more as a natural, relaxing and comfortable place, where its economic function is not significantly underlined [81,82]. Therefore, we are keen to make people aware of the considerable role of online shopping in promoting consumption equality through our research. The development of online shopping to improve the accessibility of shopping for people living in marginal and rural areas and to reduce regional consumption inequalities is the greatest benefit that online shopping gives to marginal areas. Online shopping is a significant tool for improving regional inequalities that applies equally to developed and developing countries.

Given the complexity of the various systems of e-commerce, certain issues have not yet been considered. Spatial reconstruction has been brought about by changes in technology, with counties in different regions being drawn into the digital economy to varying degrees. Research has focused more on large cities and comparatively ignored non-metropolitan areas and rural areas. Therefore, it is especially important to prioritize the analysis of the spatial diffusion characteristics of e-commerce in different areas. Research on the spatial diffusion of e-commerce is concerned not only with its impact on the spatial structure of the economy but also with its impact on changes in the spatial structure of cities and regions. Will e-commerce expand the agglomeration advantages of cities and other core areas, or will it weaken the comparative advantages of cities to bring diverse development opportunities to peripheral areas? These issues are worth discussing in the context of internet research.

7. Conclusions and Policy Insights

Retail and consumption in e-commerce are distinguished to compare the spatial diffusion patterns of online business and online shopping at the county level. This article also attempts to determine which hypotheses are fitted by online business and online shopping in the core–periphery structure at a large scale and to explore the mechanisms behind these differences in diffusion modes.

Our comparative study found that eastern China has an absolute advantage in online business. The spatial diffusion of county-level online business obeys the characteristics of a gradient distribution from east to west. Gradient transition characteristics are also the main diffusion characteristics of online shopping, but there are many continuous areas with high values in the remote areas of the northeast and northwest and the North China

Plain, breaking the pattern of gradient distribution. Online business exhibits a hierarchical diffusion of technology only from strongly urbanized areas to weakly urbanized areas, which conforms only to the innovation diffusion hypothesis. In addition to online shopping conforming to the innovation diffusion hypothesis as a whole, there is an edge rise in low-tech areas, due to their own need to actively adapt to technology. The efficiency hypothesis is also supported in peripheral areas between 1200–1800 km and 2300–3200 km from the core. Online shopping has an advantage in some remote areas in the west. However, most of the coastal areas have experienced comprehensive technical adaptation, taking into account both business and consumption technology applications.

The difference in spatial diffusion between online business and online shopping at the county level mainly exists because they are different types of internet applications. Online business is a retail-oriented internet application that requires not only internet infrastructure and user skills but also the support of macrosocial economic conditions, including the accumulation of economic factors, such as an industrial foundation and human capital. For the spatial diffusion of online business, it is not enough for residents to have internet skills alone; the online business also needs to be integrated with industry, and without the support of an industrial base, it is difficult to realize the efficiency hypothesis at scale in marginal counties. Therefore, the socioeconomic gradient between the east and west in China determines the characteristics of online business, and the structure of online business in China conforms only to the innovation diffusion hypothesis at a large scale.

Online shopping differs from online business in that it is a consumption-oriented internet application and has fewer restrictions. The role of income and savings is particularly important for online shopping, as the mobile internet infrastructure and its use are no longer issues in urban and rural areas of China. Residents' online shopping is restricted by their own budgets, and other factors have little impact. Although in the overall core–periphery structure, online shopping is associated with the innovation diffusion hypothesis, this does not prevent online shopping from achieving a breakthrough in the diffusion pattern in peripheral areas. In China, where the mobile internet is well developed, infrastructure elements and skills are no longer obstacles to online shopping. Residents in the west with high incomes and savings can easily shift their consumption patterns to overcome spatial friction.

The article also has some limitations. Time series data can provide better insight into the spatial diffusion patterns of e-commerce, but the data used in this article are not rich enough, due to the limited availability of data. In the future, it will be possible to obtain richer data and study the latest proliferation patterns of e-commerce. In addition, the data used in the regression model in this paper are cross-sectional data, and the results of the analysis are correlational rather than causal. We will look further into the panel data for regression analysis to explore the mechanism of e-commerce diffusion.

Some policy insights can be drawn from this study. For policy makers, while it is impractical to rely on online business to achieve a widespread rise of marginal areas, policy releases still need to be more tilted toward marginal areas, including the improvement of infrastructures, such as the internet, transportation and logistics, as well as the support of fiscal and taxation systems, such as microfinance and tax incentives. In particular, the layout of the infrastructure reduces the inequality of public services between the core and the periphery, bringing opportunities for the development of online business in the periphery and contributing to the equalization of consumption between the core and the periphery. At the same time, other actors, such as social groups and non-profit organizations, need to pay more attention to the ability to use digital technology in marginal areas outside of cities and contribute more to raising the level of online business in marginal areas. In addition, social groups need to create more ties for the development and exchange of e-commerce between urban and rural areas in order to facilitate integration between core and peripheral areas.

Author Contributions: Conceptualization, F.W. and M.W.; validation, M.W.; formal analysis, F.W.; methodology, F.W. and S.Y.; data curation M.W.; writing—original draft preparation, F.W. and S.Y.; writing—review and editing, M.W. and S.Y.; supervision, M.W.; project administration, M.W.; funding acquisition, M.W. All authors have read and agreed to the published version of the manuscript.

Funding: This research is financially supported by the National Social Science Foundation of China (Grant No. 19AZD007), the National Natural Science Foundation of China (Grant No. 42171207), the Innovation Program of Shanghai Municipal Education Commission (File No. 2021-01-07-00-08-E00130) and ECNU Academic Innovation Promotion Program for Excellent Doctoral Students (File No. YBNLTS2021-033).

Institutional Review Board Statement: Not applicable.

Informed Consent Statement: Not applicable.

Data Availability Statement: The data presented in this study are available on request from the first author.

Acknowledgments: We would like to express our gratitude to Chan Xu at the Sichuan Normal University for her valuable data. Thanks also to the three anonymous reviewers for their generous and helpful suggestions in shaping this paper. All errors and omissions are our own.

Conflicts of Interest: The authors declare no conflict of interest.

References

1. Anderson, W.P.; Chatterjee, L.; Lakshmanan, T.R. E-commerce, transportation, and economic geography. *Growth Chang.* **2003**, *34*, 415–432. [CrossRef]
2. Farag, S.; Weltevreden, J.; Van Rietbergen, T.; Dijst, M.; Van Oort, F. E-shopping in the Netherlands: Does geography matter? *Environ. Plan. B Plan. Des.* **2006**, *33*, 59–74. [CrossRef]
3. Harvey, D. *The Condition of Postmodernity*; Blackwell: Oxford, UK, 1990.
4. Cairncross, F. *The Death of Distance 2.0*; Texere: London, UK, 2001.
5. Friedman, T.L. *The World Is Flat: A Brief History of the Twenty-First Century*; Macmillan: London, UK, 2006.
6. Clarke, G.; Thompson, C.; Birkin, M. The emerging geography of e-commerce in British retailing. Regional Studies. *Reg. Sci.* **2015**, *2*, 371–391. [CrossRef]
7. Salemink, K.; Strijker, D.; Bosworth, G. Rural development in the digital age: A systematic literature review on unequal ICT availability, adoption, and use in rural areas. *J. Rural. Stud.* **2017**, *54*, 360–371. [CrossRef]
8. Helsper, E.J. A corresponding fields model for the links between social and digital exclusion. *Commun. Theory* **2012**, *22*, 403–426. [CrossRef]
9. Cao, X.; Douma, F.; Cleaveland, F. Influence of e-shopping on shopping travel: Evidence from Minnesota's Twin Cities. *Transp. Res. Rec.* **2010**, *2157*, 147–154. [CrossRef]
10. Norris, P. *Digital Divide: Civic Engagement, Information Poverty, and the Internet Worldwide*; Cambridge University Press: Cambridge, UK, 2001.
11. Sparks, C. What is the "Digital Divide" and why is it important? *Javn. Public* **2013**, *20*, 27–46. [CrossRef]
12. DiMaggio, P.; Hargittai, E.; Neuman, W.R.; Robinson, J.P. Social implications of the Internet. *Annu. Rev. Sociol.* **2001**, *27*, 307–336. [CrossRef]
13. Kyriakidou, V.; Michalakelis, C.; Sphicopoulos, T. Digital divide gap convergence in Europe. *Technol. Soc.* **2011**, *33*, 265–270. [CrossRef]
14. Peter, J.; Valkenburg, P.M. Adolescents' internet use: Testing the "disappearing digital divide" versus the "emerging digital differentiation" approach. *Poetics* **2006**, *34*, 293–305. [CrossRef]
15. Hollifield, C.A.; Donnermeyer, J.F. Creating demand: Influencing information technology diffusion in rural communities. *Gov. Inf. Q.* **2003**, *20*, 135–150. [CrossRef]
16. Hargittai, E. Second-level digital divide: Mapping differences in people's online skills. *Arxiv Prepr. Cs/0109068* **2001**, *7*, n4. [CrossRef]
17. Longley, P.A.; Webber, R.; Li, C. The UK geography of the e-society: A national classification. *Environ. Plan. A* **2008**, *40*, 362–382. [CrossRef]
18. Underwood, J.D. Rethinking the Digital Divide: Impacts on student-tutor relationships. *Eur. J. Educ.* **2007**, *42*, 213–222. [CrossRef]
19. Van Dijk, J.; Hacker, K. The digital divide as a complex and dynamic phenomenon. *Inf. Soc.* **2003**, *19*, 315–326. [CrossRef]
20. Ono, H.; Zavodny, M. Digital inequality: A five country comparison using microdata. *Soc. Sci. Res.* **2007**, *36*, 1135–1155. [CrossRef]
21. Hargittai, E.; Hinnant, A. Digital inequality: Differences in young adults' use of the Internet. *Commun. Res.* **2008**, *35*, 602–621. [CrossRef]
22. Van Deursen, A.; Van Dijk, J. Internet skills and the digital divide. *New Media Soc.* **2011**, *13*, 893–911. [CrossRef]
23. Van Deursen, A.J.; Van Dijk, J.A. The digital divide shifts to differences in usage. *New Media Soc.* **2014**, *6*, 507–526. [CrossRef]

24. Lu, X.; Meng, X.; Guo, J.; Huang, W. Empirical Study on Impact of New In-formation Communication Technology on Digital Divides: Beijing and Shanghai. In Proceedings of the 2013 International Conference on Information, Business and Education Technology, Beijing, China, 14–15 March 2013; Atlantis Press: Paris, French, 2013. Available online: https://www.atlantis-press.com/proceedings/icibet-13/6211 (accessed on 20 October 2021).
25. Stern, M.J.; Adams, A.E.; Elsasser, S. Digital inequality and place: The effects of technological diffusion on internet proficiency and usage across rural, suburban, and urban counties. *Sociol. Inq.* **2009**, *79*, 391–417. [CrossRef]
26. Galloway, L. Can broadband access rescue the rural economy? *J. Small Bus. Enterp. Dev.* **2007**, *14*, 641–653. [CrossRef]
27. Galloway, L.; Sanders, J.; Deakins, D. Rural small firms' use of the internet: From global to local. *J. Rural Stud.* **2011**, *27*, 254–262. [CrossRef]
28. Stenberg, P.L.; Morehart, M.J.; Vogel, S.J.; Cromartie, J.; Breneman, V.E.; Brown, D.M. *Broadband Internet's Value for Rural America*; DIANE Publishing: Collingdale, PA, USA, 2009; No. 1477-2016-121092.
29. Mehra, B.; Merkel, C.; Bishop, A.P. The internet for empowerment of minority and marginalized users. *New Media Soc.* **2004**, *6*, 781–802. [CrossRef]
30. Malecki, E.J. Digital development in rural areas: Potentials and pitfalls. *J. Rural Stud.* **2003**, *19*, 201–214. [CrossRef]
31. Lengsfeld, J.H. An econometric analysis of the sociodemographic topology of the digital divide in Europe. *Inf. Soc.* **2011**, *27*, 141–157. [CrossRef]
32. Räisänen, J.; Tuovinen, T. Digital innovations in rural micro-enterprises. *J. Rural Stud.* **2020**, *73*, 56–67. [CrossRef]
33. Cooper, J. The digital divide: The special case of gender. *J. Comput. Assist. Learn.* **2006**, *22*, 320–334. [CrossRef]
34. Hargittai, E.; Walejko, G. The participation divide: Content creation and sharing in the digital age. *Inf. Community Soc.* **2008**, *11*, 239–256. [CrossRef]
35. Atkinson, J.; Black, R.; Curtis, A. Exploring the digital divide in an Australian regional city: A case study of Albury. *Aust. Geogr.* **2008**, *39*, 479–493. [CrossRef]
36. Livingstone, S.; Helsper, E. Gradations in digital inclusion: Children, young people and the digital divide. *New Media Soc.* **2007**, *9*, 671–696. [CrossRef]
37. Servon, L.J.; Bodnar, C. Bridging the digital divide: Technology, community, and public policy. *Can. J. Commun.* **2003**, *28*, 488.
38. Rogers, E.M. *Diffusion of Innovations*. Simon and Schuster: New York, NY, USA, 2010.
39. Hägerstrand, T. A Monte Carlo approach to diffusion. *European J. Sociol.* **1965**, *6*, 43–67. [CrossRef]
40. Kortum, S. International technology diffusion: Theory and measurement. *Int. Econ. Rev.* **1999**, *40*, 537–570.
41. Keller, W. Geographic localization of international technology diffusion. *Am. Econ. Rev.* **2002**, *92*, 120–142. [CrossRef]
42. Boschma, R. Proximity and innovation: A critical assessment. *Reg. Stud.* **2005**, *39*, 61–74. [CrossRef]
43. Lagendijk, A.; Lorentzen, A. Proximity, knowledge and innovation in peripheral regions. On the intersection between geographical and organizational proximity. *Eur. Plan. Stud.* **2007**, *15*, 457–466. [CrossRef]
44. Davids, M.; Frenken, K. Proximity, knowledge base and the innovation process: Towards an integrated framework. *Reg. Stud.* **2018**, *52*, 23–34. [CrossRef]
45. Hargittai, E. Weaving the Western Web: Explaining differences in Internet connectivity among OECD countries. *Telecommun. Policy* **1999**, *23*, 701–718. [CrossRef]
46. Holt, L.; Galligan, M. Mapping the field: Retrospective of the federal universal service programs. *Telecommun. Policy* **2013**, *37*, 773–793. [CrossRef]
47. Cambini, C.; Jiang, Y. Broadband investment and regulation: A literature review. *Telecommun. Policy* **2009**, *33*, 559–574. [CrossRef]
48. Amit, R.; Zott, C. Value creation in e-business. *Strateg. Manag. J.* **2001**, *22*, 493–520. [CrossRef]
49. Guillén, M.F.; Suárez, S.L. Developing the Internet: Entrepreneurship and public policy in Ireland, Singapore, Argentina, and Spain. *Telecommun. Policy* **2001**, *25*, 349–371. [CrossRef]
50. Beilock, R.; Dimitrova, D.V. An exploratory model of intercountry Internet diffusion. *Telecommun. Policy* **2003**, *27*, 237–252. [CrossRef]
51. Lin, J.; Yu, Z.; Wei, Y.D.; Wang, M. Internet access, spillover and regional development in china. *Sustainability* **2017**, *9*, 946. [CrossRef]
52. De Blasio, G. Urban–rural differences in internet usage, e-commerce, and e-banking: Evidence from Italy. *Growth Chang.* **2008**, *39*, 341–367. [CrossRef]
53. Motte-Baumvol, B.; Belton-Chevallier, L.; Dablanc, L.; Morganti, E.; Belin-Munier, C. Spatial dimensions of E-shopping in France. *Asian Transp. Stud.* **2017**, *4*, 585–600. [CrossRef]
54. Cao, X.; Chen, Q.; Choo, S. Geographic distribution of e-shopping: Application of structural equation models in the Twin Cities of Minnesota. *Transp. Res. Rec.* **2013**, *2383*, 18–26. [CrossRef]
55. Beckers, J.; Cárdenas, I.; Verhetsel, A. Identifying the geography of online shopping adoption in Belgium. *J. Retail. Consum. Serv.* **2018**, *45*, 33–41. [CrossRef]
56. Weltevreden, J.W.; Atzema, O.A.; Frenken, K.; de Kruif, K.; van Oort, F.G. The geography of Internet adoption by independent retailers in the Netherlands. *Environ. Plan. B Plan. Des.* **2008**, *35*, 443–460. [CrossRef]
57. Song, Z.; Wang, C.; Bergmann, L. China's prefectural digital divide: Spatial analysis and multivariate determinants of ICT diffusion. *Int. J. Inf. Manag.* **2020**, *52*, 102072. [CrossRef]
58. Zhang, X. Investigation of e-commerce in China in a geographical perspective. *Growth Chang.* **2019**, *50*, 1062–1084. [CrossRef]

59. Wei, Y.D.; Lin, J.; Zhang, L. E-commerce, taobao villages and regional development in China. *Geogr. Rev.* **2020**, *110*, 380–405. [CrossRef]
60. Liu, M.; Zhang, Q.; Gao, S.; Huang, J. The spatial aggregation of rural e-commerce in China: An empirical investigation into Taobao Villages. *J. Rural. Stud.* **2020**, *80*, 403–417. [CrossRef]
61. Geng, J.; Li, C. Empirical research on the spatial distribution and determinants of regional e-commerce in China: Evidence from Chinese provinces. *Emerg. Mark. Financ. Trade* **2020**, *56*, 3117–3133. [CrossRef]
62. Shao, R.; Derudder, B.; Witlox, F. The geography of e-shopping in China: On the role of physical and virtual accessibility. *J. Retail. Consum. Serv.* **2022**, *64*, 102753. [CrossRef]
63. Coe, N.M.; Yeung, H.W. Grounding global flows: Constructing an e-commerce hub in Singapore. In Proceedings of the Conference on Global Networks, Innovation and Regional Development, UC, UC Santa Cruz, Santa Cruz, CA, USA, 11–13 November 1999.
64. Boschma, R.A.; Weltevreden, J.W. An evolutionary perspective on Internet adoption by retailers in the Netherlands. *Environ. Plan. A* **2008**, *40*, 2222–2237. [CrossRef]
65. Couture, V.; Faber, B.; Gu, Y.; Liu, L. Connecting the countryside via E-Commerce: Evidence from China. *Am. Econ. Rev. Insights* **2021**, *3*, 35–50. [CrossRef]
66. Wong, P.K. Global and national factors affecting e-commerce diffusion in Singapore. *Inf. Soc.* **2003**, *19*, 19–32. [CrossRef]
67. Farag, S.; Schwanen, T.; Dijst, M.; Faber, J. Shopping online and/or in-store? A structural equation model of the relationships between e-shopping and in-store shopping. *Transp. Res. Part A Policy Pract.* **2007**, *41*, 125–141. [CrossRef]
68. Rao, S.S.; Metts, G.; Monge, C. Electronic commerce development in small and medium sized enterprises. *Bus. Process. Manag. J.* **2003**, *9*, 11–32. [CrossRef]
69. Kabango, C.M.; Asa, A.R. Factors influencing e-commerce development: Implications for the developing countries. *Int. J. Innov. Econ. Dev.* **2015**, *1*, 64–72. [CrossRef]
70. Guercini, S.; Bernal, P.M.; Prentice, C. New marketing in fashion e-commerce. *J. Glob. Fash. Mark.* **2018**, *9*, 1–8. [CrossRef]
71. Fan, J.; Tang, L.; Zhu, W.; Zou, B. The alibaba effect: Spatial consumption inequality and the welfare gains from e-commerce. *J. Int. Econ.* **2018**, *114*, 203–220. [CrossRef]
72. Hernandez, T.; Gomez-Insausti, R.; Biasiotto, M. Non-store retailing and shopping centre vitality. *J. Shopp. Cent. Res.* **2001**, *8*, 58–81.
73. Kshetri, N.; Dholakia, N. Determinants of the global diffusion of B2B e-commerce. *Electron. Mark.* **2002**, *12*, 120–129. [CrossRef]
74. Hoggart, K.; Paniagua, A. What rural restructuring? *J. Rural. Stud.* **2001**, *17*, 41–62. [CrossRef]
75. Roche, M. Rural geography: A stock tally of 2002. *Prog. Hum. Geogr.* **2003**, *27*, 779–786. [CrossRef]
76. Leong, C.; Pan, S.L.; Newell, S.; Cui, L. The Emergence of Self-Organizing E-Commerce Ecosystems in Remote Villages of China: A Tale of Digital Empowerment for Rural Development. *Manag. Inf. Syst. Q.* **2016**, *40*, 475–484. [CrossRef]
77. Gao, P.; Liu, Y. Endogenous inclusive development of e-commerce in rural China: A case study. *Growth Chang.* **2020**, *51*, 1611–1630. [CrossRef]
78. Zhang, Y.; Long, H.; Ma, L.; Tu, S.; Li, Y.; Ge, D. Analysis of rural economic restructuring driven by e-commerce based on the space of flows: The case of Xiaying village in central China. *J. Rural. Stud.* **2018**, in press. [CrossRef]
79. Omar, C.; Anas, T. E-commerce in Malaysia: Development, implementation and challenges. *Int. Rev. Manag. Bus. Res.* **2014**, *3*, 291–298.
80. Karine, H. E-commerce development in rural and remote areas of BRICS countries. *J. Integr. Agric.* **2021**, *20*, 979–997. [CrossRef]
81. Woods, M. Engaging the global countryside: Globalization, hybridity and the reconstitution of rural place. *Prog. Hum. Geogr.* **2007**, *31*, 485–507. [CrossRef]
82. Phillips, M. Baroque rurality in an English village. *J. Rural. Stud.* **2014**, *33*, 56–70. [CrossRef]

land

MDPI

Article

Spatial Pattern and Driving Mechanism of Urban–Rural Income Gap in Gansu Province of China

Ping Zhang [1,2,*], Weiwei Li [3], Kaixu Zhao [4] and Sidong Zhao [5]

1 College of Civil Engineering and Architecture, Jiaxing University, Jiaxing 314001, China
2 College of Architecture and Urban Planning, Lanzhou Jiaotong University, Lanzhou 730070, China
3 College of Landscape and Architectural Engineering, Guangxi Agricultural Vocational University, Nanning 530007, China; wwli2021@163.com
4 College of Urban and Environmental Science, Northwest University, Xi'an 710127, China; zhaokaixu@stumail.nwu.edu.cn
5 School of Architecture, Southeast University, Nanjing 210096, China; 230189013@seu.edu.cn
* Correspondence: ppqzhangp@zjxu.edu.cn

Citation: Zhang, P.; Li, W.; Zhao, K.; Zhao, S. Spatial Pattern and Driving Mechanism of Urban–Rural Income Gap in Gansu Province of China. *Land* **2021**, *10*, 1002. https://doi.org/10.3390/land10101002

Academic Editors: Stephan Bartke and Sigrun Kabisch

Received: 1 September 2021
Accepted: 20 September 2021
Published: 23 September 2021

Publisher's Note: MDPI stays neutral with regard to jurisdictional claims in published maps and institutional affiliations.

Abstract: The urban–rural income gap is a principal indicator for evaluating the sustainable development of a region, and even the comprehensive strength of a country. The study of the urban–rural income gap and its changing spatial patterns and influence factors is an important basis for the formulation of integrated urban–rural development planning. In this paper, we conduct an empirical study on 84 county-level cities in Gansu Province by using various analysis tools, such as GIS, GeoDetector and Boston Consulting Group Matrix. The findings show that: (1) The urban–rural income gap in Gansu province is at a high level in spatial correlation and agglomeration, leading to the formation of a stepped and solidified spatial pattern. (2) Different factors vary greatly in influence, for example, per capita Gross Domestic Product, alleviating poverty policy and urbanization rate are the most prominent, followed by those such as floating population, added value of secondary industry and number of Internet users. (3) The driving mechanism becomes increasingly complex, with the factor interaction effect of residents' income dominated by bifactor enhancement, and that of the urban–rural income gap dominated by non-linear enhancement. (4) The 84 county-level cities in Gansu Province are classified into four types of early warning zones, and differentiated policy suggestions are made in this paper.

Keywords: urban–rural income gap; spatial pattern; driving mechanism; China

1. Introduction

1.1. Background

Urban and rural areas are interrelated and interdependent, and together they contribute to the sustainable development of the regional economy. The urban–rural income gap is an important basis for measuring the capacity and level of comprehensive economic development of a country or region, as well as a crucial prerequisite for comprehensive and integrated urban–rural development. A large urban–rural income gap may then have a negative impact on economic and social development. Young [1] found that national income inequality is largely caused by the urban–rural income gap, accounting for about 40% of the total, according to an empirical analysis of 65 countries. The widening urban–rural income gap poses a great challenge to cities and villages in achieving sustainable development, and it has become an economic risk and social problem that developing countries have to face and solve in the process of industrialization and urbanization [2]. Therefore, it is of great theoretical significance and practical value to study the spatial patterns and influence factors of residents' income and its changes and the urban–rural income gap and its changes, reveal their deep-seated driving mechanisms and further put forward targeted policy recommendations for narrowing the urban–rural income gap and achieving integrated urban–rural development.

As one of the countries with the largest urban–rural income gap in the world, China still has a large urban–rural income gap that is highly representative in the world. Therefore, the empirical study of China will provide inspiration or experience for other countries and regions in the world to solve the problems of residents' income increase and urban–rural income gap. China has seen accelerated industrialization and urbanization as well as great achievements in economic growth and social development since its reform and opening up, bringing a dramatic increase in the income of urban and rural residents. However, the "miracle of development" is accompanied by a significant urban–rural income gap, which is increasing in a fluctuating manner. China's urban–rural income ratio was about 2.51 in 1978, reached a historical peak of 3.33 in 2009, and remained at a high level of 2.64 in 2019, despite a decline. In the early 1990s, the income gap between urban and rural areas in China was less than $209, but in 2019 it widened to $3818. (The data comes from the China Statistical Yearbook in 1991 and 2020. According to the data released by the National Bureau of Statistics of China, the inflation rate in the same period is about 10%). As socialism with Chinese characteristics enters a new era, further narrowing the income gap between urban and rural areas has become an important part in solving the problem of "unbalanced development". In the context of poverty eradication and high-quality development, comprehensive and integrated urban–rural development is facing more complex challenges, and the formulation of scientific policies to promote the reduction in the urban–rural income gap has become a hot issue of common concern among Chinese government sectors, scholars and the public. In recent years, both the central and local governments in China have included the urban–rural income gap as a core issue to be tackled in the 13th (2011–2015) and 14th (2016–2020) Five-Year Plans. In 2021, the central government issued the No. 1 document *Opinions of the Central Committee of the Communist Party of China and the State Council on Comprehensively Promoting Rural Revitalization and Accelerating the Modernization of Agriculture and Rural Areas*, clearly requiring "fully stimulating the development vitality of the countryside to consolidate and expand the results of poverty eradication and continue to narrow the income gap between urban and rural residents".

1.2. Aim and Question

To sum up, the existing papers have provided abundant data for investigating the urban–rural income gap and lack a solid theoretical and methodological foundation for in-depth analysis of the spatial–temporal evolution law and driving mechanism of the urban–rural income gap. To address the shortcomings in the research on small cities, this paper attempts to take Gansu Province, a less developed region in western China, as an example, to systematically and quantitatively analyze the spatial and temporal patterns of the urban–rural income gap and its change patterns in 84 small county-level cities. Based on various measurement methods, such as GIS spatial analysis and the GeoDetector method, this paper tries to explore policy recommendations to promote integrated urban–rural development, so as to put forward a reference for Gansu and other similar regions in China and even the world, to coordinate urban–rural development and formulate plans or policies to narrow the urban–rural income gap.

This paper focuses on the following questions: (1) What are the regular characteristics of the spatial pattern and spatial effects of the urban–rural income gap in small cities, including the analysis and pattern identification of spatial heterogeneity and correlation characteristics of urban residents' income, rural residents' income and the urban–rural income gap? (2) What are the driving mechanisms of spatial variation in the urban–rural income gap in small cities, including the composition of influence factors, the size of direct effect and the interaction effect of multiple factors together? (3) How to create an early warning analysis model of urban–rural income gap in small cities and propose targeted response policies, including space governance risk classification and policy zoning?

2. Literature Review

The study of urban–rural income gap is a classical topic of geography, economics, sociology, planning and other subjects, and how to increase income and narrow the urban–rural income gap for residents is a hot issue of continuous concern for the government, scholars and the public. After a long-term follow-up study, academics have now achieved more fruitful research results in measurement methods, causes, coping strategies and social impacts, with continuous innovation of research fields and perspectives. For example, Attanasio [3] and Krueger [4] have extended the study of income inequality to the field of consumption inequality and analyzed the connection between the two. Additionally, Binelli [5], after an empirical study of Central and Eastern Europe, concluded that those with higher incomes are less aware of urban–rural income inequality. However, we find that there are still some shortcomings in the existing studies in research scale and methodology.

2.1. Review of Spatial Pattern

From the perspective of research scale, the research results are focused on the national and regional levels, with insufficient attention to the city scale, especially a great lack of empirical studies at the small city scale. At the national level, Burlacu [6], Tamkoc [7], Lise [8], Heshmati [9], Salvati [10], Thein [11], Bodjongo [12], Su [13] and other scholars have conducted case or empirical studies on the urban–rural income gap in Romania, Turkey, Japan, Korea, Greece, India, Myanmar, Cameroon and other countries, analyzing the gap changes. It is important to note that Zhao [14] and Gradin [15] conducted comparative analyses of China with the United States and China with India, finding that income inequality is much lower in China.

At the regional level, inter-provincial analysis is the focus. For example, Chen [16] pointed out that tourism, urbanization and fiscal decentralization all contribute to narrowing the urban–rural income gap in China. Shi [17] estimated the spillover effect of inbound tourism on the urban–rural income gap based on spatial econometric methods and concluded that inbound tourism significantly helps to reduce the income gap between urban and rural areas, but with striking differences between eastern, central and western regions. Kim [18] concluded that the interaction effect of tourism and (Foreign Direct Investment) in narrowing the urban–rural income gap is significantly larger in the autonomous regions than in other provinces, as the urban–rural income gap can be reduced through the use of FDI and the development of tourism in the autonomous regions. Li [19] found that the growth of Agricultural Environmental Total Factor Productivity further widened the urban–rural income gap in China. Jin [20] argued that the increase in social security spending helps reduce the urban–rural income gap, but there are significant regional differences in such effects. Wei [21] analyzed the effects of trade scale and mode on the urban–rural income gap at the provincial level in China and found that the scale of international import and export trade, processing trade and general trade has widened the urban–rural income gap in the eastern region, while having narrowed it in the central region. Meanwhile, in the western region, exports reduce while general trade aggravates the urban–rural income gap, but imports and processing trade have no significant effect. Hong [22] found a significant positive effect of upgrading China's industrial structure on narrowing the urban–rural income gap, and Wang [23] concluded that the increase in urbanization level and fertilizer application intensity have a significant effect on alleviating the inter-provincial urban–rural income gap in China.

At the urban level, there have been some exploratory studies, but with insufficient attention to small and medium-sized cities and cities in less developed areas. For example, Zhang [24] conducted an empirical study on 248 prefecture-level cities from 2008 to 2018 and pointed out that tourism development helps to narrow the urban–rural income gap in China. Again, Huang [25] conducted an empirical study of 278 prefecture-level cities from 2003 to 2016, and the analysis showed that highway construction has reduced the urban–rural income gap, with great regional differences in its impact—negative in western cities while positive in eastern cities. Although Thiede [26] analyzed the dynamics of urban–rural income disparity in U.S. cities and concluded that the disparity in small cities is higher than that in large cities, there is a lack of analysis of the current characteristics, changing trends and main causes of urban–rural disparity in small cities. Small and medium-sized cities account for a large proportion and hold an important position in the regional town system. In addition, large-medium-small and prefecture-county-town-level cities, impacted by the scale effect, are greatly different in spatial heterogeneity and its driving mechanisms. Insufficient studies on small and medium-sized cities at the county and town levels, especially those in less developed areas, pose certain challenges to the applicability and accuracy of existing research findings.

2.2. Review of Driving Factors

From the perspective of research methodology, existing studies are focused on econometric analysis, with weak spatial analysis. The research methods of the existing papers are dominated by time series models, panel data models, mathematical statistics, correlation analysis, regression models, Markov chains, clustering, causality tests and cointegration equations, with focus on the analysis of the current characteristics, changing trends, influence factors, countermeasures and suggestions of urban–rural income gap. For example, Kibriya [27] analyzed the dynamics and patterns of rural–urban income inequality in India based on the time series method. Oyekale [28] analyzed the determinants of rural–urban income disparity in Nigeria based on regression methods and concluded that the factors of paid work, non-farm enterprises, grants and formal letters have the greatest impact on the rural–urban income gap, with further suggestion that infrastructure development, birth control and increased access to formal education in rural areas should be accelerated in order to reduce rural–urban income inequality. Sehrawat [29], based on least squares, cointegration equation and Granger Causality test tools, analyzed that financial development and economic growth reduce poverty in South Asian countries, while urban–rural income inequality increases poverty. Borodkin [30] analyzed the wage gap between urban and rural residents in the process of industrialization in Russia based on econometric methods. Chotia [31] analyzed the connection between infrastructure development and urban–rural income inequality in Bureau of Research Information Control System countries based on least squares and cointegration tests. Vafaei [32] investigated the connection between urban–rural income inequality and health based on ecological analysis, correlation analysis and multiple linear regression, finding that population health status is a function of absolute income, but not of relative income. Sehrawat [33] investigated the impact of financial development and economic growth on urban–rural income inequality in South Asian Association For Regional Cooperation countries based on the Granger Causality analysis tool. There are significant differences in the urban–rural income gap and its changes in different cities, and such differences represent the spatial heterogeneity under the combined effect of economic, social, political and ecological factors in the region, and it is often hard to explain the impact of spatial heterogeneity on the urban–rural income gap and its changes based on the above analysis methods. Besides, the existing papers give too little care to the influence of geospatial effect, lacking practical explanatory power and presentation. Most of the papers have an insufficient application of GIS spatial analysis tools and lack the necessary quantitative empirical studies on the influence factors of spatial heterogeneity and correlation, leading to insufficient awareness of the spatial patterns, spatial relationships, spatial effects and spatial dynamic mechanisms of urban–rural income gap in different regions.

In addition, there is no comprehensive study of multiple dependent and independent variables in the existing research methods, and no sufficient attention to multiple independent variable interaction effects. Most of the current papers are empirical studies on a particular indicator that reflects or influences the urban–rural income gap, yet often it is impossible for a single indicator to accurately depict the actual level of the urban–rural income gap and the complexity of its driving mechanisms. For example, Li [34] concluded that the high-speed rail construction has effectively narrowed the urban–rural income gap in China, but the convergence effect on the urban–rural income gap in China is still weak. Wang [35] concluded that the urban-biased land development policy is the most powerful factor driving the urban–rural income gap in China. Su [36] confirmed the existence of a Financial Kuznets Curve in East China; that is, the urban–rural income gap increases and then decreases with financial development. Batabyal [37] argued that income gap affects urban–rural population distribution patterns and residential choices, and Amara [38] concluded that educational attainment and family size are the major factors affecting urban–rural income gap in Tunisia. Zhao [39] analyzed the dynamic relationship between income structure and urban–rural income gap and its driving mechanism, and the results showed that wage income is the most powerful factor widening the urban–rural income gap, followed by transfer income, with the property income at the weakest position. Zhu [40] conducted an empirical study in China and concluded that the urban–rural inequality tends to be more severe in regions that have more complex export product/destination structures, due to the concentration of export activities in urban areas and due to some barriers that inhibit the flow of input factors (e.g., capital and labor) between rural and urban areas. Chen [41] found that while FDI directly contributes to narrowing the urban–rural income gap through job creation, knowledge spillovers and contributions to economic growth, it also exacerbates urban–rural income inequality through international trade and other channels. The urban–rural income gap and its changes are influenced by many factors, and they are in a complex interaction relationship. The joint action of multiple factors may produce synergistic reinforcing effects or antagonistic constraining effects, which eventually lead to deformation or even denaturation of the driving force under the influence factors alone. However, the quantitative measurement and in-depth analysis are neglected in the existing papers. In the era of big data, comprehensive research based on multiple indicators as dependent and independent variables is imminent.

3. Research Design

3.1. Research Methods

3.1.1. Coefficient of Variation: CV

The coefficient of variation (CV) is used to compare the magnitude of dispersion of the analyzed data, which is independent of the magnitude and measurement scale. The coefficient of variation is dimensionless, and a larger value represents a greater degree of dispersion, and vice versa. According to Guan [42], Zhang [43], Ruan [44], Liu [45], Miyamoto [46] and She [47], dispersion is classified as weak, medium and strong based on the CV values. That is, the value of the coefficient of variation is weakly discrete when it is 0–0.15, reflecting the low spatial inequality of urban–rural income gap; moderately discrete when it is 0.16–0.35, reflecting the high spatial inequality of urban–rural income gap; and strongly discrete when it is greater than 0.36, reflecting the very high spatial inequality of urban–rural income gap.

The coefficient of variation is calculated according to the equation as follows:

$$C_v = \frac{1}{y}\sqrt{\frac{1}{n-1}\sum_{i=1}^{n}(y_i - y)^2} \tag{1}$$

where C_v represents the coefficient of variation, n represents the number of small cities in the study area and y_i represents the observed value of an indicator for the small city; y is the average of the observed values of an indicator for all small cities. It is important to note that when the average value is close to zero, even a tiny perturbation may have a large impact on the coefficient of variation, resulting in poor accuracy. Therefore, when the average value is close to zero, the coefficient of variation values is only of reference value and cannot be used as a basis for determining spatial differentiation.

3.1.2. Exploratory Spatial Data Analysis: ESDA

ESDA is an ideal data-driven analysis method recognized by academic circles that has been widely used in the study of spatial heterogeneity and correlation. The commonly used measures in exploratory spatial data analysis methods are the global Moran's I, the Moran's scatter plot and the Lisa agglomerative distribution plot. In this paper, the global Moran index is employed to indicate the existence of spatial autocorrelation, agglomeration and agglomeration trend in the overall space, and further explain the agglomeration types and spatial correlation characteristics in existence in terms of spatial location through the Lisa agglomeration distribution map, reflecting the spatial heterogeneity and instability within the local area. The value of Global Moran's I is in a range of [−1, 1]. At a given significance level (generally 0.05 or 0.1), the value > 0 indicates positive spatial correlation, and when the value is greater, the spatial correlation and agglomeration will be more significant; the value < 0 indicates negative spatial correlation, and when the value is smaller, the spatial variation will be larger; the value = 0 indicates random spatial distribution. According to Local Moran's I, spatial correlation patterns can be subdivided into four types, including H-H and L-L with positive spatial correlation and H-L and L-H with negative spatial correlation. The calculation equation is as follows:

$$\text{Global Moran's } I = \frac{n}{S_0} \times \frac{\sum_{i=1}^{n}\sum_{j=1}^{n} W_{ij}(y_i - y)(y_j - y)}{\sum_{i=1}^{n}(y_i - y)^2},\ S_0 = \sum_{i=1}^{n}\sum_{j=1}^{n} = W_{ij} \quad (2)$$

$$\text{Local Moran's } I_i = Z_i \sum_{i=1}^{n} W_{ij} Z_j \quad (3)$$

where n represents the quantity of cities, y_i and y_j are the observed values of cities i and j, respectively, y is the average of the observed values, W_{ij} is the spatial weight matrix in global spatial autocorrelation and the row normalized value of spatial weights in local spatial autocorrelation, S_0 is the sum of spatial weight matrices, Z_i and Z_j are the normalized values of the observed values of cities i and j. In this paper, we have conducted spatial autocorrelation analysis based on ArcGis 10.2 (Esri, Redlands CA, USA) and GeoDa 1.18 (Esri, Redlands CA, USA), where the significance level is 0.05, the spatial weight matrix is the one based on the adjacent boundaries and all parameters are those of software by default. The maximum number of neighbors is 11 and the minimum is 1, with the average of 4.64 and the median of 4.50.

3.1.3. Boston Consulting Group Matrix: BCG

BCG, also known as the four-quadrant analysis, was created in 1970 by Bruce Henderson, a leading American management scientist and founder of the Boston Consulting Group. This method is mainly applied in business management and economics, and it classifies products or markets into four types: stars, question, cows and dogs, through the interaction of two factors of "sales growth" and "market share". In this paper we use it to evaluate the spatial classification of residents' income and the risk partitioning of urban–rural income gap, depending on the average of the relative shares of the dependent variables and growth rates to classify the cities in the study area into four types of H-H, H-L, L-H, and L-L.

$$\text{Relative share} = \frac{y_i}{y_{max}} \quad (4)$$

$$\text{Growth rate} = \left(\sqrt[t]{\frac{y_i}{y_i'}} - 1\right) \times 100\% \quad (5)$$

where t represents the time, y_i and y_{max} are the observed value of city i and the maximum value of all cities, respectively, and y_i' is the observed value of city i in the base period.

3.1.4. GeoDetector

GeoDetector is a new spatial analysis model used to detect the connection between a certain geographical attribute and its explanatory factors [48] and is widely used in the study of the influence factors of natural and economic and social phenomena. This paper is devoted to exploring the spatial pattern of urban–rural income gap in small cities and the driving forces behind it, and this method is quite applicable to it due to a large number of influence factors. It should be noted that an exploratory study has been conducted in this regard by Chen [49], who empirically investigated the spatio-temporal characteristics of the urban–rural income gap and its drivers in prefecture-level cities in the Yangtze River Economic Belt from 2000 to 2017. GeoDetector consists of four functional modules of factor detection, interaction detection, risk detection and ecological detection. In this paper, we studied the magnitude of factor forces and its interaction effects that affect the spatial pattern of urban–rural income gap in county-level cities in Gansu depending on the two functional modules of factor detection and interaction detection.

Spatial differentiation is the spatial expression of natural and socio-economic processes. GeoDetector is a new statistical method to detect spatial heterogeneity and reveal its driving factors. Its basic idea is that, based on the assumption that the study area is divided into sub-regions, there is spatial heterogeneity if the sum of the variances of the sub-regions is smaller than the total regional variance, and there is statistical correlation between the independent and dependent variables if their spatial distribution tends to be the same. In other words, if independent variables have a significant influence on dependent variables, they should have similar spatial distributions [50]. The q-statistic calculated by GeoDetector can be used to measure the degree of explanation of independent variables to dependent variables and analyze the interaction between independent variables. In factor detection, GeoDetector, by calculating the q-value of each independent variable and dependent variable, quantitatively evaluates the correlation (similarity) between the two. In interaction detection, GeoDetector determines whether there is interaction between two independent variable factors, and the strength, direction, linearity or nonlinearity of interaction by calculating and comparing the q-value of the dependent variable after superposition of two independent variable factors.

Let's assume the dependent variable is Y_i and the independent variable is X_i, and use them to depict the level of urban–rural income gap and its influence factors, respectively. With the q value of the factor detection results, the level of spatial heterogeneity of Y_i and the extent to which X_i explains the spatial heterogeneity of Y_i can be measured. The value of q is in a range of [0, 1], and under the condition of passing the significance test, a larger value indicates that Y_i has a more pronounced spatial heterogeneity and X_i has a stronger explanatory power for it. In general, the threshold value for passing the significance test is 0.05 under general conditions, and 0.1 under loose conditions. With the interaction detection results we can identify interactions between different drivers X_i, i.e., to assess whether drivers X_1 and X_2, when acting together, enhance or diminish the explanatory power of the dependent variable Y_i, or whether the effects of these factors on Y_i are independent of each other. The evaluation results are classified into five categories according to the relationship between q_{12} and q_1, q_2 under the interaction of the two drivers (Table 1) [51]. The calculation equation of q is as follows:

$$q = 1 - \frac{\sum_{h=1}^{l} N_h \sigma_h^2}{N\sigma^2} = 1 - \frac{SSW}{SST},\ SSW = \sum_{h=1}^{l} N_h \sigma_h^2,\ SST = N\sigma^2 \quad (6)$$

where h is the number of strata or classifications of the independent variables, N_h and N are the number of cities in stratum h and the study area, respectively, σ_h^2 and σ^2 are the variance of the dependent variable in stratum h and the study area, respectively, SSW is the Within Sum of Squares and SST is the Total Sum of Squares.

3.2. Study Area: Gansu

The study area of this paper is 84 county-level small cities in Gansu Province, and due to a great lack of data for Anning, Jiashishan and Maqu, they are not included in this study to ensure the accuracy of the findings (Figure 1). Located in the hinterland of northwest China, Gansu Province is one of the major minority populated areas in China, and it is a typical underdeveloped province in China with backward economic and social development. In 2019, the GDP of Gansu Province was 126.4 billion US dollars, ranking fifth from the bottom among 31 provinces, autonomous regions and municipalities directly under the central government of China; during the same period, its per capita GDP was $4783, ranking first from the bottom in the country (Figure 2). At present, the problems of inadequate rural development, unbalanced development between urban and rural areas, and especially the large income gap between urban and rural areas, are still prominent in Gansu Province. In 2019, the average income of urban residents in Gansu Province was $4685.51, $1454.66 lower than the Chinese average; the average income of rural residents was $1395.80, $926.58 lower than the Chinese average; the absolute difference between urban and rural income reached up to $3289.85, $528.23 lower than the Chinese average; and the urban–rural income gap index reached 3.36, 0.7 higher than the Chinese average (Figure 3).

Table 1. Interaction between Explanatory Variables.

Graphical Representation	Description	Interaction
	$q(X_i \cap X_j) < Min(q(X_i), q(X_j))$	Weaken, nonlinear
	$Min(q(X_i), q(X_j)) < q(X_i \cap X_j) < Max(q(X_i)), q(X_j))$	Weaken, uni-
	$q(X_i \cap X_j) > Max(q(X_i), q(X_j))$	Enhance, bi-
	$q(X_i \cap X_j) = q(X_i) + q(X_j)$	Independent
	$q(X_i \cap X_j) > q(X_i) + q(X_j)$	Enhance, nonlinear

Legend: ● $Min(q(X_i), q(X_j))$ ● $Max(q(X_i), q(X_j))$ ● $q(X_i) + q(X_j)$ ▼ $q(X_i \cap X_j)$.

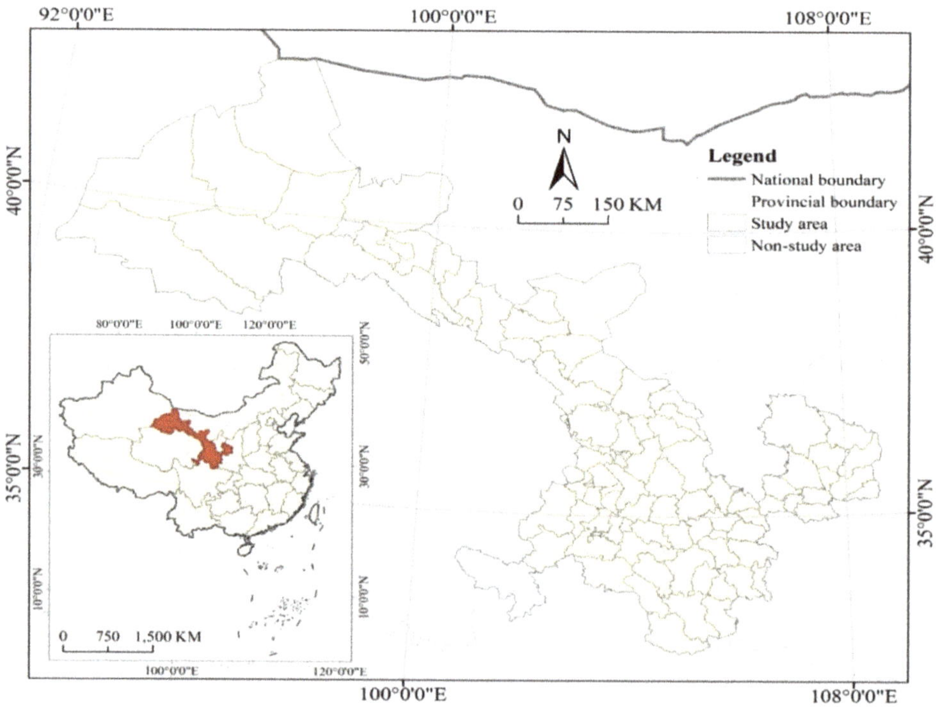

Figure 1. Study Area.

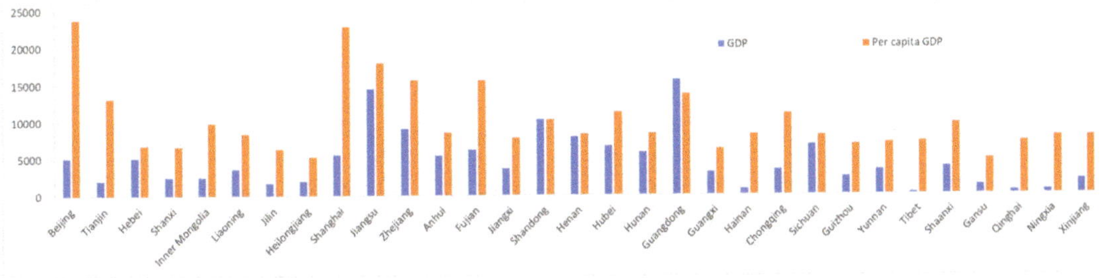

Figure 2. Comparative analysis of economic development in China's provinces.

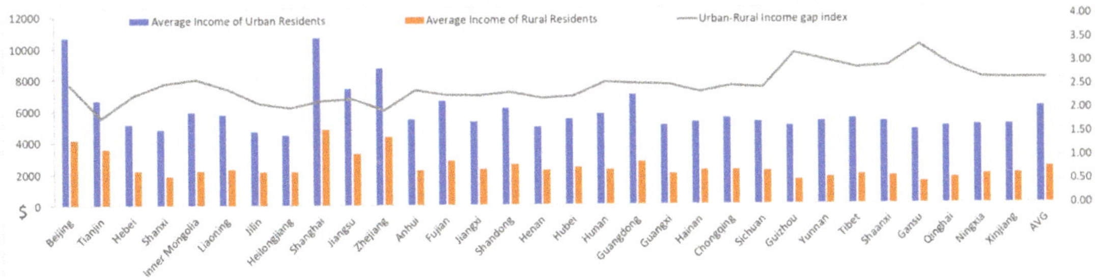

Figure 3. Comparative analysis of urban–rural income gap in China's provinces.

From the perspective of the development of Gansu Province, although the average income of rural residents in Gansu Province has been growing faster than that of urban and rural areas in recent years, and the urban–rural income gap has been gradually narrowing, the absolute gap between urban and rural incomes is still in a rapid and sustained growth, with the gap still stable at a high level and greater than the national average all the time (Figure 4). From 2013 to 2019, the average income of urban residents in Gansu Province increased by $1805, up by 7.20% annually; the average income of rural residents increased by $586, up by 8.08% annually; the absolute urban–rural income gap increased by $1219, up by 6.84% annually; the urban–rural income gap index decreased by 0.20, up by −0.82% annually. In summary, the income level of residents in Gansu province is much lower than the average of China, but its urban–rural income gap index is much higher than the national average level. It is a major task for governments at all levels in Gansu Province to increase residents' income and narrow the urban–rural income gap for a long period of time in the future. The study on Gansu province is a typical case, and it is of great reference value for other similar regions in China and the world, to solve the problems of income increase and urban–rural income gap.

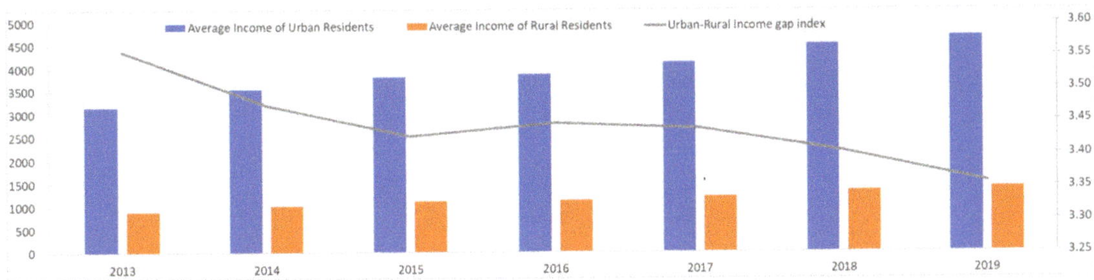

Figure 4. Analysis on the change of urban–rural income gap in Gansu Province.

3.3. Index Selection

From the perspective of dependent variable selection, the average incomes of urban residents and that of rural residents are the basic indicators, serving as the most intuitive ones for studying the urban–rural income gap and significant ones for the government to examine the coordination of regional urban–rural economic and social development. The ratio of the two can be used to construct the index of urban–rural income gap. It should be noted that narrowing the urban–rural income gap is the core of policy design, and for the government and the public, in addition to the status quo values of the three indicators above, they are also interested in the changes of these indicators. Therefore, in this paper, we finally selected six dependent variables, that is, the average income of urban residents, the average income of rural residents, the urban–rural income gap index, the change in the average income of urban residents, the change in the average income of rural residents and the change in the urban–rural income gap index (Table 2). There are many factors influencing the urban–rural income gap, and they are in a complex relationship. The analysis in part 1.2 shows that existing studies focus on urban-biased policies, a dualistic economic system, urbanization, industrialization, economic outward orientation, financial development, institutional change, natural conditions, education level and agricultural inputs [52], which are of great inspirational values for this study. The urban–rural income gap and its changes constitute a systematic problem. In line with the principles of comparability, feasibility, representativeness and accessibility, and according to the research ideas of Li [53], Zhao [54,55] and Yuan [56], this paper presents a comprehensive analysis of their influence factors based on 13 indicators from three areas of economy, society and policy (Table 2).

Table 2. Model variable description.

Variable	Index	Code	Type
Dependent Variable (Y_i)	Average Income of Urban Residents	Y_1	Situation
	Average Income of Rural Residents	Y_2	
	Urban–rural Income Gap Index	Y_3	
	Changes in Average Income of Urban Residents	Y_4	Dynamic
	Changes in Average Income of Rural Residents	Y_5	
	Changes in Urban–rural Income Gap Index	Y_6	
Independent Variable (X_i)	Gross Domestic Product (GDP)	X_1	Economic driving force
	Per Capita GDP	X_2	
	Total Retail Sales of Consumer Goods	X_3	
	Added Value of Secondary Industry	X_4	
	Added Value of Tertiary Industry	X_5	
	Total Population	X_6	Social driving force
	Floating Population	X_7	
	Urbanization Rate	X_8	
	Number of Internet users	X_9	
	Financial Expenditure	X_{10}	Policy driving force
	Amount of Bank Loans	X_{11}	
	Main Functional Area Planning	X_{12}	
	Alleviating Poverty Policy	X_{13}	

The impact of the economic development level on the urban–rural income gap is shown in the total amount and quality, and it is also in a greater connection with the industrial structure and consumption level [57]. GDP and per capita GDP are common indicators to depict the total amount and quality of urban economic development, while retail sales of social consumer goods are commonly used indicators to reflect consumption vitality. The value added by the secondary and tertiary industries is a major driver to attract population and increase income, presenting the degree of deagrarianization of the industrial structure and its employment and wealth creation effects. The key to the impact of social conditions on the urban–rural income gap lies in population size and its attribute characteristics. Population size, population mobility and the transformation of rural population to urban population have a great influence on the urban–rural income gap by differentially improving the marginal efficiency between rural and urban areas. We should note that the Internet and its applications have enjoyed a strong rise in China and have been integrated into all areas of the social economy, leading to the rapid growth of new businesses such as e-commerce live streaming and short video, as well as the size of online shopping users. The Internet has played a role as a "booster" in increasing farmers' income, selling agricultural products and transforming agriculture. As an emerging force, the Internet has reduced the cost of information search and opened up the scope of market participation for farmers, and has improved the accuracy of government policies for agricultural benefits. The popularization and development of the Internet has brought a powerful digital dividend for the development of rural areas, farmers and agriculture, and has become a major emerging factor that should not be ignored in the study of the urban–rural income gap in China in the new era.

Government initiative is the key to solving the problem of the urban–rural income gap. With direct fiscal spending, indirect bank loans and comprehensive policy design, the government can effectively macro-regulate the income of urban and rural residents. Local governments with greater fiscal strength have a greater ability to intervene directly, so we selected in this paper the size of fiscal spending to represent direct government influence. The low profitability of the agriculture-related industries makes it difficult to get loans from banks in general. For this reason, the government often indirectly intervenes in the urban–rural income gap by establishing agricultural policy banks and increasing the targeted loans related to agriculture to guide bank loans to rural areas, agriculture and farmers. The main functional area planning is a long-term strategic program in China, and it divides the space into different types of policy areas based on the resource and environmental carrying capacity, existing development density and development potential of different regions. As the main functional area planning directly determines the main function, development direction and intensity of each city, it has been a major policy that has to be considered in the study of the urban–rural income gap. In November 2015, the central government issued the *Decision on Winning the Battle against Poverty*, marking that poverty alleviation has become a core task for the central and local governments. In 2016, there were 592 national-level, poverty-stricken counties in China, including 375 in the western region. Gansu had a total of 75 poverty-stricken counties, including 58 at the national level and 17 at the provincial level, making it one of the provinces with the heaviest task of poverty eradication in China. Since the implementation of the war on poverty, the state has increased investment and policy support for Gansu Province, especially for poverty-stricken counties, all of which have now lifted themselves from poverty. In the transformation of Gansu province from a concentration of poor counties to a region with no poor counties, it can be seen that the impact of poverty alleviation policy on the urban–rural income gap in the province is obvious.

3.4. Research Steps

This study consists of three steps and seven key points (Figure 5). The first step is raw data and pre-processing. (1) Form a complete raw data table based on the data published on the relevant statistical websites. (2) Discrete the continuous data of independent variables based on Python and, to eliminate artificial influence, classify the independent variables of 84 county-level cities into nine types by the percentile method (2–10). The second step is data processing. (3) Perform a spatial analysis, including the calculation of the coefficient of variation of the dependent variable and Moran's I, and spatial analysis of the dependent variable based on ArcGis 10.2 and GeoDa1.18. (4) For Influence Factor, import the original data of the dependent variable and the discrete data of the independent variable into GeoDetector, carry out factor detection and interaction detection, and perform data review and result selection according to p-value (<0.05, <0.1 under loose conditions) and q-value. The third step is data analysis. (5) Comprehensively analyze the forces of driving factors and their acting modes, grade and classify the factors and their interaction effects. (6) Spatially classify the income level of residents, spatially classify the income gap and make adaptive and targeted policy recommendations based on the BCG model.

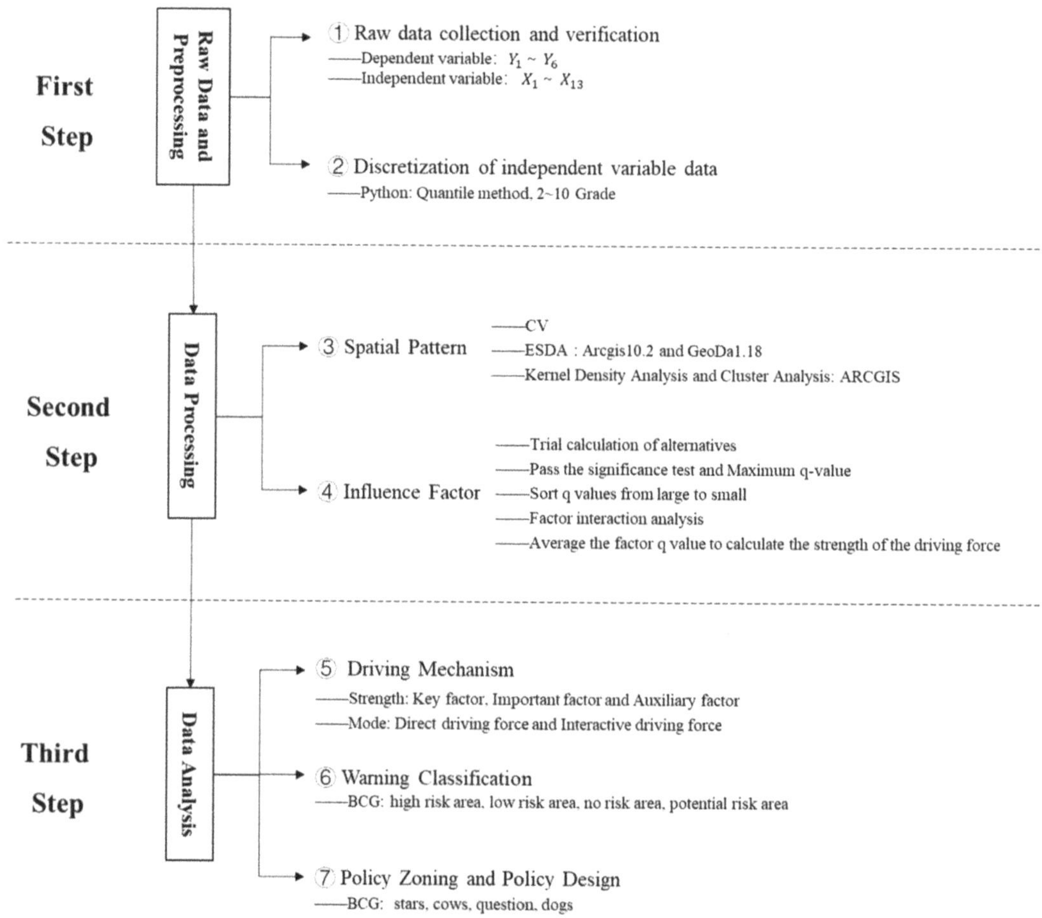

Figure 5. Research steps.

3.5. Data Sources

The dependent and independent variable indicators in this paper are mainly from the *Gansu Development Statistical Yearbook* and the *Gansu Province Rural Yearbook*, and some indicators are from the *China County Construction Statistical Yearbook*, with some missing data collected from the statistical handbooks and government work reports of each county.

The study period chosen was 2016–2019, for two main reasons. The first is to ensure data integrity. There were indeed many statistics before 2016, and lengthening the study time would affect the accuracy of the conclusions. The second is to maintain the consistency of the policy context. In November 2015, China started the battle against poverty, with the central and provincial governments strengthening support for poverty-stricken counties and impoverished people. Due to the high proportion of county-level cities in Gansu Province defined as poverty-stricken counties by the central and provincial governments, and the large impact of poverty eradication and poverty alleviation policies, the analysis of 2016 as the base year is more reasonable considering the lag in policy implementation. It should be noted that the main functional area planning divides the space into three types of ecological areas, main agricultural products production areas and key development areas, so they are assigned values of 1, 2 and 3 respectively in the processing of independent variables. In addition, the poverty-stricken counties involve both national and provincial levels, so the 84 counties in Gansu Province are classified into three types: general counties, provincial poverty-stricken counties and national poverty-stricken counties, and they are assigned values of 1, 2 and 3, respectively, in the processing of independent variables.

4. Results

4.1. Spatial Pattern

4.1.1. Spatial Heterogeneity

There is some spatial heterogeneity in residents' income and its changes and in urban–rural income gap and its changes in 84 county-level cities in Gansu Province, but it is not very prominent. The spatial heterogeneity of rural residents' income and its changes is at the highest level, followed by the spatial heterogeneity of urban residents' income and its changes, with the spatial heterogeneity of urban–rural income gap and its changes at the bottom. In 2019, the coefficients of variation for Y_1, Y_3 and Y_4 were 0.19, 0.25 and 0.24, respectively, in a range of 0.16~0.35, which were moderately heterogeneous; the coefficients of variation for Y_2 and Y_5 were 0.44 and 0.41, respectively, both greater than 0.36, which were strongly heterogeneous. Y_6 showed a large number of negative numbers, with an average value of −0.08 and a standard deviation of 0.12, indicating a low level of spatial heterogeneity. In 2016, the coefficients of variation for Y_1 and Y_3 were 0.20 and 0.26, respectively, which remained moderately heterogeneous, and the coefficient of variation for Y_3 was 0.51, which was strongly heterogeneous. The indicators related to urban–rural income gap of 84 county-level cities in Gansu Province in 2019 and 2016 were classified into high, medium and low types by nature breaks of ARCGIS 10.2.

In terms of the spatial distribution of state quantities, Y_1 and Y_2 have similar spatial patterns, and Y_3 is completely different from the first two. Besides, the spatial pattern in 2019 was generally similar to that in 2016, except for a broad contraction in the medium category of Y_1, indicating the appearance of a solidified spatial pattern of the urban–rural income gap (Figure 6). In 2019, Y_1 and Y_2 shared the same spatial pattern, with the exception of the area around the provincial capital, the three types of cities were characterized by southeast-northwest clustering and stair-step distribution. Specifically: cities of the high type are mostly distributed in the northwest corner of Gansu, including Guazhou, Jinta, Subei, Dunhuang, Sunan and Jinchang, with a small proportion in the core area of the provincial capital, including Chengguan, Anning and Xigu. Cities of the medium type are concentrated in the west corridor of the Yellow River and the edge of the provincial capital, including Shandan, Minle, Ganzhou, Yongchang, Minqin, Yongdeng, Gaolan and Yuzhong. Cities of the low type are concentrated in the eastern region of the Yellow River, including Linxia, Longxi, Huining, Xihe and Liangdang. Y_3 is completely different from Y_1 and Y_2, with high agglomeration but insignificant stepwise. Cities of the high type are mainly distributed in the west region of the Yellow River, including Xihe, Zhouqu, Longxi and Huachi, with a small proportion in the west corridor of the Yellow River, including Gulang and Tianzhu. Most of the medium type cities are concentrated in the edge of the provincial capital, including Jingtai, Baiyin, Jingyuan, Huining, Yongjing and Linxia, with a small proportion scattered in the east region of the Yellow River, including Zhengning, Lingtai, Huixian and Chengxian. Cities of the low type are mainly distributed in the west corridor of the Yellow River and the provincial capital area, including Guazhou, Yumen, Sunan, Minqin, Yongdeng, Gaolan, Chengguan and Xigu. In 2010, Y_1 changed significantly, and the number of the medium type cities increased rapidly, with the geographical scope expanding from the west region of the Yellow River to the east. Unfortunately, Y_2 and Y_3 remain unchanged, and the spatial pattern is solidified.

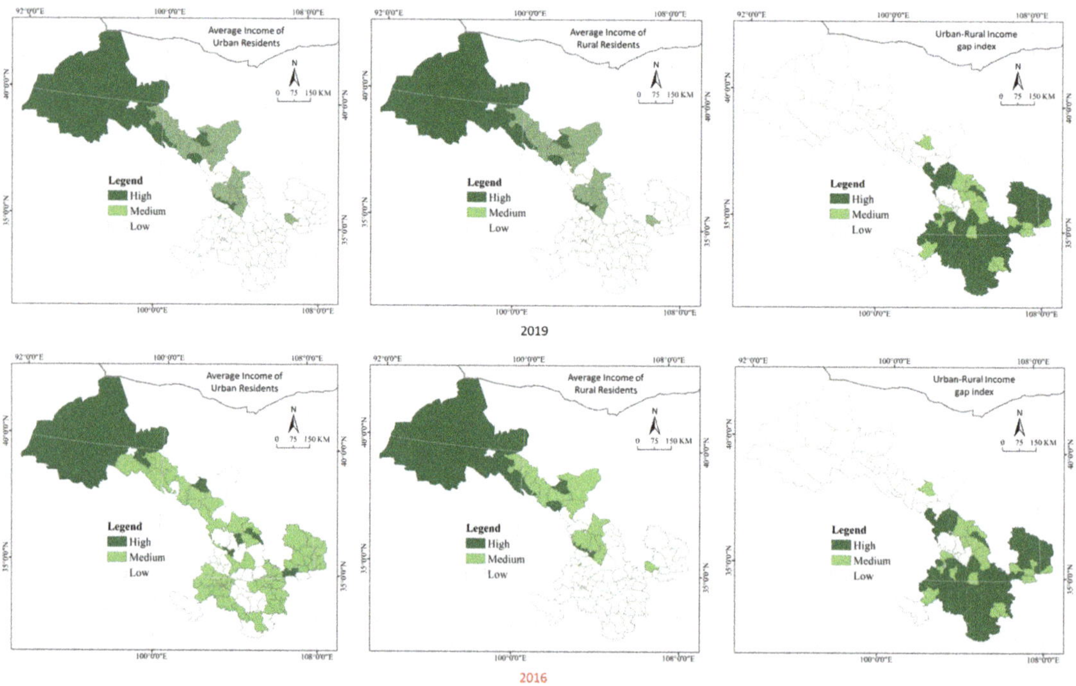

Figure 6. Analysis on Spatial heterogeneity.

From the perspective of the spatial distribution of changes, the spatial patterns of Y_4, Y_5 and Y_6 are completely different (Figure 7). For Y_4, cities of the high type are mainly distributed in regions of Lanzhou and Jiuquan, including Yongdeng, Gaolan, Chengguan, Yuzhong, Yumen, Subei, Xifeng and Jinchuan. There are agglomerations of the medium-type cities in both the west and east regions of the Yellow River, the former including desert oasis cities such as Jinta, Guazhou, Dunhuang, Sunan, Yongchang and Minle, and the latter including resource-based cities such as Huachi, Heshui, Zhengning, Zhenyuan, Lingtai, Chongxin and Jingchuan. Cities of the low type are mainly distributed in the east region of the Yellow River, including Liangdang, Diebu, Longxi, Wushan, Kangle, Xiahe, Jingyuan and Jingtai, with a small proportion in the west corridor of the Yellow River, including Minqin, Gulang, Tianzhu, Gaotai and Linze. For Y_5, cities of the high type are mainly distributed in the northwest of Gansu Province, including Subei, Yumen, Jinta, Sunan and Jinchuan. Cities of the medium type are mainly distributed in the west corridor of the Yellow River and the Lanbai metropolitan area, including Guazhou, Dunhuang, Minqin, Shandan, Ganzhou, Minle, Baiyin, Gaolan and Yongdeng. All of the low-type cities are located in the east region of the Yellow River, including Qin'an, Tongwei, Weiyuan, Longxi, Gangu and Dongxiang. For Y_6, cities of the high type are all concentrated in the provincial capital metropolitan area, including Yongdeng, Chengguan, Yuzhong and Gaolan. All the cities in the west corridor of the Yellow River and the cities in the area with intensive mineral resources in the east region of the Yellow River are of the medium type, including Heshui, Jingning, Zhengning, Huating and Lingtai. It should be noted that the spatial patterns of income growth for urban and rural residents are quite different. For urban residents, the cities with the highest income growth rate are distributed in the core area of the provincial capital, while those with the lowest income growth rate are distributed in ethnic autonomous regions, with the income growth of the urban residents in cities along the Longhai Railway, highways and the Yellow River, as well as oasis cities in the west corridor of the Yellow River, and resource-based cities in the east region of the Yellow River at a medium rate. For rural residents, the cities with the highest income growth rate are distributed in the ethnic autonomous regions, and those with the lowest growth rate in the west corridor of the Yellow River, with the income growth in most of the cities in the east region of the Yellow River at a medium rate.

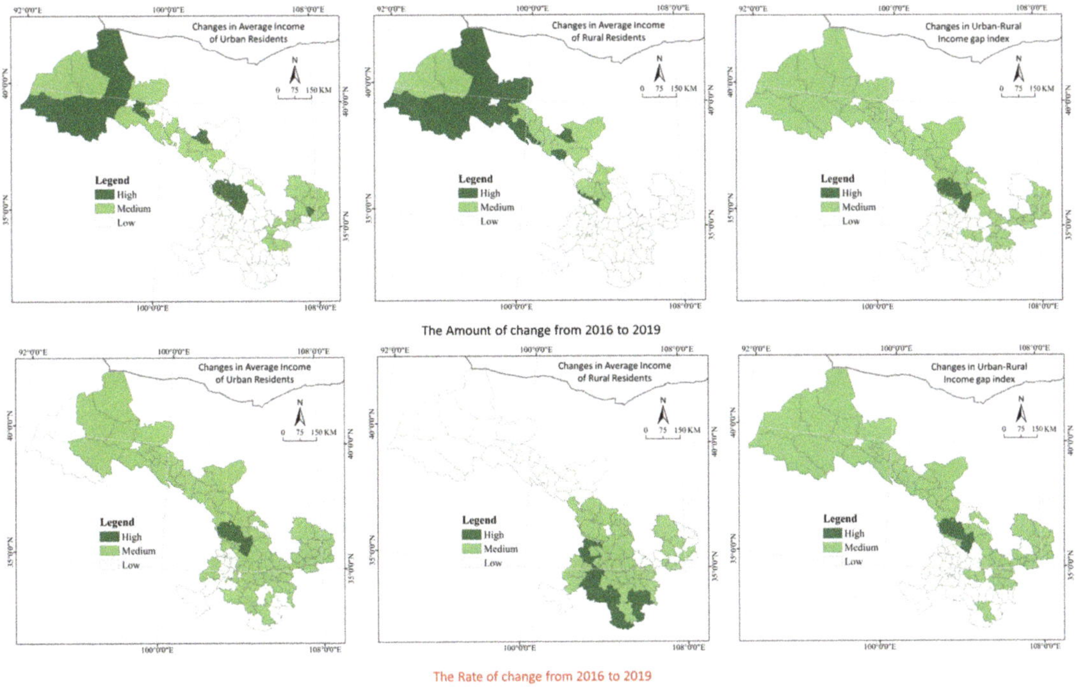

Figure 7. Analysis on Spatial heterogeneity of change.

4.1.2. Spatial Correlation

The indicators related to the urban–rural income gap in 84 county-level cities in Gansu Province are all positively spatially autocorrelated, and they are ranked as $Y_2 > Y_3 > Y_5 > Y_4 > Y_6 > Y_1$ in spatial correlation strength. From the perspective of global Moran's I, the value of Y_1 in 2019 and 2016 was 0.34 and 0.26, respectively, always at a low level; the value of Y_2 was 0.67 and 0.68; the value of Y_3 was 0.66 and 0.68; and the values of Y_4, Y_5, and Y_6 were 0.52, 0.60 and 0.43, respectively, always at the middle and high level, indicating that the urban–rural income gap and its changes in county-level cities in Gansu Province remain stable for a long time with significant global spatial autocorrelation and strong spatial agglomeration. To further analyze the types of spatial interconnections among cities, we created a Lisa diagram by means of GeoDa 1.18. Based on the spatial relationship between the sample cities and their neighboring cities, the cities were classified into four types of H-H, H-L, L-H, and L-L (Figures 8 and 9). Y_1, Y_2 and Y_3 are roughly the same in spatial pattern in 2016 and 2019, with changes only in local areas, such as the L-L type of Y_1 expanding widely in the Dingxi region, and the addition of Jingchuan and Huixian for the L-H type of Y_3 besides Kongtong. Y_4 and Y_5 are similar in spatial pattern, with the latter developing at a higher level than the former. Y_6 is obviously different from Y_4 and Y_5. It should be noted that the spatial patterns of the amount and rate of change of urban–rural income and its gap also vary widely.

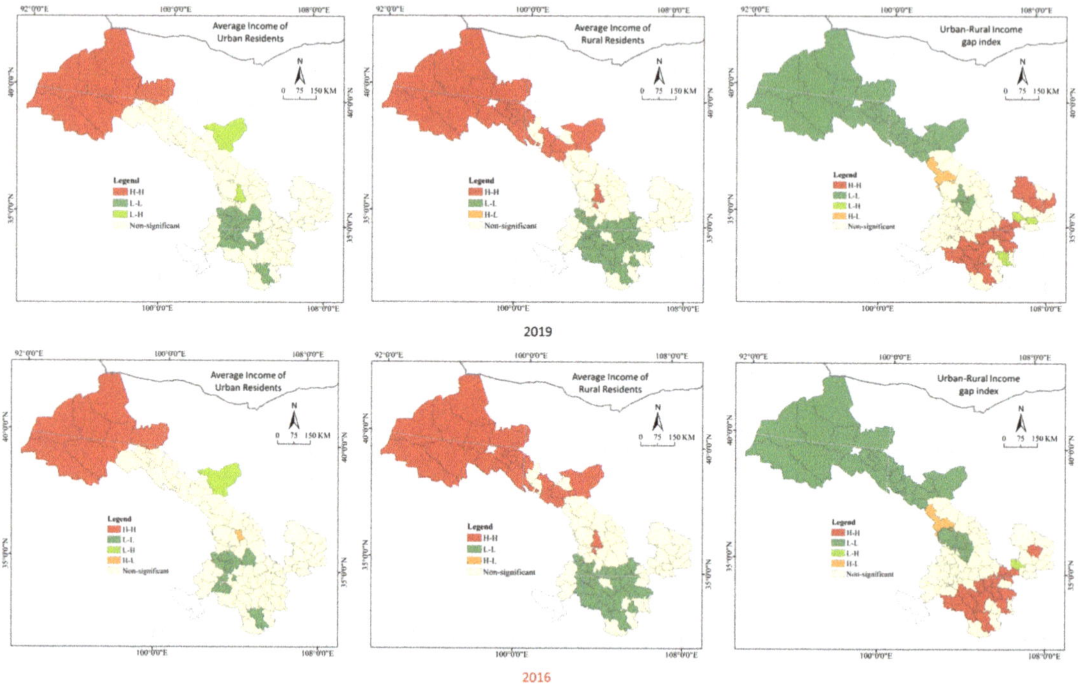

Figure 8. Analysis on Spatial correlation.

For Y_1, cities of the H-H type are concentrated in the northwest region of Gansu in a contiguous distribution, and those of the L-L type are mainly concentrated in Dingxi and Gannan regions, indicating that when the income of urban residents in the central county is high/low, that of the neighboring counties is high/low, characterized by a strong positive spatial correlation; the only cities of the L-H type (polarized) are Minqin and Gaolan, and no cities are of the H-L type (hollow), indicating that there are few cases where the income of residents in neighboring counties is low/high when that in the central county is high/low, with the negative spatial correlation quite insignificant. Y_2 is very similar to Y_1, but the H-H and L-L types have a broader geographic coverage. It should be noted that Y_2 has no L-H type, and only one city, Linxia, is of the H-L type. For Y_3, cities of the H-H type are concentrated in Longnan, Tianshui, Pingliang and Qingyang areas in the southeast of Gansu Province in a contiguous distribution; cities of the L-L type are mainly concentrated in the west of the Yellow River and the provincial capital area; cities of the L-H type are only Kongtong, Jingchuan and Huixian, and there is only one city, Tianzhu, of the H-L type. For Y_4 and Y_5, cities of the H-H type are mainly concentrated in the northwest of Gansu Province, and cities of the L-L type are mainly located in the southwest corner of Gansu. It should be noted that Y_4 has only one city of L-H and H-L types, Minqin and Qinzhou, respectively; Y_5 has no cities of the L-H type, and there is only one city, Linxia, of the H-L type. For Y_6, cities of the H-H type are mainly distributed in the provincial capital and have expanded to the northwest to oasis cities such as Sunan and

Tianzhu with minority autonomy, cities of the L-L type are also clustered in the southwest region of Gansu Province, Qinzhou and Linxia are cities of the H-L type and there is only one city, Jingtai, of the L-H type.

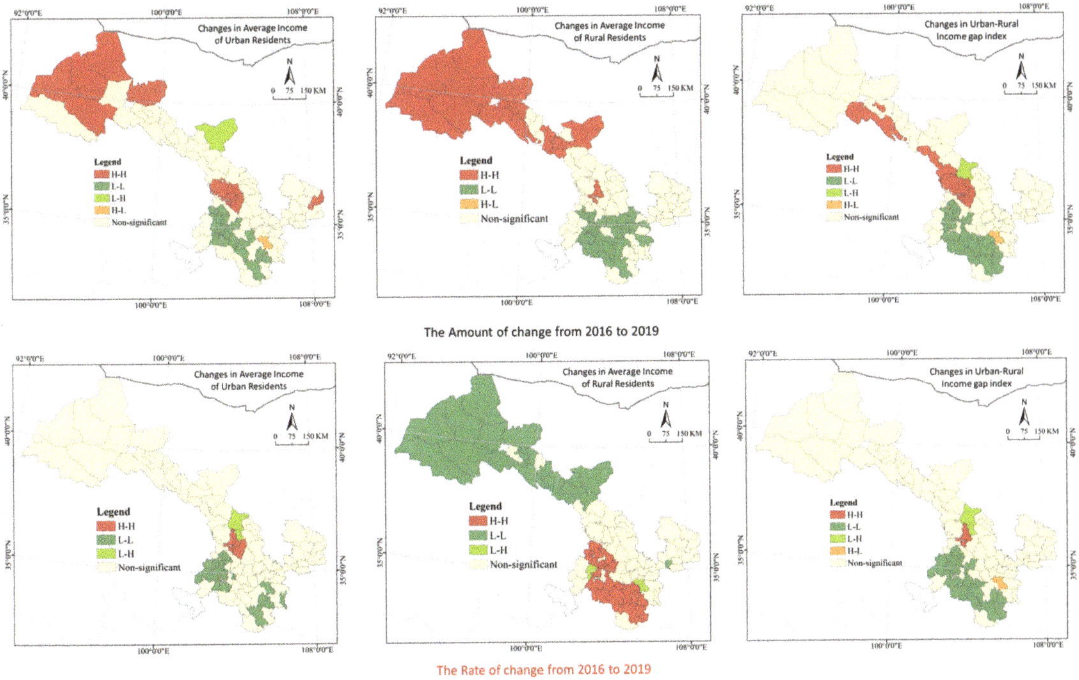

Figure 9. Analysis on Spatial correlation of change.

It is important to note that from the perspective of the spatial distribution of the rate of change, the patterns of urban residents' income, rural residents' income and urban–rural income gap are different from each other. As for the change rate of urban residents' income, cities of the H-H type are concentrated in the southeast corner of the provincial capital, cities of the L-L type are concentrated in Linxia Hui Autonomous Prefecture and the Longnan region, Baiyin and Jingtai are cities of the L-H type, and there are no cities of the H-L type. For the rate of change of rural residents' income, cities of the H-H type are concentrated in the Hedong Hui and Tibetan autonomous regions and Longnan region, cities of the L-L type are concentrated in the west corridor of the Yellow River, Hezuo and Qinzhou are cities of the L-H type and there are no cities of the H-L type. For the change rate of urban–rural income gap, there are few cities of the H-H type—only Qilihe, Xigu, Chengguan and Gaolan in the core area of the provincial capital. There are many cities of the L-L type, which are concentrated in the autonomous region for ethnic minorities in the east of the Yellow River. Baiyin and Jingtai are cities of the L-H type, and there is only one city, Qizhou, of the H-L type.

4.2. *Influence Factors*

4.2.1. Factor Detection

X_6 of Y_1 and Y_4, X_6, X_7 and X_{10} of Y_6 could not pass the significance test, while X_{10} of Y_4 and X_6 of Y_5 could only pass the significance test of 0.1. At 5% or a more stringent level of significance, the impact factors are classified as high, medium and low based on the ranking of the direct effect (q) according to top3, top7 and others (Table 3). For the average income of urban residents, per capita GDP, alleviating poverty policy, and urbanization rate are of the high type; floating population, added value of secondary industry, GDP and number of Internet users are of the medium type; total retail sales of consumer goods, added value of tertiary industry, amount of bank loans, main functional area planning and financial expenditure are of the low type. The social driving force is greater than the economic driving force in general, and the policy driving force is minimal. For the average income of rural residents, per capita GDP, alleviating poverty policy and urbanization rate are of the high type; floating population, GDP, added value of secondary industry and added value of tertiary industry are of the medium type; number of Internet users, financial expenditure, amount of bank loans, total retail sales of consumer goods, main functional area planning and total population are of the low type. The social, economic and policy driving forces are roughly equal in general, and all of them are strong. For the urban–rural income gap index, per capita GDP, urbanization rate and alleviating poverty policy are of the high type; floating population, added value of secondary industry, financial expenditure and amount of bank loans are of the medium type; GDP, total retail sales of consumer goods, added value of tertiary industry, main functional area planning, total population and number of Internet users are of the low type. The social, economic and policy driving forces are roughly equal in general, and all of them are weak (Figure 10).

Table 3. Analysis of factor detector.

		X_1	X_2	X_3	X_4	X_5	X_6	X_7	X_8	X_9	X_{10}	X_{11}	X_{12}	X_{13}
Y_1	q	0.40	0.59	0.33	0.43	0.27	0.06	0.45	0.52	0.35	0.14	0.23	0.19	0.59
	p	0.00	0.00	0.00	0.00	0.02	0.20	0.00	0.00	0.00	0.03	0.00	0.01	0.00
Y_2	q	0.31	0.76	0.24	0.30	0.28	0.10	0.47	0.68	0.25	0.25	0.25	0.11	0.76
	p	0.01	0.00	0.01	0.01	0.01	0.05	0.00	0.00	0.01	0.03	0.03	0.03	0.00
Y_3	q	0.15	0.45	0.14	0.24	0.14	0.05	0.27	0.45	0.05	0.18	0.17	0.07	0.43
	p	0.02	0.00	0.04	0.05	0.03	0.05	0.01	0.00	0.05	0.04	0.05	0.05	0.00
Y_4	q	0.46	0.59	0.33	0.58	0.32	0.05	0.32	0.37	0.34	0.08	0.34	0.25	0.37
	p	0.00	0.00	0.00	0.00	0.00	0.17	0.00	0.00	0.00	0.06	0.00	0.00	0.01
Y_5	q	0.32	0.78	0.25	0.33	0.29	0.09	0.47	0.71	0.27	0.24	0.26	0.13	0.72
	p	0.00	0.00	0.03	0.00	0.01	0.08	0.00	0.00	0.01	0.04	0.02	0.02	0.00
Y_6	q	0.30	0.33	0.26	0.34	0.25	0.04	0.02	0.08	0.21	0.02	0.26	0.09	0.09
	p	0.00	0.00	0.01	0.00	0.04	0.22	0.27	0.04	0.05	0.16	0.00	0.03	0.03

Note: The average value of influence (q) of each factor is calculated at the significance level of 5% to represent the strength of economic, social and policy driving forces.

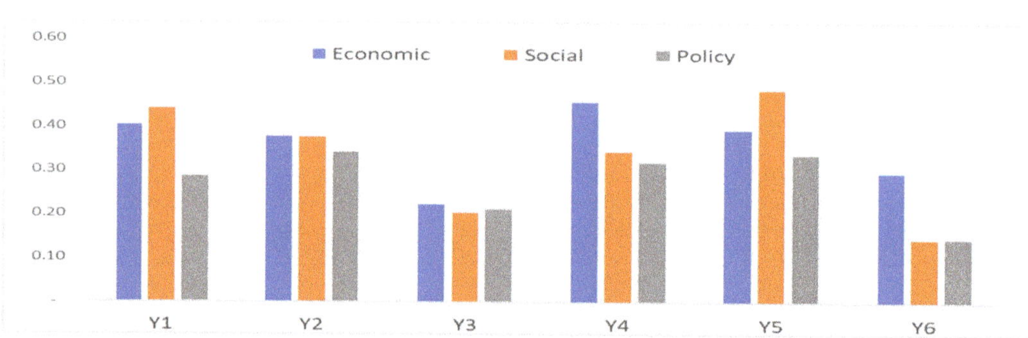

Figure 10. Analysis of driving force.

For changes in average income of urban residents, per capita GDP, added value of secondary industry and GDP are of the high type; urbanization rate, alleviating poverty policy, number of Internet users and amount of bank loans are of the medium type; total retail sales of consumer goods, added value of tertiary industry, floating population and main functional area planning are of the low type. The social and policy driving forces are roughly equal in general, lagging well behind the economic driving force (Figure 10). For changes in average income of rural residents, per capita GDP, alleviating poverty policy and urbanization rate are of the high type; floating population, added value of secondary industry, GDP and added value of tertiary industry are of the medium type; number of Internet users, amount of bank loans, total retail sales of consumer goods, financial expenditure and main functional area planning are of the low type. The social driving force is greater than the economic force in general, and the policy driving force is minimal (Figure 10). For changes in urban–rural income gap index, added value of secondary industry, per capita GDP and GDP are of the high type; total retail sales of consumer goods, amount of bank loans, added value of tertiary industry and number of Internet users are of the medium type; main functional area planning, alleviating poverty policy and urbanization rate are of the low type. The social and policy driving forces are roughly equal in general, lagging well behind the economic driving force (Figure 10).

4.2.2. Interaction Detection

All of the factor pairs are bifactor-enhanced or non-linearly enhanced with each other, and there are no independent and asymptotic relationships. The factor pairs can be classified into three types of high, medium and low based on the top10 and average value of the factor pair interaction forces (Figure 11).

Y_1 forms a total of 66 factor pairs, and the average value of the interaction forces is 0.68, with the minimum value of 0.37 and the maximum value of 0.90; the interaction effects of $X_1 \cap X_2$ and $X_2 \cap X_9$ are greater than 0.90. The factor pairs are dominated by bifactor enhancement effects, and there are only 12 non-linearly enhanced factor pairs, accounting for about 18.18%, including $X_1 \cap X_5$, $X_7 \cap X_5$, $X_{10} \cap X_5$, $X_{10} \cap X_1$, $X_{10} \cap X_2$, $X_{10} \cap X_3$, $X_{10} \cap X_4$, $X_{10} \cap X_7$, $X_{10} \cap X_9$, $X_{10} \cap X_{11}$, and $X_{10} \cap X_{12}$. X_{10} is the uppermost interaction factor.

Y_2 forms a total of 78 factor pairs, and the average value of the interaction forces is 0.71, with the minimum value of 0.29 and the maximum value of 0.95; the interaction effects of $X_1 \cap X_2$, $X_2 \cap X_3$, $X_2 \cap X_9$, $X_7 \cap X_2$, $X_4 \cap X_7$, and $X_4 \cap X_8$ are greater than 0.90. The factor pairs are dominated by bifactor enhancement effects, and there are a significantly increasing number of non-linearly enhanced factor pairs, up to 27, accounting for about 35.90%, including $X_4 \cap X_7$, $X_4 \cap X_{10}$, $X_{10} \cap X_9$, and $X_6 \cap X_8$. The uppermost interaction factors are X_4, X_6, X_{11}, X_{10}, and X_9.

Y_3 forms a total of 78 factor pairs, and the average value of the interaction forces is 0.51, with a minimum value of 0.12 and a maximum value of 0.83; the interaction effects of $X_4 \cap X_8$, $X_7 \cap X_8$, $X_4 \cap X_7$ and $X_2 \cap X_5$ are greater than 0.80. The non-linear enhanced factors dominate, up to 57, accounting for about 73.08%, including $X_2 \cap X_5$, $X_2 \cap X_4$, $X_2 \cap X_7$, $X_3 \cap X_8$, $X_7 \cap X_8$, $X_1 \cap X_8$, $X_{10} \cap X_2$, $X_{10} \cap X_{11}$ and $X_{10} \cap X_4$. All factors except X_{13} have a significant interaction effect.

Y_4 forms a total of 55 factor pairs, and the average value of the interaction forces is 0.69, with the minimum value of 0.48 and the maximum value of 0.92; the interaction effects of $X_2 \cap X_3$ and $X_2 \cap X_9$ are greater than 0.90. The factor pairs are dominated by bifactor enhancement effects, and there are only 12 non-linearly enhanced factor pairs, accounting for about 21.82%, including $X_1 \cap X_7$, $X_1 \cap X_8$, $X_3 \cap X_7$, $X_3 \cap X_8$, $X_5 \cap X_7$, $X_5 \cap X_8$, $X_8 \cap X_9$, $X_{10} \cap X_8$, and $X_{11} \cap X_9$. The uppermost interaction factors are X_7 and X_8.

Y_5 forms a total of 66 factor pairs, and the average value of the interaction forces is 0.73, with the minimum value of 0.37 and the maximum value of 0.95; the interaction effects of $X_1 \cap X_2$, $X_2 \cap X_3$, $X_2 \cap X_5$, $X_2 \cap X_7$, $X_2 \cap X_8$, $X_2 \cap X_9$, $X_2 \cap X_{10}$ and $X_4 \cap X_7$ are greater than 0.90. The factor pairs are dominated by bifactor enhancement effects, and there are only 20 non-linearly enhanced factor pairs, accounting for about 30.30%, including $X_1 \cap X_{10}$, $X_1 \cap X_{11}$, $X_3 \cap X_4$, $X_4 \cap X_{10}$, $X_4 \cap X_5$, $X_5 \cap X_9$, $X_9 \cap X_{11}$, $X_9 \cap X_{12}$ and $X_{10} \cap X_{12}$. The uppermost interaction factors are X_3, X_4, X_9 and X_{10}.

Y_6 forms a total of 45 factor pairs, and the average value of the interaction forces is 0.54, with the minimum value of 0.13 and the maximum value of 0.93; only the interaction effect of $X_2 \cap X_9$ is greater than 0.90. The factor pairs are dominated by non-linear enhancement effects, up to 13, accounting for about 73.33%, including $X_1 \cap X_2$, $X_1 \cap X_{12}$, $X_2 \cap X_3$, $X_2 \cap X_4$, $X_2 \cap X_9$, $X_3 \cap X_4$, $X_9 \cap X_4$, $X_9 \cap X_{11}$ and $X_9 \cap X_{12}$. The uppermost interaction factors are X_2, X_8, X_9 and X_{12}.

Figure 11. Analysis of interaction detector.

5. Discussion

5.1. Low Spatial Heterogeneity, High Spatial Correlation and Agglomeration, and Solidified Stepwise Spatial Pattern

The income of rural residents and its changes show strong spatial heterogeneity, but there is no high spatial heterogeneity of the income of rural residents and its changes or the urban–rural income gap index and its changes. However, the spatial correlation and agglomeration of the dependent variables of urban residents' income is high, except for the relatively low spatial correlation. From the perspective of spatial pattern and its evolution trend, the 84 county-level cities in Gansu Province are in a very stable spatial pattern, prominently characterized by high, medium and low stepwise aggregated distribution. The spatial distribution and change patterns of Y_1, Y_2, Y_4 and Y_5 show that the residents in the Lanzhou-Baiyin metropolitan area and the Jiuquan-Jiayuguan co-location area are the most affluent, with the most significant income growth of urban and rural residents, followed by the residents in the west corridor of the Yellow River (Zhangye–Jinchang–Wuwei section) with a high income growth of urban and rural residents, and then the residents of the east region of the Yellow River have the smallest income growth of urban and rural residents. Y_3 also follows the pattern of stepwise spatial distribution and change, and the urban–rural income gap is highly coupled with the poverty level, indicating that the income gap is greater in poorer places. Y_6 has a completely different spatial pattern, with the largest changes in urban–rural income gap concentrated in the provincial capital area, moderate changes in oasis cities and resource-based cities and the smallest changes in non-resource-based cities in the west region of the Yellow River. It should be noted that the spatial patterns of the amount and rate of change of urban and rural residents' income are inconsistent or even opposite to each other, reflecting that promoting increased income for urban and rural residents requires different or even opposite strategies depending on local conditions. Based on the research experience of Pan [58], Ding [59] and Fu [60] et al., the urban–rural income gap indexes are classified into four types with thresholds of 1.5, 2.5 and 3.5, and their spatial distribution reflects that the urban–rural income gap in Gansu Province has prominent spatial heterogeneity and agglomeration characteristics and that they are in a very stable spatial pattern (Figure 12). We should note that Pan [61] clearly pointed out that the urban–rural income gap in Gansu Province presents a spatial pattern of "high in the west and low in the east, high in the north and low in the south", and the spatial distribution shows a trend of "club convergence" at two poles (H-H and L-L).

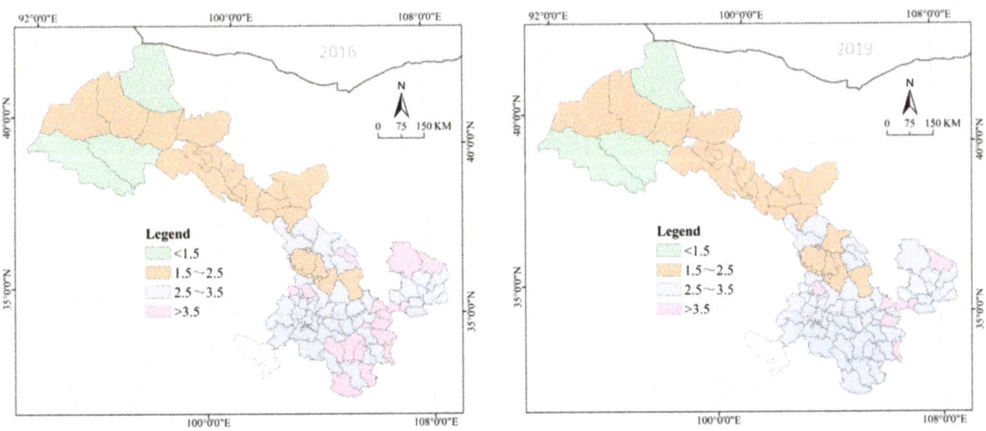

Figure 12. Map made using Pan's method.

5.2. *Diversification of Influence Factors and Complexity of Driving Mechanisms*

The factor averages of Y_1 and Y_3, Y_2 and Y_4, and Y_3 and Y_6 are calculated as the influences of urban residents' income and its changes, rural residents' income and its changes, and urban–rural income gap and its changes, respectively, at 5% or a more stringent level of significance, and the total factor averages of Y_1, Y_2, Y_3, Y_4, Y_5 and Y_6 are further calculated as the combined driving forces of all influence factors. The average of the influence factor force of urban residents' income and its changes is 0.34, the average of the influence factor force of rural residents' income and its changes is 0.37, the average of the influence factor force of urban–rural income gap and its changes is 0.21, and the combined average value of the influence factor forces of all dependent variables is 0.31. Based on the ranking and average of the factor forces, the driving factors are classified into three types of "key factors", "important factors" and "auxiliary factors" with the strength of the factor interaction effects taken into full account (Figure 13). "Key factors" are dominated by direct forces, with the strength of factor forces ranked in the top three. For urban residents' income and its changes, per capita GDP, added value of secondary industry and alleviating poverty policy are key factors; for rural residents' income and its changes, per capita GDP, added value of secondary industry and urbanization rate are key factors; for urban–rural income gap and its changes, per capita GDP, added value of secondary industry and floating population are key factors. The direct and factor interaction forces of the "important factors" work simultaneously, and the direct force must be greater than the average, otherwise a quite strong interaction force is required. For urban residents' income and its changes, urbanization rate, gross domestic product (GDP), floating population and number of Internet users are important factors; for rural residents' income and its changes, floating population, added value of secondary industry, number of Internet users and financial expenditure are important factors; for urban–rural income gap and its changes, urbanization rate, alleviating poverty policy, gross domestic product (GDP) and amount of bank loans are important factors. The direct forces of "auxiliary factors" are very weak and less than the average. It is important to note that the interaction forces of auxiliary factors, such as financial expenditure, total retail sales of consumer goods, total population, number of Internet users, main functional and area planning, are also prominent.

Figure 13. Driving mechanisms of urban–rural income gap.

From a comprehensive perspective, the factors exerting comprehensive influence on the income of urban and rural residents and its changes, and on the urban–rural income gap and its changes, are: per capita GDP, alleviating poverty policy and urbanization rate are key factors; floating population, added value of secondary industry and gross domestic product (GDP) are important factors; total retail sales of consumer goods, added value of tertiary industry, amount of bank loans, number of Internet users and financial expenditure are auxiliary factors. It should be noted that urbanization rate, added value of secondary industry, number of Internet users and financial expenditure are super interaction factors. In summary, the industrialization and urbanization stages constitute the core driving forces, with the factors of per capita GDP, urbanization rate, floating population, added value of secondary industry and GDP exerting the most prominent effect; government policies and information technology are the emerging driving forces, with the alleviating poverty policy showing a strong direct force, while the interaction effects of the number of Internet users and the financial expenditure should not be ignored; the tertiary industry and consumption vitality play a major auxiliary and supporting role, including the factors of added value of tertiary industry, total retail sales of consumer goods and amount of bank loans.

Some of the findings in this paper have corroborated certain points of view in existing papers; however, there are a few points of view that are different or even contrary to the past studies. These new findings are of great value in complementing and improving the driving mechanisms and evolutionary laws of the urban–rural income gap. Chen [62] conducted an empirical study on 31 Chinese provinces from 1978 to 2019 and found a two-way causal relationship between urbanization and the urban–rural income gap; Ma [63] found that urbanization development narrowed the urban–rural income gap after an empirical study on China from 1978 to 2014. Their findings are in agreement with the empirical results of this paper. Zhang [64] found, based on an empirical study of 31 Chinese provinces from 1978 to 2006, that per capita GDP is in an inverted U-shaped relationship with the urban–rural income gap, but financial development and the size of government spending have widened the urban–rural income gap. Liu [65] concluded that fiscal policy is negatively related to the urban–rural income gap, while positively related for the compactness of urban spatial form. Nguyen [66] investigated the effect of international integration on the urban–rural income gap in Vietnam based on a regression model and concluded that exports, GDP and urban–rural income gap are negatively related, while indicators such as FDI, per capita GDP and percentage of Internet users are positively related to income inequality. Fesselmeyer's [67] research, based on the econometric decomposition method, found that government investment policies, price manipulation incentives and education are the principal factors widening the urban–rural income gap in Vietnam, especially government policies to generate gains for urban residents at the expense of rural areas, which better supports Lipton's urban-bias hypothesis. That is, the government directs resources from rural to urban areas under strong political pressure from urban populations, without regard to efficiency or equity. Their findings are in disagreement with the empirical results of this paper. The differences between Zhang and Liu may be influenced by scale effects and geographical differences, and the differences between Nguyen and Fesselmeyer may be influenced by country context, government operating mechanisms and investment patterns.

It should be noted that we find that the Internet has become an emerging force, and its interaction forces with factors such as per capita GDP, mobile population and urbanization are particularly prominent. For all dependent variables, the interaction force of the Internet with per capita GDP is generally greater than 0.9, and the interaction force with floating population is mostly greater than 0.8. The interaction force between the Internet and urbanization is prominent in terms of residents' income and its changes; however, the interaction force between the Internet and urbanization is quite weak in terms of urban–rural income gap and its changes. Li [68] conducted an empirical test based on the panel data of 11 prefecture-level cities in Zhejiang Province, China, from 2011 to 2018, and found that there is an inverted U-shaped relationship between e-commerce development and urban–rural income gap. Zhejiang is still on the left of the inverted U-shaped curve, indicating that the development of e-commerce is gradually aggravating the urban–rural income gap. Our difference with Li may be influenced by scale effects and geographical differences. In any case, its value is to inspire the government to further create more favorable conditions for the spread of rural Internet, especially e-commerce, and to prompt the Internet to play a greater role in narrowing the urban–rural income gap. In addition, Zhao [69] concluded that the growth of productive services helps to narrow the urban–rural income gap, especially communications, transport, warehousing and logistics, which have the greatest impact, but the impact of information services and software development is insignificant. This complements our research—we find that consumption vitality and consumer services play a major auxiliary and supporting role.

5.3. Early Warning System and Differentiated Policy Design

We grade and classify spatial risks based on the BCG model and propose differentiated policy design according to the driving mechanisms identified in the previous section. Based on equation (4), the relative share of each city and their average value can be calculated, where the average value of urban residents' income is 0.63, the average value of rural residents' income is 0.39 and the average value of the urban–rural income gap index is 0.68. Based on equation (5), the growth rate and its average value can be calculated for each city from 2016 to 2019. The maximum growth rate of urban residents' income in Gansu Province was 18.95% (Yuzhong) and the minimum was 6.48% (Huixian), with the average of 8.35%; the maximum growth rate of rural residents' income was 10.37% (Dangchang) and the minimum was 7.18% (Dunhuang), with the average of 9.18%; the maximum growth rate of urban–rural income gap index was 8.62% (Yuzhong) and the minimum was −3.04% (Huixian), with the average of −0.76%. Based on the BCG matrix and taking the average value of relative share and growth rate as the threshold, we classify cities into four types: stars, cows, question and dogs (Figure 14). Due to the large difference in the spatial risk of residents' income and its changes, urban–rural income gap and its changes in county-level cities in Gansu Province, the "one-size-fits-all" approach should be avoided in spatial governance.

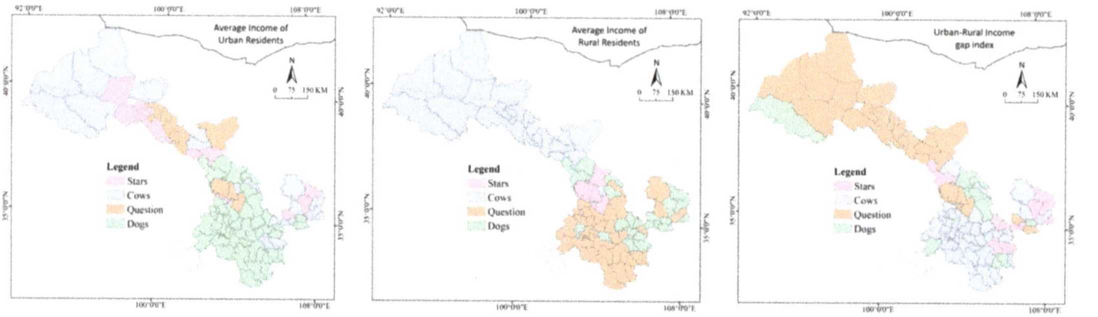

Figure 14. Policy area map of Gansu province.

In terms of urban income and its changes: The relative share and growth rate of the stars-type cities are larger than the average, characterized by "double high", indicating that these cities are in the best state—urban residents' income is very high and grows fast. The cities of the stars-type are scattered in spatial distribution, including Yumen, Sunan, Suzhou, etc. The relative share of the cows-type cities is greater than the average, while the growth rate is lower than the average, characterized by "high-low", indicating that the growth has slowed down although the income in these cities is high. Therefore, it is necessary to find the reasons for the lower growth rate and make the policy design appropriate to the situation. The cities of the cows-type are mainly concentrated in the northwest and northeast corners of Gansu in spatial distribution, including Guazhou, Subei, Dunhuang, Aksay, Heshui, etc. The relative share of the question-type cities is smaller than the average, but the growth rate is greater than the average, characterized by "low-high", indicating that the growth is fast although the income is low, with a good future development prospect in these cities. There are two areas where the cities of the question-type are concentrated: those in the provincial capital concentration area include Yongdeng, Gaolan and Yuzhong, and those in the concentration area of west corridor of the Yellow River (Zhangye section), including oasis cities such as Gaotai, Linze, Ganzhou, Shandan and Minle. The relative share and growth rate of the dogs-type cities are both smaller than the average, characterized by "double low", indicating that these cities are in the worst condition and must be the focus of policy adjustment. The cities of the dogs-type are concentrated in the east region of the Yellow River, including all counties of Baiyin, Dingxi, Linxia, Gannan, Longnan, etc. For the policy design for cows-, question- and dogs-type cities, per capita GDP, added value of secondary industry and alleviating poverty policy must be at the core, and it is required to give full play to urbanization rate, floating population and other factors.

In terms of rural income and its changes: The relative share and growth rate of the stars-type cities are larger than the average, characterized by "double high". The cities of the stars-type are concentrated in the Lanzhou-Baiyin metropolitan area, including Jingtai, Yongdeng, Gaolan, etc. The relative share of the cows-type cities is greater than the average, while the growth rate is lower than the average, characterized by "high-low". The cities of the cows-type are mainly concentrated in the spatial distribution in the west corridor of the Yellow River, including all the counties of Jiayuguan, Jiuquan, Zhangye and Jinchang, etc. The relative share of the question-type cities is smaller than the average, but the growth rate is greater than the average, characterized by "low-high". The cities of the question-type are concentrated in the area of Dingxi, Linxia, Gannan and Longnan in the east region of the Yellow River. The relative share and growth rate of the dogs-type cities are both smaller than the average, characterized by "double low", indicating that these cities are in the worst condition and must be the focus of policy adjustment. The cities of the dogs-type are mainly concentrated in the area of Pingliang, Qingyang and Tianshui in the east region of the Yellow River, and it should be noted that Tianzhu, Gulang, Pingchuan, Jingyuan and Hezuo are also of this type. For the policy design for the cities of the cows-, question-, and dogs-type, per capita GDP, alleviating poverty policy and urbanization rate must be at the core, and it is required to give full play to factors such as floating population, gross domestic product, added value of secondary industry, number of Internet users and amount of bank loans.

In terms of urban–rural income gap and its changes, the relative share and growth rate of stars-type cities are greater than the average, characterized by "double high". These cities are scattered in spatial distribution, including Aksay, Jingtai, Jingyuan and Huining, which must be the focus of policy adjustment. The relative share of the cows-type cities is greater than the average, while the growth rate is lower than the average, characterized by "high-low". Most of these cities are concentrated in the east region of the Yellow River, including Dingxi, Longnan, Linxia and Gannan. Accelerating the income growth of rural residents must be the focus of their future policy regulation. The relative share of the question-type cities is smaller than the average value, while the growth rate is higher than the average, characterized by "low-high". These cities are scattered in spatial distribution, including Aksay, Jingtai, Jingyuan, Baiyin, Huining, etc. They must find out the reasons for the rapid growth of the urban–rural income gap index and make the regulation and control policy design appropriate to the situation. The relative share and growth rate of dogs-type cities are both lower than the average, characterized by "double low", indicating that these cities are in the most ideal condition—the urban–rural income gap index is low and grows slowly. They are mainly concentrated in the provincial capital and the west corridor of the Yellow River. Lingtai and Kongtong are also cities of this type. How to maintain the current development status of these cities is the key point of future regulation policy design. For the policy design for cows-, question-, and dogs-type cities, per capita GDP, added value of secondary industry and floating population must be the core, and it is required to give full play to the factors such as urbanization rate, alleviating poverty policy, amount of bank loans, number of Internet users and main functional area planning.

6. Conclusions

The spatial patterns of residents' income and its changes, as well as urban–rural income gap and its changes, are multidimensional in nature, and different conclusions or new findings can be reached from different dimensions, such as economy, society, culture, geography and planning. Based on the geospatial variability and correlation of urban–rural residents' income and its changes and the urban–rural income gap and its changes, this paper provides an analysis of the spatial patterns of county-level cities in Gansu Province and their driving mechanisms, as well as spatial early warning and governance zoning maps, by integrating economic, social, policy and other influence factors, with the help of GIS, GeoDetector and BCG models. Our findings are as follows: (1) the urban and rural residents' income and its changes and the urban–rural income gap and its changes in Gansu Province are low in spatial heterogeneity, but high in spatial correlation and agglomeration, (2) the spatial patterns of urban and rural residents' income, income gap and their changes in Gansu Province are stepwise and solidified, with urban and rural residents' income and its changes characterized by southeast–northwest gradient escalation, and the urban–rural income gap and its changes characterized by southwest–northeast gradient changes, (3) the influence of different factors varies greatly, for example, factors such as per capita GDP, alleviating poverty policy and the urbanization rate have the most prominent force, followed by those with nonnegligible forces, such as floating population, added value of secondary industry and the number of Internet users, (4) the influence factors are increasingly diversified, and according to the ranking and average of the forces, the influence factors are divided into three types of "key factors", "important factors" and "auxiliary factors", with "key factors" being dominated by direct driving forces, "important factors" combining direct and interaction forces and "auxiliary factors" mainly depending on indirect forces, (5) the driving mechanism is becoming increasingly complex, with factor pairs dominated by bifactor enhancement effects in terms of urban and rural residents' income and its changes, and dominated by non-linear enhancement effects in terms of urban–rural income gap and its changes, (6) the suggestion of "graded spatial warning and differentiated policy design" is made and 84 county-level cities in Gansu Province are classified into four levels of high, low, potential and no risk areas, and four types of stars, cows, question and dogs. It is suggested that the government should accordingly develop adaptive and precise policy design and carry out spatial governance to narrow the urban–rural income gap while rapidly increasing the income of urban and rural residents.

Theoretically, this study provides a new research framework and method for researchers in macroeconomics, human geography, land management and spatial planning to study the spatial patterns of the incomes of urban and rural residents and the urban–rural income gap and their changing trends, which will be helpful to reveal the spatial evolution patterns of urban and rural residents' income and their governance mechanisms. Practically, this study helps urban policy makers and decision makers find scientific and reasonable measures to raise the income of urban and rural residents and narrow the income gap between urban and rural areas, provides a basis for decision making in the design of governmental urban and rural spatial management policies and provides valuable references for spatial planners in planning design, especially planning for rural revitalization. In any country or region, raising the income of urban and rural residents is the core task of the government, and narrowing the urban–rural income gap is also a challenge that the government must face in development. Small cities, located at the bottom of the pyramid in the national and regional town system, are large in number and play an important part, so they are vital hubs and key carriers to achieve the integrated development of urban and rural areas. The research methodology and findings in this paper not only provide policy recommendations for small cities in less developed regions of China to push urban–rural integration, but also offer valuable references for small cities to solve the problems of residents' income increase and the urban–rural income gap in less developed countries and regions that are undergoing rapid industrialization and urbanization, such as Pakistan, Uzbekistan, Kazakhstan, Turkmenistan, Iran, Syria, Iraq, Turkey, Egypt, Libya and Algeria. However, there are some shortcomings in our study. For example, there is a lack of comparative studies of prefecture-level cities and provincial regions at different scales, leading to the fact that the precision and applicability of some findings in this paper may be affected to some extent.

Author Contributions: Conceptualization, P.Z.; methodology, P.Z. and W.L.; software, W.L. and K.Z.; validation, P.Z. and W.L.; formal analysis, P.Z. and S.Z.; investigation, P.Z. and W.L.; resources, P.Z.; data curation, W.L. and P.Z.; writing—original draft preparation, P.Z. and W.L.; writing—review and editing, P.Z. and S.Z.; visualization, S.Z.; supervision, K.Z. and S.Z.; project administration, P.Z.; funding acquisition, P.Z. All authors have read and agreed to the published version of the manuscript.

Funding: This paper and the related research are financially supported by the National Natural Science Foundation of China (Project Number: 51768029), the National Natural Science Foundation of China (Project Number: 51768030), and the Natural Science Foundation of Gansu Province (Project Number: 20JR5RA415).

Institutional Review Board Statement: Not applicable.

Informed Consent Statement: Not applicable.

Data Availability Statement: The data used in this paper mainly comes from the Gansu Development Yearbook, Gansu Rural Yearbook, China County Construction Statistical Yearbook and China Statistical Yearbook. Most of the data can be obtained by visiting the following links: http://tjj.gansu.gov.cn/tjj/c109464/info_disp.shtml (accessed on 12 May 2021), https://data.cnki.net/yearbook/Single/N2021040103 (accessed on 12 May 2021), http://www.stats.gov.cn/tjsj/ndsj/ (accessed on 12 May 2021).

Conflicts of Interest: The authors declare no conflict of interest.

References

1. Young, A. Inequality, the urban-rural gap, and migration. *Q. J. Econ.* **2013**, *128*, 1727–1785. [CrossRef]
2. Zhang, G.S.; Wang, F.Z.; Kang, J.Y.; Yang, H.; Ding, Z.W. Spatial pattern of urban-rural income gap and its influencing factors at county level in Yangtze river economic belt. *Econ. Geogr.* **2017**, *37*, 42–51. [CrossRef]
3. Attanasio, O.; Berloffa, G.; Blundell, R.; Preston, I. From earnings inequality to consumption inequality. *Econ. J.* **2002**, *112*, C52–C59. [CrossRef]
4. Krueger, D.; Perri, F. Does income inequality lead to consumption inequality? Evidence and theory. *Rev. Econ. Stud.* **2006**, *73*, 163–193. [CrossRef]
5. Binelli, C.; Loveless, M. The urban-rural divide. *Econ. Transit.* **2016**, *24*, 211–231. [CrossRef]
6. Burlacu, M.I. Income, occupational structure and standard of living in Romania's rural areas compared to urban areas. In Proceedings of the 28th IBIMA Conference on Vision 2020: Innovation Management, Development Sustainability, and Competitive Economic Growth, Seville, Spain, 9–10 November 2016; pp. 3187–3195.
7. Tamkoç, M.N.; Torul, O. Cross-sectional facts for macroeconomists: Wage, income and consumption inequality in Turkey. *J. Econ. Inequal.* **2020**, *18*, 239–259. [CrossRef]
8. Lise, J.; Sudo, N.; Suzuki, M.; Yamada, K.; Yamada, T. Wage, income and consumption inequality in Japan, 1981–2008: From boom to lost decades. *Rev. Econ. Dyn.* **2014**, *17*, 582–612. [CrossRef]
9. Heshmati, A.; Rudolf, R. Income versus consumption inequality in Korea: Evaluating stochastic dominance rankings by various household attributes. *Asian Econ. J.* **2014**, *28*, 413–436. [CrossRef]
10. Salvati, L. The dark side of the crisis: Disparities in per capita income (2000–2012) and the urban-rural gradient in Greece. *Tijdschr. Econ. Soc. Geogr.* **2016**, *107*, 628–641. [CrossRef]
11. Thein, A.T.Z.; Akita, T. Education and expenditure inequality in Myanmar: An analysis with the 2006 and 2012 ousehold income and expenditure survey in an urban and rural setting. *Reg. Sci. Policy Pract.* **2019**, *11*, 55–70. [CrossRef]
12. Bodjongo, M.J.M.; Sikod, F. Access to the urban markets and variation of rural farmers income of the informal sector in Cameroon. *Rev. Econ. Reg. Urbaine* **2017**, *2*, 357–375.
13. Su, D.L. The Present Situation, Reasons and Countermeasures of the Income Gap between Urban and Rural Areas in China. In *Proceedings of the 2018 International Symposium on Social Science and Management Innovation (SSMI 2018)*; Atlantis Press: Washington, DC, USA, 2018; Volume 68, pp. 379–383.
14. Zhao, Q.; Lei, C.K.; Peng, L.; Gu, X. The U.S.-Sino different links between income and consumption inequality. *Appl. Econ. Lett.* **2020**, *28*, 713–716. [CrossRef]
15. Gradín, C.; Wu, B. Income and consumption inequality in China: A comparative approach with India. *China Econ. Rev.* **2020**, *62*, 101463. [CrossRef]

16. Chen, X.; Zhang, X.; Song, Y.; Liang, X.; Wang, L.; Geng, Y. Fiscal decentralization, urban-rural income gap, and tourism. *Sustainability* **2020**, *12*, 10398. [CrossRef]
17. Shi, W.; Luo, M.; Jin, M.; Cheng, S.K.; Li, K.X. Urban-rural income disparity and inbound tourism: Spatial evidence from China. *Tour. Econ.* **2019**, *26*, 1231–1247. [CrossRef]
18. Kim, J.H.; Kang, K.H. The interaction effect of tourism and foreign direct investment on urban–rural income disparity in China: A comparison between autonomous regions and other provinces. *Curr. Issues Tour.* **2019**, *23*, 68–81. [CrossRef]
19. Li, Q.; Wu, X.; Zhang, Y.; Wang, Y. The effect of agricultural environmental total factor productivity on urban-rural income gap: Integrated view from China. *Sustainability* **2020**, *12*, 3327. [CrossRef]
20. Jin, H.; He, Y.X.; Zhao, C.H. Social security expenditure, demographic structural changes, and urban-rural income gap in China: Evidence from the provincial data. *Rev. Cercet. Interv. Soc.* **2018**, *62*, 254–277.
21. Wei, H.; Li, H.B.; Guo, Y. Trade structure, trade mode and the urban-rural income gap in China. *S. Afr. J. Econ. Manag. Sci.* **2013**, *16*, 96–114. [CrossRef]
22. Hong, M.; Zhang, W. Industrial structure upgrading, urbanization and urban-rural income disparity: Evidence from China. *Appl. Econ. Lett.* **2020**, *28*, 1321–1326. [CrossRef]
23. Wang, X.; Shao, S.; Li, L. Agricultural inputs, urbanization, and urban-rural income disparity: Evidence from China. *China Econ. Rev.* **2019**, *55*, 67–84. [CrossRef]
24. Zhang, M.; Zhang, G.; Liu, H. Analysis of the impact of tourism development on the urban-rural income gap: Evidence from 248 prefecture-level cities in China. *Asia Pac. J. Tour. Res.* **2021**, *26*, 614–625. [CrossRef]
25. Huang, Z.; Xu, H.; Li, J.; Luo, N. Has highway construction narrowed the urban–rural income gap? Evidence from Chinese cities. *Pap. Reg. Sci.* **2020**, *99*, 705–723. [CrossRef]
26. Thiede, B.C.; Butler, J.L.W.; Brown, D.L.; Jensen, L. Income inequality across the rural-urban continuum in the United States, 1970–2016. *Rural Sociol.* **2020**, *85*, 899–937. [CrossRef]
27. Kibriya, S.; Bessler, D.; Price, E. Linkages between poverty and income inequality of urban-rural sector: A time series analysis of India's urban-based aspirations from 1951 to 1994. *Appl. Econ. Lett.* **2018**, *26*, 446–453. [CrossRef]
28. Oyekale, A.S. Determinants of income inequality in urban and rural Nigeria: A regression-based approach. *Asia Life Sci.* **2012**, *8*, 127–142.
29. Sehrawat, M.; Giri, A.K. Financial development, poverty and rural-urban income inequality: Evidence from South Asian countries. *Qual. Quant.* **2015**, *50*, 577–590. [CrossRef]
30. Borodkin, L.; Granville, B.; Leonard, C.S. The rural/urban wage gap in the industrialisation of Russia, 1884–1910. *Eur. Rev. Econ. Hist.* **2008**, *12*, 67–95. [CrossRef]
31. Chotia, V.; Rao, N.; Wagner, N.; Narayan, P. Investigating the interlinkages between infrastructure development, poverty and rural-urban income inequality. *Stud. Econ. Finance* **2017**, *34*, 466–484. [CrossRef]
32. Vafaei, A.; Rosenberg, M.W.; Pickett, W. Relationships between income inequality and health: A study on rural and urban regions of Canada. *Rural Remote Health* **2010**, *10*, 1430. [CrossRef]
33. Sehrawat, M.; Giri, A. Panel data analysis of financial development, economic growth and rural-urban income inequality. *Int. J. Soc. Econ.* **2016**, *43*, 998–1015. [CrossRef]
34. Li, W.; Wang, X.; Hilmola, O.-P. Does high-speed railway influence convergence of urban-rural income gap in China? *Sustainability* **2020**, *12*, 4236. [CrossRef]
35. Wang, S.; Tan, S.; Yang, S.; Lin, Q.; Zhang, L. Urban-biased land development policy and the urban-rural income gap: Evidence from Hubei Province, China. *Land Use Policy* **2019**, *87*, 104066. [CrossRef]
36. Su, C.-W.; Song, Y.; Ma, Y.-T.; Tao, R. Is financial development narrowing the urban-rural income gap? A cross-regional study of China. *Pap. Reg. Sci.* **2019**, *98*, 1779–1800. [CrossRef]
37. Batabyal, A.A.; Beladi, H. Preference matching, income, and population distribution in urban and adjacent rural regions. *Pap. Reg. Sci.* **2019**, *98*, 2201–2208. [CrossRef]
38. Amara, M.; Ayadi, M.; Jemmali, H. Rural-urban migration and income disparity in Tunisia: A decomposition analysis. *Pap. Reg. Sci.* **2018**, *98*, 1053–1083. [CrossRef]
39. Zhao, S.D. The reasons for the income gap between urban and rural residents in China. In Proceedings of the 2017 3rd International Conference on Social Science and Higher Education, Sanya, China, 28–30 September 2017; pp. 275–278.
40. Zhu, S.; Yu, C.; He, C. Export structures, income inequality and urban-rural divide in China. *Appl. Geogr.* **2020**, *115*, 102150. [CrossRef]
41. Chen, C. The impact of foreign direct investment on urban-rural income inequality. *China Agric. Econ. Rev.* **2016**, *8*, 480–497. [CrossRef]
42. Guan, X.Y.; Wang, S.L.; Gao, Z.Y.; Lv, Y.; Fu, X.J. Spatio-temporal variability of soil salinity and its relationship with the depth to groundwater in salinization irrigation district. *Acta Ecol. Sin.* **2012**, *32*, 198–206.
43. Zhang, R.F. *Theory and Application of Spatial Variability*; Science Press: Beijing, China, 2005; pp. 13–14.
44. Ruan, B.Q.; Xu, F.R.; Jiang, R.F. Analysis on spatial and temporal variability of groundwater level based on spherical sampling model. *J. Hydraul. Eng.* **2008**, *39*, 573–579.
45. Liu, X.N.; Huang, F.; Wang, P. *Spatial Analysis Principle and Method of GIS*; Science Press: Beijing, China, 2008; pp. 199–206.

46. Miyamoto, S.; Chacon, A.; Hossain, M.; Martinez, L. Soil salinity of urban turf areas irrigated with saline water I. Spatial variability. *Landsc. Urban Plan.* **2005**, *71*, 233–241.
47. She, D.L.; Shao, M.A.; Yu, S.G. Spatial variability of soil water content on a cropland-grassland mixed slope land in the Loess Plateau, China. *Trans. Chin. Soc. Agric. Mach.* **2010**, *41*, 57–63.
48. Wang, J.F.; Li, X.H.; Christakos, G.; Liao, Y.L.; Zhang, T.; Gu, X.; Zheng, X.Y. Geographical detectors-based health risk assessment and its application in the neural tube defects study of the Heshun region, China. *Int. J. Geogr. Inf. Sci.* **2010**, *24*, 107–127. [CrossRef]
49. Chen, L.; Shen, W. Spatiotemporal differentiation of urban-rural income disparity and its driving force in the Yangtze River Economic Belt during 2000–2017. *PLoS ONE* **2021**, *16*, e0245961. [CrossRef]
50. Wang, J.-F.; Hu, Y. Environmental health risk detection with GeogDetector. *Environ. Model. Softw.* **2012**, *33*, 114–115. [CrossRef]
51. Zhao, S.; Yan, Y.; Han, J. Industrial Land Change in Chinese Silk Road Cities and Its Influence on Environments. *Land* **2021**, *10*, 806. [CrossRef]
52. He, C.; Du, H. Urbanization, inclusive finance and urban-rural income gap. *Appl. Econ. Lett.* **2021**. [CrossRef]
53. Li, Y.; Wang, X.; Zhu, Q.; Zhao, H. Assessing the spatial and temporal differences in the impacts of factor allocation and urbanization on urban-rural income disparity in China, 2004–2010. *Habitat Int.* **2014**, *42*, 76–82. [CrossRef]
54. Zhao, S.; Zhao, K.; Zhang, P. Spatial Inequality in China's housing market and the driving mechanism. *Land* **2021**, *10*, 841. [CrossRef]
55. Zhao, S.; Zhang, C.; Qi, J. The key factors driving the development of new towns by mother cities and regions: Evidence from China. *ISPRS Int. J. Geo.-Inf.* **2021**, *10*, 223. [CrossRef]
56. Yuan, Y.; Wang, M.; Zhu, Y.; Huang, X.; Xiong, X. Urbanization's effects on the urban-rural income gap in China: A meta-regression analysis. *Land Use Policy* **2020**, *99*, 104995. [CrossRef]
57. Yan, D.S.; Sun, W.; Chen, D.; Tong, W.T. Relationship between urbanization rate and urban-rural income gap in the Yangtze River Delta. *China Popul. Resour. Environ.* **2021**, *31*, 28–36. [CrossRef]
58. Pan, J.H. Spatiotemporal pattern of urban-rural income gap of prefecture level cities or above in China. *Econ. Geogr.* **2014**, *34*, 60–67. [CrossRef]
59. Ding, Z.W.; Zhang, G.S.; Wang, F.Z. Spatial-temporal differentiation of urban-rural income in Central Palins Region at different scales. *Geogr. Res.* **2015**, *34*, 131–148. [CrossRef]
60. Fu, Z.H.; Mei, L.; Liu, Y.J.; Zheng, R.M. Spatial pattern of urban-rural income disparity and its differentiation mechanism in 3 provinces of Northeast China. *Sci. Geogr. Sin.* **2019**, *39*, 1473–1483. [CrossRef]
61. Pan, J.H.; Shi, P.J. Spatial difference of coordinative development of urban and rural areas and its mechanism in Gansu Provience. *Arid. Land Geogr.* **2011**, *34*, 355–362.
62. Chen, Y.; Luo, P.; Chang, T. Urbanization and the urban-rural income gap in China: A continuous wavelet coherency analysis. *Sustainability* **2020**, *12*, 8261. [CrossRef]
63. Ma, X.; Wang, F.; Chen, J.; Zhang, Y. The income gap between urban and rural residents in China: Since 1978. *Comput. Econ.* **2017**, *52*, 1153–1174. [CrossRef]
64. Zhang, H.-W.; Chen, W.-G.; Zhang, J. Urban–rural income disparities and development in a panel data set of China for the period from 1978 to 2006. *Appl. Econ.* **2011**, *44*, 2717–2728. [CrossRef]
65. Liu, Y.; Long, C. Urban and rural income gap: Does urban spatial form matter in China? *SAGE Open* **2021**, *11*, 21582440211002215. [CrossRef]
66. Nguyen, T.L.H.; Nguyen, V.C. The impact of international integration on the inequality of income between rural and urban areas in Vietnam. *J. Asian Financ. Econ. Bus.* **2020**, *7*, 277–287. [CrossRef]
67. Fesselmeyer, E.; Le, K.T. Urban-biased policies and the Increasing rural-urban expenditure gap in Vietnam in the 1990s. *Asian Econ. J.* **2010**, *24*, 161–178. [CrossRef]
68. Li, L.; Zeng, Y.; Ye, Z.; Guo, H. E-commerce development and urban-rural income gap: Evidence from Zhejiang Province, China. *Pap. Reg. Sci.* **2020**, *100*, 475–494. [CrossRef]
69. Zhao, J.; Dong, X.; Dong, K. Can agglomeration of producer services reduce urban-rural income inequality? The case of China. *Aust. Econ. Pap.* **2021**. [CrossRef]

land

Article

Experimenting with Urban–Rural Partnerships for Sustainable Sanitation in India: Learning from Practice

Shubhagato Dasgupta * and Neha Agarwal

Centre for Policy Research, New Delhi 110021, India; nagarwal@cprindia.org
* Correspondence: shubhagato@cprindia.org

Abstract: Local government partnerships for producing services are ubiquitous in many countries. However, the approach has rarely been applied in India—likely owing to a history of centralized planning and independent urban and rural governance systems. Nonetheless, the country's transforming sanitation landscape could benefit from intergovernmental partnerships for scaling services with speed and efficiency. The ongoing national sanitation program has espoused the approach in theory but the body of practice to support its wide deployment is sparse. This paper critically reviews one of the first experiments with the approach for producing sanitation services in the Dhenkanal district, Odisha, India. We ask the question: what can Dhenkanal's case tell us about the challenges and opportunities for delivering sanitation services through local-level intergovernmental urban–rural partnerships in India? As part of our practice research, we supported the district government pilot the approach. The data, consultations, and observations underpinning the experiment form the basis of our insights. We find that the urban–rural partnership increased access to sanitation services among rural households within a short period, lowered service charges, and clarified institutional responsibilities. The experiment highlighted issues relating to planning, responsibility, accountability, and financing that need tackling in order to strengthen the model going forward. We recommend that evolving a definitive model(s) of intergovernmental partnerships would require experimenting with the approach in diverse institutional contexts and granting governments the flexibility to recreate and renegotiate the form of the partnership.

Keywords: intermunicipal cooperation; India; urban–rural partnership; sanitation; faecal sludge management; wastewater management; local governance; local government partnership

Citation: Dasgupta, S.; Agarwal, N. Experimenting with Urban–Rural Partnerships for Sustainable Sanitation in India: Learning from Practice. *Land* **2022**, *11*, 1021. https://doi.org/10.3390/land11071021

Academic Editors: Stephan Bartke and Sigrun Kabisch

Received: 25 May 2022
Accepted: 1 July 2022
Published: 5 July 2022

1. Introduction

India's sanitation landscape is undergoing a rapid transformation. In 2012, India was home to a population of 2.5 billion that lacked access to improved sanitation [1]. The incumbent national government launched a large-scale program, the *Swachh Bharat Mission* (SBM), to eliminate open defecation in 2014. The program aimed to provide subsidized toilets to households lacking a toilet—a whopping 67% of all rural households and 13% of all urban households, as revealed by the Census of India in 2011. The program has reportedly enabled the construction of 110 million and 6 million new rural and urban toilets, respectively, to date [2,3]. The low and slow availability of centralized sewerage systems in urban India and their infeasibility in rural India result in a high national dependence on on-site sanitation systems (Figure 1). The increase in the number of toilets without a commensurate expansion of sewerage systems has increased the dependence significantly [4].

On-site sanitation systems prevalent in India produce the need for faecal sludge management (FSM) systems, i.e., systems to ensure the evacuation and conveyance of faecal waste from on-site sanitation systems and its treatment and disposal (or recycle) off-site. Faecal waste accumulating in septic tanks and single pits must be emptied periodically and conveyed via emptying vehicles (typically vacuum trucks). The evacuated waste must be treated at facilities, such as a Faecal Sludge Treatment Plant (FSTP) before disposal. Over time, the proliferation of centralized sewerage systems may diminish the need for FSM in the bigger cities. However, low-density small towns and peri-urban and rural settlements require FSM to be able to achieve 'safely managed sanitation' in accordance with the Sustainable Development Goal 6 [5]. The challenge of ensuring proper FSM is not unique to India; developing countries across Asia and Africa confront it [6]. What may distinguish India is the sheer scale of the challenge: developing FSM systems to serve more than 255,000 *gram panchayats*, or rural local bodies, and upwards of 4000 cities and towns. It begs the question: how to scale-up sanitation services across India with speed and efficiency?

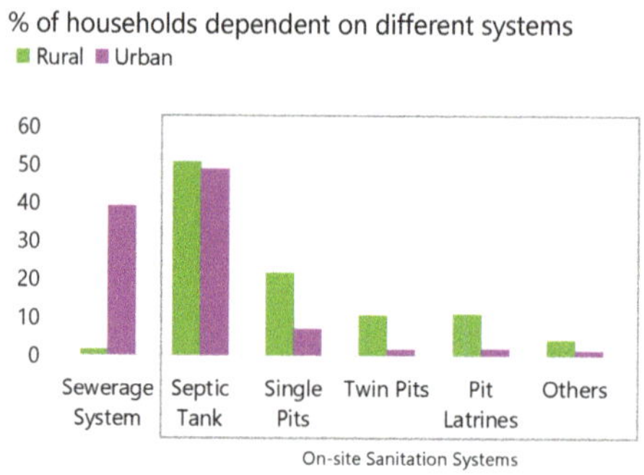

Figure 1. Dependence on different systems for managing wastewater.

Local government partnerships for the efficient production of services are common in several countries, including the United States of America (USA) and many more in Europe [7–11]. Commonly framed as 'intermunicipal cooperation', local government partnerships present an opportunity to achieve economies of scale for small local governments, especially when the alternatives of region-scale privatisation and political consolidation (e.g., in the form of metropolitan government) are infeasible or undesirable. Higher standards for services and raised citizen expectations are also proposed as imperatives for intermunicipal cooperation [12]. Cooperation can also help surmount the limitations of individual capacity and action in solving problems, such as environmental pollution, that transcend administrative boundaries. It can be formal or informal, hierarchy-based or network-based, or somewhere in the middle. For example, in the USA, cooperation can occur informally through councils of government that lack legal standing or, at its most formal, take the form of special districts for single functions (e.g., education, transport, etc.). Formal inter-municipal cooperation is also effectuated through 'empowered' counties that deliver services to all or a majority of the municipalities within their borders [7].

In contrast to its ubiquity in many countries, India has rarely applied the approach of intergovernmental partnership to produce services so far. A possible reason is that the production of many services, including sanitation, is effectuated through national programs (such as SBM). It may be that the phenomenon provides a weak incentive for local governments to partner and cooperate since national programs: (1) are bifurcated along the urban–rural divide, (2) predetermine local-level solutions, and (3) allocate financing to local governments individually. Nonetheless, in a new precedent, the national government espoused the approach of intergovernmental urban–rural partnerships (at the local level) for sanitation service delivery in 2021. The ongoing phase (2020–2025 in rural and 2021–2026 in urban) of SBM targets increasing the levels of wastewater management in both urban and rural India. The national government noted that many cities and towns already possess wastewater management systems (in the form of both FSTPs and the conventional sewage treatment plants) and that many more have such systems upcoming. It accordingly recommended that urban local governments in these cases extend their services to the neighbouring *gram panchayats* (vide its letter S-18011/6/2021-SBM-DDWS dated 14 September 2021).

The national-level mainstreaming of the intergovernmental partnership approach could be a step in the right direction. An early understanding of the challenges associated with its implementation could help evaluate its suitability and trajectory, specific to the Indian context. Towards this goal, in this paper, we discuss one of the first Indian experiments with an intergovernmental urban–rural partnership at the local level. The experiment of concern—predating the national mainstreaming—is situated in Dhenkanal district in the Indian state of Odisha. Under the experiment, the district utilized the urban–rural partnership approach to produce sanitation services for rural households.

In the following sections, we discuss our experience of supporting the government(s) conceptualize and implement a pilot model of urban–rural partnership in the Dhenkanal district. We ask the question: what can Dhenkanal's case tell us about the challenges and opportunities for producing sanitation services via local-level intergovernmental urban–rural partnerships in India? Since the case we discuss is the first-of-its-kind in India, the present paper may serve as a pioneering contribution to the body of work on urban–rural partnerships in India. We hope that it would also provide a jumping-off point for further research on, and experimentation with, the approach in India and regions with a similar institutional context and sanitation-related issues.

2. Background

2.1. Methods and Materials

The present paper is based on the authors' practice research on urban–rural partnerships in the Dhenkanal district of Odisha. The authors supported the district-level and local-level governments develop an experimental urban–rural partnership that would allow rural households to access sanitation services via urban infrastructural systems. The authors' direct participation in the entire process from the ideation and development of the partnership model to its implementation (during the period 2019–2021) has informed the paper. Over the course of the process, we interacted with and consulted officials of:

- The district government, or 'District Administration' (as it is called in India);
- The urban local government;
- The rural local governments, or '*gram panchayats*', shortlisted for forming the partnership with the urban local government(s).

The process commenced with an assessment of the sanitation landscape in the Dhenkanal district (Figure 2). To inform the assessment, we conducted a sample survey of 1000 rural households and structured interviews with political leaders, or *sarpanch(s)*, of eight rural local governments in 2020. The findings of the survey are discussed in the authors' previous work [13,14]. A variety of secondary data also informed the assessment and later steps of the process. It included:

- Geospatial data relating to administrative boundaries and the road networks;
- Demographic data from the Census of India 2011;
- Transactional data relating to emptying services and records of FSTP utilization from the urban local government (2019–2020).

The assessment was followed by analysis to identify the gram panchayats that were good fits for the urban–rural partnership. We presented the preliminary results of the analysis to the district administration and the relevant local governments for review. The local governments held further consultations to develop the terms of the partnership, including inter alia mechanisms for coordination, tariff design, and roles and responsibilities. On finalization of the terms, the urban local government and 17 *gram panchayats* signed a Memorandum of Agreement (MoA) to codify their partnership for service delivery on 28 December 2020. The signing was soon followed by multiple Information, Education, and Communication (IEC) campaigns aimed at raising awareness about services among rural households.

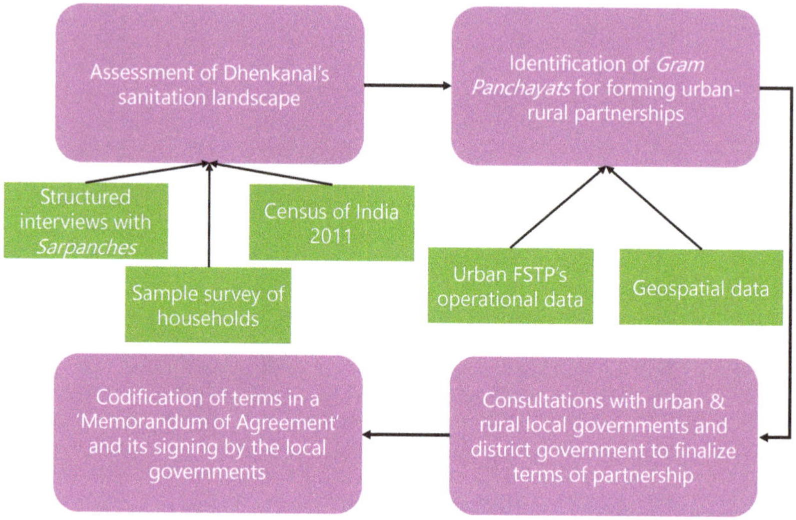

Figure 2. Process of developing the urban–rural partnership.

In addition to the communication with different government representatives, our direct and indirect observations of how the entire process unfolded forms an important basis of the paper. Complementing our practice research, we reviewed literature on more mature implementations of intergovernmental partnership to develop an understanding of what forms the partnerships can take (Figure 3). Our intent in doing so was to relate the findings from the Dhenkanal district to other cases. Given the paucity of its implementation in India, we relied primarily on literature discussing the European experience. The findings from our analysis discussed in Section 4 cluster around three main pillars: planning, responsibility and accountability, and financing.

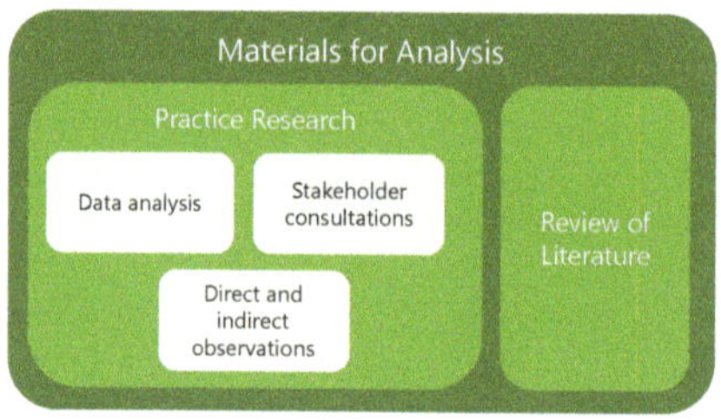

Figure 3. Materials forming the basis of analysis.

2.2. *Site of Enquiry*

Dhenkanal is one of the 30 districts in the coastal state of Odisha in India (Figure 4). A district in India is an administrative division at the state level and comprises multiple urban and rural local bodies. The Dhenkanal district has 216 local bodies-four urban local bodies, viz., the eponymous Dhenkanal, Bhuban, Hindol, and Kamkhyanagar, and 212 *gram panchayats*. These 212 *gram panchayats* cumulatively house 1237 villages. Only 9% of the district's total population of approximately 1.2 million resides in towns and the vast majority of 91% in *gram panchayats*. The four towns are small and do not exhibit urban primacy; the largest among the four, the Dhenkanal municipality, houses a population of about 67,000. Overall, the district is spread out over an area of 4452 square kilometres and has a population density of 268 persons per square kilometres.

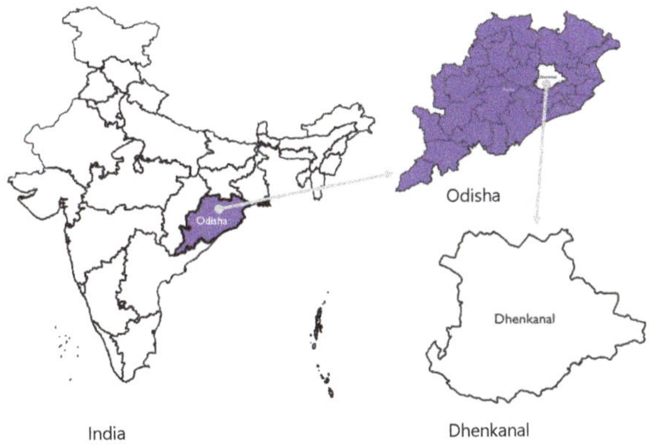

Figure 4. Site of enquiry.

The Dhenkanal municipality in the district is one of the first small towns in the country to develop an urban FSM system. The municipality's FSM system became operational in October 2018 under a pilot project focused on urban FSM called Project Nirmal. The authors' organisation had served as a knowledge partner for Project Nirmal. Under Project Nirmal, the municipality undertook town-level sanitation planning, procured emptying vehicles, and constructed a nature-based FSTP. It also systematized recordkeeping of the continuous operational data generated by the FSM system.

Following the implementation of SBM during 2014–2019, the share of rural households in the district with an individual toilet climbed from 18% to a purported 100%. Addressing the lack of proper systems for managing faecal waste beyond the toilets emerged as an important issue at the end of the program. The high share of rural population in the district lent the issue especially high importance. The second and ongoing phase of SBM also created an external imperative to solve for rural sanitation, and the Odisha Rural Sanitation Policy, issued in 2020, provided further impetus.

The availability of a functional and determinable FSM system, as well as the scope of the rural sanitation challenge made the Dhenkanal district a suitable site for piloting the approach. We discussed the pilot with the Dhenkanal's district administration and later worked with the Dhenkanal municipality and its neighbouring rural local governments to implement it.

3. Overview of Key Concepts

3.1. Local Governance in India

Most of India's population resides in rural areas, although the share has been declining steadily over time (~89% in 1901 to ~69% in 2011). As per the Local Government Directory 2021, rural and urban areas in the country comprise a little over 255,000 and 4700 local governments, respectively [15]. Local governments in India are tasked with the central role in "the provision of public services, the creation and maintenance of local public goods, and the planning and implemental of developmental activities and programs" [16]. They are part of a multi-tiered governance structure that is vertically bifurcated along the urban–rural divide. The main tiers are: national or central, state, district, and local, where the district provides a singular point of convergence between urban and rural governance.

At the national level, the Ministry of Housing and Urban Affairs steers all aspects of urban development, including sanitation and wastewater management. On the other hand, the responsibility for rural development is split between the Ministry of Rural Development and the Ministry of Jal Shakti; here, the latter administers programmes for improving rural sanitation. State-level administrations similarly house separate urban and rural departments. In Odisha's case, these are the Housing and Urban Development Department and the (rural) Panchayati Raj and Drinking Water Department.

Until the passage of the 73rd and 74th amendments in 1992, the Indian Constitution recognized states as the only official subnational units. The two amendments together accorded urban and rural local governments constitutional status [16]. The objective of the amendments was "decentralization and enduring popular participation in planning, management, and delivery of civic services" [17]. Schedule XI and XII of the Indian Constitution defined the scope of responsibilities of rural and urban local governments, respectively. The two schedules placed sanitation squarely within the purview of the local government (Figure 5). Further, the amendments directed the states to devolve powers and resources to local governments to allow the latter to fulfil their new responsibilities [16]. Article 243ZD of the 74th amendment also mandated the creation of District Planning Committees (DPCs) that would create district-level integrated development plans by consolidating plans developed by all the urban and rural local governments in a district. DPCs, at least in theory, have been tasked with identifying and facilitating joint development of projects that are of common interest to urban and rural local governments [18].

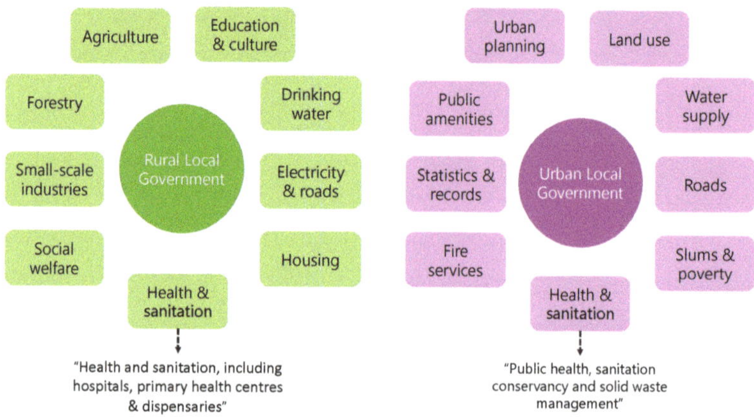

Figure 5. Indicative list of responsibilities of rural and urban local governments as per Schedule XI and XII of the Indian Constitution.

Despite its strong legal underpinnings, decentralization has been erratic in India. Studies conducted in the aftermath of the amendments observed a resistance to administrative and fiscal decentralization within rural governance. The functions for rural local governments may have been broadly specified as per the schedules, but an unclear definition of responsibilities and lack of resources for their fulfilment have inhibited local action and ownership over outcomes [16,19]. Rural local governments have played a limited role in planning, instead serving as local-level implementing agencies for programs and schemes determined and designed at the national and state levels [16]. Urban local governments purportedly confront similar challenges; state-level departments and parastatal agencies can diminish the urban local government's role in planning and, sometimes, even in the management of infrastructure and services [20]. Similarly, DPCs are absent or non-functional in most states; where they do function, they have failed at enabling urban–rural linkages and developing integrated plans [16,18]. Arguably, with local governments playing only a limited role in planning (if at all), the role of DPCs as an aggregator of local-level plans naturally diminishes too.

3.2. Sanitation in India

In the decades following independence, governance and developmental planning in India had been highly centralized [21]. Planning occurred via five-year plans that the national government's 'Planning Commission' developed and that the states were responsible for implementing. Sanitation has been articulated as an issue of national importance since the issue of the first five-year plan (1951–1956). The first plan recommended either one of individual or shared toilets, both preferably water-borne, for different types of urban housing and sanitary latrines for rural housing. Without being specific, it also emphasized the importance of "arrangements for the disposal of sewage" in relation to the latter. However, the national focus on sanitation culminated into a program only with the Central Rural Sanitation Programme (CRSP) in 1986. Subsequent efforts to improve sanitation have been channelled through large-scale national programmes underwritten by a mix of national and state-level financing.

Over several decades, successive programmes have targeted furthering access to toilets among rural households. The underlying principle of the programmes has oscillated from supply-driven and subsidized toilet construction to that which is community-led and demand-driven. Nonetheless, the spate of programs had reported low success in the past. Factors cited for toilet disuse include poor quality construction, fear of pit overflowing, and lack of knowledge about maintenance [22]. The latest programme, SBM, brought an explicit focus on Information, Education, and Communication (IEC) like its recent predecessors, but many old issues appear to persist; rural households in Odisha have cited the small size and subsequent filling up of pits as a reason to not use toilets regularly [13,23].

Regardless, in its ongoing second phase (2020–2025), SBM has shifted the focus from the construction of toilets to the safe and complete management of faecal waste beyond toilets. FSM has emerged as an important option in this regard given the high prevalence of septic tanks and single pits in rural areas, as previously noted. The recent global mainstreaming of FSM as a lower-cost alternative to sewerage systems in specific urban settings has also led to its greater acceptance in urban India [24,25]. The concomitant emergence of the two phenomenon–an increased need for FSM in rural areas and its increased adoption in urban-has produced a kind of convergence between urban and rural sanitation.

3.3. Intermunicipal Cooperation

The varied models of intermunicipal cooperation arise under differing contexts of national institutional histories, socio-techno-economic landscapes, local bodies' sizes and competencies, and goals. Four different models of intermunicipal cooperation have been proposed: (1) quasi-regional governments, (2) planning forums, (3) service delivery organisations, and (4) service delivery agreements [12]. Each model presents unique opportunities and challenges, and although it is debatable whether cooperation can sustain efficiency gains in the long run, it has been shown to at least able to improve service coverage and quality by overcoming scale-related obstacles [7,26,27]. In addition, an intensifying focus on urban sustainability and transition in recent times has provided a new impetus for shared urban governance [28].

4. Results

4.1. Planning

The main question for planning was how to size the extended service area for the urban FSM system. In the present case, it translated to: how many and which *gram panchayats* are apt for forming a partnership with the Dhenkanal municipality. We first considered supply-side constraints. Given that service delivery to the additional rural households had to utilize the existing urban FSM systems, the ability of the system to cater to additional households was one obvious factor. Although an FSM system, in general, is more modular than a centralised sewerage system, the costs of augmenting capacity may not always be insignificant. The municipality was amenable to expanding the vehicle fleet but increasing the capacity of an FSTP, requiring capital financing and land, was deemed infeasible. As a result, the available spare capacity of the FSTP provided a hard constraint for determining the service area.

A second important factor we considered on the supply side was the average distance of the gram panchayat from the urban local body. The delivery of services entails a roundtrip of the emptying vehicle and fuel has shown to be the largest cost driver of emptying services [29,30]. Even if the FSTP had infinite capacity, the exorbitant costs of longer trips would constrain the service area that is economically feasible to serve. Therefore, the identification of the suitable *gram panchayats* had to strike the right balance between the two factors.

The FSTP in the Dhenkanal municipality has a capacity of 27 kilolitres per day. Approximately 50% of the capacity (daily average) was being utilized in the first two years of its operation, with no clear year-to-year rise in utilization. Equally importantly, the records showed that the urban local body had been serving requests from rural households outside its periphery since before the commencement of the pilot. The arrangement was informal, market-led, and imposed a service charge on rural households that was 1.5–2 times of that paid by urban household per roundtrip of the vehicle. Over the period of January 2019 to February 2020, rural households constituted 13% of all households served. The relative share of the types of on-site sanitation systems emptied differed between urban and rural households (Figure 6). Overall, the analysis did establish that the FSTP in the Dhenkanal municipality was well-suited to serving rural households in addition to the urban jurisdiction.

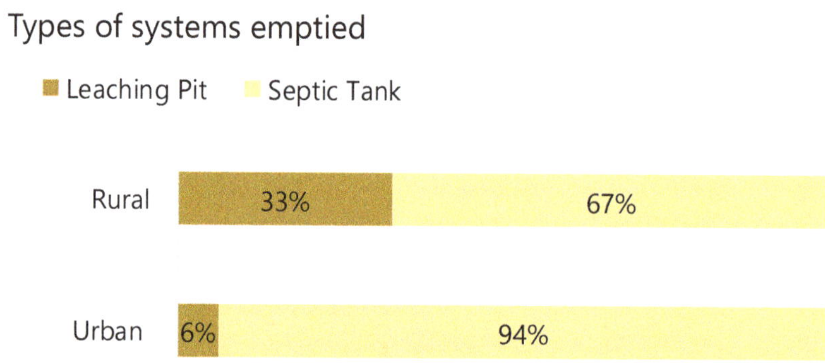

Figure 6. Relative share of different types of on-site sanitation systems emptied.

Once the magnitude of supply was established, we aimed to match it with the demand for services based on the household survey. The household survey presented three important considerations:

- The majority of households owned on-site sanitation systems (viz., septic tanks and single pits) that would require FSM over their lifetime of operation;
- A significant share of households that had emptied their on-site sanitation systems in the past reported utilizing the cheaper (almost by 50% on average) but unsafe alternative of manual emptying of the system.

The characteristics of on-site sanitation systems revealed by the survey helped estimate the number of rural households that would fit within the available spare capacity. We started with an initial arbitrary radius of 10 km and listed all *gram panchayats* falling within the perimeter. Based on the number of circumscribed rural households, the type of and characteristics of on-site sanitation systems, and the reported trends in toilet usage, we estimated that the FSTP had just enough spare capacity to serve the *gram panchayats* falling within 10 km for the next five years [31]. However, determining which *gram panchayats* 'fell within 10 km' was not straightforward.

Unlike most urban local bodies, *gram panchayats* lack a single unified landmass bound by an unbroken perimeter. One or more villages constitute a *gram panchayat*, and one or more hamlets constitute a village. The different villages and hamlets can abut each other or be separated by large tracts of land (Figure 7). Of these villages, one of the villages—

generally the largest—houses the rural local government's office and serves as the seat of the administration. The distance of a single village or hamlet from the urban local body can be greater or lesser than the average distance between the rural local body as a whole and the urban local body.

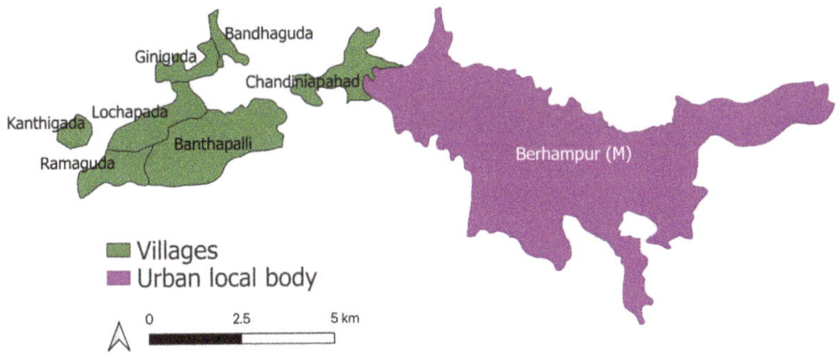

Figure 7. An example of a gram panchayat, Banthapalli (Ganjam district) with non-contiguous villages.

Therefore, we had to establish an unambiguous criterion to when a *gram panchayat* can be said to meet the distance threshold. What happens when some villages of the *gram panchayat* qualify for an urban–rural partnership by falling within the threshold and some do not? Does a *gram panchayat* only qualify if all of its constituent villages qualify? Or does it quality if a certain proportion of its villages, e.g., 50%, 75%, etc., qualify? Should the urban local body serve the entire *gram panchayat* even if one of the villages qualifies for an urban–rural partnership? These questions may appear of minor consequence but were important to consider for ensuring the long-term viability of the urban–rural partnership for three main reasons.

First, the *gram panchayat* is accountable for all the villages in its jurisdiction regardless of the distance. Although an urban–rural partnership is not the only approach to serving rural households, it is faster than the alternative of building greenfield rural FSM systems from the ground up. If only the villages individually qualifying are served under the urban–rural partnership, the remaining would be left relying on the status quo of no services or unreliable services in the short term. The *gram panchayats* could find explaining the resulting institutionally sanctioned unequal access to services difficult to justify to their jurisdictions.

Second, as previously noted, fuel is shown to be the largest cost driver of emptying services. Therefore, failing to predict how much distance the emptying vehicles would need to travel to serve a particular rural local body with reasonable accuracy could lead to a misestimation of operating costs and a failure in setting tariffs that help achieve cost recovery.

Third, the novelty of FSM as a concept in rural India has meant that *gram panchayats* have limited initial capacities for its management. The discourse on increasing access to toilets is not new to rural India owing to the long history of national sanitation programs focused on increasing access to toilets in rural areas. However, the issue of how to manage faecal waste safely beyond the toilet in the absence of centralized sewerage systems gained mainstream traction only over the last decade in India and globally [13,24,32]. Expectedly, *gram panchayats* are still learning about FSM. Therefore, holding them responsible for managing multiple FSM models within a single jurisdiction could make strenuous and unsustainable demands on their still-developing capacities.

4.2. Responsibility and Accountability

The fact that accountability improves the quality of services and increases citizen welfare is well-established [33,34]. In rural India, the *sarpanch* and their team of ward members are the politically accountable functionaries of the local rural government and are empowered to steer development in the jurisdiction they serve. However, the urban–rural partnership complicates the lines of accountability. Although rural citizens receive services from the urban local government, they cannot hold it accountable. They can hold only their own local government accountable, but rural local governments cannot, in turn, seek accountability from their urban counterparts under conventional rules of governance.

A Memorandum of Agreement (MoA) codified the terms of the urban–rural partnership early on, but later experience shows that an MoA may not be enough. As noted earlier, the MoA signed by the urban and local governments specified their roles and responsibilities at the broadest level of detail, which was enough to get the partnership off the ground. It responds to questions such as: Who will deliver services? Who will set prices? Who will pay for services? Who will maintain records? It specifies who will do what but not what processes or standards to adopt while doing something. For example, it states that "any dispute or disagreement .... shall be settled through mutual consultations and negotiations". However, in this case, the inherent form of urban–rural partnership set up a differential in leverage that could result in unfair decision-making and resource allocation. The MoA, in its existing form, does not address or correct the power imbalance.

Many other similar questions emerged. For example, if rural households are denied services despite the urban–rural partnership, who do they hold to account—the rural local government or the urban? If the quality of services delivered to rural households is poor or inferior to that received by urban households, can rural households or their local government hold the urban local government accountable? How do we define poor? What happens when the 'Project Review Committees' (set under the MoA and comprising all project partners, including representatives from different tiers of government) expire? Under what rules of engagement do the urban and rural governments convene to "coordinate and address issues and facilitate smooth delivery of FSM services" like the MoA requires? If urban and rural local governments are in dispute, who arbitrates?

4.3. Financing

Urban–rural partnerships are known to produce efficiency gains. In the present case, utilising spare capacities of existing infrastructure for the partnership has helped rural local governments avoid the expenditure resulting from the development of greenfield rural FSM systems. However, who benefits from these gains? Currently, the fee charged for emptying services only reflects the operational costs of providing the services. The tariff design does not account for the capital cost of vehicles or the capital and operating cost of treatment facilities. Whatever resultant charge urban households pay, rural households pay a distance surcharge in addition to it—although the surcharge is 50% lower than what it was before the formalization of the partnership. It was lowered because of the consultations between the urban local government and the participating *gram panchayats*, where the latter advocated for more affordable service charges.

From the point of cost recovery, the surcharge is logical. Still, it results in rural households paying more for the same level of service as their urban counterparts, despite typically earning lower incomes [35,36]. At the time of writing, rural households pay the base fee and surcharge in full and do not benefit from any subsidies. Since neither rural nor urban households pay for the recovery of capital infrastructure and assets, is it okay for the government to capture all the capital savings resulting from urban–rural partnerships? Or should the government utilise a part of the savings to subsidise the services for rural households at large? Alternatively, if the government is okay with underwriting all the capital expenditure (to an extent), leaving households to only defray operational costs, is an approach that lowers operational costs preferable over one limiting capital costs, like that of urban–rural partnerships?

Considering the former case, since expenditure on fuel is the biggest cost driver of emptying services, minimising the average travel distance over the service area might lead to lower operational costs. Therefore, if the government(s) continues to absorb capital expenditure fully, rural households could receive services at a lower fee if all rural local governments but those abutting the town or city developed their own cluster-level FSM systems (at the level of clusters to reach the optimal scale) (Figure 8). However, in the present case, the sizing of the mixed urban–rural service area only factored in the availability of infrastructure and the cost of providing services to the service provider. It presupposed that rural households could afford the services and would be willing to pay for them, no matter the distance surcharge and the total service fee.

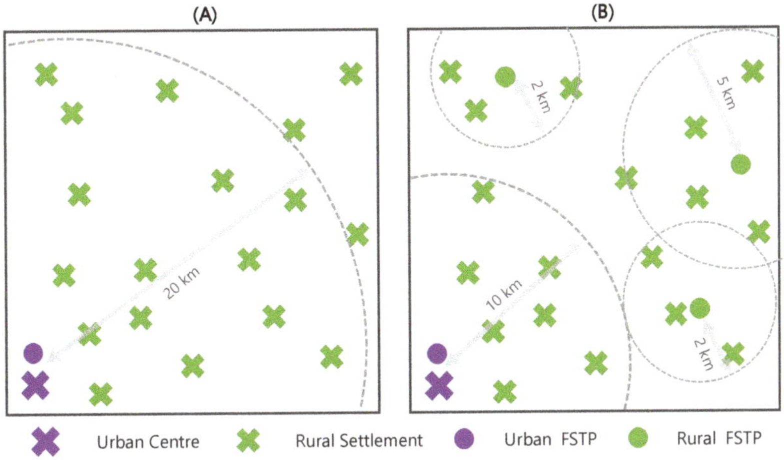

Figure 8. Two representative clustering regimes: (**A**) all gram panchayats within 20 km of a city or town clustered together for urban–rural partnership (**B**) smaller groups of gram panchayats clustered for urban–rural and rural–rural partnerships.

5. Discussion

5.1. Planning

The informal delivery of services by the urban local body to rural households even before the development of a formal intergovernmental urban–rural partnership evinced the role of the market. The phenomenon underscored that service relationships transcending administrative boundaries can emerge at the behest of market forces. Previous research has documented how similar trans-boundary relationships have emerged between small-scale independent service providers and farm owners within sanitation ecosystems of south India [37]. However, it was also clear that the informal arrangement produced negative externalities. The price for the service unilaterally determined by the urban local body rendered the service inaccessible for the share of the rural households that instead relied on manual emptying. The latter was reportedly cheaper on average, but has proved itself fatally dangerous to the lives of the service providers [38]. The formalization of the urban–rural partnership and service delivery arrangements brought down the service charge for the rural households and was accompanied by awareness-building about service availability. Therefore, at the very least, the partnership—serving as a direct government intervention—helped target the negative externalities that a market for services can produce.

In transitioning out of the market approach into one based on systematic intergovernmental partnership, the flexibility of the market had to be compromised for achieving viability. Before the partnership the municipality served rural households up to distances of 20 km, but the threshold was reduced to 10 km as part of the formalization. The motivation for defining a fixed, albeit reduced, service area was to create predictability on both the supply and demand sides. Not determining how long an FSTP can serve the increased service population created the risk that infrastructure would fail in the future or that rural households would be left stranded as demand outpaces supply. Even if augmenting the capacity of FSTPs was an option, without systematic monitoring, only a system failure would have augured the need for it if supply and demand had not been matched at the outset.

More generally, the present case represents only a limited set of opportunities for urban–rural partnerships. In the Dhenkanal district, the FSM system had been operational for an year at the time the pilot was conceptualized. It was only retrospectively that the urban–rural partnership was formalized based on the available spare capacity of urban infrastructure. Cases where an FSM system is still in the ideation or planning stages could benefit from the existing research on identifying optimal regional service areas and infrastructure locations to create service systems that are truly co-created by two or more local governments [39,40].

5.2. Responsibility and Accountability

In holding up Dhenkanal's case against the four models of intermunicipal cooperation proposed by research, the urban–rural partnership in the district most closely resembled a 'service delivery agreement'. Under 'service delivery agreement', "participating local governments enter a formal agreement to cooperate in the delivery of services without establishing a joint standing organization" and "one of the partners, frequently the largest municipality, renders and sells services to other partners" [12]. However, at present, the MoA between the urban local government and the *gram panchayat* only broadly sets out the terms of such a 'service delivery agreement'. It is not a comprehensive articulation of roles and responsibilities of the two governments and inadequately comments on issues, such as periodic reviews, service level benchmarks, grievance reporting and redressal, and dispute resolution. Moreover, it obfuscates the routes of accountability.

Past research has identified the 'short' and the 'long' routes of accountability [41]. The short route enforces direct accountability between the citizen and the service provider by allowing the former to 'vote with their feet'. When the government itself is a service provider, the short, if slow, route can still emerge in the form of political accountability over a longer time period (the years between consecutive election cycles). When the urban local government was informally serving rural households, it viewed the act as an optional practice that could help protect public goods, and not as a profit-making enterprise. Therefore, even if rural households had 'voted with their feet', the urban FSM system could have been indifferent to a shrinking rural base. At any rate, before or after the formal intergovernmental partnership, the urban local government is not politically accountable to rural households.

Under the partnership, the weak short route has transformed to a weak long route. Rural households can hold the officials of the gram panchayat that helped form or manage the partnership politically accountable. However, if the intergovernmental partnership does not meaningfully enable the gram panchayat to hold the urban local government accountable, accountability reaches an impasse. Since the urban–rural partnership is framed as an agreement and not a contract, in its current form, it does not equip *gram panchayats* with any real leverage to hold its urban counterpart accountable for poor performance. Previous research discusses the implications of poor accountability. It emphasizes that monitoring cooperative agreements—same as private contracts—is necessary for maintaining the economic benefits of intermunicipal cooperation [7].

5.3. Financing

The consultations held between the urban local government and the *gram panchayats* proved effective in lowering the price of services for rural households. However, the negotiation occurred with the eventual formalization of the partnership as a foregone conclusion. As discussed in Section 5.1, the process determining the service area aimed at matching supply with theoretical demand. It did not consider whether an urban–rural partnership was the most cost-effective way to serve the specific *gram panchayat*. A consideration of issues of financing at different levels, e.g., in determining both the macro-level approach and the micro-level operation financing, would be important as urban–rural partnerships go from controlled pilots to wider deployment.

Secondly, in Dhenkanal's case the *gram panchayats'* concerns regarding affordability caused the lowering of the price. Still, the logic of charging rural households what urban households pay plus a surcharge was not questioned. Models such as tax equalization provide a credible alternative to consider. In places such as the United States, an equalization rate is applied to municipalities to determine a fairer distribution of tax burden [42]. Tariff designing for FSM could similarly consider the prosperity of the different local bodies and index the fee to it. More generally, with an increase in the size of the service area and greater utilization of infrastructure, operating costs per service request could be expected to go down. Although variable operating costs would commensurately increase, fixed operating costs, such as the driver's monthly wages or routine maintenance checks, would now be distributed over more service requests. The resultant lowering of the fixed operating costs per trip could reduce the total operational costs of the emptying services (provided factors such as higher wear-and-tear due to higher utilization do not offset it). If it does, the service charge for urban households would come down, as would the base fee that rural households need to pay. Therefore, the urban–rural partnership in the present case hints at the need for rate rebasing.

6. Conclusions

In the present paper, we analysed the issues observed in a novel experiment with urban–rural local government partnership in India. The impetus for the experiment was the emergence of a common pathway, viz. FSM, for achieving 'safely managed sanitation' in parts of urban and rural India. Our experience in the Dhenkanal district showed that the 'producer' local government, the Dhenkanal municipality, in our case, was already convinced of transcending rigid administrative boundaries to deliver services. However, on its own, the resulting market-mode for service delivery produced failures.

We noted that shifting from the market-mode to a formal urban–rural partnership helped target issues such as unaffordability and subsequent reliance on inferior and unsafe services that are symptomatic of the market's deficiencies. The shift lowered the cost of services for rural households and compelled rural local government to target awareness-building about service availability among households as an explicit goal. The partnership approach was also faster to implement compared to the pace of development of greenfield FSM systems in our specific case. Given that the acquisition of land is the common rate-determining step for infrastructure development in urban and rural India alike, the partnership approach could offer the benefit of speed to settings such as the Dhenkanal district with underutilized infrastructure. Although not discussed in the present paper, the further implementation of the approach with other local governments in the Dhenkanal district and the neighbouring Angul district hint at its replicability. However, we cannot remark on the nature and magnitude of any scale-related efficiency gains that may have materialized since that was not within the scope of the present study. Future research could tackle this aspect.

Coming to specific issues relating to implementation, we noted that in the present setting where such speedy urban–rural partnerships are in option, a system-level analysis that explicitly accounts for the costs of service delivery to households could help determine the most sustainable approach—whether an urban–rural retrospective partnership, a rural–rural partnership for greenfield development, or something else. Since most urban and rural local governments are still in the planning phase, proactive co-creation of sanitation service delivery systems by rural and local governments could be an important area for future research. Further practice research could also explore which mechanisms are best suited for establishing clear pathways for accountability under different forms of intergovernmental partnerships. It could also evaluate if the financial sustainability and affordability of services under the partnership could benefit from application of concepts such as equalization and rate rebasing.

We conclude that both urban and rural India are striving to achieve 'safely managed sanitation' and the similarity of their infrastructural pathways is being increasingly recognized. The alignment presents an opportunity to evaluate what regional-level sanitation and wastewater management could look like in India. The present case—with its specific set of institutional and infrastructural contexts—provides one specific example. Since past research highlights how institutional histories are key determinants of the forms of the partnership, we believe that more experiments in different types of districts and states could help evolve the optimal baseline model of intergovernmental partnerships in India (Figure 9). As a result, from the outset the processes of forming, formalising, and managing partnerships would also benefit from the flexibility to re-create and renegotiate the model in a collaborative, transparent, and fair manner.

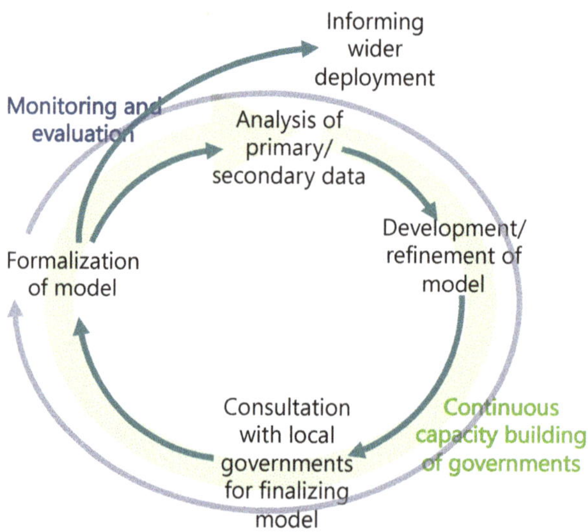

Figure 9. Proposed process cycle adopted for evolving the most contextually apt model for intergovernmental urban–rural partnerships.

Author Contributions: Conceptualisation, S.D. and N.A.; formal analysis, N.A.; writing—Original draft preparation, N.A.; writing—Review and editing, S.D. and N.A.; visualisation, N.A.; supervision, S.D.; funding acquisition, S.D. All authors have read and agreed to the published version of the manuscript.

Funding: The present work was supported by the Bill and Melinda Gates Foundation (BMGF), India (under the grant number OPP1136373).

Institutional Review Board Statement: Not applicable.

Informed Consent Statement: Informed consent was obtained from all participants involved in the study.

Data Availability Statement: The datasets generated and/or analysed during the current research may be available from the corresponding author on reasonable request.

Acknowledgments: The authors are thankful for the support of the entire project team in Delhi and Odisha; the state, district, and local governments; UNICEF India; the Bill and Melinda Gates Foundation; Arghyam; Practical Action India.

Conflicts of Interest: The authors declare no conflict of interest.

References

1. Hueso, A.; Bell, B. An Untold Story of Policy Failure: The Total Sanitation Campaign in India. *Water Policy* **2013**, *15*, 1001–1017. [CrossRef]
2. Swachh Bharat Mission-Urban. Available online: https://sbmurban.org/ (accessed on 19 June 2022).
3. Swachh Bharat Mission-Gramin. Available online: https://sbm.gov.in/sbmreport/home.aspx (accessed on 19 June 2022).
4. Dasgupta, S.; Agarwal, N.; Mukherjee, A. Moving up the On-Site Sanitation Ladder in Urban India through Better Systems and Standards. *J. Environ. Manag.* **2021**, *280*, 111656. [CrossRef] [PubMed]
5. Dasgupta, S.; Agarwal, N. Dichotomy to Continuum: Importance of Faecal Sludge Management in Shaping Future Urban–Rural Linkages in Urbanizing India. *Environ. Urban. ASIA* **2022**, *13*, 168–178. [CrossRef]
6. Harada, H.; Strande, L.; Fujii, S. Challenges and Opportunities of Faecal Sludge Management for Global Sanitation. In *Towards Future Earth: Challenges and Progress of Global Environmental Studies*; Katsumi, T., Hashimoto, S., Eds.; Kaisei Publishing: Tokyo, Japan, 2016; pp. 81–100, ISBN 978-4-87603-500-7.
7. Warner, M.E. Inter-Municipal Cooperation in the U.S.: A Regional Governance Solution? *Urban Public Econ. Rev.* **2006**, *6*, 221–239.
8. Ferrari, S. *Local Government in Brazil and Switzerland: A Comparative Study on Merger an Inter Municipal Cooperation*; Institute of Federalism: Fribourg, Switzerland, 2015.
9. Klimovský, D.; Mejere, O.; Mikolaitytė, J.; Pinteric, U.; Saparniene, D. Inter-Municipal Cooperation in Lithuania and Slovakia: Does Size Structure Matter? *Lex Localis* **2014**, *12*, 643–658. [CrossRef]
10. Haveri, A.; Airaksinen, J. Inter-Municipal Cooperation in Finland: Old Traditions and New Promises. In *Inter-Municipal Cooperation in Europe*; Hulst, R., van Montfort, A., Eds.; Springer: Dordrecht, The Netherlands, 2007; pp. 39–65, ISBN 978-1-4020-5379-5.
11. Hulst, R.; van Montfort, A. (Eds.) *Inter-Municipal Cooperation in Europe*; Springer: Dordrecht, The Netherlands, 2007; ISBN 978-1-4020-5378-8.
12. Hulst, J.; van Montfort, A.V. Institutional Features of Inter-Municipal Cooperation: Cooperative Arrangements and Their National Contexts. *Public Policy Adm.* **2012**, *27*, 121–144. [CrossRef]
13. Agarwal, N.; Mukherjee, A.; Dwivedi, A. *Towards a Sustha and Swachha Rural Dhenkanal: Institutionalizing Faecal Sludge Management for Achieving ODF Plus*; Centre for Policy Research: New Delhi, India, 2020.
14. Jain, A.; Mukherjee, A.; Agarwal, N. *Rural Sanitation Factsheet: Dhenkanal, Odisha*; Centre for Policy Research: New Delhi, India, 2020.
15. LGD—Local Government Directory, Government of India. Available online: https://lgdirectory.gov.in/ (accessed on 19 June 2022).
16. Chaudhuri, S. What Difference Does a Constitutional Amendment Make? In *Decentralization and Local Governance in Developing Countries*; Bardhan, P., Mookherjee, D., Eds.; The MIT Press: Cambridge, MA, USA, 2006; ISBN 978-0-262-26769-4.
17. Vajpeyi, D.K.; Arnold, J.M. Evolution of Local Self-Government in India. In *Local Democracy and Politics in South Asia: Towards Internal Decolonization?* Vajpeyi, D.K., Ed.; Urban Research International; VS Verlag für Sozialwissenschaften: Wiesbaden, Germany, 2003; pp. 31–43. ISBN 978-3-663-10676-0.
18. Rai, M. *Status and Functioning of District Planning Committees in India*; Participatory Research in Asia: New Delhi, India, 2009; Available online: https://www.pria.org/knowledge_resource/1547543966_DPC.pdf (accessed on 2 July 2022).
19. World Bank. *Overview of Rural Decentralization in India*; World Bank: Washington, DC, USA, 2000.
20. World Bank. *Urban Finance and Governance Review*; World Bank: Washington, DC, USA, 2004.
21. Singh, R.; Brahmankar, R.; Murty, J.V.R.; Verma, P.; Wescoat, J.L., Jr. Analyzing Rural Drinking Water Services for District Planning in Maharashtra, India. *Water Policy* **2019**, *22*, 37–51. [CrossRef]
22. Ganguly, S. India's National Sanitation and Hygiene Programme: From Experience to Policy—West Bengal and Maharashtra Models Provide Keys to Success. In *Beyond Construction: Use by All*; IRC International Water and Sanitation Centre and WaterAid: The Hague, The Netherlands, 2008; ISBN 978-9937-2-0472-9.
23. Mukherjee, A.; Dwivedi, A.; Agarwal, N. *Rural Sanitation Practices: A Rapid Assessment Study for Odisha*; Centre for Policy Research: New Delhi, India, 2020.
24. Schrecongost, A.; Pedi, D.; Rosenboom, J.W.; Shrestha, R.; Ban, R. Citywide Inclusive Sanitation: A Public Service Approach for Reaching the Urban Sanitation SDGs. *Front. Environ. Sci.* **2020**, *8*, 19. [CrossRef]
25. Ministry of Urban Development National Policy on Faecal Sludge and Septage Management (FSSM). 2017. Available online: https://smartnet.niua.org/content/8e184ef5-2232-4f0d-a5af-78876c96aff8 (accessed on 2 July 2022).
26. Sørensen, R.J. Does Dispersed Public Ownership Impair Efficiency? The Case of Refuse Collection in Norway. *Public Adm.* **2007**, *85*, 1045–1058. [CrossRef]
27. Zafra-Gómez, J.L.; Prior, D.; Díaz, A.M.P.; López-Hernández, A.M. Reducing Costs in Times of Crisis: Delivery Forms in Small and Medium Sized Local Governments' Waste Management Services. *Public Adm.* **2013**, *91*, 51–68. [CrossRef]
28. Mahmoud, I.H.; Morello, E.; Ludlow, D.; Salvia, G. Co-Creation Pathways to Inform Shared Governance of Urban Living Labs in Practice: Lessons From Three European Projects. *Front. Sustain. Cities* **2021**, *3*, 690458. [CrossRef]
29. Mukherjee, A.; Arya, P.; Dasgupta, S.; Chhabra, S.S. *Bridging the Gap Opportunities for Private Sector Participation in Faecal Sludge and Septage Management*; Centre for Policy Research: New Delhi, India, 2019; Available online: https://cprindia.org/briefsreports/%cb%9cbridging-the-gap-opportunities-for-private-sector-participation-in-faecal-sludge-and-septage-management/ (accessed on 2 July 2022).

30. Chowdhry, S.; Kone, D. *Business Analysis of Fecal Sludge Management: Emptying and Transportation Services in Africa and Asia*; Bill and Melinda Gates Foundation: Seattle, WA, USA, 2012.
31. Saxena, S.; Chouhan Singh, N.; Mojumdar, S.; Oledzka Nielsen, M.; Manchikanti, S.; Mukherjee, A.; Agarwal, N.; Rani, M.; Dutta, A.; Vignesh, R. Bridging the Rural-Urban Divide in Sanitation with a Cluster-Based Approach to Faecal Sludge Management: A Case Study from Dhenkanal District in Odisha, India. *H_2O Open J.* (Forthcoming).
32. Prasad, C.S.S.; Ray, I. When the Pits Fill up: (In)Visible Flows of Waste in Urban India. *J. Water Sanit. Hyg. Dev.* **2019**, *9*, 338–347. [CrossRef]
33. Adserà, A.; Boix, C.; Payne, M. Are You Being Served? Political Accountability and Quality of Government. *J. Law Econ. Organ.* **2003**, *19*, 445–490. [CrossRef]
34. Deininger, K.; Mpuga, P. Does Greater Accountability Improve the Quality of Public Service Delivery? Evidence from Uganda. *World Dev.* **2005**, *33*, 171–191. [CrossRef]
35. Sharma, S. India's Rural-Urban Divide: Village Worker Earns Less than Half of City Peer. 2019. Available online: https://www.financialexpress.com/economy/indias-rural-urban-divide-village-worker-earns-less-than-half-of-city-peer/1792245/ (accessed on 25 May 2022).
36. Shergill, H.S. Rural–Urban Disparity in the Standard of Living across States of India. *Econ. Polit. Wkly* **2021**, *56*, 45–46.
37. Kvarnström, E.; Verhagen, J.; Nilsson, M.; Srikantaiah, V.; Singh, K.; Ramachandran, S. *The Business of the Honey-Suckers in Bengaluru (India): The Potentials and Limitations of Commercial Faecal Sludge Recycling—An Explorative Case Study*; IRC: The Hague, The Netherlands, 2012; Volume 48, Available online: https://www.ircwash.org/OP48 (accessed on 2 July 2022).
38. Murali, M. No Deaths Due to Manual Scavenging, 161 Died Cleaning Sewers in 3 Years: Centre. 2022. Available online: https://www.hindustantimes.com/india-news/no-deaths-due-to-manual-scavenging-161-died-cleaning-sewers-in-3-yearscentre-101649270193401.html (accessed on 25 May 2022).
39. Voutchkov, N.S.; Boulos, P.F. Heuristic Screening Methodology for Regional Wastewater-Treatment Planning. *J. Environ. Eng.* **1993**, *119*, 603–614. [CrossRef]
40. Zeferino, J.A.; Cunha, M.C.; Antunes, A.P. Adapted Optimization Model for Planning Regional Wastewater Systems: Case Study. *Water Sci. Technol.* **2017**, *76*, 1196–1205. [CrossRef] [PubMed]
41. World Bank. *World Development Report 2004: Making Services Work for Poor People*; World Bank: Washington, DC, USA, 2004.
42. Equalization Rates. Available online: https://www.tax.ny.gov/pit/property/learn/eqrates.htm (accessed on 20 June 2022).

MDPI
St. Alban-Anlage 66
4052 Basel
Switzerland
Tel. +41 61 683 77 34
Fax +41 61 302 89 18
www.mdpi.com

Land Editorial Office
E-mail: land@mdpi.com
www.mdpi.com/journal/land

www.ingramcontent.com/pod-product-compliance
Lightning Source LLC
Chambersburg PA
CBHW061611220926
51850CB00005B/476
* 9 7 8 3 0 3 6 5 5 1 4 1 8 *